Business Statistics in Brief

Business Statistics in Brief

ALBERT C. OVEDOVITZ
St. John's University
Jamaica, New York

Australia • Canada • Denmark • Japan • Mexico • New Zealand • Philippines
Puerto Rico • Singapore • South Africa • Spain • United Kingdom • United States

Publisher: Dave Shaut
Senior Acquisitions Editor: Charles McCormick, Jr.
Developmental Editor: Mardell Toomey
Senior Marketing Manager: Joseph A. Sabatino
Media Technology Editor: Kevin von Gillern
Media Production Editor: Robin Browning
Production Editor: Amy S. Gabriel
Manufacturing Coordinator: Sandee Milewski
Internal Design: Ann Small, A Small Design Studio, Cincinnati, Ohio
Cover Design: Rick Moore
Cover Image: © 1999 PhotoDisc, Inc.
Photo Manager: Cary Benbow
Production House: Lifland et al., Bookmakers
Compositor: Parkwood Composition Service, Inc.
Printer: RR Donnelley & Sons Company, Willard Manufacturing Division

Printed in the United States of America
1 2 3 4 5 03 02 01 00

For more information contact South-Western College Publishing, 5101 Madison Road, Cincinnati, Ohio, 45227 or find us on the Internet at
http://www.swcollege.com

For permission to use material from this text or product, contact us by
telephone: 1-800-730-2214
fax: 1-800-730-2215
web: http://www.thomsonrights.com

Library of Congress Cataloging-in-Publication Data

Ovedovitz, Albert C.
Business statistics in brief / Albert C. Ovedovitz
p. cm.
Includes index.
ISBN 0-324-04868-8 (package)
ISBN 0-324-02434-7 (text only)
ISBN 0-324-04867-X (CD only)
1. Commercial statistics. 2. Statistics. I. Title.

HF1017.O95 2000
519.5--dc21 99-088981

This book is printed on acid-free paper.

TO RITA LEE

ABOUT THE AUTHOR

From Dix Hills, New York, Dr. Albert C. Ovedovitz holds degrees from Queens College and the City University of New York. He has taught for 34 years and spent five years as a statistician for the United Nations. Dr. Ovedovitz also works as a forensic economic consultant. A highly regarded statistician, he has received numerous awards, including the Outstanding Achievement Medal for Faculty from St. John's University and an award for Meritorious Achievement from the New York Chapter of the American Statistical Association and John Jay College of Criminal Justice.

Preface

ABOUT THE TEXT

What sets ***Business Statistics in Brief*** apart from other introductory statistics books is a systematic commitment to showing students how to convert data into meaningful information. Not only will this book teach students how to apply formulas to data and derive results; it will also build students' skills in choosing the statistical tools for their analysis, interpreting the results of their work, and expressing their findings in meaningfully written reports. This book fosters motivated learning by focusing on topics of direct interest to students who are preparing for careers in business. In writing this book, I sought out traditional business applications—statistics related to the stock market or employees' salaries, for example. In addition, I looked for applications that would appeal to students' lighter interests, such as consumer goods and sports and other recreational activities.

Text Objectives

Business Statistics in Brief was written with the following major objectives in mind.

Brevity. Most introductory business statistics courses are only one term in length. Chapters and sections in this text are therefore designed for instructional flexibility. If instructors choose to present chapters in an abbreviated form by omitting supplementary problems, this will not jeopardize the integrity of the topics in question. The *Instructor's Manual* outlines alternative syllabi.

Student Orientation. Given the difficulty that many students have in comprehending the ideas and techniques in an introductory business statistics course, this book has been carefully designed to present concepts and applications at a level students can understand. Comprehension and skill building are assured by clear, direct exposition, leading from one conceptual tool or application to the next. Examples and exercises are placed where they can provide the most help, showing in a realistic way how the tools work. The examples and exercises are intended to be motivating; that is, problems are organized to help students develop their own ability to pick the right tool and then compute, interpret, and analyze.

Applications Orientation. Today's undergraduate and MBA-level students will be exposed to statistics in their professional lives, but in a different way than are professional statisticians and other researchers. Managers are less interested in methodology than in extracting information from data and in understanding how to assess the quality of the information derived from imperfect, sample-based data. To the fullest extent possible, ***Business Statistics in Brief*** employs realistic business situations and real data in order to help students understand how inference works. Data are drawn from a variety of sources, such as *Business Week, Consumer Reports, Forbes, Fortune, Money, New York Times,* and *USA Today.*

Focus on Interpreting Output. Each chapter includes a number of large-data-set exercises that challenge students to carefully examine the outcomes of their analyses. Examples and exercises go beyond manipulation of formulas to look at problem solving, interpreting the outcomes of analytical activities, and report writing. Many exercises ask students to write a report evaluating their answers and interpretations of the problem and data. Suggestions for research provide additional opportunities for data collection, selection of statistical tools, analysis of information, and interpretation of results. These exercises are clearly marked with an identifying icon:

Chapter Organization at a Glance

Build Your Knowledge of Statistics. Unusual in a book on statistics, the chapter-opening vignettes capitalize on my teaching experience over the past 30 years. These chapter openers introduce topics to be presented in the chapter. Where appropriate, they briefly discuss how the current material relates to other chapters' topics. These interesting and inspiring introductions constitute another facet of the motivated learning approach.

Statistics Tool Kit. A boxed feature designed to encourage students' enthusiasm about statistics, the Statistics Tool Kits take students beyond the text, exploring curiosities and providing detailed explanations that make the topics more relevant.

Examples. In this example-based text, the examples are completely worked out to give students an intuitive understanding of the material and prepare them to deal with the exercises. *Every exercise is preceded by at least one example of similar type.* The examples demonstrate not just the structure and manipulation of formulas to get "correct" answers, but decision-making applications as well.

Exercises. All exercises are presented in a logical learning sequence. Most problem sections are divided into two parts: Skill Builders and Applying the Concepts. The former is dedicated to the use of equations; the latter offers opportunities to practice tool selection, analyze results, and write reports within a framework of real data and realistic business situations.

End-of-Chapter Review. Each chapter contains lists of the formulas used in that chapter and the new statistical terms introduced, as well as supplementary problems. These features serve to provide a comprehensive review of the concepts presented in that chapter.

Important Features

- Statistical process control (SPC) is presented not as an isolated statistical issue, but in connection with applications where descriptive or inferential methods of SPC arise. See Chapters 1, 8, and 9 of the text.
- Data collection and the use of sampling methods are treated as basic tools in introductory statistics. To ensure that students do not view data as a mysterious raw material that appears out of nowhere, Chapter 1 is devoted to encouraging stu-

dents to understand the sources of data. Chapter 6 emphasizes the careful management of controlled studies in a section on the use of sampling methods.

- The common software packages Minitab© and Microsoft® Excel© are featured in discussions of computer use because of their prevalence both in the classroom and in the business world. Brief instructions for using Minitab and Excel are found in the text where computer applications are illustrated. Those students using other packages or doing the work by hand may omit these sections without any loss of continuity.
- The CD-ROM found in the back of the textbook contains supplementary data sets needed for the exercises. These data sets are available in both Minitab and Excel formats. The data sets can also be found in printed form in Appendix A (pages 438–447) at the conclusion of this book.

ABOUT THE SUPPLEMENTS

Web Site. Data sets and a complete supplementary chapter on index numbers are available online at the following address: Ovedovitz.swcollege.com

The following three supplements to this text are all available to the instructor on CD-ROM (ISBN: 0-324-04867-X):

Instructor's Manual. The author-prepared *Instructor's Manual* contains recapped learning objectives, teaching suggestions, complete solutions to all exercises, and sample syllabi.

Test Bank. Prepared by Professor Sharon Neidert of the University of Tennessee and Professor Alan Smith of Robert Morris College, the *Test Bank* contains a large number of multiple choice, true/false, and essay questions, as well as quantitative problems for each chapter.

PowerPoint©. Prepared by Victor Lu of St. John's University, who collaborated directly with the author, a complete slide show in PowerPoint accompanies each chapter, highlighting the major points in each section.

ACKNOWLEDGMENTS

My thanks go to the following reviewers, whose comments and suggestions were invaluable in improving the manuscript:

Sung Ahn
Washington State University

Raid Amin
University of West Florida

Natalie Calabro
St. John's University

Satya Chakravorty
Kennesaw State College

Darrell Christie
University of Wisconsin

J. Michael Cicero
Highline Community College

Renato Clavijo
Robert Morris College

Martin Goodman
Monmouth University

Deborah Gougeon
University of Scranton

William Harman
Centennial College

Clifford Hawley
West Virginia University

Kenneth D. Lawrence
New Jersey Institute of Technology

Vamal Munshi
Sonoma State University

Roxy Peck
California Polytechnic University

Bob Reed
University of Oklahoma

Harry Reinken
Phoenix College

Vivek Shah
Southwest Texas State University

Varinder Sharma
Christopher Newport University

William Soule
University of Maine

Phillip Taylor
University of Arkansas

Dang Tran
California State University

Rudy Wuilleumier
Eastern Kentucky University

Special thanks are due to Mardell Toomey and Charles McCormick, Jr. of South-Western College Publishing for their unfailing support and encouragement. Special thanks, too, go to my colleagues at St. John's University, particularly Andrew Russakoff and Victor Lu. I offer my thanks to friends and family whose support was a source of encouragement over the course of this project. My very special thanks go to my wife, Rita, who provided moral support and encouragement.

Dix Hills, New York **Albert C. Ovedovitz**

Brief Contents

Contents

Business Statistics in Brief

1 Numbers, Management, and Statistics

OUTLINE

LEARNING OBJECTIVES

1. Understand what statistics are and how they are related to populations, variables, samples, and data.
2. Show how sample statistics are related to population parameters and understand what descriptive and inferential statistics are.
3. Identify the basic steps in statistical studies.
4. Explain the role of statistics in business and industry.
5. Differentiate between primary and secondary data and understand the uses of both.
6. Identify the different sorts of variables: quantitative, qualitative, continuous, and discrete.
7. Understand how numerical data vary in quality, from nominal data to ratio data.

BUILD YOUR KNOWLEDGE OF STATISTICS

Ignorance of statistical methods can be costly. Consider the experience of *Literary Digest,* a popular magazine of the 1930s. During the 1936 election, it sent out 10 million questionnaires. Based on 2.4 million responses, it predicted that Republican challenger Alfred Landon would defeat President Franklin Delano Roosevelt with a lead of more than 20 percent. Roosevelt won with 62 percent of the vote. The *Literary Digest* went wrong based on millions of responses, while pollster George Gallup called the election correctly based on a survey of 50,000. How could this happen?

The reason is that the *Literary Digest* failed to focus on the right *population,* the voters who would decide the election. It surveyed a group that was wealthier than the typical voter. Worse, it relied on responses from those who felt strongly enough to send in replies. Such self-selected samples are seldom representative—in this case, the responses overrepresented those who disliked Roosevelt. The *Literary Digest* passed into history, remembered for its errors. The Gallup Organization thrives today, routinely assessing national opinion to within ±3 percentage points, based on samplings of no more than 1500 people. Understanding statistics can be a key to success, while ignorance leads to failure.

In this chapter, you will find out what the discipline of statistics is all about. You will learn the basic vocabulary of the statistician. How are statistics gathered and used? How are statistics related to populations? How is a statistical study conducted? Studying statistics will teach you how to read and interpret statistical information and how to collect and analyze data. You will learn what conclusions can be drawn from numbers and where caution is needed. Your understanding will develop paragraph by paragraph, as new concepts are introduced to you. This chapter will provide you with statistical tools that you can use in business, manufacturing, accounting, and finance.

Once you have mastered the statistician's vocabulary in this chapter, the next chapters will focus on describing data sets. In Chapter 2, you will see how data are organized and portrayed. Chapter 3 will turn to the specifics of numerical description. In following chapters, you will learn how control charts are used to monitor production and how statistical tools are used in product testing, market research, financial forecasting, and profitability calculations.

SECTION 1 ▪ THE DEFINITION OF STATISTICS

An airline might figure out the average weight per passenger on its New York to Chicago run in order to estimate how much added freight its planes can carry per flight, given the number of passengers booked and the carrying capacity of the plane. This average, based on a sample of passengers on such flights, is a statistic. Every statistic is calculated by looking at some population (also called a universe), or collection of individuals or cases, and recording the values of one or several variables relating to these cases. Here the airline would define its population as all the travelers who fly this run, each one counted once for each trip made. The cases would be the individual trips made by passengers. The variable would be the weight of the traveler and bags together, perhaps along with the number of bags checked or other information.

To keep costs down, the airline would select a representative group of several hundred passengers from this population of several thousand. This representative group would be its sample. The sample statistic it would use would be the average of the weights of the individual passenger/bag totals in its sample. The airline might also note the maximum and minimum weights for passengers and baggage, to get an idea of the variability of these weights. The maximum and minimum are also sample statistics.

These sample statistics would help the airline estimate the average weight for the population as a whole. The average for the entire population is the population parameter that the sample statistic is attempting to approximate.

If a sample consists of the entire population, it is called a census. The sample statistics for a census are identical with the corresponding population parameters. Generally, it is impractical or impossible to take a census, and thus sampling is the best procedure.

The example above illustrates many of the basic concepts of statistics; the formal definitions appear below.

The **population** is the group of individuals or cases to be studied. The population is chosen for its relevance to a specific purpose; for example, we would choose registered voters as our population if we wished to predict an election.

The **cases,** or **individuals,** are the entities that collectively make up the population and whose properties, such as their preferences as voters, are of interest.

A **variable** is a property, possessed by the cases in the population, that we wish to measure or record, such as the weight or number of a passenger's bags or the declared political party of a voter.

A **sampling frame** is a system that allows us to identify a population and select a sample from it. For example, the voter registration rolls provide a sampling frame for voters. Think of a sampling frame as a list of the population.

A **sample** is a group of cases chosen from the population for study. If a sample includes the entire population, it is called a **census.**

Data consist of the cases in the sample, together with the values of the variables recorded for these cases.

A **statistic** is any number computed from sample data—for example, the average or the maximum value of a variable for all the cases in the sample. The **sample mean** is the average of the values for a sample. If the variable is denoted x, this statistic is written $\bar{x}$. The plural of *statistic* is *statistics.* **Statistics** is also the name for the field of study devoted to statistics, their properties, and their applications.

A **parameter** is a statistic derived from a sample consisting of the entire population. Suppose the variable we are looking at is a mass denoted by m. The sample statistic, the average mass for a random sample, would be denoted by $\bar{m}$. The corresponding population parameter, the average mass for the whole population, would be denoted by μ (the Greek letter mu, corresponding to a Roman m).

Simple **random sampling** is a procedure for taking samples of a given size, say k, in such a way that each set of size k in the population is equally likely to be selected. A completely random procedure is free from systematic bias in the sense that it neither favors nor avoids any special group of individuals in the population.

One way to generate a random sample of k individuals is to write the names of the individuals in your population on cards and then, after thoroughly shuffling, deal out the top k cards as your sample. In most situations, this method will be impractical, but your sampling frame will permit some way of assigning numbers to the individuals. Computer software will then let you pick k of these numbers at random, somewhat like spinning a giant wheel of fortune.

Random sampling avoids systematic biases. Sample statistics (such as the average, or arithmetic mean) for simple random samples will almost surely approach the corresponding population parameters as larger and larger samples are taken.

The principles above form the basis for statistical techniques. Statistical methods allow us to say how likely a sample statistic is to be within a certain range of the corresponding population parameter.

Random sampling is the best approach we have, but not every random sample is representative of the population. Dealing poker hands from a well-shuffled deck gives each player a random sample of size five. But the lucky player who has just received a royal flush in spades should not conclude that the deck consists of nothing but spades nor that it is more than half face cards simply because his hand has these properties. What is true is that on average, over the long run, the information gleaned from random sampling will be representative of the population at large.

Sampling is used when a census is impractical, impossible, or too costly. If you wanted to know what proportion of Superbowl attendees live near the site of the Superbowl, you would sample points, not attempt a census—there are too many people to ask. But if a third-grade teacher wanted to know what proportion of his students had brothers or sisters, he would ask them all, for taking a census is easy in this case.

STATISTICS TOOL KIT

Notation That Makes Things Easier

Common names are easy to remember, so you would use "COST" to head a spreadsheet column in a cost survey. Likewise, you would choose c to represent the cost variable. The usual practice is to use Roman letters for variables. The average cost for your sample would then automatically be $\bar{c}$. The overbar denotes a sample average. In a study where only one variable is being looked at, the population average is indicated by the Greek letter μ, which corresponds to the Roman m (the leading letter of mean, another word for average). The general pattern is to shift from Roman for a sample statistic to the corresponding Greek letter for the population parameter. For example, if the sample statistic were s^2, the population parameter would be σ^2, where σ is the Greek letter sigma, which corresponds to the Roman s. (Note that the capital sigma, Σ, also plays a role in statistical formulas; it is used to indicate the summation of a number of terms.)

The following table contains some rules for statistical notation. Following these conventions will make things clearer for everyone.

SOME RULES FOR STATISTICAL NOTATION

Concept	Notation
Constants	Letters from the beginning of the Roman alphabet: a, b, c, etc.
Variables	Letters from the end of the Roman alphabet: u, v, w, x, y, and z
Subscripts	Letters from the middle of the Roman alphabet: i, j, k, l, m, and n
Sample statistics	Letters from the Roman alphabet—for example, $\bar{x}$, s, and s^2 for the sample mean, standard deviation, and variance, respectively
Population parameters	Letters from the Greek alphabet—for example, μ, σ, and σ^2 for the mean, standard deviation, and variance (which is the square of the standard deviation), respectively

SECTION 2 ▪ STEPS IN A STATISTICAL INVESTIGATION

The largest statistical studies in the United States are the U.S. Census, used to apportion seats in the House of Representatives, and the Current Population Survey, which gives us employment figures, among other things. At the other end of the spectrum, studies may be as small as testing of a few dozen mice. The budgets for the large-scale projects are staggering. The projected cost for the next U.S. Census is four to five billion dollars, and the annual budget for the Current Population Survey runs forty to fifty million dollars. Large-scale national studies have complex designs, but the business examples we will begin with are small and easy to understand.

Suppose a tire manufacturer needs to convince buyers that his products will last. He decides to advertise 60,000-mile tires and to support this claim by selecting a sample of tires from each batch at random and testing them on a machine that simulates road wear. An average sample mileage of $\bar{x}$ = 59,957 miles would be consistent with his claim of a population average of about 60,000 miles per tire, as some variation in mileage is to be expected. Here, statistical testing monitors quality, and statistics help to sell the product. Sampling and statistical analysis are necessary, because the tires are tested until they fail. If they were all tested, there would be none left to sell. Statistical methods tell the manufacturer and the buyer what variations of x are consistent with the claim that this is a population of 60,000-mile tires.

Now suppose that you work for a taxicab company that has a fleet of 20 cars, which the company maintains and fuels. Your boss thinks that the different vehicles are getting quite different mileages. Your boss wonders how much she might save if she could improve the mileage on the less efficient cars. Your job is to answer this question. First you must decide what data to collect. This means deciding how mileage should be defined, so that everyone participating will gather numbers in the same way.

You decide to begin by looking at a small set of numbers, to get a feeling for what is going on. For this pilot study, you ask the drivers to keep logs of miles driven and gas put into their tanks. You decide to look at two successive fillings for each of the twenty vehicles. Two sorts of information are involved. The information for the separate vehicles gives cross-sectional data on the fleet's mileage, while the information on successive fills for a given vehicle gives longitudinal data on the performance of that particular vehicle over time. You realize that the successive figures for the same vehicle depend, in part, on how full the attendant gets the tank each time, but you decide to sort that problem out later.

The important thing is to look at the data and see what is going on. In looking at the data, you use the tools of descriptive statistics: graphing and looking at averages and other summary statistics. Some of the steps you will need to take to complete your study are shown in Figure 1.1.

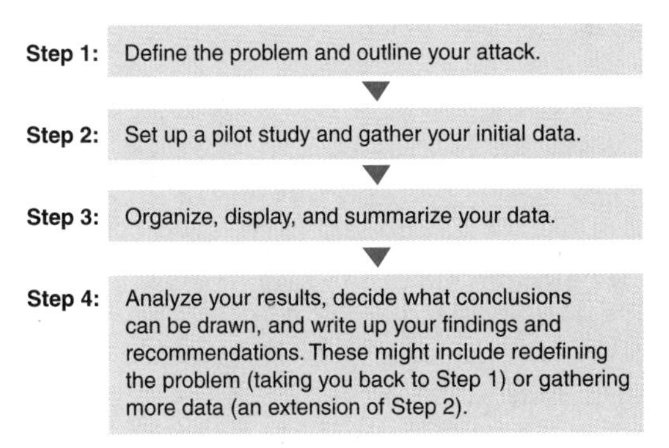

Figure 1.1
Steps in a statistical investigation

For this mileage study, you would ask each participant in the fleet to keep a record like the one illustrated in Figure 1.2. After recording the date, mileage, and amount of gasoline at each fill-up, each person would compute the number of miles between fill-ups and provide data on mileage (miles per gallon).

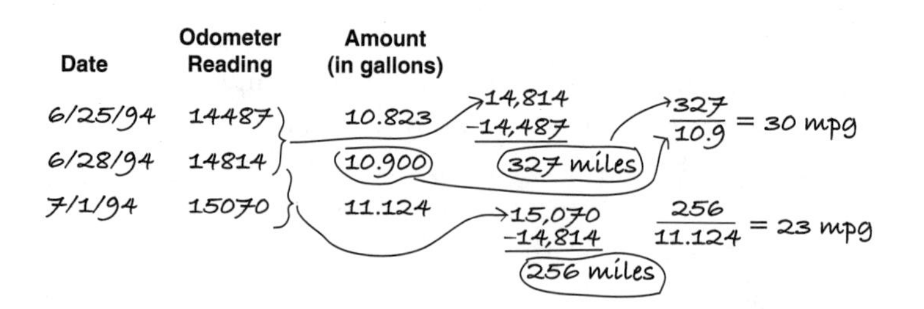

Figure 1.2
Forms for collection of data on gas mileages and transformation of the data into a set of observations

The collected data in their raw form would not tell you too much. They might look like this:

Gasoline Mileages for a Sample of Cars

30	32	28	22	17	19	20	34	23	31
23	29	22	18	20	21	30	35	32	27
24	26	23	19	22	31	31	23	28	33
15	22	24	27	26	39	28	32	31	34

The second step would consist of putting the data into some organized format. A table would help to make some sense of the data. We shall explore the methods of organizing data in Chapter 2; for the moment, let's simply organize the data in intervals of 5 miles per gallon (mpg), as in Table 1.1. This table shows how the cars are distributed according to their gas mileage. (We will return to this idea of a frequency distribution in Section 1 of Chapter 2.)

Table 1.1

A FREQUENCY DISTRIBUTION OF GASOLINE MILEAGES

Mileage Category	Number of Cars
15–19 mpg	5
20–24 mpg	13
25–29 mpg	8
30–34 mpg	12
35–39 mpg	2

Organizing the data reveals patterns. We see that the data fall between 15 mpg and 39 mpg. The majority of the cars (33) fall in the 20–34 mpg range.

Tables are useful when further computations are going to be made with the entries. Charts are more useful in presentations, as they allow patterns to be seen directly. Figure 1.3 contains a bar graph (or bar chart) of the data. Notice that the highest bar is for 20–24 mpg and the lowest bar is for 35–39 mpg.

The third step, summarizing the numerical data, consists of reducing the sample to a number of summary statistics.

> **Summary statistics** describe a data set with a few key numbers. These numbers may indicate where the center of the data set falls, as well as the extremes.

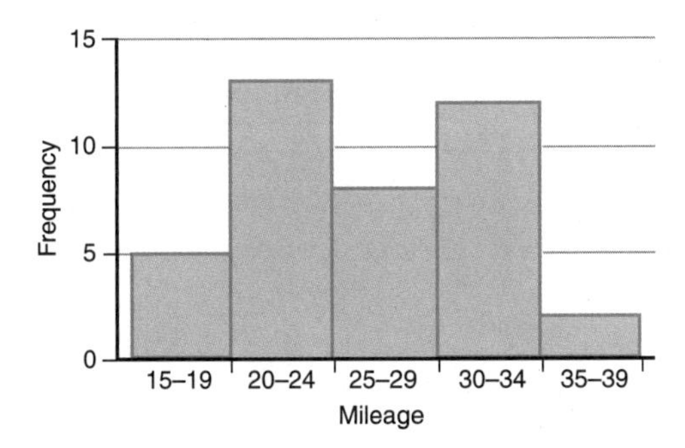

Figure 1.3
Graph of gasoline mileages in Table 1.1

Summary statistics tell a story. With today's electronic computational tools, a set of summary statistics can be generated very quickly. We will explore how this is done in more detail in Chapter 3. For now, however, here's the story:

> *In a sample of 40 cars, it was found that the average mileage was about 26 miles per gallon. Cars did as poorly as 15 mpg and as well as 39 mpg. The middle 50 percent of the vehicles tested fell between 22 and 31 mpg.*

The analysis comes after the summary of numerical data. For the analysis, it is necessary to consider the context of the data. Is 26 mpg the desired goal? Is there a reason why some vehicles perform so poorly? What explains the superior performance of other cars? What policy initiatives do the data suggest? Once you have answered all these questions, you are ready to make your presentation. Typically presentations are made in the form of reports and/or exhibits. Reports, including charts and graphs, might be presented to an employer, a board of directors, or a client.

The example above involved use of the techniques of descriptive statistics. The other main branch of statistics is devoted to the study of inferential statistics.

> **Descriptive statistics** are numbers—such as averages, maxima, or minima—that are used to describe the samples taken from a population. The field of descriptive statistics is the study of sample variables using sample statistics, graphs of the distribution of the variables, and other measures.
>
> **Inferential statistics** provide quantitative answers to questions such as whether the observed difference in two sample statistics is more likely the result of chance variation or the result of systematic differences in the underlying populations. For example, we would turn to inferential statistics for a scientific approach to determining whether the fact that the Red Sox edged out the Blue Jays in the Series was more likely a matter of luck or the result of superior skill.

SECTION 3 ▪ STATISTICS AT WORK IN BUSINESS AND INDUSTRY

How might an automobile maker use statistical methods in manufacturing? On the input side of the assembly line are parts such as starters, shipped to the manufacturer by various suppliers. The manufacturer will take samples of these starters and determine whether they meet its specifications. The same will be done for windshields, tires, air bags, and so on. This process is known as acceptance sampling.

> **Acceptance sampling** is a statistical procedure for testing input items or materials used in manufacturing, construction, or other processes in order to ensure that the necessary specifications are met.

Acceptance sampling can be carried out at the start of production or as the product moves from one stage to the next. Notice that the early stages of production serve as sources of supply for the later ones. Acceptance sampling is part the general process of control used to assure the quality of products and the proper operation of industrial processes. For instance, when a tire manufacturer tests a sample from a batch of tires to assure itself and its customers of the quality of its products, it is engaged in statistical quality control. During the manufacturing process, the same company will sample to monitor variables such as tread depth and thickness in order to detect and correct potential problems. This statistical process control tells workers when to halt and readjust machinery that is going awry.

> **Statistical quality control (SQC)** provides statistical procedures for testing manufactured items to assure that user specifications will be met.
>
> **Statistical process control (SPC)** provides statistical techniques for monitoring ongoing processes so that the final results will meet set specifications.

These techniques apply equally to products such as cars and to services such as health delivery. Will your car's air bag deploy when needed and not before? Does your HMO deliver an acceptable level of care? Such issues are monitored by methods of SQC and SPC.

Sampling is used by everyone from the IRS (in checking taxpayers' returns), to air and rail lines (in splitting up fares or other tariffs between passengers and freight), to marketers (in assessing what fraction of a given population might buy a new product).

- ***Accounting and Sampling.*** Beginning in the 1950s, companies such as the Chesapeake and Ohio Railroad studied the possibility of using sampling (rather than tracking all shipping bills) to allocate revenues to various railroads jointly hauling cargo. The Chesapeake and Ohio study, which ran for six months, showed that $4000 in expenses could be saved by sampling and that the discrepancy between the estimates from sampling and an exact accounting of freight bills was only $83. The overwhelming advantage of sampling was clear, and sampling methods were adopted for allocating both freight and passenger revenues. Sampling methods are widely used in accounting to deal with problems such as allocation of revenues or expenses.
- ***Market Research and Sampling Surveys.*** When a new food product is being taste tested, statistical methods are used, along with a scale such as the one shown in Figure 1.4. Such testing is a part of market research, which includes

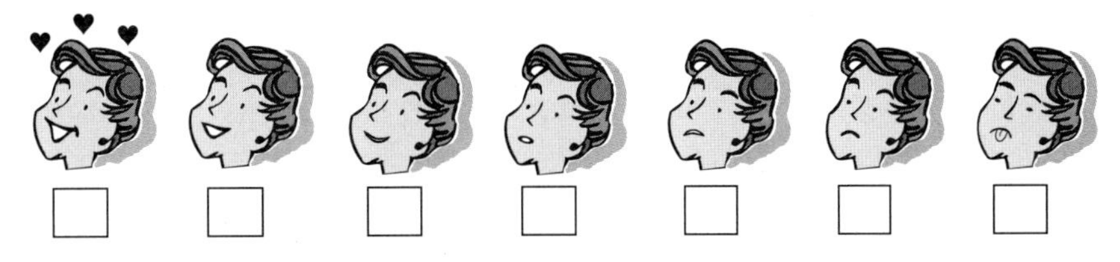

Figure 1.4
Response item for a taste test
Adapted by permission from Mavis B. Carroll.

surveys of existing and potential customers to assess their preferences and to see how one's products might be made more attractive to consumers or more effectively marketed. Market research allows comparisons between existing products and alternatives, permitting a company to find new potential in existing markets and to discover new markets. It also provides a guide for developing such markets. The ingenuity needed to design surveys is suggested by the response item shown in Figure 1.4.

- ***Forecasting and Model Building.*** A tire manufacturer may notice that the lifetime of his batches of tires depends on such variables as the composition of the rubber and the temperature at which the tires are processed in the presses on the production line. The relationship between these variables can be worked out by the general statistical method of regression, which yields the equation that best fits the conditions and data. Using this relationship, the manufacturer can optimize production. Forecasting involves estimating future sales, costs, income, and the like, by using historical data together with related statistics, such as government projections of population and personal income. No matter how sound the methods used for such forecasts, uncertainties remain and caution is in order.

Statistics such as the gross domestic product (GDP) or market averages give an overall view of where the national economy is headed. These numbers plus forecasting methods form the basis for estimating tax revenues and governmental outlays. Annual reports give reviews of current results and projections of future returns for companies. Your monthly report to your boss and the balances in your accounts do the same on a personal level. A knowledge of statistics will help you understand numbers and will show you how a systematic approach can improve results in businesses large and small, in industries from manufacturing to service.

EXERCISES 1.1–1.9

Applying the Concepts

1.1 Your employer, a large corporation, has undertaken to study the bad debt the company has incurred.
   a. Following the steps of a statistical investigation, start by defining the population of interest.
   b. Would you look at all accounts payable or just those in arrears?
   c. How would you choose the accounts for the sample?
   d. Exactly what information would you try to gather from each observation in your sample?

1.2 The United States government regularly publishes statistics on unemployment.
   a. What is the population of interest?
   b. Why not conduct a census to collect data from this population?
   c. How would you draw a random sample from this population?

1.3 The consumer price index (CPI) is a well-known measure of inflation. Each month the Bureau of Labor Statistics compiles data making up the CPI.
   a. Describe the population of interest.
   b. What problems could you anticipate in measuring this population?
   c. Would random sampling be an appropriate procedure for collecting data in this instance? Why or why not?

1.4 Suppose you work for Ford Motor Company as a market researcher and you wish to estimate the average age of Mercury lease holders.
   a. What is the population of interest?
   b. How would you draw a random sample from the database in Appendix A (and on the CD in the back of this book)?
   c. Would random sampling be an appropriate procedure for collecting data in this instance? Why or why not?

1.5 The Social Security Administration conducts a monthly Medicare survey to obtain estimates of hospital and medical care services used and charges incurred by persons covered by the SSA's hospital insurance and supplementary medical care insurance programs. This is done to provide an up-front estimate of hospital and medical bills on their way through the system.
   a. What is the population of interest?
   b. How would you draw a random sample from this population?
   c. What data would you wish to collect?

1.6 Countries often conduct household budget surveys in order to decide what to include in the "market baskets" used to determine their consumer price indices. Suppose you were in charge of designing such a survey for the Bahamas.
   a. What is the population of interest?
   b. How would you draw a random sample from this population?
   c. What data would you wish to collect?

1.7 Tourism is a major source of foreign exchange for many countries. The government of India conducted a survey at major air, sea, and land checkpoints from which tourists leave the country. Answer the following as if you were doing the planning for such a survey.
   a. Describe the population of interest.
   b. How would you select a random sample from this population?
   c. What questions would you ask?
   d. Put together a sample of five cases of data you would collect.
   e. Summarize what your findings might be.

1.8 A company has developed a new product: a mini-CD player. Before going whole-hog into production, the president of the company would like to know if this product is going to be successful. The president of the company has asked you to do some market research. You've decided to conduct a survey.

a. What is the population of interest?
b. How would you go about selecting a random sample?
c. How would you collect the data?
d. What questions would you ask?
e. Generate five sample responses.
f. What might your recommendations look like?

1.9 The reliability of a certain product depends upon the tensile strength of a certain component—say, the parachute pull cord. As the quality control supervisor, you have been asked to ensure that specifications are met. Components are made on an assembly line.

a. What is the population of interest?
b. How would you select a random sample?
c. What data would you collect?
d. What conclusions could you reach?
e. Is it possible to reach a wrong conclusion?

SECTION 4 ▪ SOURCES AND TYPES OF DATA

Every business, large or small, uses data to control, improve, and forecast performance. Data are the grist for statistical mills, the raw material for statistical analysis. In Sections 2 and 3, you saw the steps in launching a statistical study and some of the sorts of data that might be gathered. Understanding the different sources and types of data and their uses is essential for using statistics.

Data Sources

One fundamental way of categorizing data is based on the sources of the data: by whom, how, and why the data were gathered. The Bureau of the Census gathers income, cost-of-living, and job statistics in the government's Current Population Survey. These numbers are used to determine cost-of-living increases and to estimate governmental costs and tax receipts. Comparative figures from different parts of the country are used by businesses in planning where to expand or to cut back.

For the government, these surveys yield primary data, gathered by the government for its planning and public information services. For businesses, the statistics produced are secondary data, the sort of thing you might look up in publications such as the *Wall Street Journal*. The advantage of primary data is that the information is specific to your purpose and you know how trustworthy it is. The disadvantages are the cost, time, and trouble of gathering it. The advantages of secondary data are availability, relatively low cost to the user, and, under the best of circumstances, high quality. The disadvantages are that you must look at how the information was gathered and what definitions were used; in addition, it may not provide exactly the facts that you need.

Any study is likely to use several types of data, and what constitute primary data to the gathering agent may later become secondary data for others.

Primary data are data collected by the user to answer specific questions set out at the start of the study. Primary data are often difficult and costly to gather, but if the study is properly designed, the information will be directly suited to answering your questions and of known dependability.

Secondary data are data you make use of that were collected for other purposes. Secondary data may have been gathered by you or by others in your organization for other purposes or to provide general background information, or such data may come from commercial providers, arms of the government, publications, the Internet, or wherever. Although relatively inexpensive and widely available, secondary data may be of unknown accuracy and the information may not be on target for your purposes.

Primary Data Gathered by Businesses. The marketing department of Beauté Garden distributed a brief questionnaire with one of its products, Scent of Lilly perfume. This questionnaire asked about the age, income, and employment of the customer, where the customer bought the product, and whether this was a first-time purchase. It asked how the customer had learned of the product: in an advertise-

Figure 1.5
The questionnaire used by Beauté Garden

We'd like to ask you a favor:
We value the information you can give us about Scent of Lilly.

1. (Check One)
   ☐ I purchased this product for myself.
   ☐ I received this product from a man.
   ☐ I received this product from a woman.
2. Have you ever used Scent of Lilly before?
   ☐ Yes ☐ No
3. Have you seen any advertising for Scent of Lilly in the past six months?
   (Check One) Where? (Check all that apply)
   ☐ Yes ☐ No ☐ TV ☐ Catalogues
   ☐ Magazines ☐ Radio
4. List 3 magazines you read regularly.
   1. ____________ 2. ____________ 3. ____________
5. Employment:
   ☐ Full Time ☐ Part Time ☐ Not Employed Outside the Home
6. Age Group:
   ☐ Under 18 ☐ 18–24 ☐ 25–34 ☐ 35–44 ☐ 45+
7. Annual Household Income:
   ☐ Under $10,000 ☐ $10,000–19,999 ☐ $20,000–29,999
   ☐ $30,000–49,999 ☐ $50,000 +

Name: ______________________________

Address: ______________________________

City: ______________ State: __________ Zip Code: __________

Telephone: ______________

Thank you for your assistance.

ment or from a friend. The purpose was to gather primary data on customers and their buying choices in order to map a more effective selling campaign. The company wanted to know what groups to target and what media to use for advertising. Figure 1.5 reproduces the Beauté Garden questionnaire.

Secondary Data in Newspapers, Magazines, Books, and Databases. The *Wall Street Journal,* the *New York Times,* and a host of general and specialized newspapers and journals make a business of providing basic information, statistics, and data to subscribers. Daily stock reports are an example, as are periodic reports on the money supply and interest rates around the nation. Weekly newsletters and monthly magazines track other statistics for different industries and trades. A wealth of information can be found in the *Statistical Abstract of the United States,* published yearly by the government. Specialized information is available on various industries, some from industry groups (for example, the *Life Insurance Fact Book*). For financial information, you can look in reports by *Moody's, Standard and Poor's,* or *Value-Line* in most libraries. Some data are available from databases in electronic form, either from government sources or through commercial services at a fee.

Primary and Secondary Data Provided by the Government. The primary data the government gathers in its Decennial Census constitute secondary data for other users. Information gathered includes a person's age, sex, marital status, race, and place of residence. See the 1990 Individual Census Report in Figure 1.6 on page 16. For the Current Population Survey, about 60,000 households are interviewed each month about employment status. For the Consumer Price Index (CPI), the Bureau of Labor Statistics gathers information yearly on roughly 1,300,000 food items, 462,000 nonfood items, 80,000 rental fees, and property taxes on 13,000 titles.[1] Data from census forms are presented in tables such as Table 1.2 on page 17.

All aspects of the economy receive government attention, including agriculture, mining, energy, health care, and education. For a baseline number or the record of growth over time, you would do well to start with government sources. The Bureau of the Census has a useful Internet site at http://www.census.gov/. Another interesting site for government statistics is the site of the Bureau of Labor Statistics, at http://stats.bls.gov/blshome.htm.

In addition to primary data, governments have access to large collections of secondary data, such as statistics gathered by other governments and international bodies, including the United Nations (UN) and the World Health Organization (WHO). Beyond these sources, trade associations such as petroleum refiners or producers regularly gather statistics, though these may not always be available to outsiders. United Nations publications are a good place to look for general worldwide baseline figures.

The Sources, Methods, and Motives Behind Secondary Data. The sources and methods used to gather the data are an important consideration in deciding what use you can make of data or statistics. Knowing how the *Literary Digest* was gathering its data on the 1936 election, George Gallup had reason to believe the

---

[1]Bureau of Labor Statistics, *Handbook of Methods,* 1976.

Figure 1.6
Questionnaire used by the Bureau of the Census

1990 INDIVIDUAL CENSUS REPORT

1. Please print your name:
   Last name First name Middle initial

   ____

2. a. Are you: (Mark (X) in the box that applies.)
   (1) ☐ a person WHO USUALLY LIVES HERE or who stays here most of the week while working? — Please continue with question 3.
   (2) ☐ a person with NO USUAL PLACE OF RESIDENCE?
   (3) ☐ a person AWAY FROM YOUR USUAL HOME FOR A SHORT TIME, such as on a vacation or business trip? — Print your home address in b, and continue with question 3.
   b. House number, street name, apartment number

   ____

   Rural route number Box number

   ____

   City

   ____

   County or foreign country

   ____

   State Zip code

   ____

   Telephone number (Include area code.)

   ____

   Names of nearest intersecting streets or roads

   ____

3. Sex: (Mark (X) ONE box.)
   1. ☐ Male 2. ☐ Female
4. Race: (Mark (X) ONE box for the race you consider yourself to be.)
   1. ☐ White
   2. ☐ Black or Negro
   3. ☐ Indian (Amer.) (Print the name of the enrolled or principal tribe.)

   ____

   ____

   4. ☐ Eskimo
   5. ☐ Aleut

   <u>Asian or Pacific Islander (API)</u>

6. ☐ Chinese	11. ☐ Japanese
7. ☐ Filipino	12. ☐ Asian Indian
8. ☐ Hawaiian	13. ☐ Somoan
9. ☐ Korean	14. ☐ Guamanian
10. ☐ Vietnamese	15. ☐ Other API

   ____

   ____

5. Age and year of birth
   a. Age b. Year of birth
   _____ 1 _____

Figure 1.6 (continued)

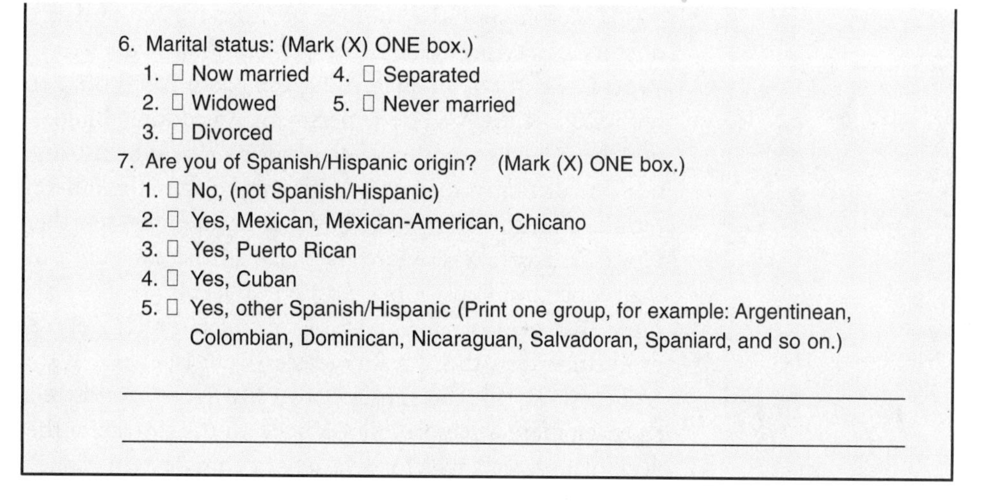

6. Marital status: (Mark (X) ONE box.)
   1. ☐ Now married 4. ☐ Separated
   2. ☐ Widowed 5. ☐ Never married
   3. ☐ Divorced
7. Are you of Spanish/Hispanic origin? (Mark (X) ONE box.)
   1. ☐ No, (not Spanish/Hispanic)
   2. ☐ Yes, Mexican, Mexican-American, Chicano
   3. ☐ Yes, Puerto Rican
   4. ☐ Yes, Cuban
   5. ☐ Yes, other Spanish/Hispanic (Print one group, for example: Argentinean, Colombian, Dominican, Nicaraguan, Salvadoran, Spaniard, and so on.)

____

____

magazine's predictions would be wrong. Gallup correctly predicted the results of the *Digest's* survey from his own mail sample of size 3000 from that magazine's lists. This example shows the importance of understanding the source and methods used to collect data if you are to judge the accuracy of the information or make use of it.

Table 1.2

POPULATION DATA FROM THE GOVERNMENT OF THE UNITED STATES OF AMERICA

No. 19. Resident Population, by Race and Hispanic Origin: 1980 to 1991

[In thousands. As of July, except as indicated. These data are consistent with the 1980 and 1990 decennial enumerations and have been modified from the official census counts; see text, section 1 for explanation]

YEAR	Total	RACE				Hispanic origin[1]	NOT OF HISPANIC ORIGIN, BY RACE			
		White	Black	American Indian, Eskimo, Aleut	Asian, Pacific Islander		White	Black	American Indian, Eskimo, Aleut	Asian, Pacific Islander
1980 (April)	226,546	194,713	26,683	1,420	3,729	14,609	180,906	26,142	1,326	3,563
1980	227,225	195,185	26,771	1,433	3,837	14,869	181,140	26,215	1,336	3,685
1981	229,466	196,635	27,133	1,483	4,214	15,560	181,974	26,532	1,377	4,022
1982	231,664	198,037	27,508	1,537	4,581	16,240	182,782	26,856	1,420	4,367
1983	233,792	199,420	27,867	1,596	4,909	16,935	183,561	27,159	1,466	4,671
1984	235,825	200,706	28,212	1,656	5,249	17,640	184,243	27,444	1,512	4,986
1985	237,924	202,031	28,569	1,718	5,606	18,368	184,945	27,738	1,558	5,315
1986	240,133	203,430	28,942	1,783	5,978	19,154	185,678	28,040	1,606	5,655
1987	242,289	204,770	29,325	1,851	6,343	19,946	186,353	28,351	1,654	5,985
1988	244,499	206,129	29,723	1,923	6,724	20,786	187,012	28,669	1,703	6,329
1989	246,819	207,540	30,143	2,001	7,134	21,648	187,713	29,005	1,755	6,698
1990 (April)	248,710	208,704	30,483	2,065	7,458	22,354	188,300	29,273	1,796	6,988
1990	249,415	209,150	30,620	2,075	7,570	22,554	188,559	29,400	1,806	7,096
1991	252,177	210,899	31,164	2,117	7,996	23,350	189,588	29,910	1,845	7,504

[1]Persons of Hispanic origin may be of any race.

Source of Data: U.S. Bureau of the Census, *Current Population Survey,* p. 25-1095.

Understanding the motives of those collecting data can be important too. Spin doctors shape questions so as to get the answers they want. Here is an example. Pepsi-Cola™ ran an advertising campaign based on a comparison of their product with Coca-Cola™. The experiment was double blind—neither the experimenters nor the subjects were told which drink was which. To give the appearance of fairness, the drinks were served in identical glasses; the different types were coded by letter in a way known to neither the experimenters nor the subjects, all of whom were Coca-Cola drinkers. More than half of the subjects said they preferred the glasses that contained Pepsi-Cola. So here it seems we have a scientific test showing that more than half of all Coca-Cola drinkers prefer Pepsi-Cola over Coca-Cola. The problem was in the way the glasses were labeled—"Q" for the glasses with Coca-Cola and "R" for the glasses with Pepsi-Cola. Given the motives of those collecting the data, there is reason to look beneath the surface of the experiment. That look reveals that subjects will, for a variety of complex emotional and aesthetic reasons, favor glasses marked "R" over those marked "Q," no matter which drink is in the glasses.

Data Types

Cross-Sectional versus Time Series Data. Like the distinctions between primary and secondary data, the distinctions between cross-sectional and longitudinal data are tied to the type of study and the sources of the data. Cross-sectional and time series data give different pictures even when we look at the same population of individuals.

> **Cross-sectional data** give a snapshot of a population at a given time.
>
> Cross-sectional data stand in contrast to **time series data,** or **longitudinal data,** which track the evolution of individuals or populations over time.

The type of data collected will depend on the purposes of the study, and the terms *cross-sectional* and *longitudinal* refer to the type of study as well as the nature of the data. If we wanted to know about the educational development of children, we would plan a longitudinal study. If we wanted to know how the current graduating class stands, we would gather cross-sectional data on that class.

Consider the sizes of children in a gym class. To pick out the right-sized uniforms for this year, you need cross-sectional data. But if you are a parent of one of these same children and are wondering how long your child will be able to wear clothes that fit now, you want time series data on children's growth. Following are examples from business and sports that illustrate the difference between these two types of data.

Figure 1.7 is a bar graph of sales of the Hudson Bay Company from 1986 through 1994. This picture packs a surprising punch. If your three-year-old cousin were to shrink by almost a quarter of her height between her third and fourth birthday, the decline would be similar to that shown for 1988. Time series, or longitudinal, data reveal both regular patterns of growth and departures from such patterns—departures that should be looked into and understood. What *did* happen to Hudson Bay between 1987 and 1988? Mired in debt, the company was forced to sell its oil and gas operations and 179 of its northern province stores. So the company did shrink, for a reason, and then resumed its growth.

Figure 1.7
Time series data

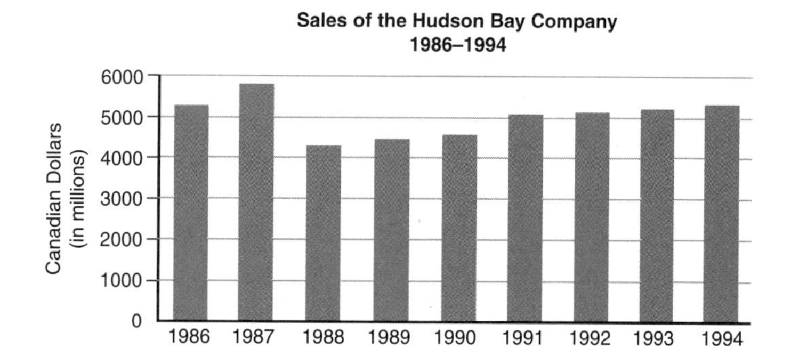

Table 1.3 gives the batting statistics for a recent season of the American League. The cross-sectional data tell you how the teams compared in batting in 1993. They tell you nothing about previous years or about future ones.

Table 1.3

CROSS-SECTIONAL DATA: AMERICAN LEAGUE 1993 BATTING STATISTICS

Team	Average	Team	Average
New York	.279	Minnesota	.264
Toronto	.279	Boston	.264
Cleveland	.275	Kansas City	.263
Detroit	.275	Seattle	.260
Texas	.267	California	.260
Baltimore	.267	Milwaukee	.258
Chicago	.265	Oakland	.254

Source of Data: M. Simpson, ed., *Baseball America's 1994 Almanac.*

Quantitative versus Qualitative Data. Some sorts of data can be usefully averaged (for example, sales per customer at a gas station), but other sorts of data can't be averaged (for example, the states represented by the license plates on the cars pulling into the gas station). So, in statistics, we distinguish quantitative variables, which can be averaged, from categorical variables, which can't. Quantitative variables, which must take on numerical values that can be added together or averaged, include weights and heights. Categorical (qualitative) variables might tell us where individuals are from or whether they are male or female, Republicans or Democrats. You can't average categorical variables, even if they are in the form of numbers, such as ZIP codes.

Quantitative variables are those taking on numerical values that can be meaningfully averaged and subjected to other arithmetic operations. Data involving such variables are called **quantitative data.** (See Figure 1.6, question 5(a).)

Qualitative, or **categorical, variables** are variables (such as political party or marital status) that specify which of various categories the individual belongs to or what qualities he or she possesses. Arithmetic operations can't be applied to such values. Data involving such variables are called **qualitative data** or **categorical data.** (See Figure 1.5, question 2.)

Because quantitative and qualitative variables often are studied together, the data arising from a given study may be partly quantitative and partly qualitative.

This question is from the Scent of Lilly survey:

2. Have you ever used Scent of Lilly before?
☐ yes ☐ no

Here the variable is qualitative, as are the data.

The following question is from a survey of people leasing Lincoln-Mercury vehicles:

24. What make of car did you previously drive?
☐ Ford ☐ GM ☐ Chrysler ☐ European
☐ Japanese ☐ Other ☐ None

Again, the variable and the data are qualitative, or categorical.

A health questionnaire might ask a woman for the number of children she has had; this is a quantitative variable. The responses can be averaged, though one needs to keep in mind what such an average means.

Continuous versus Discrete Data. Some numerical variables can, in principle, take on any value in a given range. These variables, such as your height or weight, are continuous variables. Other numerical variables, such as the number of siblings you have, are limited, in principle, to a set of possible values, each separated from the others with an intervening gap. These are discrete variables. Your age is a continuous variable. Your age at your most recent birthday is a discrete variable. Note that despite the fact that you can only weigh yourself to plus or minus a few pounds on your bathroom scale, your weight actually varies continuously over a range. The situation with a variable such as age at most recent birthday is quite different. It cannot be 21.563425 or π. It must be 21 or 22 or 3. This is what distinguishes a continuous from a discrete variable.

> A **continuous variable** is a numerical variable that can, in principle, take on all values within a certain interval or intervals. Examples are weights, heights, and speeds. Data involving such variables are called **continuous data.**
>
> A **discrete variable** is a numerical variable that can take on only a limited set of values, each separated from the others by a certain interval. Typically discrete variables count things, such as the number of pages in a book, the number of unemployed persons, or the number of angels that can dance on the head of a pin. Data involving such variables are called **discrete data.**

Consider the two variables asked for in the following question from the 1990 census:

5. Age and year of birth:
a. Age b. Year of birth
_____ 1____

The second variable is discrete, but the first may be either discrete or continuous, depending on the precise definition given by the Bureau of the Census. If the Bureau's definition is your age in years as of your most recent birthday, then this is a discrete variable. But if the definition is your age in years, months, days, hours, seconds, etc., at the instant the question was answered, then the variable is continuous. Note that categorical variables such as "Likes," "Neutral," and "Dislikes" may be coded as "1," "2," and "3." This makes sense because there is a clear progression both in the range of liking and in the numerical range. But there is not a strict numerical relation between the categorical variables and the numbers used to code them, nor can there be such a strict correspondence. This brings us to our next subject—the quality of data.

The Quality of Data. Data range from categorical data, which are purely descriptive, to ratio data, which are numerical and subject to all arithmetic and mathematical operations. Within this range, we have data that merely name some aspect of the individuals in our population, such as their political party. Such data are nominal.

In the Scent of Lilly questionnaire, question 3 provides an example of a nominal variable.

3. Have you seen any advertising for Scent of Lilly in the past six months?
(Check One)
☐ Yes ☐ No

The question asks the respondent for a "yes" or "no" response. With nominal variables, information concerning the percentage of respondents falling into a particular category may be calculated. Nominal data are associated with classifications.

Next come data that have a natural order. For example, questions that give alternatives such as "Likes," "Neutral," and "Dislikes" provide ordinal data. The structure of such data allows further analysis.

The Windsor Shirt Company conducted a survey at its manufacturer's outlet stores. The survey asked respondents to rate various factors. These ratings are an example of an ordinal variable. A series of statements was followed by four ratings categories, such as the following:

	Excellent	Good	Fair	Poor
Good buys for your money	☐	☐	☐	☐

Not only do we have categories of ratings, but we can say that one rating is better than another. We can order responses. "Excellent" is preferable to "Good," and so on. We could not say that with nominal variables.

Next come data that allow meaningful comparisons of the intervals between data points, without necessarily allowing us to fix a zero point. For example, both Muslims and Christians will agree that Charlemagne ruled for an interval of 43 years, although they may not agree on how these years should be numbered.

Again, question 5 from the Census Bureau questionnaire illustrates the point.

5. Age and year of birth:
a. Age b. Year of birth
_____ 1____

Year of birth is measured from an arbitrary zero point, which depends on whose calendar is used. We can consider differences in interval data, but we cannot consider one observation as a multiple of another.

Finally, we come to ratio data, which have a fixed zero point and can be compared or combined in any way you wish. Ratio data allow the most extensive and detailed analysis.

In the Scent of Lilly questionnaire, the ages asked about in the "Age Group" question are ratio variables. Of course, we lose some of the information in these ratio variables when we record the information only by groups.

6. Age Group:
☐ Under 18 ☐ 18–24 ☐ 25–34 ☐ 35–44 ☐ 45+

However, the actual ages, like all ratio data, have the characteristics of interval data plus one more: a meaningful zero point. When we are born, we are at age zero. Zero income is very meaningful and very depressing!

Nominal data are categorical data. If numerical coding is used, it is only a substitute for the name of the category. There is no natural ordering or way of associating an interval with the values of nominal variables.

Ordinal data admit a natural ordering of the values of the variable, and these may be consistently coded by numbers. For example, the grade of "A" may be coded with a "4" and the grade of "D" with a "1," but a "2" is not strictly twice a "1" nor are intervals necessarily meaningful.

Interval data are quantitative data for which the differences between the variables have a definite meaning, but for which there need not be a fixed zero point. With interval data (such as dates), it is not meaningful to say one is twice another.

Ratio data are quantitative data to which the full range of mathematical operations can be meaningfully applied. Examples include weights or costs measured on a common scale.

EXERCISES 1.10–1.20

Applying the Concepts

1.10 Name some activities that generate primary data.

1.11 What distinguishes primary data from secondary data?

1.12 In the 1994 *Berlitz Complete Guide to Cruising and Cruise Ships* by Douglas Ward, cruise ships are evaluated and described extensively. Some of the factors considered are
a. tonnage
b. casino
c. gymnasium
d. country of registry
e. length
f. outside cabins
g. cinema/theater
h. cruise line

Classify each of the factors mentioned with respect to the characteristics of data mentioned in this chapter.

1.13 Classify the following as primary or secondary sources of data:
a. a survey to measure unemployment in the United States, conducted by the Bureau of Labor Statistics and Bureau of the Census for government uses

b. *Compton's Interactive Encyclopedia*
c. an experiment measuring the effectiveness of a new product
d. the *Wall Street Journal*
e. the Scent of Lilly questionnaire

1.14 Tell whether the data below are quantitative or qualitative; discrete or continuous; nominal, ordinal, interval, or ratio.
a. employment status
b. age
c. gender
d. magazines read regularly
e. gross domestic product
f. household income

1.15 Mannington Mills sent out a survey to all customers who purchased a Mannington floor covering. Part of their questionnaire appears below:

1. Which brand of Mannington floor covering did you purchase? (Check one box only.)
   ☐ Beau Flaire ☐ Acclaim ☐ Boca
   ☐ Architect's Choice ☐ Aristocon ☐ Vega
2. How old was the flooring you replaced or covered?
   ☐ Less than 1 year
   ☐ 1 to 3 years
   ☐ 3 to 5 years
   ☐ 5 to 7 years
   .
   .
   .
   ☐ Over 15 years

a. Are the data from the first question qualitative or quantitative? Explain.
b. Are the data from the first question cross-sectional or time series?
c. Are the data from question 2 qualitative or quantitative?
d. Are the data from question 2 discrete or continuous?
e. Describe the quality of the data in each question—nominal, ordinal, interval, or ratio.

1.16 A question in the Mercury Lease Survey asks:

	Definitely Would		Probably Would		Maybe Would		Probably Would Not		Definitely Would Not	
Based on your dealership experience, would you recommend the salesperson?	__	__	__	__	__	__	__	__	__	__
	10	9	8	7	6	5	4	3	2	1

(This is what is called a Likert scale.)
a. Are the data from the questionnaire qualitative or quantitative?
b. Are the data discrete or continuous?
c. Characterize the quality of the data.

1.17 Classify the variables "ease of shopping," "most important ranking," and "marital status" according to the dimensions discussed in this chapter. Are they quantitative or qualitative? Are they nominal, ordinal, interval, or ratio?

1. Ease of shopping in store
   ☐ Excellent ☐ Good ☐ Fair ☐ Poor
2. Rank from 1 to 3 (1 being the highest) what is most important in making your Windsor shopping decision.
   Value ____ Fashion ____ Brand ____
3. Marital status
   ☐ Now married
   ☐ Widowed
   ☐ Divorced
   ☐ Separated
   ☐ Never Married

1.18 The May 1994 issue of *Forbes* provided data on CEO compensation. One partial observation from their table appears below:

Company/ Exec	Industry Rank	Age	Birthplace	Education	Tenure with Firm
Advanced Micro/ Sanders	2	57	Chicago, Ill.	BSEE	25

Classify each variable with respect to all the dimensions discussed in this chapter—that is, qualitative or quantitative, continuous or discrete, nominal, ordinal, interval, or ratio.

1.19 *Consumer Reports* rates hotels with respect to five variables:
a. satisfaction score
b. price
c. value (ranging from better to worse)
d. staff (ranging from better to worse)
e. condition (ranging from better to worse)

Describe the type of data each of these variables represents.

1.20 To compare patient satisfaction in fee-for-service and HMO health plans, a cross-sectional study is in order. To see how the health of subscribers evolves over time, a longitudinal study following the same subscribers would be needed. Describe five situations where cross-sectional studies would be appropriate and five where longitudinal studies would be needed.

NEW STATISTICAL TERMS

acceptance sampling
cases
categorical data
census
continuous variable
cross-sectional data
data
descriptive statistics
discrete variable
inferential statistics
interval data
longitudinal data
nominal data
ordinal data

parameter
population
primary data
qualitative data
quantitative data
random sampling
ratio data
sample
sampling frame
secondary data
statistic
statistical process control (SPC)
statistical quality control (SQC)
statistics
summary statistics
time series data
variable

2

The Organization and Display of Data

OUTLINE

Section 1 **Organizing and Presenting Data**
Beginning the Study | Gathering the Data | Organizing and Presenting the Data | Recommending Action

Section 2 **Stem Plots and Dot Plots**

Section 3 **Graphs and Charts**
Line Graphs | Ogives | Density Curves | Bar Graphs | Pareto Charts | Pie Charts

Section 4 **Using the Computer**
A Line Graph | A Bar Graph | A Pareto Chart | A Pie Chart

LEARNING OBJECTIVES

1. Organize numerical data and categorical data.
2. Make "quick and dirty" assessments of the characteristics of a data set.
3. Present data in graphical form.
4. Distinguish among different types of graphs and their uses.
5. Understand how graphs show the distribution of data.
6. Use a computer to produce graphs.

BUILD YOUR KNOWLEDGE OF STATISTICS

"... of all methods for analyzing and communicating statistical information, well-designed data graphics are usually the simplest and at the same time most powerful."
—Edward R. Tufte, *The Visual Display of Quantitative Information*

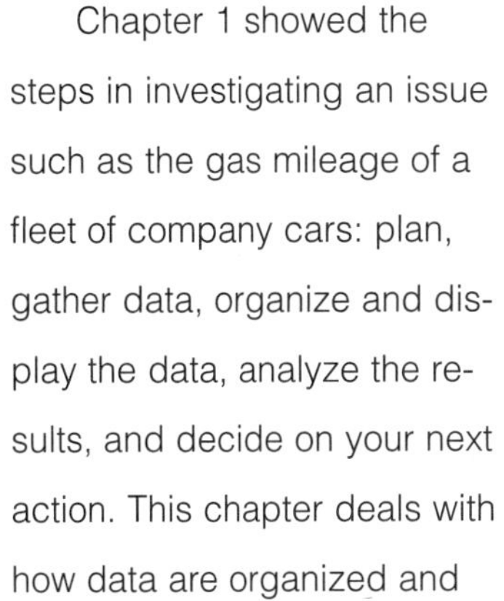

We live in a society dominated by the electronic media. Pictures and graphics deliver much of the information we encounter every day. Data in graphic form, such as tables and charts, are commonly used to communicate quantitative information. In the workplace, the masses of data that confront someone like a stock analyst or a purchasing agent for a large corporation must be organized into chunks before they can be digested. Once data have been organized, they can be presented in the form of tables and charts, which are more immediately understandable. Good graphics reveal patterns such as trends or clusters, and they make a clear and lasting impression. There is nothing like an attractive chart to bring home a point at a meeting. The creation of good graphic displays is a combination of art and science. However, that should not worry the poorest of artists. Even the beginning analyst can construct professional-looking graphs and tables with presentation or word-processing software. This text assumes that you will use such software, which is now universally available in business settings.

Chapter 1 showed the steps in investigating an issue such as the gas mileage of a fleet of company cars: plan, gather data, organize and display the data, analyze the results, and decide on your next action. This chapter deals with how data are organized and grouped by size or type to reveal patterns. This step builds on the data already gathered. You will see how to construct meaningful, attractive, and user-friendly tables and charts. Although computer packages will be used to generate the tables and charts electronically, you must decide what to present and how to present it. You will see that paper and pencil can still play a useful role in quickly roughing out ideas based on small sets of data. If you have questions about producing any type of graph, you may look ahead to Section 3 for guidance.

The pictures introduced in this chapter will help you understand and interpret the analytic statistics in future chapters. In turn, the analytic tools developed later will allow you to produce a number of interesting analyses. These include identifying and plotting trend lines to show how company profits vary over time, as well as setting up quality control charts to monitor production processes.

SECTION 1 ▪ ORGANIZING AND PRESENTING DATA

After data collection, organization is the next step in a statistical investigation. Preparing a simplified table is essential to understanding complex data sets. The following discussion illustrates how a properly devised table can show the distribution of a set of observations, reveal the overall pattern, and indicate which values are typical and which values are outliers. Tables and charts are both important tools in looking at data.

Consider an investment advisor who wants to show a client the characteristics of various mutual funds. Her point of departure is an annual listing in *Money* magazine, a small part of which is shown in Table 2.1. This list has information on the type of fund, such as growth, total return, or income. The investment advisor would like to show her client how the ranges of returns compare for different

Table 2.1

THE 2698 MAJOR FUNDS

				% Gain (or Loss) to July 1, 1996		% Annualized Return to July 1, 1996		Performance Analysis (percentile ranking by type)				% tax	Expense Analysis		Net
Fund name	Type	Style	Risk level	Six months	One year	Three years	10 years	1996	1995	1994	1993	effic-iency	% max sales charge	% annual charges	assets ($ millions)
AARP Balanced Stock & Bond	TR	Lg/Bl	—	4.1	14.4	—	—	32	48	—	—	—	None	1.01	358.6
AARP Capital Growth	Gro	Lg/Val	10	9.5	21.4	12.3	11.6	44	44	6	69	89	None	0.95	801.3
AARP Growth & Income	G&I	Lg/Bl	6	8.7	24.0	16.6	12.7	41	51	89	78	90	None	0.72	3,893.2
► Accessor Growth	Gro	Lg/Bl	4	9.7	21.0	18.8	—	46	63	88	61	91	None	1.26	58.1
◄ Accessor International Equity	Intl	Md/Gro	—	11.5	23.8	—	—	63	36	—	—	—	None	1.83	61.1
Accessor Small Cap	Agg	Sm/Val	10	10.2	27.6	15.0	—	35	53	27	34	90	None	1.31	58.7
Accessor Value & Income	TR	Lg/Val	7	8.6	22.5	13.7	—	88	94	59	65	86	None	1.40	31.8
Achievement Balanced Ret. A	TR	Lg/Bl	—	2.7	12.1	—	—	18	—	—	—	—	4.5	1.15	2.2
Achievement Equity Ret. A	Gro	Lg/Bl	—	5.7	17.6	—	—	13	—	—	—	—	4.5	1.15	2.7
◄ Acorn	Agg	Sm/Bl	10	15.2	24.8	13.4	15.0	67	20	11	96	86	None	0.57	2,856.3
◆ Acorn International	Intl	Sm/Gro	10	17.3	24.6	14.9	—	88	44	42	86	99	None	1.20	1,581.1
Addison Capital Shares	G&I	Md/Val	8	9.2	25.9	14.1	—	51	98	5	59	82	None	2.06	50.6
◄ Advantus Cornerstone A	G&I	Md/Val	—	13.1	30.2	—	—	97	59	—	—	—	5.0	1.35	41.4
Advantus Enterprise A	Agg	Md/Gro	—	7.2	25.3	—	—	15	56	—	—	—	5.0	—	38.4
Advantus Horizon A	G&I	Lg/Gro	9	9.2	22.4	13.8	11.5	52	30	68	10	91	5.0	1.41	35.8
Advantus Intl. Balanced A	TR	Lg/Val	—	6.7	10.8	—	—	72	6	—	—	—	5.0	—	37.9
Advantus Spectrum A	TR	Lg/Gro	7	5.4	15.3	9.4	—	54	50	56	7	82	5.0	1.33	56.3
Aetna Advisor	TR	Lg/Val	—	6.0	17.4	—	—	63	58	—	—	—	1.0[3]	2.04	3.1
Aetna Ascent Select	TR	Sm/Val	—	8.0	21.3	—	—	82	—	—	—	—	None	—	23.9

sorts of funds so that she and her client can settle on a strategy for investing the client's money. They have already decided to look at just growth funds and specialty funds (funds investing in particular segments of the economy, such as chemical companies).

Here we see the main theme of statistical work in business or finance. A choice is to be made as to where resources—in this case, cash assets—are to be directed. The analyst must choose the variables relevant to her client's purposes; in this case, the critical variable is the fund's rate of return. The analyst must gather and organize data and present the client with a recommended investment plan that is supported by the data. If she is successful, she will hold onto this client's business and draw in more customers. In business, finance, or industry, statistical work begins with clear practical goals and ends with operational decisions based on hard data and clear analysis.

Beginning the Study

Let us look at how the analyst proceeds. First, she notes that the full table is complicated and contains data not immediately useful. All she needs is the data on growth funds and specialty funds and their rates of return. She must organize this information and display it in a table or graph to help guide the client's decisions.

Gathering the Data

If the data were in electronic form, the analyst might use a computer to deal with the entire data set. If she was working by hand and had to enter the data herself, she would use random sampling to get a representative sample. Here we will assume that she uses sampling. She wants a sample of 25 of the 718 growth funds in this list, or roughly one in every 30 (as 718/25 is roughly equal to 30). Her sampling frame is the ordered list of 718 growth funds. To take her sample, she picks a growth fund at random from the first 30 growth funds in the list and then takes every 30th growth fund in the list, starting with her random selection. This creates a systematic random sample.

Systematic random sampling

> A **systematic random sample** is obtained by deciding on an interval k between individuals that will give a sample of the desired size, selecting at random a first individual from the first k individuals listed in the sampling frame, and then completing the sample by taking every kth individual in the listing thereafter.

Systematic random samples are easy to draw, but they can be subject to bias. The random choice of the first individual helps to avoid bias, but there can be other problems. For example, if your sampling frame is a year's worth of ridership data on a subway system and k is 7, all the data in your sample will be for the same day of the week, when ridership varies quite a bit weekdays to weekends. So caution is in order with systematic samples.

Following are the two samples our analyst draws. Consisting of 25 observations each, one is a sample of growth funds and one is a sample of specialty funds. Because they give the average annual rate of return over ten years, these values are classified as continuous quantitative ratio data (see Chapter 1).

Rate of Return for Growth Funds

8.8	10.2	11.2	12.3	13.6
9.1	10.6	11.7	12.7	14.1
9.3	10.7	11.8	12.8	14.8
9.4	10.9	12.0	13.0	16.9
9.7	11.1	12.1	13.6	17.4

Rate of Return for Specialty Funds

7.1	10.7	11.8	13.7	15.4
8.3	10.7	11.8	13.8	16.9
8.7	11.2	11.9	14.3	17.1
8.7	11.3	11.9	14.8	17.5
10.6	11.5	12.3	15.2	19.5

Organizing and Presenting the Data

Now the sample data must be organized into groups so that the advisor and her client can easily see how they are distributed. The method described here can be applied to all quantitative data. To begin with, we will work with the growth fund data. The data will be grouped into k classes of equal size, determined as follows:

1. Find the **range,** r, of the data set:

$$\text{Range} = \text{High} - \text{Low}$$

Classifying numerical data

2. Determine the approximate number of **classes** (groups), k, for the data. As a rough guide,

$$k = \sqrt{n}$$

where n is the number of observations.

3. Calculate the **class width** (c.w.):

$$\text{c.w.} \approx r/k$$

where $\approx$ means "approximately equal."

4. Set preliminary **lower class limits (LCLs)** and **upper class limits (UCLs).**
5. Tally the data by classes.
6. Present the final table.

Example 2.1 will show exactly how to carry out these steps.

Generally speaking, no table should have more than 15 classes or fewer than 5 classes. A display with more than 15 classes is visually too busy, whereas a display with fewer than 5 classes is not generally informative enough to reveal a pattern in the data.

EXAMPLE 2.1 ▪ ORGANIZING QUANTITATIVE DATA: RATES OF RETURN ON GROWTH FUNDS

Let's work through the steps our advisor would follow in organizing the data in her sample of growth funds.

Step 1. We find the range:

$$\text{Range} = 17.4 - 8.8 = 8.6$$

Step 2. We determine the approximate number of classes, k:

$$k = \sqrt{25} = 5$$

Step 3. We calculate the approximate class width, c.w:

$$\text{c.w.} = 8.6/5 = 1.72$$

Because 1.72 is not a pretty number, we make it an even 2.0. This will make further calculations easier without any loss of accuracy.

Step 4. We set up preliminary class limits. For the LCLs we start at 8.0 and add the class width. The LCLs are 8, 10, 12, 14, and 16. For the initial UCLs, we subtract one-tenth from the LCL above, which is 10. To get the other UCLs, we add the class width. The UCLs are 9.9, 11.9, 13.9, 15.9, and 17.9. We now have our preliminary class limits (in descending order):

LCLs UCLs
16.0–17.9
14.0–15.9
12.0–13.9
10.0–11.9
8.0–9.9

Our data are rounded to the nearest tenth, so the class 8.0–9.9 includes all data between 8 and 10, including 8 but excluding 10.

Step 5. We tally up the data. We go down each column in the data, entering first 8.8, then 9.1, etc. The end result is shown below:

16.0–17.9	16.9, 17.4
14.0–15.9	14.1, 14.8
12.0–13.9	12.0, 12.1, 12.3, 12.7, 12.8, 13.0, 13.6, 13.6
10.0–11.9	10.2, 10.6, 10.7, 10.9, 11.1, 11.2, 11.7, 11.8
8.0–9.9	8.8, 9.1, 9.3, 9.4, 9.7

Step 6. We present the final table, shown in Table 2.2.

Table 2.2

ANNUAL RATES OF RETURN OVER 10 YEARS FOR A SAMPLE OF 25 GROWTH STOCK FUNDS

Presentation of a table

Rate of Return	Number of Funds
16.0 to 17.9%	2
14.0 to 15.9%	2
12.0 to 13.9%	8
10.0 to 11.9%	8
8.0 to 9.9%	5

Source of Data: *Money*, August 1996.

Table 2.2 allows us to see the pattern in the data. The organization into classes has converted raw data into more useful information, or classified data.

Data organized into classes or groups are known as **classified data.**

Organization makes it possible to interpret data. We see that a majority of the funds exhibited average rates of return of between 10% and 13.9%. Four funds did better than the majority of funds, and five did worse. If your fund experienced a rate of return of between 10% and 13.9%, you're doing OK. If your growth fund did worse than 10%, perhaps it's time to seek a new investment. The investment managers for that fund were either unlucky or poor managers.

Classified data may be displayed in a **histogram,** as shown in Figure 2.1. We will return to histograms later; for now it is enough to know that the areas of the bars in a histogram are proportional to the frequency with which data fall into the range covered by the horizontal base of the bar.

Figure 2.1
Histogram of growth fund performance

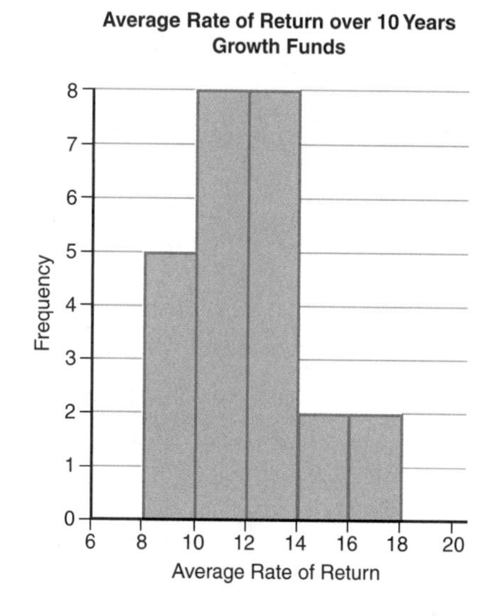

Presentation of a histogram

Following the same steps as in Example 2.1, our advisor would then carry out an analysis of the systematic random sample of 25 specialty funds, to get a comparable histogram. Figure 2.2 shows a histogram for the sample of specialty funds. Note that, in order to facilitate comparison, the same classes are used.

Figure 2.2
Histogram of specialty fund performance

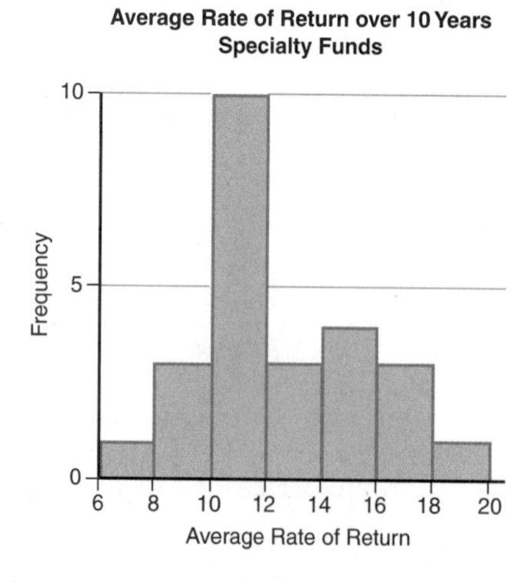

Recommending Action

Comparing the histograms, we see that the specialty funds show greater variability in returns. It may not be possible to argue that one sort of fund is absolutely better than the other, but the histograms provide a clear picture of the relative variability of the two types of investments. Based on this information, the advisor and her client can select a strategy they wish to follow. Perhaps they would put 60% of the client's investment capital in growth funds, with their somewhat greater stability; put 20% in specialty funds, with their greater variability (that is, higher risk); and hold 20% as a cash reserve in a money market account or bond fund.

Unlike the rate of return, which is quantitative, some of the data in the *Money* table are categorical, like the type of fund. To show the distribution of such categorical data, follow these steps:

1. List all the categories.
2. Count the number of individuals in each category.
3. Present the results in a table, a bar graph, or a pie chart.

EXAMPLE 2.2 ■ ORGANIZING CATEGORICAL DATA: TYPES OF FUNDS

Organizing categorical data

What could our advisor do if she wished to tempt her client to look into other types of funds? First, she could list all the categories and tally the number of funds in each, as shown in Table 2.3.

Table 2.3

CATEGORIES OF FUNDS

Fund Type	Count
Total Return	577
Growth	718
International	270
Aggressive	343
Specialty	437
Miscellaneous	353

Then she could present the results to her client in the form of a bar graph, such as Figure 2.3.

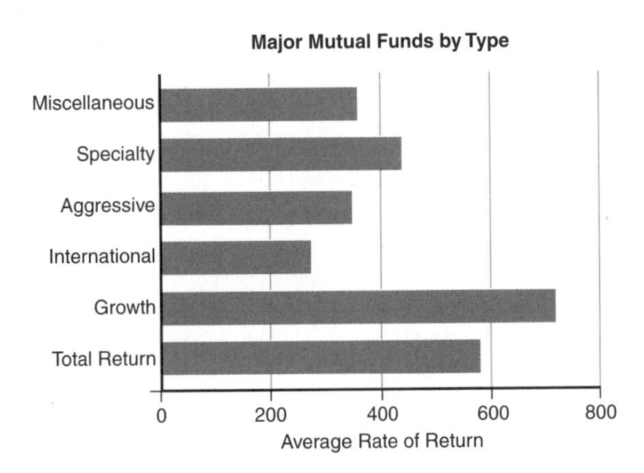

Figure 2.3
Bar graph of types of mutual funds

Now that her client can see at a glance the range of choices for each type of mutual fund, she may persuade him to commission her to do further studies for him.

EXERCISES 2.1–2.10

Skill Builders

2.1 The range of numbers in a data set is 100 units. There are 25 observations. Work out a reasonable class width for these data. If the minimum were 43 and the maximum 143, what lower and upper class limits would you use?

2.2 Organize the following data into classes, and present your results in a table and a histogram.

53	79	72	31	2
4	87	52	28	8
29	16	20	80	41
3	56	72	56	24
28	32	91	90	97

2.3 Organize the categorical data below on blood types into a table. Then show the distribution of these cases using a bar graph.

A	AB	B	O	O	O
B	O	O	AB	AB	A
AB	O	O	O	O	O
B	AB	AB	AB	A	A
A	A	A	B	B	B

Applying the Concepts

2.4 A market researcher collected 50 observations concerning people's expenditures on dining out. Observations ranged from $7.59 to $101.35 per week. Set up the class limits to organize your data. Explain how you made your choices.

2.5 Examine the following table concerning weekly paperboard produced (in thousands of tons) over a given year.

Tonnage Produced	Number of Weeks
800– 850	12
850– 900	13
900–1000	14
1000–1050	11
1100–1149	2

a. What is wrong with this table?

b. If you reconstructed this table, what would be appropriate lower and upper class limits? What further information would be necessary to construct a proper table?

2.6 The prices of two-bedroom homes in Staten Island, a borough of New York City, in June 1994 are listed below. A real estate agent wanted to summarize these prices in a concise and informative manner to show his customers.

87,500	109,900	119,900	129,500	139,500
89,900	114,900	119,900	129,900	149,000
92,000	119,500	124,500	134,900	149,900
96,900	119,900	128,000	135,000	150,000
98,000	119,900	128,500	139,000	153,000

a. Set up class limits and construct a table of the data, explaining the choices you have made.

b. Construct a histogram from the data in your table.

c. If you were in the market for a two-bedroom home in Staten Island in 1994, what should you expect to pay? Explain your reasoning.

2.7 *Consumer Reports' Buying Guide* for 1992 rated umbrellas. As insignificant as umbrellas may seem, some 50 million are sold in the United States each year. The prices of umbrellas vary, as indicated by the following sample data (in dollars):

7	9	12	14	15	17	20	25	30	36
8	10	12	15	15	17	22	26	30	55
8	11	13	15	15	18	24	30	34	65
8	12	13	15	16	20	25	30	35	196

a. Determine the class limits and construct a table for these data.

b. What would you consider a reasonable price for an umbrella? What might constitute a bargain?

c. Construct a table without the highest value. Which table would you use to present a picture of the umbrella market? Why?

2.8 Many passenger vessels cruise the Nile River. Their capacities vary markedly. A travel agent wishing to better understand the Nile River trade collected the following data on passenger capacity:

12	25	33	39	42	50	58	65	75
17	25	33	40	44	51	58	67	75
20	27	33	40	44	51	59	68	75
20	27	35	40	44	52	60	72	80
20	28	36	40	45	52	60	74	80
20	28	37	41	45	54	62	74	81
21	30	38	42	45	54	62	75	85
22	31	38	42	46	55	62	75	92
23	32	38	42	48	56	65	75	105

Source of Data: *Berlitz Complete Guide to Cruising and Cruise Ships*, 1994.

a. Construct a table summarizing passenger capacity data for Nile River vessels. (*Hint:* Organize the table in intervals of 10—that is, 10–19, 20–29, etc. Sometimes data lend themselves to a natural breakout rather than the step-by-step procedure described above.)

b. What is the typical passenger capacity for a Nile River vessel?

2.9 The Grand Prix Formula-One Road Race began back in 1906. Winners have come from all over the world. Data on winners for the period between 1950 and 1994 appear below:

Italy	Austria	United States	Brazil	Australia
Australia	Italy	France	Australia	Brazil
New Zealand	United States	Argentina	Brazil	Finland
Austria	Great Britain	Great Britain	Argentina	Brazil
Brazil	Austria	Great Britain	Great Britain	Great Britain
Argentina	France	South Africa	Great Britain	Great Britain
Australia	Italy	Brazil	Brazil	Brazil
Great Britain	Great Britain	Argentina	France	Great Britain
Great Britain	Austria	Great Britain	Argentina	Germany

Source of Data: *The Universal Almanac,* 1996.

a. Organize the data into a table.

b. Which country has had the most winners? Which of the countries listed has had the fewest winners?

2.10. Average number of weekly labor hours in manufacturing is a leading economic indicator. Data sampled over the period between 1952 and 1992 appear in the table below, in order by year:

39.8	39.6	38.8	39.4	39.7	39.6	39.4	39.4	39.8	40.0
39.7	39.2	38.8	39.7	39.6	39.9	39.6	39.4	39.8	39.7
38.8	38.5	39.3	40.2	38.9	39.4	39.8	38.8	39.9	39.8
39.5	39.3	39.2	39.6	39.1	38.9	39.6	39.3	39.9	40.4

Source of Data: New York State Department of Labor, *Employment Review,* May, 1994.

a. Construct a table summarizing the movement in the average number of weekly hours in manufacturing.

b. Over the past 40 years, what would you describe as typical?

SECTION 2 ■ STEM PLOTS AND DOT PLOTS[1]

Building a stem plot

When you are out in the field, far from your desk and computer, it is often necessary to take a quick look at a small data set in order to get a rough feeling for a situation. Stem plots and dot plots are excellent tools for such times, although you will also find them in most software packages.

We begin with a set of data on the prices of fluoride toothpastes. (We will reexamine these data using analytic tools in Chapter 3.)

Raw Data on Fluoride Toothpaste Prices

1.53	2.04	1.99	1.97	2.53
2.23	3.18	2.01	2.04	1.70
2.96	2.00	2.03	3.29	2.65
2.53	1.53	2.55	1.97	1.79
1.99	2.06	1.94	2.24	3.26

To see a pattern in such a sea of numbers, you need to organize them. Working by hand, you might arrange the toothpaste data in a **stem plot,** which is an arrangement like the one in Figure 2.4.

[1]This is an optional section.

Figure 2.4
Stem plot of the toothpaste data

Stem	Leaf
1	53 53
1	70 79
1	94 97 97 99 99
2	00 01 03 04 04 06
2	23 24
2	53 53 55
2	65
2	96
3	18
3	26 29

On the left of a stem plot is the vertical list of base numbers known as the stem. These establish the class limits for the display. Stretching to the right of the stem are the leaves, usually given in increasing numerical order. John Tukey, a pioneer in the use of such graphs, called the plots stem-and-leaf plots. Each leaf represents one data point. Thus, for instance, the two 53s in the top line in Figure 2.4 represent the two toothpastes in this survey priced at \$1.53. Notice that a stem plot not only captures all the information in the original data, but also puts the data in order and breaks the data into class intervals. These class intervals are determined by the numbers we choose for the base values on the stem and can be varied at will to create a more informative plot.

Many variations are possible. The values along the stem can be arranged in increasing or decreasing order to suit your purpose. It is common for the leaves to be indicated by single digits and written next to each other; this is the format software will generally produce. So as to avoid the complications associated with special rules for abbreviating data, in this book we use a stem-and-leaf design in which the leaves are tied directly to the data.

Notice that a stem plot gives us just the information we need to draw a rough histogram for the data by hand, as shown in Figure 2.5. The heights of the bars of the histogram correspond to how far the leaves would stretch out at each level if

Figure 2.5
Histogram of the toothpaste data

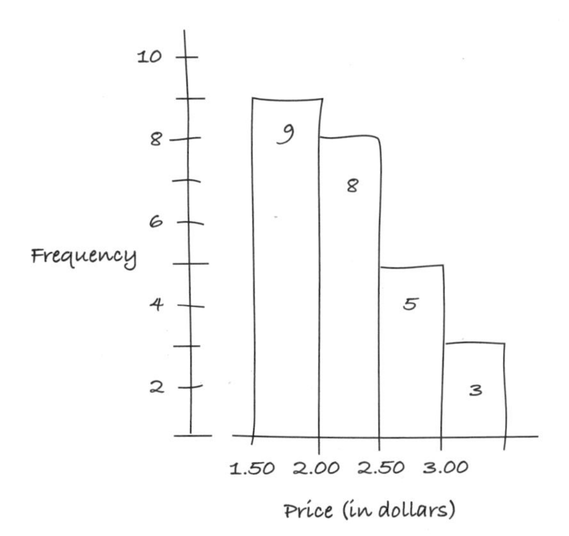

Constructing a histogram

you spaced them consistently. The equal intervals between the base numbers on the stem translate into equal widths for the class intervals of the histogram. The bars above these intervals therefore have areas proportional to the frequencies they represent, as required in a histogram.

There are 25 data points in the toothpaste data set—the same number of data points as there were in the samples of stock funds our analyst presented to her client. The analyst could easily have roughed out her whole report on the train as she commuted to work, and then left it with her secretary to put into final form. He could have created the graphics on his computer while she went on to other tasks. Indeed, the list in step 5 of Example 2.1 is essentially a stem plot for the rates of return of growth funds.

A comparison of growth and specialty funds can be made by means of **back-to-back stem plots,** as shown in Figure 2.6, as well as the pair of histograms in Figures 2.1 and 2.2. In a back-to-back stem plot, a common stem is used for right-hand and left-hand sets of leaves, and the values on any line increase as you go away from the stem.

Growth Funds		Specialty Funds
	18	19.5
17.4 16.9	16	16.9 17.1 17.5
14.8 14.1	14	14.3 14.8 15.2 15.4
13.6 13.6 13.0 12.8 12.7 12.3 12.1 12.0	12	13.3 13.7 13.8
11.8 11.7 11.2 11.1 10.9 10.7 10.6 10.2	10	10.6 10.7 10.7 11.2 11.3 11.5 11.8 11.8 11.9 11.9
9.7 9.4 9.3 9.1 8.8	8	8.3 8.7 8.7
	6	7.1

Figure 2.6
Back-to-back stem plot comparing percent return for two types of funds, based on two samples of size 25

If you make regular use of stem plots and other hand methods, you will gradually develop many shortcuts and refinements to speed your work. John Tukey's 1977 book *Exploratory Data Analysis* brought stem plots and other informal graphical methods into the mainstream, and others have since simplified and extended his ideas. The brief outline here is intended to help you to analyze situations without the aid of machines or software.

Like the stem plot, the **dot plot** is easy to understand and to draw by hand. Dot plots are also a common option for computer display of data. The dot plot for the toothpaste data is shown in Figure 2.7. A dot plot is constructed by drawing a scale for the variable and then representing each data point by a dot above the corresponding value. The dots may need to be stacked, as in Figure 2.7.

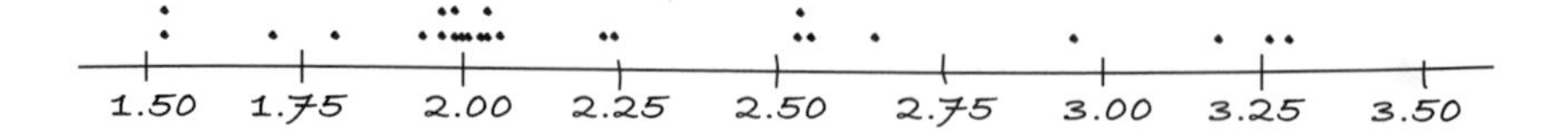

Figure 2.7
Dot plot of the toothpaste data

Stem plots, histograms, and dot plots show patterns not evident in the original blocks of data. What they help make evident is the distribution of the data by size—either grouped, as in a stem plot or histogram, or spread out on a line, as in a dot plot. The way data are distributed or spread out, among various classes or along a line is an important concept in statistics.

A **distribution** is a display of data organized by size, possibly in groups, so that patterns may be observed.

EXERCISES 2.11–2.16

Skill Builders

The following exercises will give you practice in organizing and displaying data and in using your judgment to reach conclusions. The stem plots, dot plots, and bar graphs may be done by hand or using software. Exercises 2.11–2.16 contain sets of test scores from actual small classes. For each of these sets:

a. Make a stem plot of the data.
b. Make a dot plot of the data.
c. Decide how you would award grades of A, B, C, D, and F for these tests. Use your judgment; *many* answers are possible.
d. Make a bar graph showing the distribution of grades in each class.

2.11 Test scores: 78, 72, 91, 90, 91, 96, 75, 43, 43

2.12 Test scores: 77, 94, 99, 44, 51, 97, 73, 82, 79, 74, 73, 63, 96, 60, 71, 96, 53, 94, 31, 88, 81, 34

2.13 Test scores: 57, 48, 81, 84, 86, 67, 67, 64, 67, 87, 90, 66, 72, 96, 85, 79, 76, 92, 86, 79, 98

2.14 Test scores: 97, 51, 54, 96, 70, 70, 93, 61, 38

2.15 Test scores: 40, 99, 94, 77, 81, 81, 76, 70, 70, 78, 60, 0, 37, 83, 44

2.16 Scores on a 200-point final: 136, 163, 129, 176, 191, 160, 93, 126, 61, 104, 125, 141, 74, 187, 186, 184, 93, 109, 65, 115, 134

Notes: The stem plots and dot plots in parts a and b preserve all the information in the data. In part c, hard decisions must be made, and a considerable amount of information is lost in going from scores to grades. Even earlier, much information may have been lost, because two sets of answers receiving a score of 87 may reveal very different patterns of understanding on the part of the students who wrote them. These exercises show that, although graphics and statistical analysis can help us to grasp the essentials of what is going on, there is no way to avoid loss of information when you boil down a complicated situation to a few numbers or letters.

SECTION 3 ▪ GRAPHS AND CHARTS

Numbers that appear in text and tables provide detailed information. A well-organized table, as you have seen, can tell a great deal about a set of data. Organized into a table, the data on stocks provided much information to guide investing. This information can have even greater impact when presented graphically. Because they are visual displays, graphs may allow you to notice trends and patterns more quickly than tables do. In this section, we shall examine line graphs, ogives, density curves, bar graphs, Pareto charts, and pie charts.

Line Graphs

Line graphs are connected points that exhibit trends in the values of variables. They are especially useful in illustrating monthly, quarterly, or yearly data. Every graph should have a title, which tells the reader what the graph is about. It should be brief, but informative. The vertical axis, usually on the left, gives the scale and units of measurement, such as percentages, dollars, or counts. The horizontal axis often reflects units of time—weeks, months, or years, for example. The source of the information graphed should be indicated somewhere toward the bottom left.

Figure 2.8 is a graph of McDonald's sales data, presented to stockholders as part of the company's annual report. The graph shows strong growth, but the reader would have to go to the narrative to see how this growth was tied to company strategy and future prospects. In business, such graphs are simply a means to a larger end; they are not the end of the analytic and planning process.

Figure 2.8
Line graph showing trends in sales

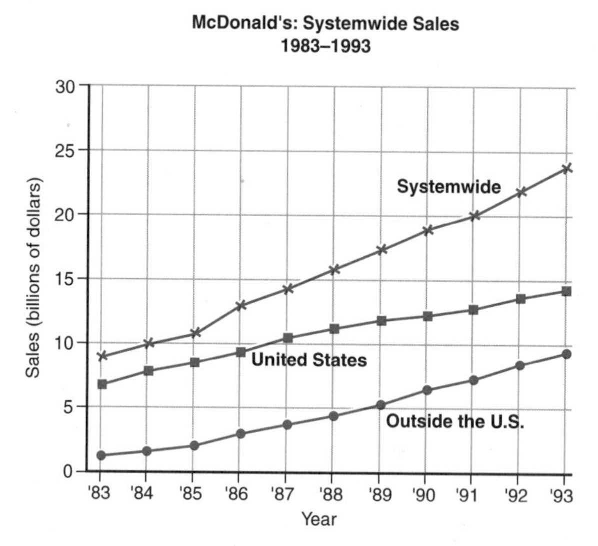

Source of Data: *Annual Report,* 1993.

Drawing a line graph

In Figure 2.8, the first line of the title indicates the subject matter of the graph. The time period covered is indicated on the second line of the title. To the left is the scale, in billions of dollars, divided into even segments of $5 billion. If the scale does not start at 0, there should be a small break at the bottom to indicate the discontinuity. The time period is indicated along the horizontal axis. For ease of presentation, the first two digits of the year have been omitted.

The top graphed line illustrates systemwide sales. The line indicates a steady increase in sales from just under $10 billion to about $25 billion. The second graphed line shows the course of sales in the United States. We see that at the beginning of the time period studied, U.S. sales made up about 80% of McDonald's revenues. Over the time period, sales outside the United States increased in importance. (Notice the increasing space between systemwide sales and U.S. sales.) The bottom graphed line shows sales outside the United States. The line increases slowly at first and then more quickly in the 1990s. It would be difficult to see these trends and relationships in a table.

EXAMPLE 2.3 ■ USING A LINE GRAPH TO SHOW A TREND IN SALES OF ELECTRICITY

The "Statistical Review" section of Detroit Edison's 1992 *Annual Report* contained the data shown in Table 2.4 on system sales. Using these data, we will construct a graph to show the trend in total system sales.

Table 2.4

SALES OF ELECTRICITY

Year	Sales (millions of kWh)	Year	Sales (millions of kWh)
1982	31,995	1987	39,492
1983	34,299	1988	41,144
1984	35,887	1989	40,684
1985	36,695	1990	41,049
1986	38,040	1991	41,049
		1992	40,697

Source of Data: Detroit Edison, *Annual Report,* 1992.

A line graph is appropriate for showing a trend, because it makes growth or decline easy to see. The line will rise more steeply if growth is rapid than if growth is slow. A good title is "Detroit Edison: Total System Sales, 1982–1992." Time is often shown increasing from left to right horizontally. In keeping with this convention, we put years on the horizontal axis. To avoid clutter in the axis labels, we use only the last two digits of each year. Here sales are represented on the vertical axis. The completed line graph is shown in Figure 2.9.

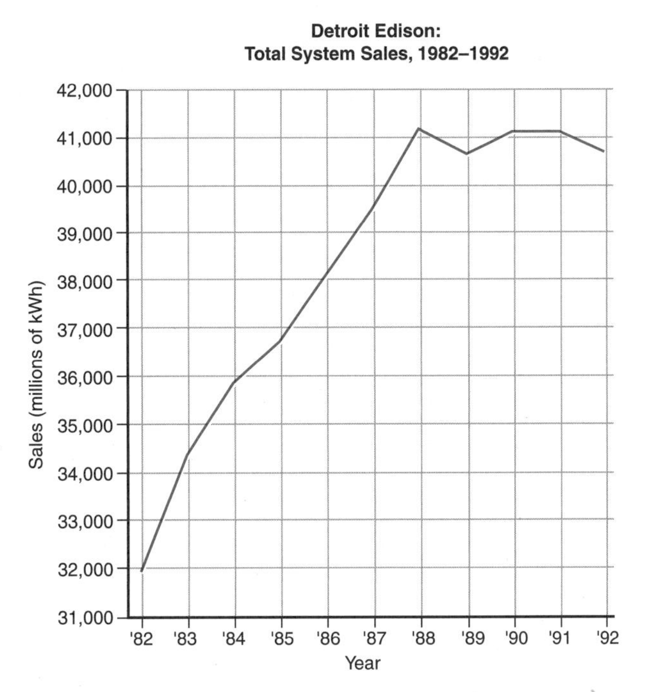

Figure 2.9
Line graph showing a trend in sales

The graph clearly illustrates the growing demand facing Detroit Edison. It shows that consumption peaked for the first time in 1988 and once again in 1990. It appears that the trend may be leveling off. This leveling is something the company will have to deal with.

Ogives

An ogive (pronounced o-jive) is a special type of line graph. Used with classified data, it shows the cumulative frequency distribution (CFD) for a variable.

> A **cumulative frequency distribution (CFD)** shows the total number of values that fall below an upper class limit. It is the statistical equivalent of a "running total."

Consider the distribution of household income in the United States in 1996, given in Table 2.5. Information of this sort would be useful in estimating the size of the potential market for various upscale products and services.

Table 2.5

HOUSEHOLD INCOME IN THE UNITED STATES, 1996

Income Category	Households (000)	Income Category	Households (000)
Under $10,000	11,880	60,000–69,999	6,717
10,000–19,999	16,526	70,000–79,999	5,072
20,000–29,999	14,887	80,000–89,999	3,447
30,000–39,999	12,475	90,000–99,999	2,408
40,000–49,999	10,552	100,000 & over	8,293
50,000–59,999	8,763		

Source of Data: U.S. Bureau of the Census, *Current Population Survey,* Table H-1: Selected Characteristics of Households, by Total Money Income in 1996.

The table is interesting in and of itself, but it does not easily lend itself to answering questions, such as the following: What percentage of households earn less than $20,000 (the approximate poverty level for a family of six)? What is the median level of income (a concept that will be explored in more detail in Chapter 3)? An ogive helps us answer these questions.

> An **ogive** is the line graph of the cumulative frequency distribution for classified data.

To construct an ogive, we must expand upon Table 2.5. The worksheet for this process appears in Table 2.6. The first data column contains the actual number of observations in each category. The second data column contains the cumulative frequency distribution (running total) for the data. The cumulative relative distribution consists of the cumulative frequency distribution divided by the total number of observations. The cumulative percentage distribution is 100 times the cumulative relative distribution.

Drawing an ogive

Table 2.6

WORKSHEET FOR DRAWING AN OGIVE

Income Category	Households (000)	Cumulative Freq. Dist.	Cumulative Relative Dist.	Cumulative Pct. Dist.
Under $10,000	11,880	11,880	.12	12
10,000–19,999	16,526	28,406	.28	28
20,000–29,999	14,887	43,293	.43	43
30,000–39,999	12,475	55,768	.55	55
40,000–49,999	10,552	66,320	.66	66
50,000–59,999	8,753	75,083	.74	74
60,000–69,999	6,717	81,800	.81	81
70,000–79,999	5,072	86,872	.86	86
80,000–89,999	3,447	90,319	.89	89
90,000–99,999	2,408	92,727	.92	92
100,000 & over	8,293	101,020	1.00	100
Total	101,020			

Thus, the **cumulative relative distribution** is given by

$$\text{CRF} = \frac{\text{Cumulative frequency distribution}}{\text{Total number of observations}}$$

and the **cumulative percentage distribution** is given by

$$\text{CPD} = \text{CRF} \times 100$$

An ogive is constructed by plotting the values in the last column of the table against the class limits. The ogive for Table 2.6 is illustrated in Figure 2.10.

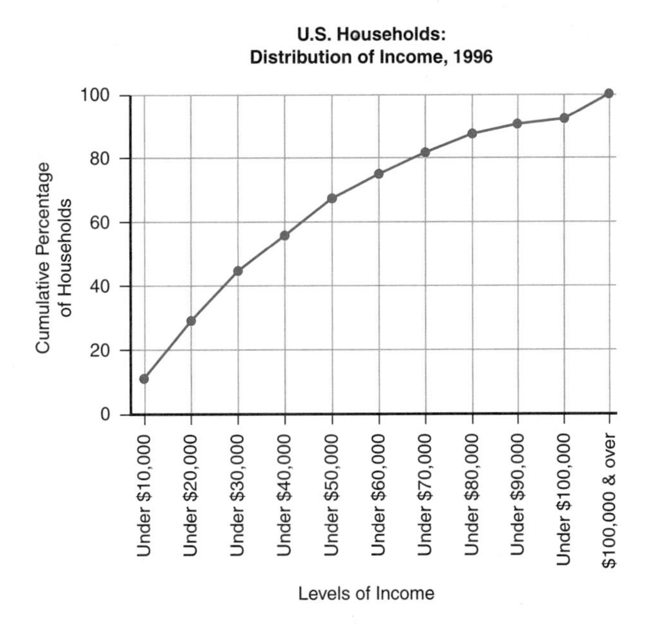

Figure 2.10
Ogive of U.S. household income distribution

Now we can answer the question "What percentage of households have an income under $20,000?" We turn to Figure 2.10, find the point on the graph above

$20,000, draw a horizontal line from that point, and observe where that line intersects the vertical axis. We get a figure in the 30% range. Now to the question "What is the median income?" The median income is the income that divides the distribution in half. To answer this question, we draw a horizontal line at 50% of the households and drop a vertical line where it intersects the graph. We get a figure in the mid-thirties. (The median is actually $35,492, according to the Census Bureau.) Although an ogive gives a good approximation of the median or measure under study, it can give only an approximation because some information is lost when the data are gathered into intervals, such as the income categories used here. With a direct plot of the cumulative percentage distribution function (CPF), we would capture all the information in the original data and could graphically work out the exact level of income corresponding to the median. The analytic details are given in Chapter 3.

EXAMPLE 2.4 ■ USING AN OGIVE TO FIND A DIVIDING LINE

Data compiled by the Bureau of the Census in the *Current Population Survey* reveal the distribution of household income for high school graduates. These data appear in Table 2.7.

Table 2.7

HOUSEHOLD INCOME FOR HIGH SCHOOL GRADUATES IN THE UNITED STATES, 1996

Income Category	Households (000)	Income Category	Households (000)
Under $10,000	3,408	$60,000–69,999	1,939
10,000–19,999	5,423	70,000–79,999	1,316
20,000–29,999	5,258	80,000–89,999	799
30,000–39,999	4,206	90,000–99,999	472
40,000–49,999	3,470	100,000 & over	1,192
50,000–59,999	1,308		

Source of Data: U.S. Bureau of the Census, *Current Population Survey,* Table H-1: Selected Characteristics of Households, by Total Money Income in 1996.

The next step is to construct a worksheet with the cumulative frequency distribution, cumulative relative distribution, and cumulative percentage distribution.

Table 2.8

WORKSHEET FOR DRAWING AN OGIVE

Income Category	Households (000)	Cumulative Freq. Dist.	Cumulative Rel. Dist.	Cumulative Pct. Dist.
Under $10,000	3,408	3,408	.08	8
10,000–19,999	5,423	8,831	.31	31
20,000–29,999	5,258	14,089	.49	49
30,000–39,999	4,206	18,295	.64	64
40,000–49,999	3,470	21,765	.76	76
50,000–59,999	1,308	23,073	.80	80
60,000–69,999	1,939	25,012	.87	87
70,000–79,999	1,316	26,328	.91	91
80,000–89,999	799	27,127	.94	94
90,000–99,999	472	27,599	.96	96
100,000 & over	1,192	28,791	1.00	100
Total	28,791			

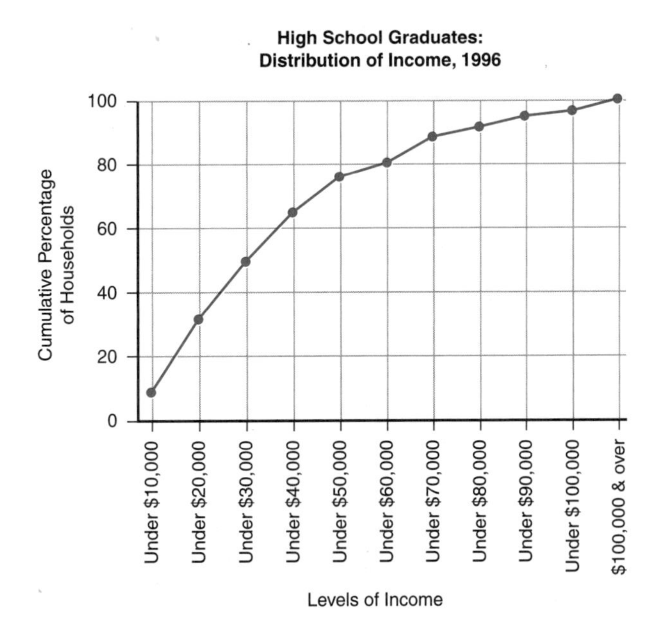

Figure 2.11
Ogive of income distribution of high school graduates

The values in the last column of Table 2.8 are plotted against the class limits to generate the ogive. The ogive is illustrated in Figure 2.11.

To divide the top 25% from the remaining 75%, we find the point on the vertical axis corresponding to 75% of the households. From that point, we draw a horizontal line intersecting the cumulative distribution function. We then drop a vertical from this point of intersection. From Figure 2.11, it appears that the value we seek is around $50,000. This value is known as the 75th percentile.

For the reasons noted previously, an ogive can only give a good approximate answer. We will look at how to find the exact dividing value for a given percentile in the next chapter.

> **Percentiles** divide a data set into one hundred equal parts. The *n*th percentile divides the data such that *n* percent of the data fall below it and 100 − *n* percent of the data fall above it.

Density Curves

The histogram in Figure 2.12 on page 46 shows the distribution of the data in Table 2.5 on U.S. household incomes. The horizontal scale is in thousands of dollars of household income. The income interval used is $10,000 per marked horizontal unit. The vertical scale is the percentage of households found per $10,000 interval on the horizontal scale.

Drawing a density curve

Notice that the base of the bar in the histogram representing incomes of $100,000 and up has been extended, to give a more realistic picture. This interval is so long that it cannot be shown accurately to scale. But the area of this bar accurately represents the 8.2% of all households that have income in this range. Remember that, in a histogram, the area of the bar must be proportional to the frequency (or relative frequency or percent) that the bar represents.

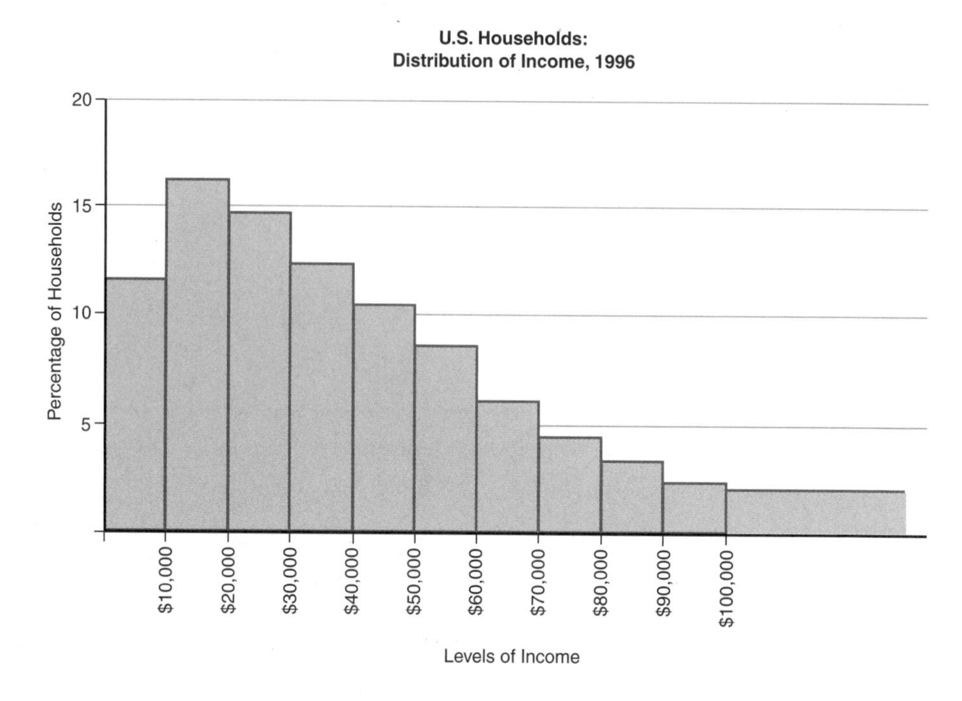

Figure 2.12
Distribution of household incomes in the United States

If we wanted to look at household incomes in more detail, we could keep the same horizontal and vertical scales but greatly reduce the widths of the class intervals—say, down to \$1–\$10,000 or even the nearest cent. This would give us Figure 2.13, where the data are presented in such detail that the upper boundary line appears as a smooth curve. This curve is the **density curve,** or density function, for household incomes in the United States. The horizontal scale is in thousands of dollars of household income. The income interval used is \$10,000 per marked horizontal unit. The vertical scale is the percentage of households found per \$10,000

Drawing a bar graph

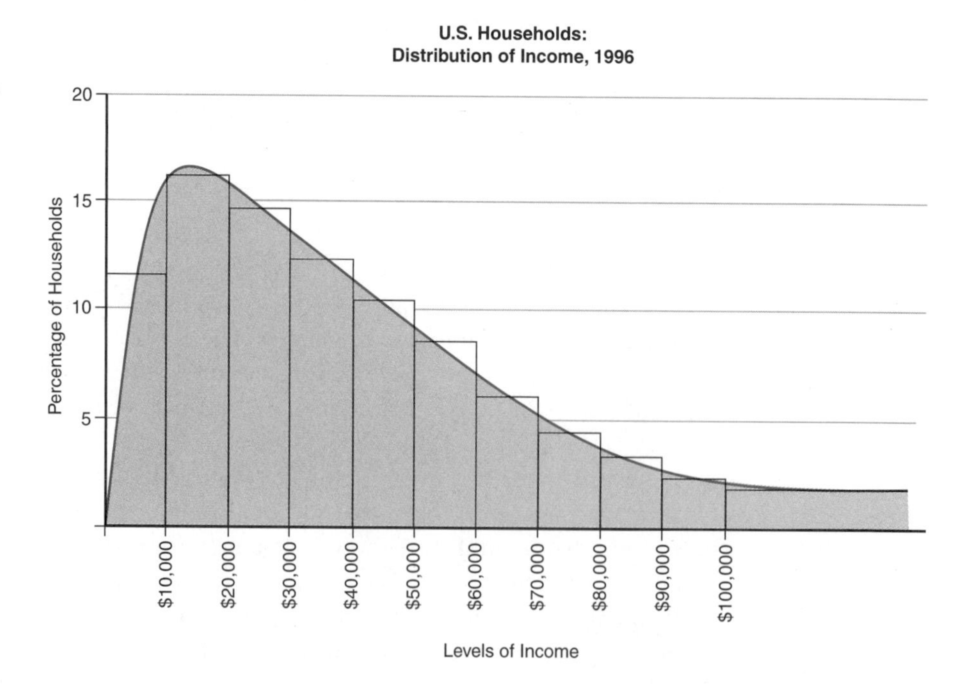

Figure 2.13
Density curve for distribution of household incomes in the United States

interval on the horizontal scale. The percentage of households with incomes falling within a given interval along the horizontal axis is proportional to the area below the curve up to the given point. The total area under the curve corresponds to 100% of households. There are some households with incomes so large that they can't be included in this picture, so the shaded region representing the population must be thought of as continuing to the right to include incomes of millions of dollars.

For comparison, the outline of the earlier histogram also appears in Figure 2.13. You can see that the area under the curve matches that of the bar of the histogram for any of our original class intervals.

Density curves provide a general way of picturing the distribution of data. In a density curve such as the one shown in Figure 2.14, the fraction or percentage of the population with values between limits A and B is proportional to the shaded area. Equivalently, the probability that a randomly selected observation from the population will have values between A and B is proportional to the shaded area.

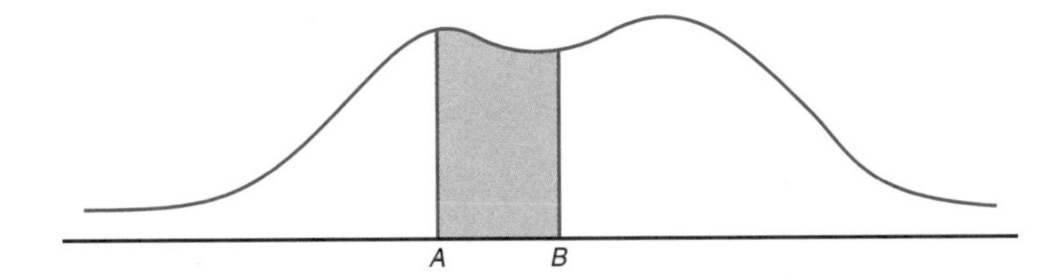

Figure 2.14
Density curve picturing a distribution

Density curves are useful in describing data and in describing the outcomes of experiments—particularly experiments involving statistical sampling. They play an important role in understanding the quantitative aspects of sampling, estimation, and hypothesis testing.

The height of the cumulative relative distribution (CRD) at x is just the fraction of the population with values less than x. So the value of the cumulative function at x is represented by the area under the density function lying to the left of x on the horizontal axis. Thus, ogives, histograms, CRDs, and density functions are all alternative ways of picturing various aspects of frequency distributions. In the next chapter, we will consider how these pictures of statistics relate to the hard numbers used in analyzing populations.

Bar Graphs

Bar graphs use rectangles of different lengths to depict quantitative or qualitative data. In the case of quantitative data, bar graphs display frequency counts or percentages.

Bar graphs are often used to compare characteristics of different places or groups. If we wished to compare the percentages of older homes in various American cities, we could use a bar graph such as the one shown in Figure 2.15 on page 48. Such a graph would be useful to someone in the home remodeling business.

The top line of the title tells us what the graph is about. The second line of the title expands upon the meaning of "older homes." The vertical axis is a percentage scale. The names of the cities with high percentages of older homes are on the horizontal axis. Each bar represents the percentage of older homes in the given city. The source of the data is indicated at the bottom left of the chart.

Figure 2.15
Bar graph comparing older homes

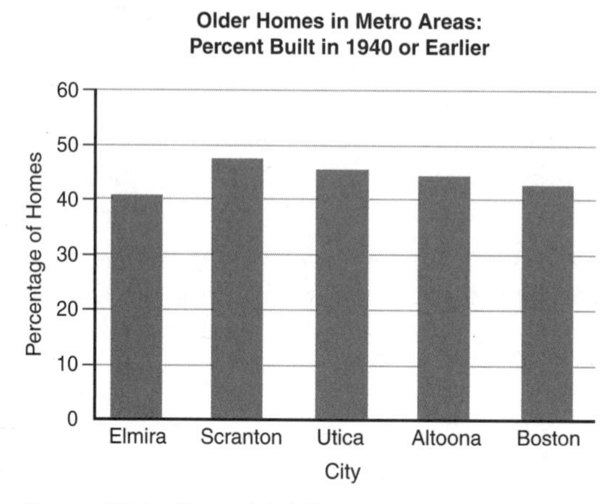

EXAMPLE 2.5 ▪ **USING A BAR GRAPH TO SHOW PERCENTAGES OF DOMESTIC CONTENT**

In order to put itself in a more favorable light with American car consumers, Ford Motor Company distributed a brochure indicating the percentage of domestic content in each of its models. Using the data in Table 2.9, we can prepare a graph comparing Mercury models.

Table 2.9

MERCURY MOTOR CARS

Car	Percent Domestic Content
Grand Marquis	72
Sable	94
Capri	19
Cougar	94
Tracer	81
Topaz	94

Because we are comparing percentages, we will create a bar graph. The title will be "Percent Domestic Content: Mercury Motor Cars, 1993." Model labels will be placed on the horizontal axis, and percentages on the vertical axis. The graph is shown in Figure 2.16.

The graph illustrates that the majority of the models have a substantial domestic content. Half the models have a domestic content of over 80%.

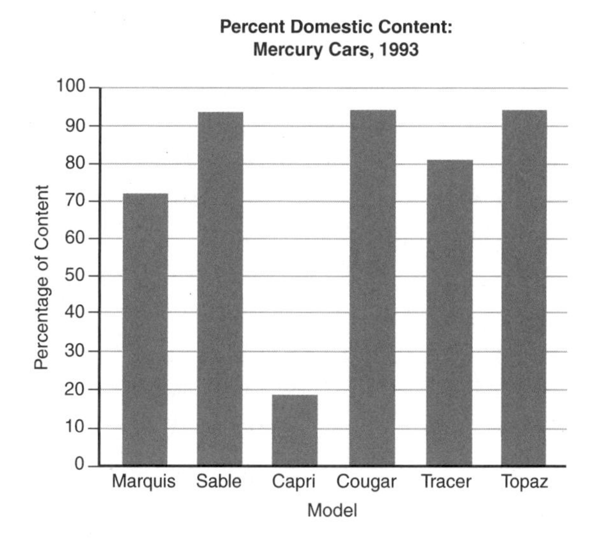

Figure 2.16
Bar graph showing percentages of domestic content

Pareto Charts

While bar graphs are most often used to present quantitative data, as in Figures 2.15 and 2.16, they may also be used to represent qualitative data. The **Pareto chart** is a bar graph used as the basis for the quality control technique of Pareto analysis.

Pareto analysis is a procedure that helps to identify important problem areas in an operating process by isolating the significant factors contributing to a problem. In a Pareto chart, the heights of the bars reflect the frequency of problems. The line rising from the top of the first bar represents the cumulative percentage of complaints. The left vertical axis represents frequency, the horizontal axis indicates the category of the problem, and the right vertical axis represents cumulative percentage. This double scale can be used anywhere counts and percentages apply. Figure 2.17 shows the frequency of complaints by type for a hotel chain.

Constructing a Pareto chart

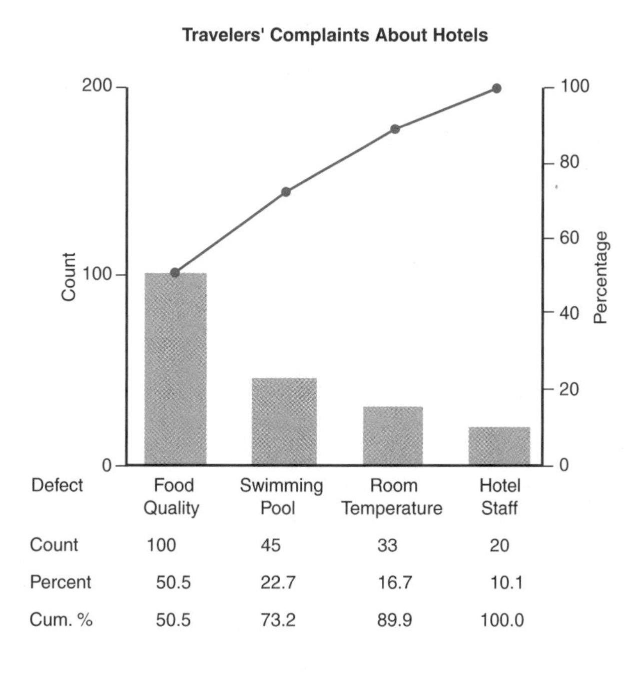

Defect	Food Quality	Swimming Pool	Room Temperature	Hotel Staff
Count	100	45	33	20
Percent	50.5	22.7	16.7	10.1
Cum. %	50.5	73.2	89.9	100.0

Figure 2.17
Pareto chart for travelers' complaints

You see immediately from the chart that the vast majority of complaints (73.2%) about this hotel chain concern food quality and the swimming pool. The remaining complaints are but a small part of the hotel's problems. If there is one lesson in quantitative work it is to *go after the big numbers first.* The hotel's management would ultimately face the question of whether complaints in these areas have a significant impact on revenues, but we do not have enough information in our data to address this question directly. At least with this chart, management has a good starting point.

EXAMPLE 2.6 ■ USING A PARETO CHART TO ANALYZE HOMEOWNERS' INSURANCE CLAIMS

Suppose an insurance company specializing in homeowners' insurance conducted a survey to examine the nature of claims made by its customers. The results in Table 2.10 were reported.

Table 2.10

HOMEOWNERS' INSURANCE CLAIMS

Category	Frequency
Water	24
Earthquake	3
Theft	40
Fire	6
Wind	15
Hail	12

To create a Pareto chart, we put these results in order of decreasing frequency. The resulting chart appears in Figure 2.18.

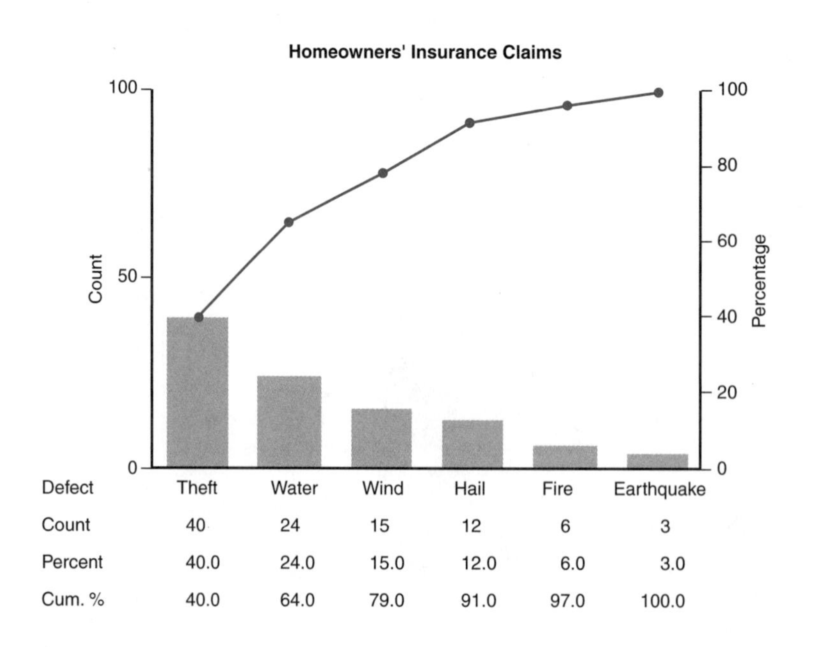

Figure 2.18
Pareto chart for homeowners' insurance claims

The major categories of claims are theft, water, and wind. These categories account for 79% of the claims. This information shows that claims adjusters should be particularly well prepared to discuss theft and water damage claims. It suggests, as well, that specialists be called in to deal with other types of damage. If we were to look at the total dollar amounts of claims, fires, which can be very costly, would play a larger role. But this is an even greater reason for a company to rely on highly trained specialists to deal with fire claims.

Pie Charts

Pie charts, or circle graphs, shine at showing how things are allocated percentagewise when the total must equal 100%. Government agencies use pie charts to show how tax dollars are divided among spending categories. Companies use them to show which products contribute the most to company sales. Whenever we wish to show relative importance, a pie chart is useful. As with other charts, a title is essential. There is, however, no vertical or horizontal scale.

Constructing a pie chart

Figure 2.19 is a pie chart of the sources of energy used by Con Edison during 1990. Each section of the pie is produced with a different type of shading so that one can easily distinguish one source from another. Next to each area is a label, accompanied by the percentage of energy derived from this source.

The whole circle is 100% and represents all the sources of power. It shows that nearly equal percentages of energy are derived from gas, oil, and nuclear sources and that coal, hydro, and refuse provide relatively little power. Notice that oil is relatively more important than coal. The chart does not show trends or the direction of changes, because it presents a snapshot at a given moment in time.

A pie chart such as the one in Figure 2.19 can be produced using Microsoft Excel's Chart Wizard. Given today's technology, it is unlikely that you would ever produce any chart without a computer. All the major computer software packages have graphing software. Graphing tools are commonly part of presentation software and are also provided by some word processing software.

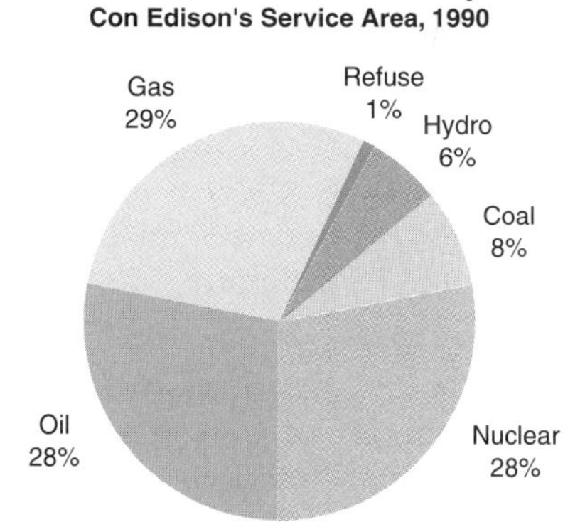

Figure 2.19
Pie chart for percentage share

Pie charts are suitable for visual presentation of market share in situations where it is natural to think of each product as getting its "slice of the pie." But the eye can more easily compare quantities if information is presented in a bar graph. In the bar graph in Figure 2.20, we easily see that coal beats out hydro; this is not so clear in the pie chart in Figure 2.19. You might want to use pie charts to dazzle your audience in live presentations, but favor bar graphs for more careful written work.

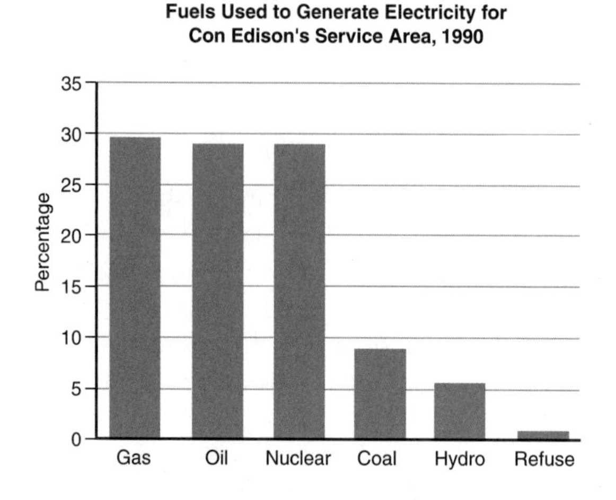

Figure 2.20
Bar chart for comparing percentages

SECTION 4 ■ USING THE COMPUTER

Computers are an integral part of the data analyst's tool kit. Calculations that once would have taken hours can now be done in nanoseconds. Computers don't get tired, and if they are fed correct information, they will perform complex calculations correctly over and over again. Vast amounts of data can be stored and accessed electronically. Once a data set has been entered into a computer file, it can be revised or updated easily. The result is that statistical analyses can be performed much more easily and efficiently than ever before.

Charts can be produced with Excel, Harvard Graphics, Lotus, Microsoft, Minitab, SAS, SPSS, and other packages. All these programs produce professional-quality tables and graphs. The best feature of these software packages is that they let you (the analyst) create a chart, look at it, and modify it. You need only produce a copy when you are perfectly satisfied with the chart's appearance.

A Line Graph

The first step in producing a line graph is to enter your data, either by keyboarding the data or by retrieving the data from a data file. Today's programs rely on graphical user interfaces (GUIs) as opposed to command line programs. (See Statistics Tool Kit: Inside Your Statistical Package.) To use Minitab to produce a line graph for the data on total system sales by Detroit Edison, you go to the GRAPH menu. Using your mouse, click TIME SERIES PLOT. A dialog box will appear. Select SALES, as measured in kilowatt-hours, as the variable to be plotted. Now go to the section marked CALENDAR and select YEARS as the period of measure. Next turn to OPTIONS and type in 82 (as 1982 was the first year for the data). To save space at the bottom of the graph, it is better to use 82 than to use 1982. Lastly, turn to ANNOTATIONS and type in the title and footnote. The result appears in Figure 2.21.

Figure 2.21
A Minitab line graph

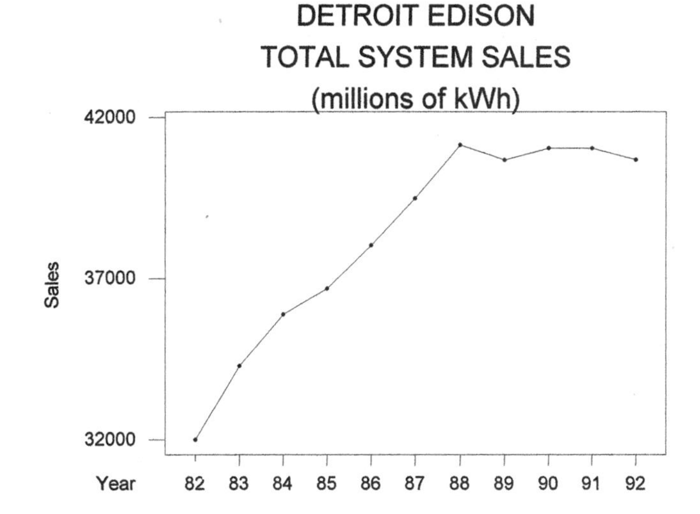

Source:Annual Report, 1992.

A Bar Graph

Suppose you want to produce a bar graph, or chart, next. Using Excel to produce a bar chart for the same Detroit Edison data is very easy. Enter the data into a worksheet in rows, as shown below:

	1982	1983	1984	1985	1986	1987	1988
Sales	31995	34299	35887	36695	38040	39492	41144

Data in this worksheet extend from cell A1 to cell B13. Select data in the specified range, and then click on the CHART WIZARD TOOL. This tool will give you a selection of charts to choose from. Selecting the first column subtype will produce the chart in Figure 2.22.

Figure 2.22
Excel bar chart

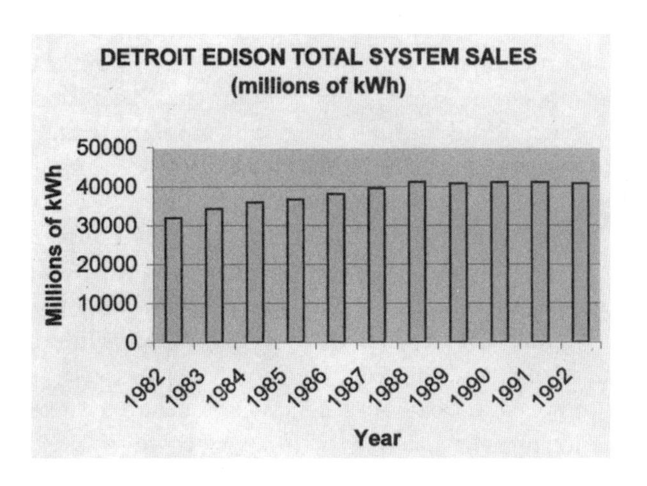

A Pareto Chart

A Pareto chart is complex, containing both count and percentage data. It is with such a chart that the computer exhibits its versatility. With Minitab, the approach is to go to the SPC (statistical process control) menu. The required tools may be on

STATISTICS TOOL KIT

Inside Your Statistical Package

With a graphical user interface, you can call up the analysis or display you want with a few clicks of your mouse. Most packages have a tutorial to familiarize you with the system so that you can get what you want without studying a manual or seeking help. The tutorial is ideal for beginners or occasional users. Heavy users may prefer to enter all commands through the keyboard; this is where the package's command line language comes in.

Packages like Minitab and Excel provide a window in which you can view the command line instructions corresponding to your mouse actions. Be sure to use any extensions to your software needed to solve statistical problems and produce graphics. For example, if you are an Excel user, you will want to have its Data Analysis Tool Pak available. You can keep up with the latest information on such products by checking with your software company on the Internet or asking the computer resources people at your school. Shown below is part of the window for the session used to create the Detroit Edison line graph in Figure 2.21.

```
MTB> RETR 'A:\DETROIT.MTW'.
Retrieving worksheet from file: A:\DETROIT.MTW
Worksheet was saved on 9/11/1997
MTB> TSPlot 'Sales';
SUBC> Year;
SUBC> TDisplay 11;
SUBC> Start 82;
SUBC> Symbol;
SUBC> Connect;
SUBC> Title "DETROIT EDISON";
SUBC> Title "TOTAL SYSTEM SALES";
SUBC> Title "(millions of kWh)";
SUBC> Footnote "Source:Annual Report, 1992."
```

These commands will be in the language your package uses to describe data manipulation, statistical procedures, and the form of its output. By looking at what you did through the graphical user interface of your computer, you will be able to understand what most of these computer commands do, even without consulting a manual. For instance, in this session, when we pointed to OPTIONS with our mouse, a dialog page opened on which we typed 82, indicating to the program that the chart was to begin with 1982 and increase in increments of 1 year. The 1 does not appear in the dialog box because it is the "default value" for the increments—that is, the value the computer assumes if no overriding value is given.

If you need to carry out a complex procedure or to write a program for routinely carrying out special procedures, you will want to learn how to use the command line instructions for your system. *For most people and most situations, the mouse-driven approach is easier and, for that reason, better.*

other menus for other packages, and some hunting or consulting may be needed to find them. The procedure is outlined below.

Consider, once again, the insurance claim data in Table 2.10. After the data have been entered from the worksheet, it is a good idea to verify that they were properly entered. This can be done by going to the menu bar at the top of the screen and selecting "Manip," followed by "Display Data . . . ." The computer will respond:

Minitab Pareto defects chart

Data Display

Row	CLAIM	VOLUME
1	Water	24
2	Earthquake	3
3	Theft	40
4	Fire	6
5	Wind	15
6	Hail	12

Remember that, in Minitab, the Pareto defects chart is found not under the GRAPH menu, but under the SPC menu. By following the menu commands, you arrive at the Pareto defects chart and can fill in the details. The package will take it from there. The result will be a chart similar to the one in Figure 2.18.

A Pie Chart

A pie chart, as noted earlier, is good for presenting categories that add up to 100%. As always, you should begin by verifying your data with the print command.

Economists divide the unemployed into four categories: job losers, job leavers, re-entrants, and new entrants. No one wants to be a job loser. A job leaver is someone who leaves voluntarily. A re-entrant is someone coming back to the labor force. Most students will become new entrants. Policy makers must understand the makeup of the unemployed population in order to deal with the problem. A pie chart calls attention to the various categories and their *relative importance.* Figure 2.23 shows the pie chart for the following data:

	NUMBER
JOB LOSERS	4.8
JOB LEAVERS	0.9
RE-ENTRANTS	2
NEW ENTRANTS	0.8

REASONS FOR UNEMPLOYMENT

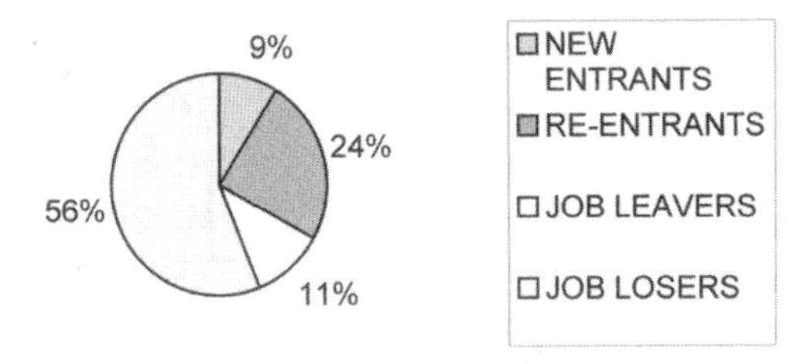

Figure 2.23
Excel pie chart

Each software package has different strengths and weaknesses. This section was designed just to serve as a general guide to graphing.

EXERCISES 2.17–2.30

Applying the Concepts

2.17 The *Places Rated Almanac* for 1993 reports that the five metro areas in the United States with the highest household income were as follows:

Stamford–Norwalk, Connecticut	$96,760
Long Island, New York	88,415
Northern New Jersey	84,416
Lake County, Illinois	84,060
Central New Jersey	81,411

Using these data, prepare a chart (suitable for a presentation) on prospective locations for three luxury clothing stores.

2.18 In 1988, the U.S. labor force consisted of 118.1 million individuals. A total of 23.6 million had 1 to 3 years of college. A total of 26.0 million had 4 years of college or more. The remainder had 4 years of high school or less. Using these data, prepare a chart that might be used to indicate the size of the market for continuing education programs.

2.19 Between 1980 and 1990, the total number of telephone access lines grew as summarized below.

Year	Lines	Year	Lines
1980	102,216,367	1986	118,345,686
1981	105,559,222	1987	123,010,150
1982	107,519,214	1988	127,087,323
1983	110,612,689	1989	131,623,290
1984	112,550,739	1990	136,184,917
1985	116,042,281		

a. Prepare a graph portraying the trend in telephone access line growth.
b. Describe the trend and how it might bear on the market for switching and line equipment.

2.20 Between the 1890s and the mid-1920s, wages in manufacturing rose considerably. Union wages in manufacturing rose from $.32 per hour (1890) to $.99 per hour (1925). Non-union wages in manufacturing rose from $.15 per hour to $.49 per hour over the same period.

Year	Union	Non-union
1890	$.32	$.15
1895	.33	.14
1900	.34	.15
1905	.38	.17
1910	.40	.19
1915	.44	.21
1920	.88	.56
1925	.99	.49

Source of Data: U.S. Department of Commerce, *Historical Statistics of the United States.*

a. Construct a chart showing growth in both these series.
b. Which group experienced greater percentage wage growth?

c. Write a caption for this chart that might be suitable for (1) a union brochure or (2) a National Association of Manufacturers publication.

2.21 A summary of mutual fund performance published by the *New York Times* (Oct. 3, 1994) compared the performance of different types of funds. The average total returns over 3 months, 1 year, and 5 years for general stock funds are shown below:

Type of Fund	3 months	1 year	5 years
Aggressive growth	8.22%	−.83%	11.12%
Growth	5.48	2.01	9.39
Growth and income	4.14	2.65	8.67
Equity income	4.08	2.24	8.19
Small income	8.52	2.73	11.69

a. Construct a chart comparing growth for each of these types of funds over 3 months.
b. Construct a chart containing the data for 3 months, 1 year, and 5 years.
c. Suppose you work for an advertising company. Using these data, write a short sales pitch aimed at a manager of one of these sorts of funds.

2.22 When the 1990 census appeared, the baby boomer generation (ages 26–40) became the subject of demographers', economists', and business community interest. Mortgage status is one of the variables on which data were compiled. The data below represent the distribution of monthly payments on the mortgages held by the baby boomer generation.

Monthly Owner Cost	Number of Baby Boomer Households
Less than $300	373,637
$300–$499	2,341,692
$500–$699	3,524,682
$700–$999	4,469,902
$1,000–$1,499	3,253,878
$1,500–$1,999	1,127,353
$2,000 and more	772,482

a. Construct an ogive for the data.
b. Find the approximate median for monthly owner cost.
c. Complete this statement: A total of 90% of baby boomer homeowner households pay $_____ or less per month.

2.23 Consider the data on Monthly Owner Cost in Exercise 2.22.
a. Make a histogram of the data. (Note the unequal lengths of intervals.)
b. On top of your histogram, sketch your best guess as to what the density curve would be.

2.24 When drugs are developed, they are tested for bad reactions. A certain drug developed for the treatment of high cholesterol exhibited a variety of bad reactions (as all drugs do). In a sample of 500 individuals, 72 suffered from respiratory problems, 60 experienced muscular problems, 100 suffered from digestive problems, 15 suffered from psychiatric disorders, and 25 had headaches.
a. Prepare a Pareto chart for the data.
b. What problems account for approximately 80% of all disorders?

c. Go to the *Physician's Desk Reference* (PDR), look up drugs for high cholesterol problems, and compare this drug with those you find. Does this drug exhibit more adverse reactions than other competing drugs in this category?

2.25 Videocassette recorders (VCRs) have become more and more popular. The number of U.S. households with VCRs grew from 200,000 in 1978 to over 58 million in 1989.

Year	U.S. Households with VCRs (000)	Year	U.S. Households with VCRs (000)
1978	200	1984	8,800
1979	400	1985	17,600
1980	840	1986	30,920
1981	1,440	1987	42,560
1982	2,530	1988	51,390
1983	4,580	1989	58,400

a. Using the data given here, prepare a chart showing the trend in VCR ownership.
b. Write a paragraph that might accompany this chart in a VCR industry trade magazine.

2.26 Average family income in Canada rose from $10,113 to $53,131 between 1971 and 1991. It rose faster in some provinces than in others. In 1991, the province with the highest family income was Ontario, whereas the province with the lowest family income was Newfoundland. Data for the country as a whole and these two provinces are presented below (all figures are in Canadian dollars).

Year	Canada	Newfoundland	Ontario
1971	$10,113	$ 6,855	$11,154
1975	16,368	12,359	17,772
1980	27,246	20,374	28,313
1985	37,981	29,022	41,291
1991	53,131	41,645	58,634

Source of Data: Statistics Canada.

a. Construct a chart illustrating the trend in income growth in these areas.
b. Which area experienced greater growth—Ontario or Newfoundland?
c. For Canada as a whole, when was the greatest growth experienced?

2.27 Sales of Anheuser-Busch rose from $3.8 billion in 1980 to $10.3 billion in 1988. Advertising expenditures rose from $428 million to $1.9 billion over the same period.

Year	Sales (millions)	Advertising Expenditures (millions)
1980	$3,822.4	$428.0
1981	4,435.9	518.6
1982	5,251.2	758.8
1983	6,714.7	1,226.4
1984	7,218.8	1,338.5
1985	7,756.8	1,498.2
1986	8,478.8	1,709.8
1987	9,705.1	1,824.5
1988	10,238.6	1,876.5

a. Using the data, prepare a graph showing the trends in these time series.
b. Do you believe the data support increasing advertising expenditures as a way to increase sales? Why or why not?

2.28 F.W. Woolworth was the original Five and Dime store. Over most of the 20th century, it grew tremendously. Just before it failed in the 1990s, it consisted of a number of subdivisions. Profits for each of its subdivisions in the mid-90s are shown below:

Subdivision	Profits (in millions)
Kinney	$328
Richman Bros.	10
Woolworth	229
Other specialty operations	12
Total	$579

Prepare a chart showing the percentage share of profit from each division.

2.29 Many members of the baby boomer generation rent their housing units. Below are data from the 1990 census concerning renter-occupied housing units.

Monthly Rental	Number of Units
Less than $200	773,960
$200–$299	1,544,076
$300–$499	6,015,868
$500–$749	4,860,763
$750–$999	1,578,038
$1,000 or more	716,559

a. Draw an ogive of the renter-occupied housing unit data for the baby boomer generation.
b. Find the 75th percentile.
c. Find the 25th percentile.
d. Complete this statement: The middle 50% of baby boomer renters pay between $____ and $____ per month.

2.30 Consider the data on monthly rental cost in Exercise 2.29.
a. Make a histogram of the data. (Note that some intervals may be larger than others.)
b. On top of your histogram, sketch your best guess as to what the density curve would be.

FORMULAS

Approximate Number of Classes (Groups):

$$k = \sqrt{n}$$

Approximate Class Width:

$$\text{c.w.} = r/k$$

NEW STATISTICAL TERMS

bar graph
class width
classes
classified data
cumulative frequency distribution (CFD)
cumulative percentage distribution (CPD)
cumulative relative distribution (CRD)
density curve
distribution
dot plot
histogram
line graph
lower class limits (LCLs)
ogive
Pareto chart
percentile
pie chart
stem plot
systematic random sample
upper class limits (UCLs)

EXERCISES 2.31–2.40

Supplementary Problems

2.31 Energy consumption may be divided into three categories.

End-Use Sector	Quadrillion BTUs
Residential & commercial	29,560
Industrial	29,658
Transportation	22,286
Total	81,504

a. Using the data, prepare a chart illustrating the breakdown of energy consumption by end-use sector.

b. Based on this chart, which sectors would you target for conserving energy and why?

2.32 Each year IBC/Donoghue publishes a *Mutual Funds Almanac.* In its 22nd edition, a beta coefficient was provided for each mutual fund. "Beta" is a risk evaluation tool. It is a measure of the relative volatility of a portfolio of stock with respect to the market as a whole. The beta of the market is 1.0. More volatile investments have a beta greater than 1.0; less volatile investments have a beta less than 1.0. The betas for a sample of 60 funds are shown below.

.89	.06	1.42	.90	1.00	.15	.16	.28	.11	.77
.72	.86	.26	1.37	.99	.20	.26	1.05	.23	.47
1.02	.69	.23	.28	.79	.17	.97	1.12	.20	.25
.21	.62	.26	.28	.22	.36	1.12	.16	.88	.42
.31	1.22	.82	.24	.18	.24	.40	.51	.16	.19
.45	.27	.57	1.26	.34	.58	.18	.10	.79	.20

a. Construct a table or histogram of these data.

b. What is the typical beta coefficient for these mutual funds?

c. Suppose an investment advisor wanted to use the data to plan a conservative investment strategy. How might she use the data in developing this strategy with a client?

2.33 In 1990, the salaries of U.S. governors ranged from $30,000 to $130,000. The table below summarizes the distribution of governors' salaries for that year.

Range	Number of Governors
$30,000–49,999	1
50,000–69,999	10
70,000–89,999	29
90,000–109,999	8
110,000–129,999	1
130,000–149,999	1

Source of Data: *The Universal Almanac,* 1991.

a. Draw an ogive of governors' salaries.
b. What is the median value?
c. Complete this statement: Ninety percent of all governors earn less than $_____.

2.34 If you had a table of the 50 salaries of the governors, which of the following graphic tools would you use to picture the distribution in more detail? Select one and give the reasons for your choice.
a. A stem plot
b. A dot plot
c. A histogram
d. A density curve

2.35 Daimler-Benz, the world's oldest automobile manufacturer, is renowned for its luxury cars. Sales and net income data for the 10-year period between 1984 and 1993 appear below.

Year	Sales (millions of DM)	Net Income (millions of DM)
1984	43,505	1,145
1985	52,409	1,735
1986	65,498	1,805
1987	67,475	1,787
1988	73,495	1,675
1989	76,392	1,700
1990	85,500	1,684
1991	95,010	1,872
1992	98,549	1,418
1993	97,737	602

Source of Data: *Hoover's Handbook of World Business,* 1995–96.

a. Construct a line graph showing the growth in sales.
b. Construct a line graph showing the change in net income.
c. Has growth in sales been accompanied by an increase in net income?
d. What might explain what you observe?

2.36 According to the *Brewer's Almanac,* the United States is the biggest beer producer in the world, while France ranks 10th in production. Data on beer production in 1986 for the ten biggest beer producers are listed here. Prepare a bar chart comparing beer production in the various countries.

Country	Barrels
United States	195,123,000
Germany	102,670,000
Soviet Union	57,949,000
England	52,580,000
Japan	42,032,000
Brazil	24,713,000
Czechoslovakia	21,267,000
Mexico	20,121,000
Canada	19,259,000
France	18,821,000

2.37 The U.S. Industrial Outlook for 1978 indicated that U.S. tableware shipments were $628 million in 1970 and $1.537 in 1977. A breakdown is given in the accompanying table.

	1970	1977
Glass	$285	$659
Plastics	213	692
Earthenware	51	85
China	79	101

a. Construct the appropriate chart for 1970.
b. Construct the appropriate chart for 1977.
c. What happened to the tableware market between 1970 and 1977?

2.38 A radio talk show host asked the opinions of listeners concerning national political issues. During one afternoon, 25 of the callers were concerned with inflation, 16 were worried about unemployment, 20 said health care was an issue, 9 were concerned about a balanced budget, 2 said Social Security was an issue, and 5 saw a problem with welfare.
a. Prepare a Pareto chart of these issues.
b. What issues should the talk show host pursue?

2.39 Statistics Canada, the national statistical agency of Canada, compiles a variety of data on its population. One of the tables it compiles concerns age of mother at birth of first child. Data for 1931 and 1991 appear below.

Age of Mother	1931	1991
Under 15	14	262
15–19	9,639	20,192
20–24	25,224	47,580
25–29	13,826	66,535
30–34	4,802	31,818
35–39	1,580	7,963
40–44	342	934
45 and over	27	27
Totals	55,454	175,311

a. Draw an ogive for 1931.
b. What is the approximate 90th percentile value?
c. Repeat the process for 1991.
d. Write a brief paragraph concerning your findings and their implications.

2.40 Refer to the data from Statistics Canada in Exercise 2.39.

a. Make a histogram for 1931. (Note that all intervals are not equal.)

b. Sketch your best guess as to a compatible density curve.

c. Repeat the process for the 1991 data.

d. According to your results, has the probability of teenage pregnancies increased?

3

Central Tendency and Variation

OUTLINE

LEARNING OBJECTIVES

1. Calculate and interpret quantitative measures of central tendency.
2. Compute and explain measures of variation.
3. Understand how quantitative measures and graphic presentations are related.
4. Calculate and interpret relative measures.
5. Identify meaningful patterns in data.
6. Select the most appropriate measure to use in a given situation.
7. Use the computer to perform precise calculations on large data sets.

BUILD YOUR KNOWLEDGE OF STATISTICS

Despite legislation, a gender gap still exists in the job market. How wide is the gap? In "All Employees Are Created Unequal," Dana Kaplan provided the following statistics (www.miningco.com, dateline January 12, 1998):

- In 1996, the median income for a woman who was the sole provider for the family (without a spouse present) was $19,911, and the median income for a man who was the sole provider for the family (without a spouse present) was $31,600.
- In 1996, women high school graduates without college working full-time on a year-round basis earned an average of $21,893 a year, while fully employed men who had completed 9th to 12th grade *with no high school diploma* earned an average of $25,283 a year.
- It would take the average woman four months longer to earn what the average man earned in 1997.
- In 1979, women in the workforce were only earning 59.7% of what men earned, while in 1996 the figure was 73.8% and in 1997, 74.4%.
- In the top twenty occupations for women in 1997, there was not even one occupation that hit the 95th percentile. The highest percentile rank went to female bookkeepers—who earned 93.7% of what their male counterparts earned in 1997.

These statistics tell an important story. They illustrate the type of data collected by labor economists to assess the status of members of the labor force. As you saw in Chapter 1, there are many types of data, which come from many different sources. In Chapter 2, you saw how these data could be organized to provide an overview of an event or group of interest. In this chapter, we will explore the methods used to summarize data and create a set of statistics that reveals their essence.

You will learn about three types of numerical measures: measures of central tendency, measures of variation, and relative measures. Each tells something different about a population. Together, these measures provide a quantitative description of the data center and the degree of scatter of the values about that center. Methods for finding the data center and dealing with the ever-present variations that show up in data shape this chapter. Here, you will learn how the organizing and profiling of raw data can turn the data into useful information.

SECTION 1 ▪ MEASURES OF CENTRAL TENDENCY

The common **average** is a single number indicating the center of an array of values. It is one example of a measure of central tendency.

> **Measures of central tendency** are numbers that indicate where the data cluster. The arithmetic mean (or average), median, and mode are the most popular measures of central tendency.

Measures of central tendency serve as rough guides. Suppose you are planning to open up a manufacturing plant. How much should you expect to pay your employees, if you want to pay them an average wage? In 1990, weekly earnings in manufacturing in New York averaged $429.55. So, if you were to have 20 employees, you would need to allow for a payroll of about $8600 per week.

Numerical data sets can be characterized in a number of ways. Single numbers falling toward the center of a data set are a significant source of information. Consider the characteristics of students in a given class—specifically, their ages, levels of education, and grade point averages. Suppose the average age is 21. Suppose further that all are sophomores and that their grade point average is 3.5. These numbers tell a lot about the students in that class. In this section, you will learn about the three measures of the center of a data set: the arithmetic mean, the median, and the mode.

But first let's briefly discuss terminology. We all have a good idea what the "average" of a column of numbers is. It is the result of adding the column and dividing by the number of summands. More properly, this result is called the arithmetic average. But because the word *average* has many informal meanings and statistical terminology must be precise, statisticians prefer to use the more specific terms **mean** and **arithmetic mean.** While we will use common terminology for numbers such as batting averages, we will use more specific statistical terminology for measures of center and range.

The Arithmetic Mean

Averages tell a story. Consider the salary data for a labor-intensive organization such as the 1990 Baltimore Orioles. Examination of Table 3.1 yields a broad impression of the range of salaries. The minimum salary, paid to Mesa, Segui, and Gallagher, was $100,000. Cal Ripken received the maximum salary of $1,367,000. What single number best represents this data set?

The arithmetic mean is a single number that characterizes a data set. In this case, it conveys information about what the typical Oriole player made in 1990. If you were comparing salaries of the Baltimore Orioles with those of another baseball team or comparing baseball salaries with those in another sport, the arithmetic mean would provide useful information. The arithmetic mean is the most fre-

Table 3.1

BALTIMORE ORIOLES' SALARIES, 1990

Player **Pitchers:**	Salary (thousands of dollars)	Player **Infielders:**	Salary (thousands of dollars)
Ballard	290	Gonzales	206
Bautista	120	Halett	208
Harnisch	120	Horn	120
Johnson	113	Milligan	155
McDonald	262	Ripken, B.	215
Mesa	100	Ripken, C.	1367
Milacki	245	Segui	100
Mitchell	111	Worthington	208
Olson	305	**Outfielders:**	
Price	400	Anderson	120
Schilling	103	Deveraux	145
Telford	100	Finely	130
Williamson	285	Gallagher	100
Catchers:		Kittle	575
Melvin	350	Orsulak	610
Tettleton	825		

Source of Data: *USA Today,* October 31, 1990, p. 4C.

quently used measure of central tendency because it is easy to understand and easy to calculate. The formula is presented below:

$$\text{Arithmetic mean} = \frac{x_1 + x_2 + \cdots + x_n}{n} \tag{1}$$

Our first observation, Ballard's salary, is represented by x_1, and our second observation, Bautista's salary, is represented by x_2. The last observation in a set of n observations is represented by x_n. Here, it is Orsulak's salary. The "$+\cdots+$" represents all the other values in the series, from x_3 to x_{n-1}, and is simply shorthand used to avoid writing out the other terms. We may also use a summation operator (see the Appendix at the end of the chapter). The summation operator is the uppercase Greek sigma, Σ. And so Equation (1) becomes

$$\text{Arithmetic mean} = \frac{\sum_{i=1}^{n} x_i}{n} \tag{2}$$

where n = number of observations and $\sum_{i=1}^{n} x_i$ is read as "sigma x sub i, as i goes from 1 to n."

Because we take into account the entire population of the Baltimore Orioles in 1990, the result obtained in this example is a parameter called the **population mean,** represented by the Greek mu, μ. (See the Statistics Tool Kit on notation in Chapter 1.) We rewrite equation (2) as follows:

$$\text{Population mean,} \quad \mu_X = \frac{\sum x_i}{N} \tag{3}$$

where N = the population size.

Typically, the subscripts and superscripts on the summation operator are omitted for simplicity. In Equation (3), an uppercase N is used to represent the number in the population. Think of Equation (3) as telling you that, in order to find μ (the population mean), you must add up all the x's, and then divide by the number of x's.

EXAMPLE 3.1 ■ FINDING THE AVERAGE OF THE 1990 BALTIMORE ORIOLES' SALARIES

What is the average of the 1990 Baltimore Orioles' salaries? For the population mean, we have

Calculating the population mean

$$\mu_X = \frac{\sum x_i}{N} = \frac{290 + 120 + \cdots + 610}{29} = \frac{7988}{29} = 275$$

The average salary of a Baltimore Orioles player in 1990 was $275 thousand.

Had we drawn just a sample of players, the process would have been the same, but the notation would have changed. The **sample mean** is represented by $\bar{x}$. The sum of the sample values is divided by the number in the sample, which is represented by a lowercase n.

$$\text{Sample mean,} \quad \bar{x} = \frac{\sum x_i}{n} \tag{4}$$

EXAMPLE 3.2 ■ SAMPLING TO ESTIMATE AN AVERAGE

To examine the cost of preparing a complex tax return, *Money* (March 1992) asked a number of accounting firms what they would charge to prepare a tax return for the Butlers, a hypothetical family. Because of the large number of firms involved, we have chosen a representative random sample of 10 from which to calculate an estimate for the overall average.

The fees for these 10 firms are listed below:

725	1175	900	650	1500
1350	1825	700	724	2880

Calculating the sample mean

Then

$$\bar{x} = \frac{\sum x_i}{n} = \frac{725 + 1175 + \cdots + 2880}{10} = 1242.90$$

Thus, we may conclude that preparation of the Butler family's tax return would cost an average of $1243 (rounding to the nearest dollar). This single summary number—the arithmetic mean—serves as a guideline for what such a family should pay. It indicates whether a family is paying too much or getting a good deal. There may be reasons for paying more than $1243, but the family should know them.

Graphics and the Mean. The arithmetic mean closely relates to the graphic displays discussed in Chapter 1. Look at Figure 3.1, which is a dot plot of the baseball salaries given in Table 3.1. Imagine the horizontal axis as a weightless but rigid plank and each dot as an equal weight sitting on the plank. If we place a wedge under the plank with its point exactly at the mean, the weights will balance perfectly. The dot plot of the data is balanced about the mean.

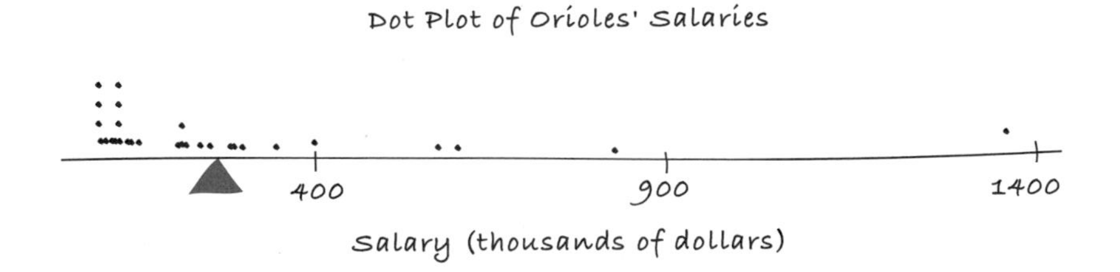

Figure 3.1 *Geometric interpretation of the mean with a dot plot*

If the distribution of the population is represented by a density curve, the mean again provides the balance point. As shown in Figure 3.2, the region under the density curve balances at the mean

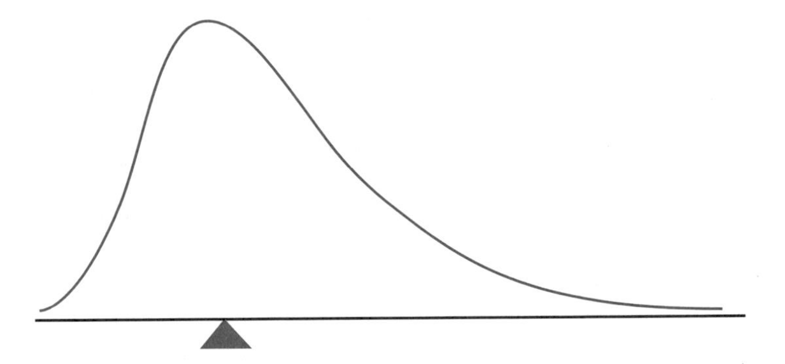

Figure 3.2 *Geometric interpretation of the mean with a continuous distribution*

Figures 3.1 and 3.2 clearly illustrate that the arithmetic mean is a measure of central tendency. They will also help you understand the properties of the mean that are about to be discussed.

Properties of the Mean. An important property of the arithmetic mean is that it is influenced by extremes. Imagine a community of 20 families in which the average income is $50,000. What would happen if a family with a multimillion-dollar annual income moved in?

EXAMPLE 3.3 ▪ ADDING A HIGH VALUE TO A DATA SET

Suppose the annual incomes for 20 families living in a community are as shown below, where numbers are in thousands of dollars:

Calculating the effect of extreme values on the mean

45	30	53	96	29	72	55	70	4	79
47	71	28	21	91	9	61	56	44	39

We can find the average income of these 20 families as follows:

$$\mu = \frac{\sum X}{N} = \frac{1000}{20} = 50$$

For these 20 families, average income is $50 thousand.

The average income of these 20 families and Arnold Schwarzenegger, who earned $55 million in 1989–1990, is

$$\mu = \frac{1000 + 55{,}000}{21} = 2666.7$$

As you can see, average community income is greatly influenced by extremes. Adding Arnold Schwarzenegger to the community raises the average income from $50 thousand to about $2.7 million. If the community levies a flat tax of 1% of income, its average receipts will be 1% of $56 million (or $560,000). So, the arithmetic mean is useful in estimating quantities like tax receipts. But this large average income is no assurance that the 21 residents of this town are potential customers for new Rolls Royces with price tags of $95,000 each.

Extreme values that lie far from the pack and are not representative of the data set, known as *outliers,* are sometimes removed before the mean is found in order to get more stable and representative results. Some teachers will drop a student's lowest grade before averaging scores for the final grade, believing that the low value is nonrepresentative. Why should you prefer this practice? Let's see analytically what happens to the average when lower values are dropped.

EXAMPLE 3.4 ■ DROPPING LOW VALUES FROM A DATA SET

Calculating the effect of low values on the mean

If we take into account all 29 players on the team, the average salary of a 1990 Baltimore Orioles player is, in thousands of dollars,

$$\mu = \frac{7988}{29} = 275$$

Keeping the geometric picture in mind, let's look at the effects of dropping the lower tail of the distribution. Dropping the combined salaries of Mesa, Telford, Segui, and Gallagher from the Orioles data will significantly shift the mean:

$$\mu = \frac{7588}{25} = 304$$

The mean increases by $29 thousand, showing again that when a low figure is dropped, the mean rises. When finding a mean, be sure that the full range of the data is represented and that you have not left out the lower or upper tail of the data. This makes sense both analytically and graphically.

Although the mean salary for the entire Orioles team was $275 thousand in 1990, no one player actually earned $275 thousand. A total of 21 Orioles players earned a good deal less than \$275 thousand; only 8 on the roster earned more. It is important to remember that *sometimes the arithmetic mean conveys the wrong impression, because it is influenced by extremes.*

The Median

A measure of central tendency that is less influenced by extremes than the mean is the median. The median is a positional measure. When data are placed in an ordered array—that is, from lowest to highest—the median is the middle value. In Chapter 2, we looked at percentiles in connection with the ogive; the median corresponds to the 50th percentile.

To find the center of the distribution of a variable x, you order the data in a spreadsheet by ascending values of x. If there are ties, simply write all the tied entries one after the other. Multiply the number of entries in the data set by ½. We shall call the resulting number k. If k is an integer, average the values in positions k and $k + 1$. If k is not an integer, then n is odd; round k upward, to get $(n + 1)/2$. The median is given by $x_{(n+1)/2}$.

Calculating the median

> The **median** is the center of a distribution of data found by arranging the n data items by ascending values of the variable x. If n is odd, the median is $x_{(n+1)/2}$. If n is even, the median is the average of the values found in the central rows, $\frac{1}{2}(x_{n/2} + x_{n/2+1})$. At least half the values are greater than or equal to the median, and at least half of them are less than or equal to it.

In Example 3.3, only one of the 21 families whose *mean* income was above \$100,000 would have been a potential customer for services aimed at families with incomes of over \$100,000. But when a group has a median income of \$100,000, at least half of those in the group have incomes of \$100,000 or more. That is why advertisers are interested in the *median* income of a magazine's subscribers, not just their *mean* income. Let's see how the median works for the Orioles pitchers.

EXAMPLE 3.5 ■ CALCULATING THE MEDIAN WITH AN ODD NUMBER OF OBSERVATIONS

In Table 3.2 on page 72, the data have been arranged in rows by ascending values of salary. With six observations above it and six observations below it, the seventh value in this set of observations is in the center and, thus, is the median. Harnisch earned the median salary for pitchers: \$120,000.

Here the median is clearly evident by inspection. When there are numerous observations, the median is often not so obvious. Although the position of the median is very clear, let's go through the steps necessary to find it.

$$k = .50 \times n$$
$$= .50 \times 13 = \text{Position } 6.5$$

Since k is not an integer, we round up. So, 6.5 becomes 7. The seventh and median value is \$120,000. Since the median is \$120,000 per year, at least half of those in this group might have use of services aimed at higher income individuals.

Table 3.2

BALTIMORE ORIOLES PITCHERS' SALARIES, 1990

Observation Number	Player	Salary (thousands of dollars)
1	Mesa	100
2	Telford	100
3	Schilling	103
4	Mitchell	111
5	Johnson	113
6	Bautista	120
7	Harnisch	120
8	Milacki	245
9	McDonald	262
10	Williamson	285
11	Ballard	290
12	Olson	305
13	Price	400

EXAMPLE 3.6 ■ CALCULATING THE MEDIAN WITH AN EVEN NUMBER OF OBSERVATIONS

The salaries of the six outfielders on the 1990 Baltimore Orioles team, in ascending order, can be found in Table 3.3.

Table 3.3

BALTIMORE ORIOLES OUTFIELDERS' SALARIES, 1990

Observation Number	Player	Salary (thousands of dollars)
1	Gallagher	100
2	Anderson	120
3	Finley	130
4	Deveraux	145
5	Kittle	575
6	Orsulak	610

$$\begin{aligned} k &= .50 \times n \\ &= .50 \times 6 = \text{Position } 3 \end{aligned}$$

Since k is an integer, we average the values at positions k and $k + 1$. The average of the third and fourth values is

$$\frac{130 + 145}{2} = 137.5$$

Thus, the median value is $137,500.

In Example 3.5, exactly half the other pitchers earned more than Harnisch, and exactly half earned less. In Example 3.6, 50% of the salaries for outfielders lie above $137,500, and 50% fall below that figure. If there are ties in the data—for example, $x_1 = 1$, $x_2 = 2$, $x_3 = 2$, $x_4 = 2$—the median is defined as usual. For these data, the median is $\frac{1}{2}(2 + 2) = 2$. Only x_1 is less than the median, and none of the values are greater than the median. However, at least half the data lie at or below the median and at least half lie at or above the median. The median—in this case, 2—is still the best measure of the center of the distribution. More typically, half the data fall below the median and half fall above it, with the possible exception of the median itself.

Recall from Chapter 2 that the *n*th percentile divides the data, with *n* percent of the data falling below it and $100 - n$ percent of the data above it. Thus, as Figure 3.3 shows, the median and the 50th percentile are the same.

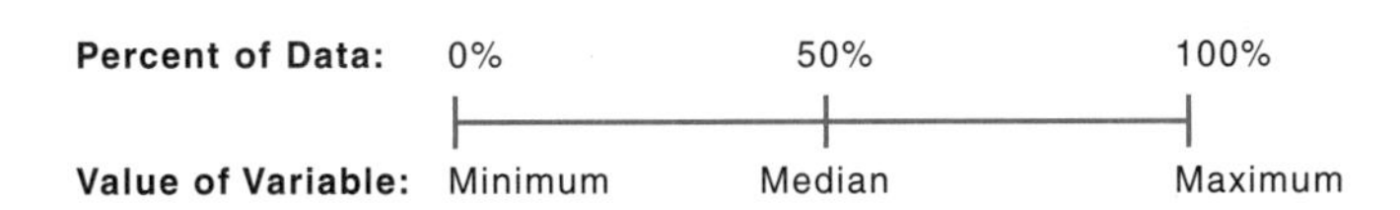

Figure 3.3
The relative position of the median in the data set

The median is unaffected by extremes, as the median takes into account only the order of the values. Suppose Gallagher received a salary of $1 million, while all other outfielders' salaries remained the same. Compute the median. You will see that it remains the same. Moving an extremely high or a low value farther from the center does not change the median, as long as there are data points near the center left unaffected. Therefore, the median is a useful tool if there is concern that extremes may distort the value given by the arithmetic mean.

The Mode

If there are many repeat observations, the mode is a useful measure of central tendency. The mode, or typical value, is the value that occurs most frequently, and it can be very revealing. When someone says of a professor "He gives lots of C's," she is saying that C appears to be the typical grade in his class. It is the grade given most frequently. It is the mode.

> The **mode** is the value or class of values that occurs most frequently in a data set.

EXAMPLE 3.7 ▪ FINDING THE MODE FOR DENTAL VISITS

When a group of individuals were asked, "How often did you visit the dentist last year?" they gave the following responses:

0	1	2	1	0	0	3	1	1	1	1
0	1	0	0	0	2	1	1	1	1	

What is the typical number of dental visits made by members of this group during the year?

Examining the data and arranging them in an ordered array from fewest to most visits gives Table 3.4 on page 74.

Table 3.4

TYPICAL NUMBER OF DENTAL VISITS PER YEAR

Number of Visits	Frequency
0	7
1	11
2	2
3	1

Finding the mode

Clearly, the typical number of visits is the number that occurs the most frequently. Eleven individuals visited their dentist once. Therefore, 1 is the mode. In this case, it is also the median. Note that the mode is more useful in profiling individuals than the arithmetic mean; in this instance, the mean is .86 visit.

The mode is a useful tool for summarizing numerical data when frequency is an important feature—for example, visits to the dentist, vacations per year, peak sales days. Figure 3.4 shows a bar graph in which the distribution of the number of dental visits peaks at the mode. Sometimes, a distribution has two peaks. If a distribution has two peaks, even when those peaks are not identical in height, the distribution may be referred to as **bimodal.** Sales of certain products may exhibit a bimodal distribution. For example, sales of perfumes peak around Valentine's Day and Mother's Day; cigar sales tend to peak around Christmas and Father's Day.

Figure 3.4
Spotting the mode

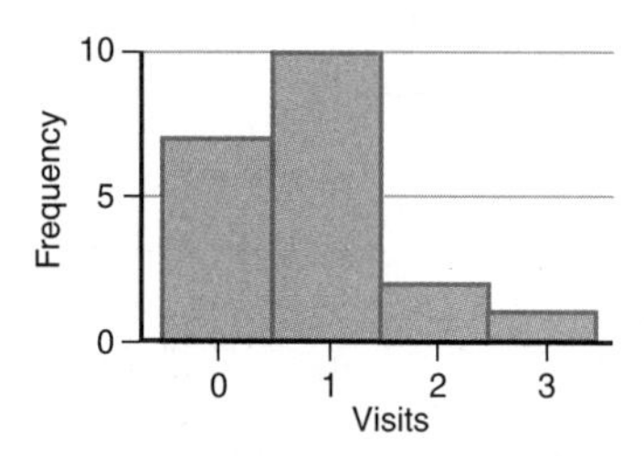

EXERCISES 3.1–3.16

Skill Builders

3.1 Consider the following data set:

53 4 29 29 79 87 16 56 32

a. Find the arithmetic mean.
b. Find the median.
c. What is the mode?
d. Suppose you dropped the 87 and added 722. How would this affect the mean? the median? the mode?

3.2 Consider the numbers below:

518 202 715 2 910 202 202

a. Find the arithmetic mean.
b. Find the median.
c. What is the mode?

3.3 The grades of statistics students on an examination were as follows:

75	80	85	50	72	82	91	83
15	42	97	74	79	57	84	66
86	82	100	77	82	51	78	92

a. Compute the arithmetic mean.
b. Construct a dot plot of the data.
c. What is the median grade?
d. Find the mode.
e. If 100 is replaced by 80, what is the effect on the mean? the median? the mode?
f. Is there an outlier that should be removed? If so, identify it and explain why it should be eliminated from the set.
g. How does removal of the outlier affect the results of parts a, c, and d?

3.4 Sharon has been walking a mile each day. Her times, in minutes, appear below:

20.5	21.4	19.8	22.1	22.7	18.0	17.2	17.8	20.0	30.4
17.9	18.1	21.5	20.0	19.6	22.3	21.6	19.7	20.0	20.1

a. Calculate Sharon's mean time.
b. What is her median time?
c. Find the mode.
d. Make a stem plot of the data.
e. What happens to the measures of central tendency if the highest time is removed from the data set?

3.5 Consider the data set of large numbers below:

52,920	4,109	4,070	56,156	51,816	178
4,358	28,483	86,910	31,575	20,190	91,161
29,492	78,687	15,608	72,232	71,533	30,624

a. Compute the arithmetic mean.
b. Find the median.
c. Remove the smallest value and the largest value at the same time. What happens to the mean and the median?
d. Now, remove just the largest value. What happens to the mean and the median?

3.6 Consider this data set:

.28	−.80	.56	−.90	−.02
−.08	.41	.24	−.97	1.05
.95	−.43	.62	.99	−.01
.00	.10	.59	.18	.84

a. Compute the mean value.
b. Find the median value.

c. Replace the value .84 with 3.84. What happens to the mean? What happens to the median?

Applying the Concepts

3.7 *Consumer Reports* (January 1991) evaluated brands of coffee. Data for a dozen popular brands appear below:

Brand Name	Package Size (in oz)	Price (per oz)	Caffeine (in mg/oz)
Yuban 100% Colombian	13	$0.27	6.38
Brown Gold 100% Colombian	16	0.28	5.38
Chock Full O'Nuts	13	0.17	8.38
Folgers Special Blend	11.5	0.19	8.78
Albertsons	13	0.15	8.62
Maxwell House			
Regular	13	0.20	6.62
Master Blend	11.5	0.20	7.91
Hills Brothers	13	0.19	8.69
Martinson	16	0.20	5.50
A&P Special Roast	13	0.17	6.84
Chase and Sanborn	11.5	0.16	10.26
Savarin	16	0.15	7.56

a. Find the mode for the standard size can of coffee.
b. Find the average price.
c. What is the median price?
d. Find the average amount of caffeine in these brands.
e. What is the median level of caffeine?
f. Which coffee would you choose and why?

3.8. In its "1991 Investment Guide" (December 17, 1990), *U.S. News & World Report* rated a dozen stocks as falling in the 99th percentile for EPS (earnings per share)—that is, 99% of the stocks examined by *U.S. News & World Report* had a lower EPS than these dozen stocks. The return on equity (R.O.E.) varies considerably, as do the price/earnings (P/E) multiple and current dividend yield. The data for these stocks appear below. The R.O.E. is the return on invested capital. The P/E multiple is the ratio of share price to per share corporate earnings. Current yield is the ratio of current annual dividends to share price.

Company Name	R.O.E. (%)	P/E Multiple	Current Yield (%)
Novell, Inc.	24	26	0
Conner	27	13	0
Costco	16	29	0
Microsoft	38	26	0
Blockbuster	27	28	0
Banco Central	17	8	3.2
LVMH Moet	22	19	1.1
Pacific Dunlap	26	8	4.1
De Beers	16	6	3.4
Intel	17	12	0
Banco Bilbao	23	7	4.8
Fannie Mae	31	7	2.9

a. What is the mean R.O.E.?
b. What is the median R.O.E.?
c. Find the mean and median P/E multiple.
d. What is the modal (most frequent) dividend yield?
e. What strategy might lie behind this yield?
f. If you were discussing price/earnings multiples, which measure of central tendency would you choose? Why?

3.9 The number of crew members on operational space shuttle flights varied considerably between 1982 and 1990. The numbers are given below:

4	5	7	5	5	5	5	4
7	7	5	6	5	5	5	7
5	5	7	8	5	5	7	5
7	5	7	5	6	7	5	

a. What is the average number of crew members on a shuttle flight?
b. Find the median number.
c. What is the mode?
d. There were four shuttle test flights with two crew members each. If you factor these values into your calculations, how do they influence the mean? the median?
e. If you were describing the typical shuttle crew, which measure of central tendency would you use? Explain your choice.

3.10 In a survey of popular brands of frozen yogurt, *U.S. News & World Report* (November 5, 1990) provided the following information on an 8-ounce serving:

Brand	Calories	Total Fat (grams)	Saturated Fat (grams)
Baskin-Robbins Low Fat	240	8	4.9
I Can't Believe Original	216	6	3
TCBY Original	240	6	4
Dairy Queen	200	0	0
Dannon Low Fat	220	4	2.4
Elan Premium	260	6	4.8
Penguins' Place Regular	208	5.6	3.2
Stonyfield Farm Nonfat	200	0	0

a. Find the arithmetic mean caloric value.
b. What is the median caloric value?
c. Is there a modal caloric value? If so, what is it?
d. Find the mean, median, and mode for total grams of fat.
e. Find the mean, median, and mode for saturated fat.
f. If Baskin-Robbins Low Fat Frozen Yogurt were taken off the list, how would this change your answers in parts a through e?
g. Which brand of frozen yogurt (if any) would you choose from the list and why?

3.11 More than two dozen strong earthquakes occurred in the 1980s. *Universal Almanac 1991* gave the magnitudes of these earthquakes and their locations as follows:

Date	Location	Magnitude
Oct. 10, 1980	Algeria	7.3
Nov. 26, 1980	Italy	7.2
July 28, 1981	Iran	—
Dec. 13, 1982	Yemen	6.0
Mar. 31, 1983	Colombia	5.7
Oct. 28, 1983	Idaho	6.9
Oct. 30, 1983	Turkey	7.2
Mar. 3, 1985	Chile	7.8
Sept. 19, 1985	Mexico	8.1
Jan. 31, 1986	Ohio	4.9
Mar. 5, 1987	El Salvador	7.5
Aug. 20, 1988	India/Nepal	8.5
Nov. 6, 1988	Burma/China	—
Nov. 25, 1988	Quebec	6.0
Dec. 3, 1988	Pasadena, California	5.0
Dec. 6, 1988	Iran	5.6
Dec. 7, 1988	Armenia	6.9
Jan. 22, 1989	Tadzhikistan	5.5
May 10, 1989	Malawi	6.2
May 23, 1989	Macquarie Ridge (300 miles SE of New Zealand)	8.2
Aug. 1, 1989	Indonesia	5.9
Oct. 17, 1989	San Francisco, California	7.1
Oct. 18, 1989	China	7.1
Oct. 19, 1989	Algeria	6.0
Dec. 15, 1989	Philippines	7.3
Dec. 27, 1989	Australia	5.4

a. Examine the data for the first half of the decade. Describe the data. What was the mean? the median?

b. Do the same for the second half of the decade.

c. If these data are representative, did earthquake activity become more severe over the course of the decade?

d. According to these data, did earthquake activity become more frequent over the course of the decade?

e. Prepare a brief paragraph describing your results in parts a through d.

3.12 The numbers of unemployed in the state of New York between 1970 and 1989 appear below:

Year	Number (thousands)	Year	Number (thousands)
1970	331	1980	597
1971	491	1981	612
1972	501	1982	684
1973	404	1983	688
1974	482	1984	583
1975	727	1985	544
1976	789	1986	526
1977	705	1987	412
1978	605	1988	358
1979	571	1989	442

Source of Data: New York State Department of Labor, *Employment Review* (January 1991), Volume 44, No. 1.

a. Find the average number of unemployed during the 1970s.
b. What was the average level of unemployment in the 1980s?
c. Average unemployment in New York was 451,000 in 1990. How does this compare with the averages for the 1970s and 1980s?
d. Some of the variation in the numbers of unemployed may be due to changes in the population. Could this explain all of the variation in unemployment seen here?

3.13 In Fall 1990, *Money* published a "Shoppers' Guide to 1,000 Schools." A random selection of 20 of the schools is given below.

College	Tuition	College	Tuition
Appalachian State (No. Carolina)	$5,788	Hobart (New York)	$15,346
Bloomsburg University (Pennsylvania)	4,480	Iona (New York)	8,260
California Lutheran	9,450	Knoxville (Tennessee)	5,290
Clark (Georgia)	6,170	Long Island University–C.W. Post (New York)	8,520
Cumberland (Tennessee)	4,400	Murray State (Kentucky)	3,760
Delaware (Mississippi)	3,170	Ohio Wesleyan	12,328
East-West (Illinois)	5,360	Radford University (Virginia)	4,760
Georgia Institute of Technology	6,054	Anshelm (New Hampshire)	9,640
St. John's University (New York)	6,860	Seton Hall (New Jersey)	9,790
Greensboro (No. Carolina)	6,201	University of Florida	4,630

a. What is the average tuition?
b. Find the median value for tuition.
c. What is the average tuition in New York and New Jersey?
d. Are the colleges in New York and New Jersey more expensive than colleges elsewhere? On what do you base your conclusion?

e. Check the most recent "College Guide" issue of *Money*. Repeat parts a through d.
f. Using your results from parts a through e, write a summary of your findings.

3.14 The New York State Department of Labor Report *Tomorrow's Jobs, Tomorrow's Workers, 1995–1996* gave the following average weekly wages for professional, technical, and managerial "jobs with favorable prospects" in the Long Island region.

Occupation	Wage	Occupation	Wage
Accountant/Auditor	$688	Occupational therapist	$612
Biomedical electronics engineer	942	Pharmacy technician	306
Biomedical equipment technician	604	Physical therapist	757
Computer programmer	755	Physician's assistant	837
Computer systems analyst	1161	Psychologist	548
Electronic engineer	635	Quality control engineer	974
Electronic publisher/editor	610	Registered nurse	761
Health information manager	753	Speech-language pathologist/audiologist	753
Industrial production manager	813	Statistician	1040
Lawyer	818	Technology manager	1249
Marketing manager	790		
Media/graphic artist	762		

a. Find the average weekly wage for professional, technical, and managerial "jobs with favorable prospects" in the Long Island region.

b. What is the median wage in these high demand occupations?
c. If the average weekly wage of statisticians is removed from the analysis, what is the new average? the new median?

3.15 Boxing's top 10 earners in 1989, as reported in *Inside Sports* (February 1990), are listed below:

Ray Leonard	$29,500,000	Frank Bruno	$3,200,000
Mike Tyson	14,000,000	Michael Nolan	3,000,000
Thomas Hearns	11,000,000	Ray Mancini	1,400,000
Roberto Duran	8,165,000	Carl Williams	1,000,000
Evander Holyfield	3,500,000	Roger Mayweather	1,000,000

a. What was the mean amount earned?
b. Find the median amount.
c. If Ray Leonard is not considered, what happens to the average amount? the median amount?

d. Who are boxing's top 10 earners today? Repeat parts a and b. Compare these results with the 1989 figures.

3.16 An evaluation of telephone answering machines, which appeared in *Consumer Reports* (November 1991), revealed the following prices for models without phones:

100	100	70	100	89	149	60	55
110	150	100	58	88	60	70	40

a. What is the mean price for a telephone answering machine?
b. Find the median price.
c. Is there a modal price? If so, what is it?
d. Based on these prices, what would you budget for such a purchase? Explain your reasoning.

SECTION 2 ■ MEASURES OF VARIATION

We now turn from measures of central tendency to measures of variation. Variability lies at the heart of statistics. It exists everywhere in quantitative life. Students' grades on an examination will vary around the average grade. The amount of coffee in cans filled on a production line will vary around the specified amount. We need ways to measure this variation so that a manager can distinguish between normal fluctuations and unusual variations that require corrective action.

> **Measures of variation** reflect the spread of a data set. The most popular measures of variation are the range, interquartile range, and standard deviation.

Measures of variation tell us about the ever-present variability of data sets. When constructing a portfolio for a client, a financial planner wants to have some knowledge about the volatility (or stability) of various investments, as the client will be concerned about variability. If the client purchases IBM stock, how much can she expect it to move in a year's time? On September 20, 1997, the *New York Times* reported that the 52-week high and low prices of IBM stock were $109\frac{7}{16}$ and $60\frac{3}{4}$;

the stock closed at 99¼. The variability of its price is of considerable importance, and knowledge that a particular stock is trading near the top of its range (range being a measure of variability) is certainly useful to an investor.

Once data on an issue such as rainfall have been characterized by a single central number, that information may be qualified by a number that shows how the rainfall data vary. This number is useful in many situations. For example, when moving to an area, you'll want to know not only the average rainfall, but the highs and lows over a year's time. The highs will determine whether your basement is likely to flood, and the lows how much you will have to water the lawn. You will also want to know when the rainfall occurs. But for now, let's concentrate on the variation in rainfall. This variation is as important as the average. The average monthly rainfall in San Francisco is 1.64 inches. That fact tells you something about the climate, but the fact that the monthly averages range from a high of 4.65 inches in January to a low of .05 inch in July tells you much more. Considering these highs and lows brings us to the first and simplest measure of variation that we will discuss—the range.

The Range

The **range** is defined as the difference between the maximum value (x_{max}) and the minimum value (x_{min}) for the data.

Calculating the range

$$\text{Range} = x_{max} - x_{min} \quad (5)$$

Look again at the monthly rainfall in San Francisco. The measure of center is the mean, 1.64 inches, and the measure of variation is the range, 4.6 inches (from .05 inch to 4.65 inches). How do these measures allow us to compare patterns in various cities? Let's consider two cities separated by thousands of miles: New York and Seattle. The mean monthly rainfall for New York is 3.68 inches and that for Seattle is 3.21 inches, so they are about the same. Now let's look at the ranges. For New York, the range is from 4.22 inches to 3.13 inches, or 1.09 inches, for Seattle, the range is from 6.33 inches to .74 inch, or 5.59 inches. So, we can see that although the measures of center are almost the same, month-to-month variations in the rainfall are far greater in Seattle than in New York.

Dot plots provide a natural way to picture the range. Figure 3.5 shows the dot plots for monthly rainfall in New York and Seattle. The means are indicated by the balance points for these distributions—the colored wedges. Although the balance points are about equal, the ranges, given by the lengths of the colored arrows are quite different.

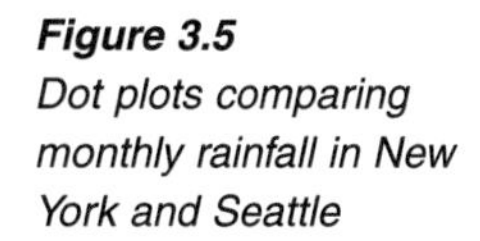

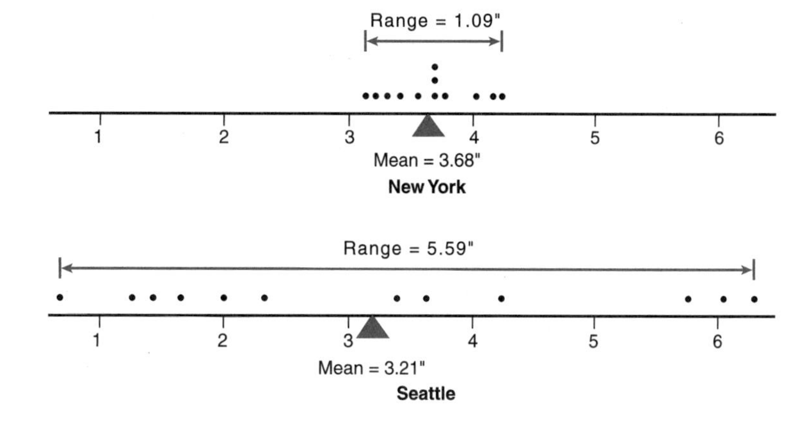

Figure 3.5
Dot plots comparing monthly rainfall in New York and Seattle

EXAMPLE 3.8 ▪ FINDING THE RANGE FOR CD RATES

As of September 30, 1991, *Newsday* reported that the average interest rate on a 3-month CD was 5.19%. An important thing to know is the spread of interest rates on such investments. Would a rate of 6.00% be an unreasonable expectation? Could an investor end up with an interest rate of 5.00%? To answer these questions, we need to determine the range of the rates shown in Table 3.5.

Table 3.5

THREE-MONTH CD RATES FOR SELECTED LONG ISLAND BANKS

Bank	Rate
Apple Bank	5.25%
Bank of New York	5.10
Bowery Savings	5.26
Chase Manhattan	4.75
Crossland	5.41
Long Island Savings	5.54
Marine Midland	4.90
Queens County Savings	5.44
Riverhead Savings	5.40
Westbury Federal	5.30

The range for this data set is calculated as follows:

$$\text{Range} = 5.54\% - 4.75\% = .79\%$$

The range is .79%. Here, the limits of the range may be more important than the range itself. The minimum value is 4.75%; the maximum value is 5.54%. This answers the question "Would a rate of 6.00% be an unrealistic expectation?" A rate of 6.00% is slightly above the range. Could an investor end up with 5.00%? You bet! This rate is well above the minimum value, but within the range.

The range is used widely in industry. It's easy to compute, and the extremes needed to compute the range are themselves useful numbers. However, all the values in the data set could be changed with the exception of the high and the low, and the range would remain the same. Returning to the graphs of Chapter 2, we can draw dot plots that illustrate this point. Figure 3.6 shows two sets of data points. Both sets have the same range, but set 1 is more spread out than set 2. We will now look at measures of the range that capture such differences.

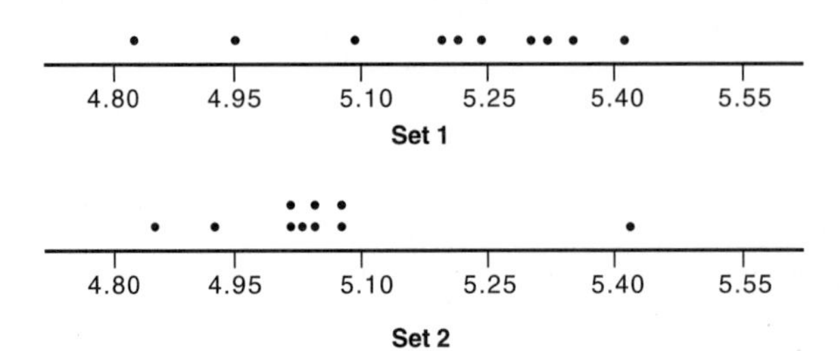

Figure 3.6
Two data sets with the same range

The Variance and Standard Deviation

The variance and standard deviation are measures of spread that depend on all the observations in a data set. Even a small change in one number can influence the result.

> The **variance** is a measure of the average variability in a data set. It depends on the difference between each observation and the arithmetic mean of the data set. It is the mean square deviation from the mean.

To understand the variance, consider a set of observations in which each observation is represented by the symbol x_i. The **deviation,** or **difference from the mean,** for each observation is calculated by subtracting: $x_1 - \mu, x_2 - \mu, \ldots, x_n - \mu$. If these deviations are averaged, the result *will always equal zero.* Negative values cancel out positive ones, as shown in Example 3.9.

EXAMPLE 3.9 ▪ UNDERSTANDING VARIANCE

The average Baltimore Orioles infielder earned a salary of $322,375 in 1990. Salary data for all Orioles infielders in 1990 are shown in Table 3.6.

Table 3.6

BALTIMORE ORIOLES INFIELDERS' SALARIES, 1990

Player	Salary	Deviation (salary − mean)
Gonzales	206,000	−116,375
Horn	120,000	−202,375
Hulett	208,000	−114,375
Milligan	155,000	−167,375
Ripken, B.	215,000	−107,375
Ripken, C.	1,367,000	+1,044,625
Segui	100,000	−222,375
Worthington	208,000	−114,375

A property of the mean

The sum of the deviations from the arithmetic mean equals 0. The mean of this set of deviations is the sum, 0, divided by $n = 8$, or $0/8 = 0$.

The fact that the positive and negative deviations from the mean balance out to zero is simply the arithmetic expression of the fact that the dot plot of the data is perfectly balanced when pivoted about the mean of the data (see Figure 3.2).

Represented symbolically, the mean deviation from the mean looks like this:

$$\frac{\sum (x_i - \mu)}{N} = 0$$

The capital Greek sigma, Σ, tells us to add up all the terms of the form $(x_i - \mu)$. These are $(x_1 - \mu)$, $(x_2 - \mu)$, ..., $(x_N - \mu)$. The resulting sum divided by N is the mean deviation from the mean. Since the sum of the deviations from the mean is always zero, it cannot be the measure of variation. To avoid this problem and to eliminate negative numbers, the deviations are squared and the average is taken. This result is called the variance. Because finding the variance is a step on the path to finding the standard deviation (which will be defined shortly), the variance is usually denoted by the square of the standard deviation—that is, σ^2 (σ is the Greek lowercase sigma) for a population or s^2 for a sample. The formula for population variance, with N being the size of the population, is given below:

$$\text{Population variance,} \quad \sigma_X^2 = \frac{\sum (x_i - \mu_X)^2}{N} \tag{6}$$

The formula for sample variance is similar:

$$\text{Sample variance,} \quad s_x^2 = \frac{\sum (x_i - \bar{x})^2}{n - 1} \tag{7}$$

where $\Sigma(x_i - \bar{x})^2$ is the sum of the squared deviations from the sample mean.

The sample variance is a statistic calculated without use of the population mean, μ. On average, $\Sigma(x_i - \bar{x})^2$ will be smaller than $\Sigma(x_i - \mu_X)^2$. To ensure that the statistic s^2 provides the best possible estimate for σ^2, we must use the smaller denominator, $n - 1$, in calculating s^2.

The denominator $n - 1$ is built into many statistical calculators; it should be used unless you are looking at a small population and have easy access to a census. In that case, you can calculate μ and σ^2 using Equation (6), rather than the estimate s^2.

The variance formula yields results in squared units of whatever is measured. As a consequence, if you double all your measurements, you will multiply the variance by 4. Since you would expect a reasonable measure of spread to double (rather than quadruple) if all the measurements were doubled, the variance is not a good measure of spread. However, its square root, called the **standard deviation,** is. The results of the standard deviation formula are in the same units as the original data—that is, not squared. The standard deviation is a measure of "average scatter."

$$\text{Population standard deviation,} \quad \sigma_X = \sqrt{\frac{\sum (x_i - \mu_X)^2}{N}} \tag{8}$$

where N = population size.

$$\text{Sample standard deviation,} \quad s_x = \sqrt{\frac{\sum (x_i - \bar{x})^2}{n - 1}} \tag{9}$$

where n = sample size.

Figure 3.5 showed that while New York and Seattle have roughly the same average monthly rainfall, Seattle's rainfall is much more variable from month to month than is New York's. How is this greater variability reflected in the standard deviations of monthly rainfall for these two cities? Computation shows that for New York, s is .364, while for Seattle, s is 1.968. Thus, a comparison of these standard deviations offers quantitative confirmation of the observation that rainfall in Seattle is more variable.

EXAMPLE 3.10 ■ FINDING THE AVERAGE SCATTER OF CD RATES

Consider the CD rate data once again. The data have been put into a worksheet in Table 3.7.

Calculating the sample standard deviation

Table 3.7

WORKSHEET FOR COMPUTING THE SAMPLE STANDARD DEVIATION FOR CD RATES FOR LONG ISLAND BANKS

Bank	Rate x_i	Deviation $x_i - \bar{x}$	(Deviation)² $(x_i - \bar{x})^2$
Apple	5.25%	+.015	.000216
Bank of New York	5.10	−.135	.018225
Bowery Savings	5.26	−.025	.000625
Chase Manhattan	4.75	−.485	.235625
Crossland	5.41	+.175	.030625
Long Island Savings	5.54	+.305	.093025
Marine Midland	4.90	−.335	.112225
Queens County Savings	5.44	+.205	.042025
Riverhead Savings	5.40	+.165	.027225
Westbury Federal	5.30	+.065	.004225
Total	52.35	.000	.564041

After a total has been obtained for the rate column, the sample arithmetic mean can be calculated:

$$\bar{x} = \frac{\sum x_i}{n} = \frac{52.35}{10} = 5.235$$

Once the mean has been calculated, the next column (deviations from the mean) can be completed. Note that the sum of this column is zero, as discussed in Example 3.9. The individual results from the deviations column are squared in the last column. This column is totaled, and the result appears in the numerator of the sample standard deviation formula:

$$s = \sqrt{\frac{\sum (x_i - \bar{x})^2}{n - 1}} = \sqrt{\frac{.564041}{10 - 1}} = .25\%$$

The standard deviation for interest rates on three-month CDs is .25% (or one quarter of one percent). For many data sets, the majority of observations fall between $\mu - 2\sigma$ and $\mu + 2\sigma$ (this is part of the empirical rule, which we will consider later in this chapter). Using $\bar{x}$ to estimate μ and s to estimate σ and applying this rough rule, we would expect most CD rates to fall between 4.735 (which is 5.235 - 2 × .25) and 5.735 (which is 5.235 + 2 × .25). Indeed, in September 1991, this was the case.

EXAMPLE 3.11 ▪ FINDING THE AVERAGE SCATTER OF EMPLOYEES' HOURLY WAGES

Following are the hourly wages for the employees of the ABC Company:

4.55	3.00	4.10	5.35	7.05	9.00	4.75
12.65	3.35	11.02	3.00	13.00	4.65	

We can compute s using a calculator. Putting the calculator in statistical mode and entering the data, we find that the sample data have an average of $6.57 (found by hitting the $\bar{x}$ key) and a standard deviation of $3.64 (found by hitting the s key or σ_{n-1} key). Spreadsheet programs, such as Excel or Lotus, also have statistical capabilities. However, there are dozens of statistical packages especially designed for this type of job. We shall explore the use of one such package, Minitab, later in this chapter.

The Coefficient of Variation

The **coefficient of variation** is the percentage of variation around the arithmetic mean. You'll find it very useful when comparing data sets that differ vastly in scale. Suppose you wish to compare unemployment in the United States with unemployment in the state of New York during the 1980s. Over the decade, unemployment in New York averaged 544.6 thousand with a standard deviation of 111.6 thousand. During that same period, unemployment in the United States averaged 8278.2 thousand with a standard deviation of 1536.5 thousand. Where was there more relative variability? In absolute terms, the standard deviation for the United States was nearly 14 times that for New York. In relative terms, the variation about the mean was 18.6% for the nation as a whole, but 20.5% for New York, suggesting that New York may experience more severe cycles of employment and unemployment than does the United States as a whole. These percentages are coefficients of variation.

$$\text{Population coefficient of variation} = \frac{\sigma_X}{\mu_X} \times 100 \tag{10}$$

where σ_X is the population standard deviation and μ_X is the population mean.

$$\text{Sample coefficient of variation} = \frac{s_x}{\bar{x}} \times 100 \tag{11}$$

where s_x is the sample standard deviation and $\bar{x}$ is the sample mean.

Example 3.12 illustrates the procedure for calculating the coefficient of variation.

EXAMPLE 3.12 ▪ COMPARING THE DOW JONES AND NIKKEI INDICES

Calculating the coefficient of variation

Let's compare the Dow Jones Industrial Index for the week of November 27–December 2, 1991 with the Nikkei Index for the same period. The data are given in Table 3.8. Which index shows greater volatility? The degree of volatility of a market is a measure of the relative percentage gains or losses one is likely to experience in that market.

Table 3.8

DOW JONES INDUSTRIAL INDEX AND NIKKEI INDEX, NOV. 27, 1991–DEC. 2, 1991

	Dow Jones Industrial Index	Nikkei Index
Average	2910.03	22,858.34
Standard deviation	22.11	148.84
Coefficient of variation	$\frac{22.11}{2910.03} \times 100$	$\frac{148.84}{22858.34} \times 100$
	$= .0076 \times 100 = .76\%$	$= .0065 \times 100 = .65\%$

Over this period of time, the Dow was more volatile than the Nikkei.

EXERCISES 3.17–3.30

Skill Builders

3.17 Consider the following data set, treating these data as a complete population.

31 28 80 56 90 41 24

a. Find the range for this data set.
b. Calculate the average, μ, for this data set.
c. What is the standard deviation, σ, for this data set?
d. Calculate the coefficient of variation.

3.18 Which of the following data sets has a greater coefficient of variation? Treat both sets of data as samples.

Set A	Set B
10	642
4	820
6	380
1	443
2	337
8	225

3.19 The population data below represent a complete census:

66,194	43,195	67,299	58,036	43,782
76,396	17,469	87,286	49,023	34,710
9,325	63,360	4,117	72,121	41,377
75,884	1,223	51,986	38,012	69,813

a. Calculate the arithmetic mean, μ.
b. Find the range for this data set.
c. Compute the standard deviation, σ.
d. Calculate the coefficient of variation.

3.20 Consider these data as samples:

.28	−.80	.56	−.90	−.02
−.08	.41	.24	−.97	1.05
.95	−.43	.62	.99	−.08
.00	.01	.59	.18	.84

a. Calculate the range.
b. Find the mean, $\bar{x}$.

c. Compute the standard deviation, s.
d. Find the coefficient of variation.

Applying the Concepts

3.21 Al tries to do some walking every day to keep physically fit. His record of distances walked (in miles) for the past month is below:

2.4	2.4	2.0	2.4	2.4	2.4	2.4	2.4	2.4	2.4
2.4	2.4	2.4	2.4	2.4	2.4	0.0	4.0	2.4	4.0
0.0	2.4	0.0	2.4	2.4	4.0	2.4	2.4	2.4	2.4

a. Find the range.
b. Calculate the arithmetic mean.
c. Compute the standard deviation.
d. What is the coefficient of variation for this data set?

3.22 The number of customers walking into a local deli each hour for several days is found below. Treat these data as samples.

5	4	5	6	5	3	10	10	8	7
12	5	21	23	10	12	15	18	42	31
23	14	20	9	53	54	35	27	22	21
30	43	35	30	23	57	22	22	15	15
18	13	9	11	16	6	8	6	3	6
4	2	7	3						

a. What is the range?
b. Calculate the arithmetic mean.
c. Find the standard deviation.
d. Compute the coefficient of variation.

3.23 The tuitions listed below are a random sample taken from "Shoppers' Guide to 1,000 Schools," in *Money* (Fall 1990).

Colleges in the Northeast (New York, New Jersey, Pennsylvania, and New Hampshire)

$4,480	6,860	15,346	8,260
8,520	9,640	9,790	

Colleges Elsewhere in the United States

$5,788	9,450	6,170	4,400
3,170	5,370	6,054	6,201
5,290	3,750	12,328	4,706
4,630			

a. Calculate the standard deviation for the entire sample (both groups). (The mean is $7,010.15.)
b. Calculate the average tuition for each group.
c. Calculate the standard deviation for each group.
d. Calculate the coefficient of variation for each group.
e. Which group in this sample exhibits higher tuition?
f. In which group in this sample was there more variation in tuition?

3.24 Returning to the list of strong earthquakes that occurred during the 1980s *(Universal Almanac 1991)*, we find that we can divide the data into two regions: the Americas and the rest of the world.

The Americas

5.7	6.9	7.8	8.1	4.9
6.0	5.0	7.1	7.5	7.0

The Rest of the World

7.3	7.2	6.0	7.2	6.5
5.5	5.5	8.2	5.9	7.1
5.4	5.6	6.9	6.0	7.3

a. Calculate the mean and standard deviation for each region.
b. Calculate the coefficient of variation for each region.
c. Which region exhibits a more unpredictable record (that is, a greater coefficient of variation)?

3.25 *Consumer Reports'* "1992 Buying Guide" rated car radio/tape players. Players can be classified into two types by mounting—removable models and shaft-mounted models for older cars and trucks. They are organized by type below. Treat the data as samples.

Removable		Shaft-Mounted	
Sony	$550	Pioneer	$350
Kenwood	449	Blaupunkt	350
Pioneer	340	JVC	300
Alpha	430	Sony	320
Clarion	350	Audiovox	300
JVC	350	Clarion	300
Panasonic	380		

a. Find the average cost for a removable radio/tape player.
b. Find the standard deviation for removable radio/tape players.
c. Repeat parts a and b for shaft-mounted players.
d. Calculate the coefficient of variation for each category.
e. Which category shows greater relative variation?

3.26 Salary data for the population of the top 20 earners in baseball for 1990 and 1991 are in the tables below.

1990		1991	
Robin Yount	$3,200,000	Daryl Strawberry	$3,800,000
Kirby Puckett	2,700,000	Will Clark	3,750,000
Roger Clemens	2,600,000	Kevin Mitchell	3,750,000
Don Mattingly	2,500,000	Joe Carter	3,666,667
Eddie Murray	2,492,091	Mark Davis	3,625,000
Paul Molitar	2,433,333	Eric Davis	3,600,000
Kent Hrebk	2,300,000	Willie McGee	3,562,000
Will Clark	2,250,000	Mark Langston	3,550,000
Ricky Henderson	2,250,000	José Canseco	3,500,000
Mike Scott	2,125,000	Tim Raines	3,500,000
Tom Browning	2,125,000	Dave Stuart	3,500,000
Mark Davis	2,125,000	Bob Welch	3,450,000
Ted Higuera	2,125,000	Don Mattingly	3,420,000
Dave Winfield	2,122,800	Doug Drabek	3,350,000
Eric Davis	2,100,000	Dennis Martinez	3,333,333
Andre Dawson	2,100,000	Andre Dawson	3,300,000
Jack Morris	2,100,000	Nolan Ryan	3,300,000
Pedro Guerro	2,083,333	Dave Winfield	3,300,000
Kevin Mitchel	2,083,333	Glen Davis	3,275,000
Mark Gubicza	2,066,667	Danny Darwin	3,250,000

Source of Data: *Sport,* June 1990 and June 1991, Peterson Publishing Co., Los Angeles, CA.

a. Calculate the average, μ, for 1990.
b. Calculate the standard deviation, σ, for 1990.
c. Calculate the coefficient of variation for 1990.
d. Repeat parts a through c for 1991.

3.27 In 1991, the New Jersey Casino Control Commission for Atlantic City Casinos compiled the following figures on winnings as a percentage of wagers. Treat the data as a census.

Casino	Gamblers' Win % Quarter Slots	Gamblers' Win % Dollar Slots
Bally's Grand	86.6	90.0
Bally's Park Place	87.2	90.9
Caesar's	85.4	91.5
Claridge	86.7	89.0
Harrah's	88.7	91.1
Resorts	87.3	90.6
Sands	85.7	89.3
Showboat	88.2	91.5
Tropworld	87.9	91.2
Trump's Castle	88.4	90.6
Trump Plaza	86.8	91.2
Trump's Taj Majal	87.8	90.7

a. Find the average percentage win, μ, on quarter slots.
b. Compute the standard deviation, σ, on quarter slots.
c. Repeat parts a and b for dollar slots.
d. What is the coefficient of variation for quarter slots?
e. What is the coefficient of variation for dollar slots?
f. Which type of slot machine shows greater relative variation?

3.28 Commercial glass cleaners sell for between \$.01 and \$.15 per fluid ounce. The data below, in cents per fluid once, are from *Consumer Reports* (October 1989). Treat the data as samples.

13	8	7	15	9	6	8
6	6	1	14	12	1	6
6	6	1	7	8	9	9

a. Calculate the range.
b. Find the average cost, $\bar{x}$.
c. Compute the standard deviation, s.
d. Calculate the coefficient of variation.
e. Based on the data, what shopping strategy would you follow?

3.29 A random sample of runners in the Women's 4 Miler held in Central Park in October 1990 (*New York Running News*, December/January 1991) gave the following running times (in minutes) for the competition:

41.52	32.97	32.28	39.66	45.62	36.30	35.92
49.08	38.33	26.62	38.38	38.65	29.17	32.93
38.43	24.60	27.43	41.02	35.90	25.53	51.03

a. Compute the range.
b. Calculate the sample mean, $\bar{x}$.

c. Find the sample standard deviation, s.
d. Calculate the coefficient of variation.

3.30 *U.S. News & World Report* (December 1990) contained the 1990 percentage returns for top-performing mutual funds. Results for a random sample of 25 funds are as follows:

−1.89%	6.60	−6.12	5.95	5.00
−3.49	−6.76	7.37	−3.31	3.50
4.92	1.97	4.10	−21.37	2.78
−5.01	−5.49	−16.49	1.04	.55
15.68	15.50	10.73	−.19	.44

a. What is the range of percentage returns for this sample of mutual funds?
b. Find the mean return, $\overline{x}$
c. Calculate the sample standard deviation, s.
d. Calculate the coefficient of variation.
e. If you invested $1,000 in each of these funds at the beginning of the year, what would your return be at the end of the year?

SECTION 3 ■ PERCENTILES, QUARTILES, AND THE INTERQUARTILE RANGE

In Chapter 2, we used percentiles in connection with the ogive of a distribution. In this chapter, we will treat percentiles analytically.

The percentile is the most popular relative measure. Relative measures indicate where a single point falls within the entire set of data. You have probably seen percentiles at work in SAT scores, which are reported in two ways: a score (for example, 690) and a percentile (for example, the 95th percentile). If your score is in the 95th percentile, 95% of the scores fall at or below it and 5% lie at or above it.

Relative measures indicate the position of an observation in a data set. They provide a basis of comparison between data points. Relative measures include the maximum (which none of the data exceed), percentiles, and quartiles.

Percentiles and Quartiles

Table 3.9 shows a sample of scores and their corresponding percentiles for the verbal portion of the 1996 SAT. The median is the 50th percentile. At least half of those taking this test scored 510 or better, and at least half scored 510 or worse.

Table 3.9

SELECTED PERCENTILES FOR THE 1996 VERBAL SAT

	Scaled Score	Percentile
	690	95th
	650	90th
	590	75th
Median	510	50th
	440	25th
	370	10th
	330	5th

While calculation of any percentile may be useful, the 25th and 75th percentiles play especially important roles. The 25th percentile is also known as the 1st **quartile;** the 75th percentile is also known as the 3rd quartile. Together, they define the interquartile range, or IQR, our third measure of variation.

Interpretation of the 25th percentile on a dot plot is obvious: 25% of the data points lie to the left of it, and 75% of the data points lie to the right of it. We will return to this graphical interpretation later in this section, after looking at some numerical examples.

> A **percentile** is a measure of relative position. The *i*th percentile, denoted P_i, is determined so that *i* percent of the data lie at or below it and 100 − *i* percent of the data lie at or above it.

Finding the *i*th percentile

To find the position of any percentile *i*, we calculate

$$\text{Position of the } i\text{th percentile} = \frac{i}{100} \times n$$

where n = number of observations.

EXAMPLE 3.13 ■ FINDING PERCENTILES FOR BAR EXAM DATA

In order to qualify to become lawyers, students take a multi-state bar examination. According to PMBR, a company that specializes in preparing students for the bar examination, different states require different passing scores. Minimum passing scores, as given in a law school brochure, are as follows:

120	126	130	131	137
120	127	130	132	140
120	128	130	133	142
125	128	130	135	145
125	130	130	135	152

The mean of the data is 131.24, and the median and mode are both 130. Therefore, the center of the distribution is around 130—but how tightly are the data clustered about 130? An approximate answer is given by the range, which is 32 points (from 120 to 152). To get a sharper estimate, we must look at how close the middle 50% of the data lie to 130; so we need to find the 25th percentile (P_{25}) and the 75th percentile (P_{75}), between which 50% of the data fall.

The points P_{25}, P_{50} (the median), and P_{75} divide the data as nearly as possible into four equal quarters by the size of the variable. These points, called the 1st, 2nd, and 3rd quartiles, respectively, are commonly used in describing the distribution of data. Figure 3.7 shows the relationship of a data set to its percentile and quartile scales.

We can calculate the position of P_{25}, the 25th percentile, as follows:

$$\text{Position for } P_{25} = \frac{25}{100} \times 25 \text{ observations} = \text{Position } 6.25$$

But we need a rank, not a position. To get at least 25% of the data, we must round up to rank 7. When the data are arranged in ascending order (see Table 3.10),

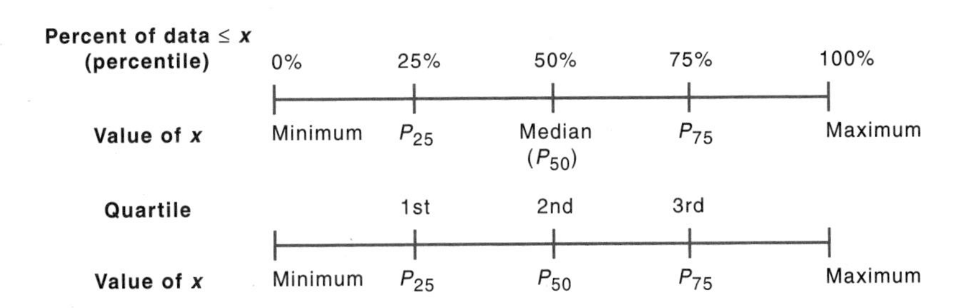

Figure 3.7
Percentiles and quartiles for a data set

we see that P_{25}, which is at rank 7, has a value of 127. This means that 25% of the states in the sample are satisfied with a score of 127 points or less, and 75% of the states in the sample require 127 points or more.

Table 3.10

MINIMUM PASSING SCORES RANKED IN ASCENDING ORDER

Rank	Score	Rank	Score	Rank	Score
1	120	11	130	21	137
2	120	12	130	22	140
3	120	13	130	23	142
4	125	14	130	24	145
5	125	15	130	25	152
6	126	16	131		
7	127	17	132		
8	128	18	133		
9	128	19	135		
10	130	20	135		

We can apply the same procedure to calculate P_{75}:

$$\text{Position for } P_{75} = \frac{75}{100} \times 25 \text{ observations} = \text{Position } 18.75$$

Position 18.75 rounds up to rank 19, so $P_{75} = 135$. This means that at least 75% of the states require scores no higher than 135 while 25% require scores above 135.

To determine the minimum passing score required by 80% of the states on the multi-state bar examination, we can apply a similar procedure to find P_{80}.

$$\text{Position of } P_{80} = \frac{80}{100} \times 25 \text{ observations} = \text{Position } 20$$

The position is an integer, so we do not need to round; 80% of the data lie at or below the number at the 20th rank, which is 135, and 20% of the data lie at or above the number at the 21st rank, which is 137. When the position for P_i is an integer, k, we define P_i to be the average of the kth and $(k + 1)$st values. Here, P80 = ½(135 + 137) = 136.

You should not worry too much about carrying out the types of calculations shown in Example 3.13 on large data sets. Large data sets are always stored and processed on computers, using methods described in Section 6 of this chapter. The important thing is to understand the ideas behind and the uses of the measures. This understanding will come with practice on the small data sets in the exercises.

The Interquartile Range

The range from P_{25} to P_{75}, from the first to third quartile, covers the central "half" of the data and is called the **interquartile range (IQR).** In Example 3.13, a total of 14 of the 25 observations fall in the range from $P_{25} = 127$ to $P_{75} = 135$. Because of ties and rounding, this is more than half of 25, but it is the closest we can come to the central 50% of the data.

Calculating the interquartile range

$$\begin{aligned} \text{IQR} &= P_{75} - P_{25} \\ &= Q_3 - Q_1 \\ &= 135 - 127 = 8 \end{aligned}$$

where Q_1 stands for the first quartile and Q_3 for the third quartile.

Figure 3.8 shows how Q_1 and Q_3 and the interquartile range relate to a dot plot of the data from Example 3.13. The positions of Q_1 and Q_3 are marked in color, and the colored arrow shows the length of the interquartile range.

Figure 3.8
Interquartile range

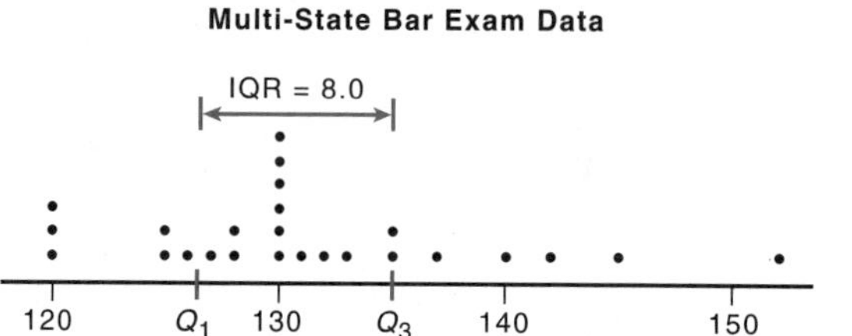

EXERCISES 3.31–3.40

Skill Builders

The purpose of these skill-building problems is to provide practice in making ranked lists and doing calculations.

3.31 Consider the following data set:

28	19	41	22	67	25
44	100	49	57	62	60

a. List these numbers in ascending order.
b. Find the 25th percentile.
c. Find the 75th percentile.
d. Find the interquartile range.
e. Find the 90th percentile.

3.32 Consider the numbers below:

81,147	87,208	69,493	283,545	97,670
87,641	80,527	91,506	246,905	113,918
110,465	79,944	59,053	140,097	107,441

a. List these numbers in ascending order.
b. Find the first quartile, Q_1.
c. Find the third quartile, Q_3.

d. Find the interquartile range, IQR.
e. Calculate the median.

3.33 Consider the numbers below:

−.03	47.00	4.79	9.00	−8.12
−10.05	17.01	.23	−5.16	170.00
7.18	11.74	6.23	2.12	.10
19.44	−19.45	1.23	44.07	18.45

a. List these numbers in ascending order.
b. Find the 17th percentile.
c. Calculate the interquartile range, IQR.
d. Find the median.
e. Complete this paragraph: Over 83% of the values fell above _____. The middle 50% of the values fell between _____ and _____. The middle value in this set of data is _____.

Applying the Concepts

3.34 Look once again at Sharon's walking times from Exercise 3.4:

17.2	18.1	20.0	20.5	22.1
17.8	19.6	20.0	21.4	22.3
17.9	19.7	20.0	21.5	22.7
18.0	19.8	20.0	21.6	30.4

a. Calculate the first quartile, Q_1.
b. Calculate the third quartile, Q_3.
c. Find the interquartile range, IQR.
d. Calculate the 65th percentile.
e. Complete this paragraph: Fifty percent of the time, it takes Sharon between _____ and _____ minutes to complete her walk. She can complete her walk in a maximum of _____ minutes 65% of the time.

3.35 Take another look at the data from Exercise 3.22, on customers entering a deli each hour for several days:

5	4	5	6	5	3
10	10	8	7	12	5
21	23	10	12	15	18
42	31	23	14	20	9
53	54	35	27	22	21
30	43	35	30	23	57
22	9	15	15	18	13
9	11	16	6	8	6
3	6	4	2	7	3

a. Calculate the interquartile range.
b. Find the 90th percentile.
c. Complete this paragraph: Ninety percent of the time, there are fewer than _____ customers entering the market. About half the time, we may expect between _____ and _____ customers to enter the store.

3.36 The following data on the cost of laptop PC systems is from *PC Portables Magazine* (July 1997):

999	1599	1899	1999	2299	2499
1099	1600	1970	2095	2346	2499
1199	1749	1995	2099	2399	2510
1299	1799	1999	2100	2399	2595
1399	1799	1999	2170	2399	2999
1399	1799	1999	2199	2399	
1499	1850	1999	2199	2499	
1549	1899	1999	2299	2499	
1595	1899	1999	2299	2499	

a. Find the interquartile range.

b. Find the 90th percentile.

c. Complete this paragraph: Price data on laptop systems reveal that a laptop system can be obtained for as little as $____. The highest price paid, however, was $____. The lowest price was $____. The middle 50% of systems are priced between $____ and $____. Ninety percent of laptop systems can be obtained for less than $____.

d. Prices and capabilities in the laptop market change rapidly. Check on current prices, and then write a paragraph summarizing the current market and making recommendations on the purchase of six laptops for the staff of an accounting firm.

3.37 The following data, on percentage change in sales for the chemical industry for the first quarter of 1989, are from *Business Week* (August 1989).

Firm Name	Percentage Change	Firm Name	Percentage Change
Arco	7	Great American	24
Air Products & Co	10	Great Lake Chem	39
American Cyanamid	4	Hanna	11
Aristech Chemical	0	Hercules	12
Betz Laboratory	17	Hinmont	−7
Cabot	11	Lubrizol	14
Crompton & Knowles	39	Monsanto	4
Dexter	4	Nalco Chemical	8
Dow Chemicals	8	Olin	14
Dupont	11	Pennwalt	11
Engelhard	14	Quantum	4
Ethyl	22	Rexene	−14
Ferro	6	Rohm-Haas	−5
Freeport-McMoran	6	Schulman	0
Fuller	14	Sterling	−27
Georgia Gulf	6	Union Carbide	7
Goodrich	8	Vista	7
Grace	4		

a. Find the interquartile range for the data.

b. Find the 85th percentile.

c. Complete this paragraph (written as background for analysts looking at the performance of chemical companies): During the first quarter, sales in the chemical industry increased between ____ and ____% for the middle 50% of the firms. The lowest change in sales was ____% for ____, while two firms vied for the largest percentage change in sales:____ %. These firms were ____ and ____. The median percentage change in sales was ____. Fifteen percent of the firms experienced an increase of over ____ %.

d. In the most recent issue of *Business Week* that has first-quarter percentage change sales data, find the latest figures for the companies listed above. Compute the results for parts a through c, and write a paragraph describing your results. Compare these results with the results for 1989.

3.38 Average circulation for the 25 U.S. newspapers with the highest circulation, both daily and Sunday, appears below:

Newspaper	Average Daily Circulation	Average Sunday Circulation
Wall Street Journal	1,795,448	NA
USA Today	1,418,477	1,785,310
Los Angeles Times	1,177,253	1,529,609
New York Times	1,114,830	1,700,825
Washington Post	791,289	1,143,145
Newsday	763,972	875,239
New York Daily News	762,078	911,684
Chicago Daily Tribune	723,178	1,107,938
Detroit Free Press	598,414	1,202,604
San Francisco Chronicle	553,433	705,260
New York Post	552,227	NA
Chicago Sun-Times	531,462	537,169
Boston Globe	504,675	798,057
Philadelphia Inquirer	502,603	974,697
Atlanta Constitution	474,578	688,175
Newark Star-Ledger	470,672	700,237
Detroit News	446,831	1,202,604
Houston Chronicle	439,574	622,608
Cleveland Plain Dealer	413,678	544,362
Minneapolis Star Tribune	408,365	527,153
Miami Herald	398,067	510,549
Dallas Morning News	393,511	618,283
Denver Rocky Mountain News	355,940	425,443
St. Louis Post-Dispatch	350,350	562,700
Orange County Register	347,675	400,375

Source of Data: *USA Today,* Nov. 15, 1991.

a. Find the 90th percentile for daily circulation.

b. Repeat part a for Sunday circulation.

c. Compare the 90th percentiles and medians for daily and Sunday circulation. (*Hint:* See the format suggested in Exercises 3.36 and 3.37.)

d. How have daily and Sunday circulation changed since 1991 for the newspapers listed above? Write a brief paper summarizing your findings.

3.39 The various Apollo Space Missions (1967–1972) lasted from as little as 7 hours to as much as 301 hours. The data on mission duration (in hours) appear below:

9	241	216	7	192	295	10
195	301	260	244	147	142	

a. Find the median duration of a flight.

b. Calculate the mean.

c. Find the 60th percentile.
d. Complete this paragraph: Apollo Space missions lasted for an average of _____ hours. Fifty percent of the missions lasted longer than _____ hours. Sixty percent were no longer than _____ hours.

3.40 Recycling is becoming increasingly popular, as people become more aware of how fragile our environment is. A study reported in the *New York Times* (May 27, 1992) indicated that the percentage of households participating in recycling by district varied from 6.4% to 62.7%. Data from districts in the New York City area appear below:

12.7	6.4	31.5	47.4	47.6
19.2	8.0	32.4	53.0	47.2
24.0	15.2	41.8	25.0	59.2
41.5	18.3	44.0	38.1	54.6
43.5	18.8	45.1	43.8	59.5

a. Find the interquartile range.
b. Calculate the median.
c. Find the 95th percentile.
d. As a principal research analyst for the Department of Environmental Control, you have been asked to prepare a report for the commissioner. He would like to know where the program stands and what improvements might be made. Complete this paragraph: Participation in recycling programs has varied from _____ to _____%. The middle 50% of the districts have participation rates between _____ and _____%. The median level of participation is _____%. Ninety-five percent of the districts exhibit a participation rate of less than _____%.

SECTION 4 ▪ THE SHAPE OF A DISTRIBUTION

Chapter 2 showed how histograms and density curves can be used to describe the distribution of data. Using these types of graphs, we can put geometrical ideas, such as symmetry, to work in describing distributions. In this section, we will look at the geometry of distributions and relate these ideas to the measures of center developed earlier in this chapter.

If we could draw an imaginary line through the center of a distribution and fold one side over the other with a perfect fit, we say that the distribution is perfectly symmetrical. Symmetry is an important condition for the use of some statistical tools, and your eye can quickly spot whether or not a distribution is symmetrical. Figure 3.9 is an example of a symmetrical distribution. Note that μ equals the median for a symmetrical distribution. If we cannot fold one side to fit over the other, we have a distribution that is not symmetrical. Nonsymmetrical distributions are often skewed.

Assessing skewness

You probably have utilized the concept of **skewness** without even realizing it. If your friends tell you that Professor Julia Deadman is a low marker, what does this say about her grade distribution? Professor Deadman rewards high performers, at the expense of average and low performers. The enhancement of the right tail

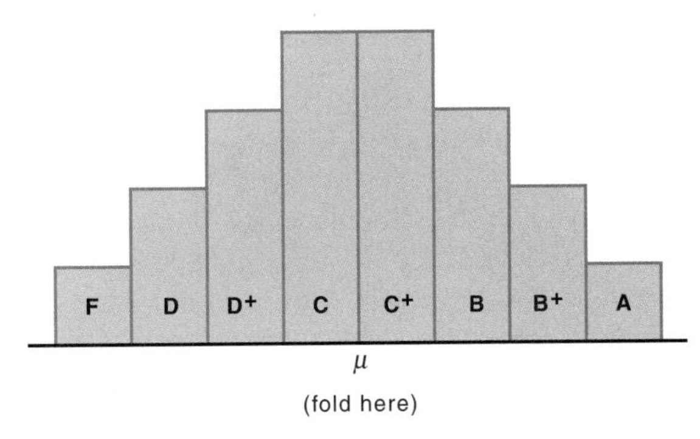

Figure 3.9
Perfectly symmetrical distribution

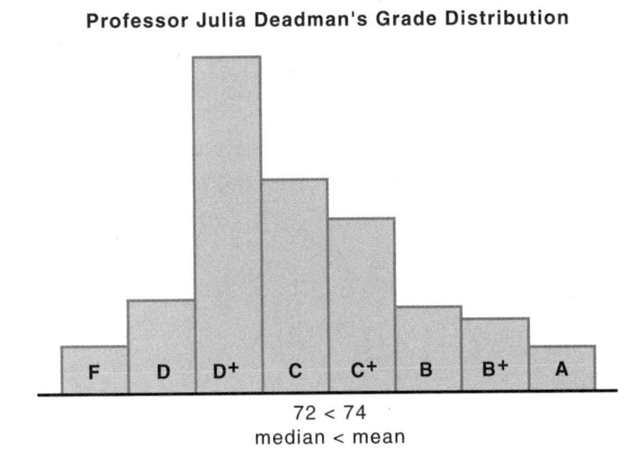

Figure 3.10
Positively skewed distribution (mean shifted to the right)

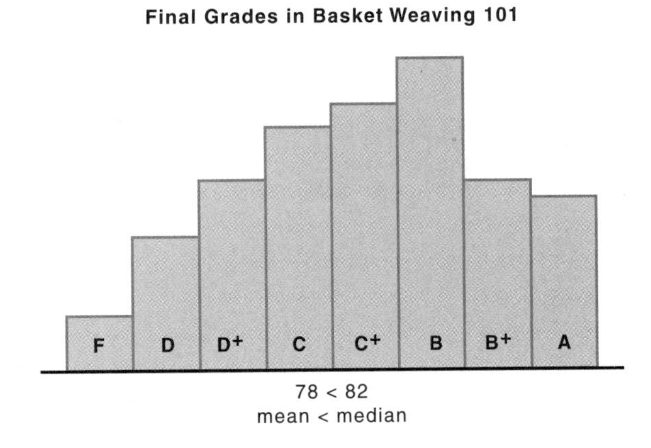

Figure 3.11
Negatively skewed distribution (mean shifted to the left)

causes the distribution to be skewed to the right. The bulk of her grades are shifted toward the low side, as shown in Figure 3.10.

Figure 3.10 illustrates a distribution that is skewed to the right (positively skewed). On the other hand, the grade distributions of professors known as "easy A's" are skewed to the left (negatively skewed). See Figure 3.11, where the left tail of the distribution is enhanced.

In addition to the visual impression, there is another indication of skewness. If the distribution is symmetrical, the mean equals the median. However, if the distribution is skewed to the right, the mean exceeds the median. Conversely, if the distribution is skewed to the left, the mean is less than the median. It is important to note that the distribution is skewed to the side of the mass of data with the enhanced tail:

Identifying Skewness

Mean > Median	Positively skewed Skewed to the right
Mean < Median	Negatively skewed Skewed to the left

EXAMPLE 3.14 ▪ EXAMINING THE DISTRIBUTION OF SALARIES AND BONUSES

Each year, *Business Week* examines chief executive officers' salaries and bonuses. Data for 1991 for food-processing industry executives appear below. The data (in thousands of dollars) are in ascending order.

729	921	1198	1431	1661	1852
815	1036	1361	1444	1751	2000
883	1067	1405	1514	1806	4000

a. Calculate the arithmetic mean.
b. Find the median.
c. Is this distribution skewed? If so, how?

The mean for the data is $1493 thousand. The median is $1418 thousand. The mean exceeds the median, so the distribution is skewed to the right (that is, positively skewed), suggesting that there are some relatively high compensation figures. While the average CEO in food processing earned approximately $1.5 million, one CEO received compensation of $4 million. The situation may be illustrated as shown in Figure 3.12.

Figure 3.12
Distribution of salaries and bonuses

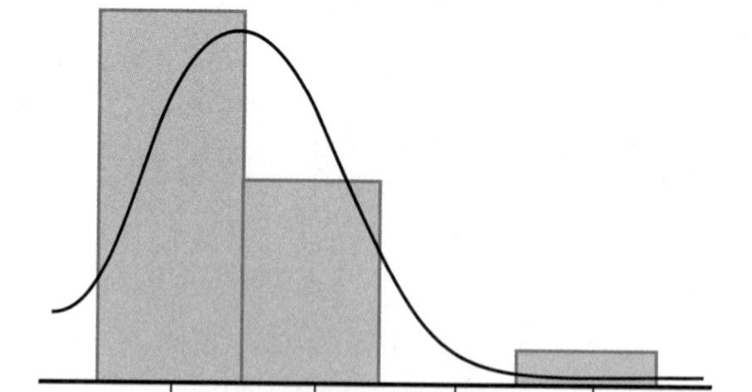

The tools intended for use with a symmetrical distribution cannot be used with a distribution that is skewed. This concept will be discussed further later.

EXERCISES 3.41–3.50

Skill Builders

3.41 Consider the following data set:

2	3	3	3	5	6	8	9

a. Calculate the arithmetic mean.
b. What is the median value?
c. Based on your results in parts a and b, is the distribution skewed? If so, how?

3.42 Consider the data set below:

2	202	306	435	621	797	950
22	245	316	518	715	899	965
79	276	415	562	722	912	991

a. Find the arithmetic mean.
b. Compute the median value.
c. Based on your answers to parts a and b, is this distribution positively or negatively skewed? Why?

Applying the Concepts

3.43 The data on Sharon's times for walking the mile appear once again below:

17.2	18.1	20.0	20.5	22.1
17.8	19.6	20.0	21.4	22.3
17.9	19.7	20.0	21.5	22.7
18.0	19.8	20.1	21.6	34.4

a. Using your calculations on the median and the mean, can you conclude that the distribution is skewed? If so, in what way?
b. If 34.4 were considered an outlier and removed from the data set, would this change your conclusions?

3.44 The grades of students on their most recent homework assignment appear below:

4.0	8.0	8.5	9.0	9.5	10.0
6.5	8.0	9.0	9.0	9.5	10.0
7.0	8.0	9.0	9.0	10.0	10.0
7.0	8.5	9.0	9.0	10.0	10.0
7.5	8.5	9.0	9.5	10.0	10.0

a. What is the median grade?
b. Calculate the arithmetic average.
c. Is this distribution skewed? If so, indicate whether it is skewed to the right or to the left and why.

3.45 In its "1991 Investment Guide," *U.S. News & World Report* compiled data on a dozen stocks rated as falling in the 99th percentile for earnings per share. Included were the following data on current dividend yield:

0	0	0	0	0	0	1.1	2.9	3.4	4.1	4.8

a. Calculate the mean dividend yield.

b. Find the median dividend yield.
c. Is this distribution positively or negatively skewed? Give your reasoning.

3.46 Below are the numbers of ounces in various standard-size cans of coffee:

11.5	11.5	13.0	13.0	16.0	16.0
11.5	13.0	13.0	13.0	16.0	

a. Compare the mean and median for this data set.
b. Is this distribution skewed? to the right or to the left? Why?
c. State your conclusions in layman's English without using *any* statistical terms. What does skewness mean in this context?

3.47 In 1991, when the president "suggested" lowering consumer interest rates to stimulate the economy, interest rates charged by the largest credit card issuers became a major subject of discussion. The lowest annual percentage rates offered by the largest card issuers appear below:

Issuer	Lowest Rate Offered
Bank of New York	13.40
First Chicago	14.40
Discover/Sears	14.80
Citibank	14.90
Optima/American Express	15.25
MBNA/America	17.80
Household Bank	18.90
Chase Manhattan	19.80
Bank of America	19.80
Manufacturers Hanover	19.80

Source of Data: *New York Times,* November 13, 1991.

a. Calculate the mean.
b. Compute the median.
c. Is this distribution skewed? If so, how? Explain your reasoning.
d. What conclusions would you draw from this sample about shopping for rates as a credit card user?

3.48 The September 1990 issue of *Consumer Reports* classified hotel chains by expense into four categories: economy, moderate, high-priced, and luxury. Chains rated "economy" received overall satisfaction index (OSI) ratings of between 63 and 85. They are listed below:

Scottish Inn	63	Comfort Inn	79
Red Carpet	66	Red Roof Inn	80
Motel 6	71	Super 8	80
Roadway Inn	73	Shoney's Inn	81
Econo Lodge	73	La Quinta	82
Royal Inn	74	Budgetel	82
Day's Inn	74	Drury Inn	83
Knights Inn	79	Hampton Inn	85

a. What is the mean OSI?
b. Find the median OSI.
c. Group the data by tens—that is, 60–69, 70–79, etc.—and prepare a histogram of the data.

d. Is this distribution symmetrical?

e. What conclusions can you draw concerning the shape of the distribution of overall satisfaction index ratings?

3.49 *Journal of Business Forecasting* (Summer 1992) published the following forecasts by a number of banks of the consumer price index (CPI) for the third quarter of 1993:

144.7	145.1	145.6	146.2	146.4
146.9	147.2	147.2	147.5	147.6

These numbers are relative to the base years of 1982–1984, for which the CPI was 100.

a. Find the median forecast.

b. What is the mean forecast?

c. Is this distribution skewed?

d. What was the level of the CPI for the third quarter of 1993? How accurate were the banks' forecasts?

3.50 The freshman retention rate of so-called quartile three universities, as classified by *U.S. News & World Report* (September 1992), varies from 55% for the University of Idaho to 89% for Marquette. The data for all these schools appear below:

55	68	71	76	77	79	81	84	87
58	68	71	76	77	79	81	84	88
60	69	71	76	77	80	81	85	89
61	69	73	76	78	80	81	85	89
63	70	74	77	78	80	82	85	
67	70	75	77	79	80	82	85	

a. What is the average retention rate?

b. Find the median rate.

c. Is this distribution skewed or symmetrical?

d. Group the data by tens—that is, 50–59, 60–69, etc.—and prepare a histogram of the data.

e. Based on your findings, write a short paragraph about retention rates. How might this information figure in the choice of a college?

SECTION 5 ▪ THE EMPIRICAL RULE

Applying the empirical rule

In this section, we will look at how the numerical summary provided by the mean and standard deviation relates to the general shape of the distribution. Many data sets fall in a symmetrical bell-shaped pattern around the arithmetic mean. Graphing the distribution of net asset values (NAVs) of municipal bond fund shares provides an example of a symmetrical bell-shaped pattern. A sample of 80 observations of such bonds exhibited a mean of \$11.13 and a standard deviation of \$3.20. Figure 3.13 on page 104 shows a histogram for the sample of 80 observations, superimposed on a normal density curve for the population of NAVs.

The distribution of data often shows a pattern described by the empirical rule for the percentage of data within a given number of standard deviations from the

Figure 3.13
Distribution of net asset values

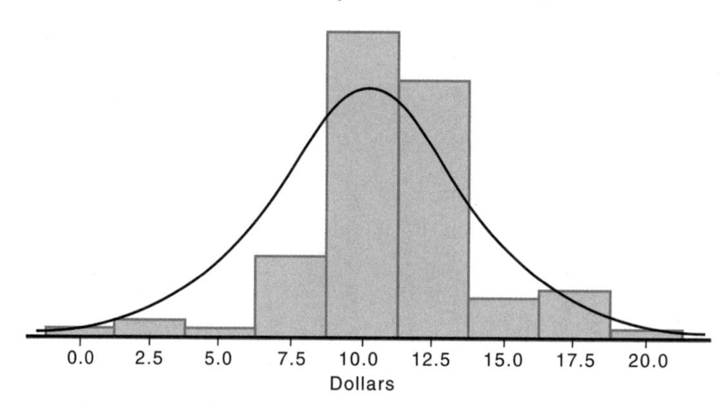

mean. The empirical rule is derived from the percentages for a fixed type of distribution called a normal distribution. A **normal distribution** has as its graph the familiar bell curve, and its values are as tabulated in Appendix Table IV.

The Empirical Rule

When data follow a bell-shaped distribution, we may expect that:

- 68% fall within 1 standard deviation of the mean.
- 95% fall within 2 standard deviations of the mean.
- 99.7% fall within 3 standard deviations of the mean.

Figure 3.14 illustrates a normal distribution.

Figure 3.14
Normal distribution

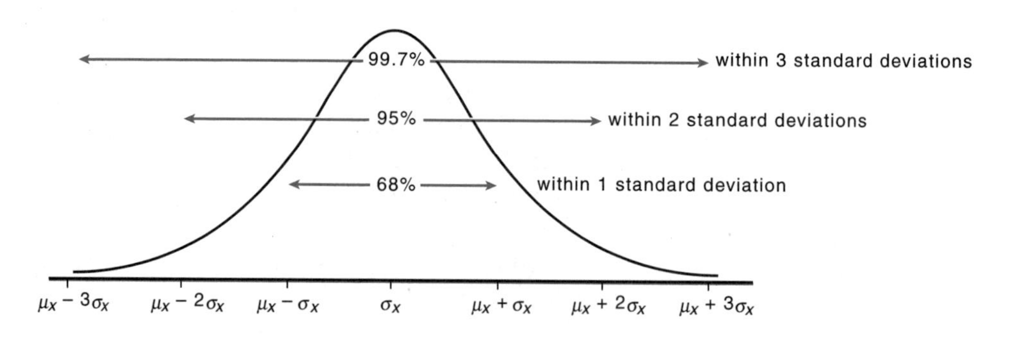

EXAMPLE 3.15 ▪ FINDING THE DISTRIBUTION OF AGES

The empirical rule specifies the minimum percentage of data elements that will fall within a given distance of the mean, using the standard deviation as the unit of measure. How closely do the ages of professional basketball players fit the empirical rule? Ages of a sample of basketball players are presented below:

25	23	23	29	29	28	26	27	28
26	26	34	25	27	24	29	33	28
25	23	25	33	28	33	32	24	28

Calculating, we find that the mean is 27 years of age and the standard deviation is 3. If the distribution is normal, the data should fit the pattern described by the empirical rule. Table 3.11 compares the actual data to the empirical rule.

Table 3.11

THE FIT OF ACTUAL DATA TO THE EMPIRICAL RULE

Range	Number of Players	Actual Percentage	Empirical Rule
Within 1 standard deviation (24 to 30 years)	20	74%	68%
Within 2 standard deviations (21 to 33 years)	26	96	95
Within 3 standard deviations (18 to 36 years)	27	100.0	99.7

Based on the sample, it is reasonable to conclude that, for practical purposes, the data appear to be distributed in a roughly normal way.

SECTION 6 ▪ USING THE COMPUTER

Computers are essential for handling massive amounts of data quickly and can generate a variety of useful displays. Computers are also useful for exploratory data analysis—that is, answering "what if" questions. In this section, you will learn how to interpret computer-produced results by looking at sample outputs.

Many different computer packages are available for these purposes—for example, Minitab, Excel, Statistical Analysis System (SAS), JMP IN, and Statistical Package for the Social Sciences (SPSS). Although the details of their programs vary, their capabilities and output are generally similar.

To give focus to our discussion of the use of the computer to tabulate and display data, let's consider the situation of a school district superintendent. Confronted with the perennial complaint that taxes are too high, the school board has suggested that schools reduce their expenses by increasing their student–teacher ratios. The ratio in the superintendent's district is 11 to 1. His opponents claim that this is too low. To prove that this ratio is not too low, he has collected data from all the districts in the county. His data appear in Table 3.12 on page 106.

The Data Analysis Add On feature of Microsoft Office Excel produces all of the typical descriptive statistics. Although the exact results will depend on which version of Excel you are using, here is a typical Excel printout for the student–teacher ratio data set:

Using Excel

RATIOS	
Mean	11.65672
Standard Error	0.38957
Median	13
Mode	13
Standard Deviation	3.188769
Sample Variance	10.16825
Kurtosis	1.08877
Skewness	−1.26927
Range	14
Minimum	2
Maximum	16
Sum	781
Count	67

Table 3.12

STUDENT–TEACHER RATIOS IN SUFFOLK COUNTY

District	Ratio	District	Ratio	District	Ratio
Amagansett	8	Hampton Bays	11	Rocky Point	14
Amityville	12	Harborfields	15	Sachem	14
Babylon	16	Hauppauge	3	Sag Harbor	10
Bay Shore	13	Huntington	13	Sagaponack	2
Brentwood	15	Islip	13	Sayville	12
Bridgehampton	5	Kings Park	12	Shelter Island	9
Brookhaven	14	Laurel	11	Shoreham	10
Center Moriches	11	Lindenhurst	14	Smithtown	13
Cold Spring	11	Longwood	15	So. Huntington	14
Commack	13	Mattituck	13	So. Country	13
Connetquot	13	Middle Country	15	So. Manor	15
Copiague	15	Miller Place	15	Southampton	14
Deer Park	13	Montauk	10	Southold	11
East Hampton	10	Mount Sinai	14	Springs	14
East Islip	14	New Suffolk	8	Three Village	13
East Moriches	13	North Babylon	13	Tuckahoe	8
East Quogue	14	Northport	12	Wainscott	4
Eastport	10	Oysterponds	9	W. Babylon	13
Elwood	13	Patchogue	14	W. Islip	12
Fire Island	4	Port Jefferson	10	Wm. Floyd	15
Fishers Island	4	Quogue	5	Wyandanch	14
Greenport	9	Remsenburg	9		
Half Hollow	11	Riverhead	14		

Excel provides all the statistics we have discussed thus far, as well as statistics for standard error, kurtosis, and skewness. Standard error is a concept we shall explore further in Chapter 6. **Kurtosis** refers to the degree of peakedness in a distribution. Skewness is measured in a similar way to variance. However, it may result in a positive or negative result, corresponding to a distribution that is positively or negatively skewed.

Is a student–teacher ratio of 11 to 1 too low? The arithmetic mean for the county is 11.65672 (rounded to 12). The median value is 13. While the ratio for this school district falls below either measure of central tendency, it is still within the typical range. There are several districts with much lower ratios. These districts may, indeed, be operating uneconomically or may face special circumstances requiring lower ratios.

Minitab presents many of the same statistics as Excel. While it omits kurtosis and skewness, it adds the TRMEAN (trimmed mean) and Q1 and Q3 (the 1st and 3rd quartiles). Here is the Minitab printout for the student–teacher ratio data set:

Using Minitab

	N	MEAN	MEDIAN	TRMEAN	STDEV	SEMEAN
RATIOS	67	11.657	13.000	11.885	3.189	0.390
	MIN	MAX	Q1	Q3		
RATIOS	2.000	16.000	10.000	14.000		

The TRMEAN, or **trimmed mean,** is arrived at by deleting, or trimming away, the lowest 5% and highest 5% of the data. Thus, it is an average of the central 90% of the data. If the trimmed mean is close to the mean, it suggests that the data set is, for the most part, symmetrical. The trimmed mean may be a more reliable figure in situations where the data are subject to an occasional erratic variation, such as inputting 10003 for 1003. In the case of the school district data, fifty percent of the values fall between 10 and 14 (the interquartile range).

When observations tend to repeat themselves and fall within a limited range, as these student–teacher ratios clearly do, a chart can reveal some interesting information. Figure 3.15 illustrates one of the charting features of Minitab.

Figure 3.15
Bar chart produced by Minitab

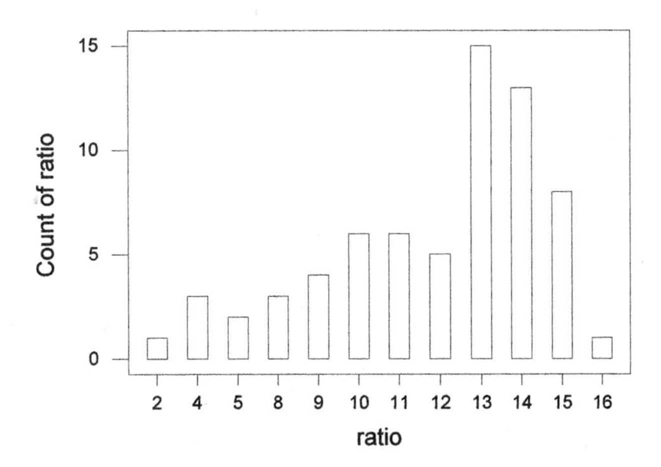

JMP (produced by SAS) yields statistics similar to those produced by Minitab, but in a different output format. The following array of descriptive statistics was produced using JMP IN, the student version of the JMP software:

Using JMP IN

Quantiles

maximum	100.0%	16.000
	99.5%	16.000
	97.5%	15.300
	90.0%	15.000
quartile	75.0%	14.000
median	50.0%	13.000
quartile	25.0%	10.000
	10.0%	7.400
	2.5%	3.400
	0.5%	2.000
minimum	0.0%	2.000

Moments

Mean	11.68657
Std Dev	3.14406
Std Error Mean	0.38411
Upper 95% Mean	12.45346
Lower 95% Mean	10.91967
N	67.00000
Sum Weights	67.00000

Figure 3.16 on page 108 consists of a histogram and a box-and-whisker plot produced by JMP IN for the student–teacher ratio data. The histogram shows frequency on the vertical axis and the student–teacher ratios on the horizontal axis. Because it gives a quick snapshot of the distribution, the histogram easily shows the concentration of ratios between 9 and 15. The box-and-whisker plot indicates the

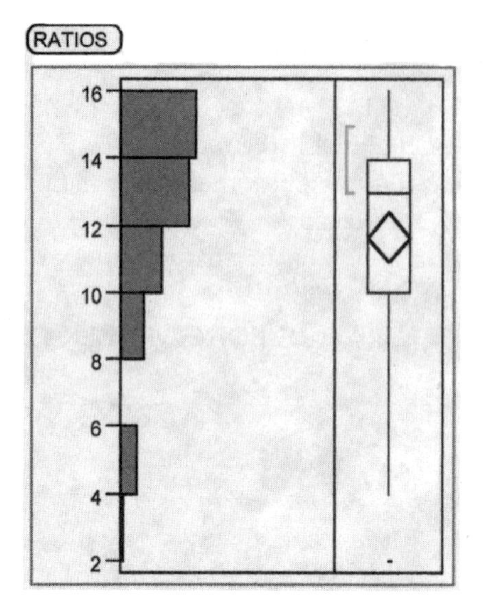

Figure 3.16
Histogram and box plot produced using JMP

distribution of the data by showing where the extremes lie (typically, although not necessarily, at the ends of the whiskers), where the quartiles fall, and where the median falls. It is an essential tool for visualizing and comparing distributions. Side-by-side displays of box-and-whisker plots for two distributions allow viewers to compare various measures of central tendency and range at a single glance. Were a distribution perfectly symmetrical, the box-and-whisker plot would appear so. The plot focuses on what is called the five-number summary.

> The **five-number summary** consists of five locational measures: the minimum, the first quartile, the median, the third quartile, and the maximum.

The box-and-whisker plot in Figure 3.16 shows the student–teacher ratio on the horizontal axis. Starting at the bottom of the diagram, we see an outlier at a student–teacher ratio of 2. The bottom edge of the rectangle in the center of the plot defines the first quartile. The enclosed diamond represents a 95% confidence interval for the mean (a concept we will examine in Chapter 7). The line above the diamond defines the median, and the top edge of the rectangle represents the third quartile. The bracket on the top left indicates the most concentrated 50% of the data. The maximum of the data is at 16. If the data are sufficiently extreme, whiskers are drawn to a distance of 1.5 IQRs beyond the quartiles to define an **inner fence.** In this box-and-whisker plot, 16 and 4 are inner fences. If there are more extreme values, the whisker extends 3 IQRs beyond the quartile to an **outer fence.** The value 2 may be considered a "real outlier."

> **Outliers** are observations that fall outside the usual range—for some purposes, 1.5 IQRs below Q_1 or 1.5·IQRs above Q_3. An unusually high or low figure in a data set may require special attention, as outliers often indicate special circumstances—for example, reduced production at a plant due to flood, fire, or strike or an error in data entry.

Side-by-side box-and-whisker plots allow a quick comparison of measures of center and spread. Shown in Figure 3.17 are box plots for the monthly rainfall data for Seattle and New York. They show you immediately that Seattle's median rainfall is a bit lower than that of New York, but also that the variability of Seattle's monthly rainfall averages is far greater than that of New York's, as indicated by the interquartile range.

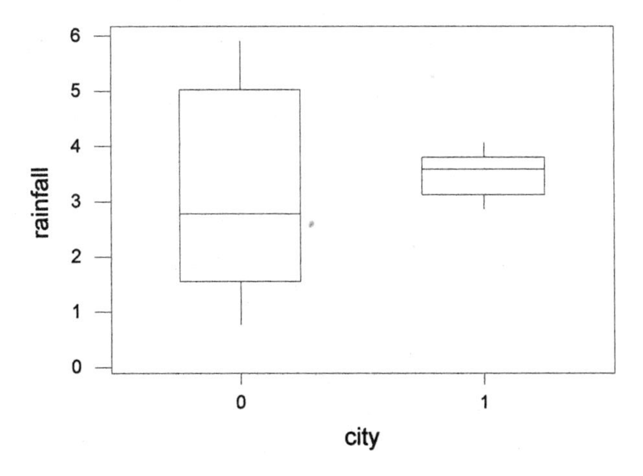

Figure 3.17
Box plots for comparing data sets

As you can see, the computer can easily calculate a set of summary measures and produce graphics to reflect these measures. The computer enables statisticians to focus their skills on interpreting the output obtained rather than dealing with computational details.

EXERCISES 3.51–3.57

Applying the Concepts

3.51 The 1990 incomes (in millions of dollars) for the largest charities involved in social services appear below:

Pan American Development	10.4
World Concern	11.9
American Foundation for the Blind	12.3
Direct Relief International	12.5
World Relief	17.4
Armenian General Benevolent Union	19.5
Habitat for Humanity	20.7
Braille Institute	21.1
Amnesty International	23.9
Christian Appalachian Project	25.4
National Urban League	26.0
American Friends Service Committee	28.5
Christian International/Holy Land Christian Mission	29.3
PLAN International, USA	30.7
Children's Aid Society	30.7

The Lighthouse	31.1
Population Council	35.6
Mennonite Central Committee	36.0
Food for the Hungry	36.2
Girl Scouts of the USA	37.0
International Rescue Committee	37.6
Compassion International	40.7
Father Flanagan's Boys Home	40.9
Church World Service	43.0
Girls Inc.	43.3
Camp Fire Girls	45.2
Mothers Against Drunk Driving	49.3
Jewish Board of Family and Children Services	56.7
Larry Jones Ministries/Feed the Children	73.9
National Benevolent Association of Christian Churches	74.2
Big Brothers/Big Sisters	76.9
Save the Children Federation	90.8
Special Olympics International	90.9
Lutheran World Relief	92.6
Rotary Foundation of Rotary International	95.0
Covenant House	98.2
Christian Children's Fund	103.0
World Vision	215.5
Volunteers of America	239.2
Boys & Girls Clubs of America	239.6
CARE	293.5
Jewish Community Centers Association	380.0
Planned Parenthood	383.7
United Jewish Appeal	426.6
Boy Scouts of America	430.0
Association for Retarded Citizens	464.0
UNICEF	821.0
Salvation Army	1215.5
Young Men's Christian Association	1438.5
American Red Cross	1465.6
Lutheran Social Ministry Organization	1486.7
Catholic Relief Services	1538.6

Source of Data: *Money,* December 1991.

a. Use a computer to generate a printout and histogram of the data.

b. Have your program produce the descriptive measures, and print or record them.

c. Draw a box-and-whisker plot of the data.

d. Using the empirical rule, estimate the range of the central 95% of the data set. Compare this figure with the actual range from the 2.5th percentile to the 97.5th percentile, which represents the central 95% of the data. Does the empirical rule fit this data set well?

e. Complete this paragraph: In 1990, there were _____ major social service charities operating in the United States. They received an average income of _____. The smallest income of _____ was reported by _____. The largest income was reported by _____, which reported an income of _____. The middle 50% of the charities reported income between _____ and _____.

3.52 Baseball fans have always been interested in home run leaders. The National League and American League home run leaders for the 1966–1990 period are listed below:

Year	National League Leader	Home Runs	American League Leader	Home Runs
1966	Hank Aaron	44	Frank Robinson	49
1967	Hank Aaron	39	Harmon Killebrew	44
1968	Willie McCovey	36	Frank Howard	44
1969	Willie McCovey	45	Harmon Killebrew	48
1970	Johnny Bench	45	Frank Howard	44
1971	Willie Stargell	48	Bill Melton	33
1972	Johnny Bench	40	Dick Allen	37
1973	Willie Stargell	44	Reggie Jackson	32
1974	Mike Schmidt	36	Dick Allen	32
1975	Mike Schmidt	38	George Scott	36
1976	Mike Schmidt	38	Craig Nettles	32
1977	George Foster	52	Jim Rice	39
1978	George Foster	40	Jim Rice	46
1979	Dave Kingman	48	Gorman Thomas	45
1980	Mike Schmidt	48	Reggie Jackson	41
1981	Mike Schmidt	31	Bobby Grich	22
1982	Dave Kingman	37	Gorman Thomas	39
1983	Mike Schmidt	40	Jim Rice	39
1984	Mike Schmidt	36	Tony Armas	43
1985	Dale Murphy	37	Darrell Evans	40
1986	Mike Schmidt	37	Jesse Barfield	40
1987	Andre Dawson	49	Mark McGwire	49
1988	Darryl Strawberry	39	Jose Canseco	42
1989	Kevin Mitchel	47	Fred McGriff	36
1990	Ryne Sandberg	40	Cecil Fielder	51

Source of Data: *The World Almanac and Book of Facts,* 1991, pp. 940–941.

a. Use a computer to draw a histogram for each league.
b. Print out the summary statistics for each data set.
c. Draw a box-and-whisker plot for each data set. How do they compare?
d. Based on your statistical analysis, write a brief article for your school's sports column.

e. How do recent home run leaders compare with those of the past?

3.53 A computer analysis of the prices of gift-quality fountain and ball-point pens (*Consumer Reports,* November 1991) yielded the printouts below. As a research analyst for a prominent retailer, you have been asked to prepare a summary of your findings concerning these two types of pens.

VARIABLE	N	MEAN	MEDIAN	TRMEAN	STDEV	SEMEAN
fountain	18	98.5	72.5	92.8	79.6	18.8
ballpt	14	53.8	40.0	48.6	38.6	10.3

VARIABLE	MIN	MAX	Q1	Q3
fountain	13.0	275.0	47.5	122.2
ballpt	20.0	150.0	29.5	63.8

a. Based on your examination of the chart, which type of pen is generally more expensive?

b. Describe the middle 50% of the range for ball-point pens, and compare it with that for fountain pens.
c. Which distribution appears to be more skewed? Explain your answer.
d. Use your findings to prepare a brief paragraph comparing the prices of the two types of pens.

3.54 The November 11, 1991 issue of *Forbes* presented a list of the 200 best small companies. These companies, with sales between $5 million and $350 million, had consistently increasing sales, earnings, and returns on equity. Data on price/earnings ratios for a random sample of 25 of these firms appear below. The price/earnings ratios indicate what investors are willing to pay per dollar of reported profits. Higher P/E ratios are associated with firms with good growth prospects; low ratios are associated with firms in more fully exploited markets, where current results are good but prospects for growth seem limited.

Firm (area of activity)	P/E Ratio
Luiski International (PC distributor)	9.0
SCOR US (property and casualty insurance)	9.2
Advanced Logic Research (PCs)	9.5
Vertex Communications (satellite antennas)	11.2
Dataflex (PC retailer)	11.6
P. Leiner (nutritional products)	12.3
Frederick's of Hollywood (sensual gifts)	15.7
Serv-Tech (oil and chemical plant maintenance)	16.4
Laser Precision Corp. (precision instruments)	16.5
RehabCare (rehab clinics)	16.7
Sealright (food and dairy products packages)	16.8
Selectron (circuit boards)	18.0
American Management Systems (consulting)	18.5
Nelicor (patient monitoring equipment)	19.6
CR Gibson (gifts and stationery)	19.9
Nature's Sunshine Products (personal care products)	20.9
Tootsie Roll Industries (candy)	23.4
Hach (water analysis equipment)	26.2
Dreyer's Grand Ice Cream (ice cream)	28.2
Linear Technology (integrated circuits)	28.5
Mid-Atlantic Waste Systems (solid waste disposal)	28.9
VeriFone (transaction automation systems)	33.2
Fastenal (fastener retailer)	33.5
Paychex (payroll accounting)	35.9
Diagnostek (pharmacy by mail service)	51.2

a. Use a computer to construct a histogram of the data.
b. Find the mean and standard deviation.
c. Draw a box-and-whisker plot.
d. Describe your findings concerning P/E ratios. Then write a sentence about each statistic calculated and a sentence about the histogram.

e. How has this portfolio of potential investments behaved recently? Have these companies maintained their P/E ratios? Have any worsened? Have any improved? What does a change in the P/E ratio imply about the company?

3.55 The principal bodies of salt water throughout the world include the Caspian Sea, covering an area of 170,000 square miles, and the Salton Sea, covering an area of only 266 square miles. Kuku-nor is 10,000 feet above sea level, whereas the Dead Sea is 1290 feet below sea level. Below is a list of the principal bodies of salt water, with their areas and elevations.

Body of Water	Area (square miles)	Elevation (feet above sea level)
Aril Sea (Russia)	26,000	155
Balkash (Russia)	8,600	780
Caspian Sea (Europe)	170,000	−86
Dead Sea (S.W. Asia)	340	−1,290
Eyre (Australia)	3,600	−35
Great Salt Lake (U.S.)	1,750	4,218
Issuk-Kul (Russia)	2,230	5,400
Kuku-nor (China)	2,300	10,000
Maracaibo (Venezuela)	8,000	0
Salton Sea (U.S.)	266	−280
Urumia (S.W. Asia)	1,795	4,100
Van (Armenia)	1,400	5,214

Source of Data: Lincoln Library of Essential Information, 1961.

a. Use a computer to find the mean and standard deviation for the area.
b. Calculate the five-number summary for the area.
c. Draw a box-and-whisker plot for the area.
d. Describe your findings concerning the areas of principal bodies of salt water.
e. Repeat parts a through d for the elevation.

3.56 Presidents of the United States have varied in age from 42 for Theodore Roosevelt to 69 for Ronald Reagan. Below are the ages when inaugurated for all presidents through Ronald Reagan:

President	Age at Inauguration	President	Age at Inauguration
1. George Washington	57	21. Chester A. Arthur	50
2. John Adams	61	22. Grover Cleveland	48
3. Thomas Jefferson	57	23. Benjamin Harrison	55
4. James Madison	57	24. Grover Cleveland	55
5. James Monroe	58	25. William McKinley	54
6. John Quincy Adams	57	26. Theodore Roosevelt	42
7. Andrew Jackson	61	27. William H. Taft	51
8. Martin Van Buren	54	28. Woodrow Wilson	56
9. William Henry Harrison	68	29. Warren G. Harding	55
10. John Tyler	51	30. Calvin Coolidge	51
11. James K. Polk	49	31. Herbert Hoover	54
12. Zachary Taylor	64	32. Franklin D. Roosevelt	51
13. Millard Fillmore	50	33. Harry S. Truman	60
14. Franklin Pierce	48	34. Dwight Eisenhower	62
15. James Buchanan	65	35. John Kennedy	43
16. Abraham Lincoln	52	36. Lyndon Johnson	55
17. Andrew Johnson	56	37. Richard Nixon	56
18. Ulysses S. Grant	46	38. Gerald Ford	61
19. Rutherford B. Hayes	54	39. Jimmy Carter	52
20. James A Garfield	49	40. Ronald Reagan	69

a. Use a computer to calculate the mean and standard deviation.
b. Draw a histogram of the data.
c. What is the interquartile range, IQR?
d. Write a sentence about each of your findings in parts a through c.

3.57 According to *U.S. News & World Report* (March 22, 1993), average starting salaries for MBAs from the top 25 business schools ranged from $42,000 for the University of Texas at Austin to $66,000 for Stanford University. The list appears below:

School	Starting Salary
Stanford	$66,000
Harvard	62,500
Sloan/MIT	60,000
Wharton/Penn	57,000
Columbia University	54,500
Kellogg/Northwestern	54,000
University of Chicago	54,000
University of Michigan	54,000
Tuck/Dartmouth	53,000
New York University	53,000
Darden/University of Virginia	53,000
Cornell University	53,000
University of California, Los Angeles	53,000
Fuqua/Duke University	52,530
University of California, Berkeley	52,000
Yale University	52,000
Carnegie Mellon	51,150
University of North Carolina	50,000
University of Southern Califfornia	48,500
Purdue University	48,000
Indiana University	46,800
University of Rochester	46,100
University of Pittsburgh	44,000
Vanderbilt University	43,100
University of Texas, Austin	42,000

a. Use a computer to find the mean and median.
b. Produce a histogram.
c. Calculate the interquartile range.
d. Using your results, write a brief report, suitable for your alumni newsletter, about starting MBA salaries.

e. What are the current starting MBA salaries? How do these salaries compare with the data presented above? How do starting salaries compare with tuitions? What are the implications, if any, for one's choice of schools?

FORMULAS

Arithmetic Mean

For the population:

$$\mu_X = \frac{\sum x_i}{N} \quad \text{where } N = \text{population size}$$

For a sample:

$$\bar{x} = \frac{\sum x_i}{n} \quad \text{where } n = \text{sample size}$$

Median for an Ordered Array

Calculate $k = .50 \times n$.
If k is an integer, average the values at k and $k + 1$.
If k is not an integer, round up to the next value.

Range

$$\text{Range} = x_{\text{maximum}} - x_{\text{minimum}}$$

Variance

For the population:

$$\sigma_X^2 = \frac{\sum (x_i - \mu_X)^2}{N}$$

For a sample:

$$s_x^2 = \frac{\sum (x_i - \bar{x})^2}{n - 1}$$

Standard Deviation

For the population:

$$\sigma_X = \sqrt{\frac{\sum (x_i - \mu_X)^2}{N}}$$

For a sample:

$$s_x = \sqrt{\frac{\sum (x_i - \bar{x})^2}{n - 1}}$$

Coefficient of Variation

For the population:

$$CV = \frac{\sigma_X}{\mu_X} \times 100$$

For a sample:

$$CV = \frac{s_x}{\bar{x}} \times 100$$

Percentile Position

$$\text{Position of the } i\text{th percentile} = \frac{i}{100} \times n \qquad \text{(rounded up as necessary)}$$

NEW STATISTICAL TERMS

arithmetic mean
average
bimodal distribution
coefficient of variation
deviation
difference from the mean
empirical rule
five-number summary
inner fence
interquartile range (IQR)
kurtosis
mean
measures of central tendency
measures of variation
median
mode
normal distribution
outer fence
outlier
percentiles
population mean
population variance
quartile
range
relative measures
sample mean
sample variance
skewness
standard deviation
summation operator
trimmed mean
variance

EXERCISES 3.58–3.60

Supplementary Problems

3.58 The number of players attending a bingo game determines the level of revenue. During a recent 21-week period at the Suffolk Jewish Center, attendance at bingo varied between 90 and 160 people. The data are below:

98	98	110	122	128	110	153
112	100	125	150	105	107	90
118	160	109	128	140	119	139

a. Find the mean and standard deviation for the data.
b. Calculate the median.
c. Is this distribution skewed? If so, how?
d. Draw a box-and-whisker plot of the data.
e. Construct a histogram of the data.
f. Each player spends approximately $50 each time he or she attends bingo. Bingo is held once a week, 40 weeks a year. At each session, $3000 in prizes is awarded. What is the expected profit each year for the center?

3.59 Orlando, Florida, the home of DisneyWorld, has many motels and hotels. According to the *1987 Mobil Travel Guide,* prices for single-room occupancy vary considerably. The prices (in dollars) for a random sample are below:

24	38	41	45	49	58	65	79	100
28	39	42	46	49	60	66	85	105
34	40	44	48	49	63	67	85	185
35	40	44	48	51	63	67	89	
35	40	44	48	55	65	75	95	
36	41	44	48	56	65	78	99	

a. What is the average price of a motel room in Orlando?
b. Calculate the standard deviation.
c. Find the interquartile range.
d. Compute the coefficient of variation.
e. Calculate the 90th percentile.
f. If a person budgeted $50 per night per room, would he or she be under-budgeting?

g. Using at a current *Mobil Travel Guide,* repeat your calculations for parts a through f. Prepare a brief report on the change in the cost for a single-occupancy room in Orlando.

3.60 Data for 1990 for some health-related charities are given below.

Charity	Income (millions of dollars)	Expenses: Fundraising (% of income)	Expenses: Administration (% of income)
AmeriCare	78.4	0.9	1.3
MAP International	35.1	2.5	1.2
Hadassah	32.0	3.1	6.3
Project Hope	47.8	4.8	5.0
Interchurch Medical Assistance Incorporated	15.6	1.1	1.9
Juvenile Diabetes	27.7	19.8	5.1
United Cerebral Palsy	359.9	2.6	12.7
Cystic Fibrosis	62.2	6.8	9.1
Easter Seal	288.5	7.7	9.9
Muscular Dystrophy	118.8	18.8	6.8
National Mental Health Association	72.4	2.0	12.3
City of Hope	157.0	3.2	10.2
American Diabetes	56.3	11.0	8.5
Epilepsy Foundation	31.9	10.4	11.4
Leukemia Society	32.5	11.3	12.7
March of Dimes	124.4	15.1	9.3
American Heart	264.4	12.2	7.7
American Cancer	365.5	16.1	7.0

American Institute for Cancer Research	12.3	18.4	9.9
American Lung	121.0	18.5	6.2
National Kidney	20.7	9.4	6.8
National Multiple Sclerosis	76.4	16.4	9.1
Alzheimer's	36.4	13.9	10.9
Joslin Diabetes	31.2	1.8	12.8
Arthritis Foundation	76.6	12.2	6.8
St. Jude's	124.0	13.6	9.5
Shriner's Hospital	413.0	0.5	2.1

Source of Data: *Money,* November 1991.

a. Find the mean and standard deviation for each variable.

b. Produce a histogram for 1990 income.

c. Suppose you were an investigator for the attorney general's office. What would you make of the data? Are there any charities you would single out for special investigation? Why? (See "State Gathers Data to Show Who Profits from Fund Raising Dollars," *New York Times,* September 15, 1995, Section 1, p. 46.)

APPENDIX: THE SUMMATION OPERATOR

There are four basic operational symbols in arithmetic: +, −, ×, and ÷. Calculating statistics frequently requires repeated use of these operations—particularly addition. The Greek uppercase sigma, Σ, is the symbol for the process of summation; it tells you to add the values following it. If the variable is X, then its specific values are $x_1, x_2, x_3, \ldots$, and so on. Similarly, for the variable Y, the specific values are $y_1, y_2, y_3, \ldots$, and so on. Consider the data in Table 3.22.

Table 3.22

Variable *X*	Value
Observation 1, x_1	26.1
Observation 2, x_2	25.7
Observation 3, x_3	26.2
Observation 4, x_4	25.9
Observation 5, x_5	29.3

Σx_i is read as "sigma x sub i." There are five observations in the table, and their sum is 133.2. To arrive at this sum, we add the observations: $x_1 + x_2 + x_3 + x_4 + x_5$. Here, the sample size is $n = 5$, for the five observations. Often, Σx_i is expressed more briefly as ΣX, where the variable X tells us what is to be added up and the sample size and observations are understood from the context.

To calculate the average, we divide the sum of the values by the number of values:

$$\bar{x} = \frac{\sum x_i}{n} = \frac{\sum X}{n} = \frac{133.2}{5} = 26.64$$

In some cases, expressions involving a constant may follow the Σ symbol. For example, in the expression $\Sigma(X - \bar{x})$, the value $\bar{x}$ is a constant. For the values in Table 3.22,

$$\begin{aligned}\sum (X - 26.64) &= (26.1 - 26.64) + (25.7 - 26.64) + (26.2 - 26.64) \\ &\quad + (25.9 - 26.64) + (29.3 - 26.64) \\ &= 133.2 - (5 \times 26.64) = 0\end{aligned}$$

Here, $\Sigma\, 26.64 = 5 \times 26.64$, and, in general, $\Sigma c_i = n \times c$, where c is any constant. Using the compressed form, we may write C to represent variables whose values are all equal to the constant C. Then ΣC is equivalent to observation 1 + observation 2 + ⋯ + observation n. The summation process looks like this:

$$\begin{aligned}\sum 3 &= 3_1 + 3_2 + 3_3 + 3_4 + 3_5 \\ &= 5 \times 3\end{aligned}$$

Thus, we say $\Sigma C = n \times C$, where C is a constant.

EXAMPLE 3.16 ■ CALCULATING THE VALUE OF AN EXPRESSION WITH A SUMMATION SYMBOL

Using the data set in Table 3.23, let's calculate the value of $\sum XY$ and the value of $(\sum X)(\sum Y)$ and see what the difference is between the two.

Table 3.23

X	Y
5	9
6	8
7	3
8	2

$\sum XY$ means $x_1y_1 + x_2y_2 + x_3y_3 + x_4y_4$, so

$$\sum XY = (5 \times 9) + (6 \times 8) + (7 \times 3) + (8 \times 2)$$
$$= 45 + 48 + 21 + 16 = 130$$

$(\sum X)(\sum Y)$ means $(x_1 + x_2 + x_3 + x_4) \times (y_1 + y_2 + y_3 + y_4)$, so

$$\left(\sum X\right)\left(\sum Y\right) = (5 + 6 + 7 + 8) \times (9 + 8 + 3 + 2)$$
$$= 26 \times 22 = 572$$

The difference between the two is that the first is the sum of the products and the second is the product of the sums.

EXERCISES 3.61–3.71

Skill Builders

Use the following data set in the exercises below.

X	Y	C
3	10	2
7	4	2
9	8	2
11	6	2
13	12	2

3.61 Calculate $\sum X$.

3.62 Calculate $\sum X^2$.

3.63 Calculate $(\sum X)^2$.

3.64 Interpret the calculations done in Exercises 3.61, 3.62, and 3.63 in words.

3.65 Calculate the values of $\sum XY$ and $(\sum X)(\sum Y)$.

3.66 Calculate the value of $\sum CX$.

3.67 Calculate the value of $\sum (X + Y)$.

3.68 Calculate the value of $\Sigma (Y - 2)$.

3.69 Calculate the value of $(\Sigma Y) - 2$.

3.70 Calculate the value of ΣC.

3.71 Calculate the value of $\Sigma (X - C)$.

4

The Laws of Chance

OUTLINE

LEARNING OBJECTIVES

1. Speak the language of probability.
2. Identify sample spaces.
3. Understand the classical approach and the empirical approach to probability.
4. Explain the rules of probability.
5. Apply the rules of probability to real-life examples.

BUILD YOUR KNOWLEDGE *OF STATISTICS*

It's the bottom of the ninth; there's a man on first and a man on third. Your team is losing by one run. There are two outs. What's the probability of scoring? According to Hal Stern of Iowa State (Chance Lecture, Dartmouth College, December 13, 1997), the probability of scoring at least a single is .37. Baseball is one of the games where the laws of probability can be readily applied. In the dugout, these laws are often applied informally when, for example, a manager makes a decision to call for a sacrifice bunt or a steal. Each call is based on a perceived probability that it will be successful. Probability plays a role in business strategy as well. The role of probability is perhaps most obvious in the casino and gaming industry. The concept of insurance against risk is based on the rules of mathematical expectation, a derivative of probability.

Up to this point, we have dealt with descriptive statistics. We discussed how and where data are collected. In Chapter 2, we examined the presentation and organization of data. In Chapter 3, we reviewed a variety of numerical measures of data sets. Frequently, however, we do not have the luxury of complete data about a population because it is too large. When it is impossible or impractical to take a complete count (census) of every population, we use data drawn from samples. Consequently, we must rely on statistical inference. Generalizations about the characteristics of a population or predictions about future events depend on partial information. With the aid of statistical methods—specifically, the tools of probability—we can use samples to derive accurate information about populations.

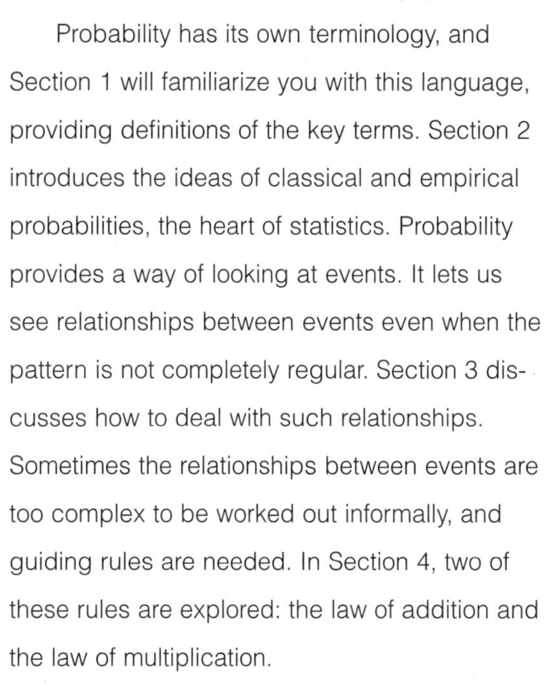

Probability has its own terminology, and Section 1 will familiarize you with this language, providing definitions of the key terms. Section 2 introduces the ideas of classical and empirical probabilities, the heart of statistics. Probability provides a way of looking at events. It lets us see relationships between events even when the pattern is not completely regular. Section 3 discusses how to deal with such relationships. Sometimes the relationships between events are too complex to be worked out informally, and guiding rules are needed. In Section 4, two of these rules are explored: the law of addition and the law of multiplication.

SECTION 1 ▪ THE LANGUAGE OF PROBABILITY

Take a deck of cards, shuffle it seven or eight times, and then record what the top card is. This is an experiment. It can be repeated any number of times. Each repetition is a trial, and the set of all possible outcomes for a given trial is the sample space for this experiment. Probability theory cannot say what particular sequence of results your experiment will yield, but it can predict that roughly a quarter of the cards you turn up will be hearts. The classical definition of probability tells us that, if 13 of the 52 cards are hearts, the chance that a heart will be the top card is 1/4. Then, the relative frequency definition of probability says that an event having a probability of 1/4 should happen about a quarter of the time in a long run of trials.

An **experiment** is a process that produces observations. In order to apply statistical or probabilistic methods in analyzing experiments, we must think of them as repeatable, so that our reasoning applies to the behavior that would be observed in a long series of such experiments.

Tossing a coin is an experiment, as is rolling a die or drawing a card from a well-shuffled deck. Each toss of the coin or roll of the die or draw of a card is a trial.

Each repetition of an experiment is a **trial.** A trial is one output of the process that defines the experiment.

For every experiment there is a sample space.

A **sample space** for an experiment consists of all possible outcomes of the experiment.

The sample space for the experiment of tossing a coin consists of two elements, which may be either pictured or listed.

Picture: $\Omega = \{\ \ \}$

List: $\Omega = \{\text{Heads, Tails}\}$

The Greek capital omega is frequency used to represent the sample space for an experiment.

Like the sample space for tossing a coin, the sample space for throwing a single die may be represented either by showing its possible outcomes or by listing the numbers that can turn up.

Picture: $\Omega = \{\ \ \}$

List: $\Omega = \{1, 2, 3, 4, 5, 6\}$

STATISTICS TOOL KIT

Picturing Sample Spaces and Sets

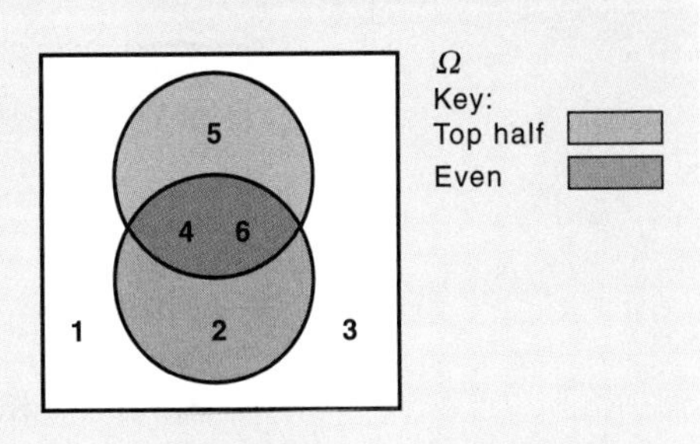

If we represent the set $\Omega = \{1, 2, 3, 4, 5, 6\}$ for the throw of a single die by a box in the plane, we can picture various subsets of Ω by drawing curves around them. For the set Ω, the subsets Even = $\{2, 4, 6\}$ and Top half = $\{4, 5, 6\}$ might be pictured as shown to the right.

The **intersection** of Even and Top half is the dark region common to both sets. It is written Even $\cap$ Top half and is equal to $\{4, 6\}$. The **union** of Even and Top half is the entire shaded region, containing the elements belonging to either of these sets. It is written Even $\cup$ Top half and is equal to $\{2, 4, 5, 6\}$.

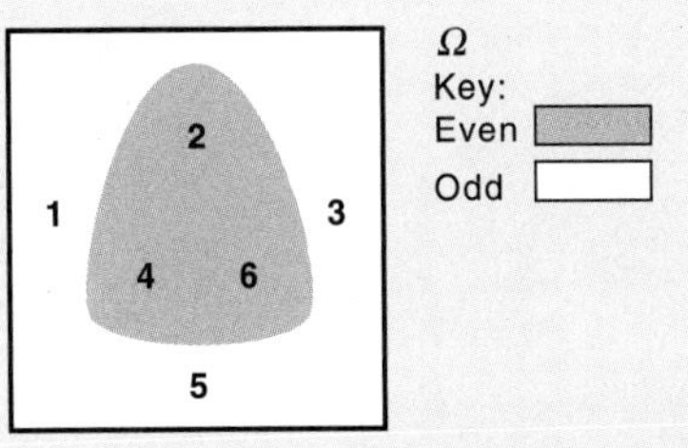

Within a fixed set Ω that contains all the sets we wish to consider, sometimes it is useful to refer to the complement of a set. The **complement** of a set A in Ω is all of the elements of Ω that are not in A. The complement of A is denoted $\overline{A}$. Here, $\overline{\text{Even}} = \{1, 3, 5\}$. In this case, a natural name for $\overline{\text{Even}}$ would be Odd. In the picture to the right, the set Odd is the unshaded region.

Now consider the event of throwing two different-colored dice at once. The sample space can be represented by the picture or array of numbers shown below:

$\Omega =$

$$\Omega = \begin{array}{cccccc} 11 & 21 & 31 & 41 & 51 & 61 \\ 12 & 22 & 32 & 42 & 52 & 62 \\ 13 & 23 & 33 & 43 & 53 & 63 \\ 14 & 24 & 34 & 44 & 54 & 64 \\ 15 & 25 & 35 & 45 & 55 & 65 \\ 16 & 26 & 36 & 46 & 56 & 66 \end{array}$$

These are the possible outcomes for the first roll in a game of craps. The points in the sample space where the sum is 7 or 11, giving the player a win, are circled.

Probability methods give us a way of talking about events and finding their probabilities.

> **Events** are subsets of a sample space Ω. A **simple event** is a subset of Ω containing a single element. The event A is said to occur on a given trial if the outcome of the trial is a point in A.

In craps, a player wins on the first roll by rolling a 7 or an 11. This is the event Win1 = $\{16, 25, 34, 43, 52, 61, 56, 65\}$. Likewise, other outcomes in the game correspond to other subsets of Ω, all of which are events.

If we take three samples at random from a production batch and record S (for satisfactory) or D (for defective) for each one in the order sampled, the sample space is

$$\Omega = \{(SSS), (SSD), (SDS), (DSS), (SDD), (DSD), (DDS), (DDD)\}$$

Let's suppose we accept this batch if at least two of the three samples are satisfactory. Then acceptance corresponds to the event Accept = {SSS, SSD, SDS, DSS}. In this way, all questions about probabilities of various outcomes of trials in an experiment can be turned into questions of calculating the probabilities of a trial's yielding a certain event. The probability of a certain event is simply the probability that a trial's result will fall in some specific subset of the sample space Ω.

SECTION 2 ▪ TWO DEFINITIONS OF PROBABILITY

There are two basic definitions of probability: the classical definition, based on the assumption of equal probabilities of simple events, and the relative frequency definition.

The **classical definition of probability** is based on the ratio of the ways a particular event can occur to the total number of events that can occur in the sample space.

Probability of a simple event: classical approach

Classical Definition of Probability

If an experiment can result in N outcomes, each an equal chance of occurring, then the probability of event A is the number of outcomes that make up event A, denoted by $n(A)$, divided by the total number of outcomes in the sample space, denoted by N.

$$\text{Probability of event } A, P(A) = \frac{n(A)}{N}$$

You will see how this definition works in Examples 4.1 and 4.2. Each of these examples describes an experiment that could result in a known number, N, of simple events—for example, getting a heart on drawing a card from a standard deck (13 hearts out of 52 cards) or obtaining a 4 on the roll of a die (one 4 out of six sides).

EXAMPLE 4.1 ▪ FINDING THE PROBABILITY OF ROLLING A 4

Let's consider the simple experiment of rolling a single die. A simple experiment consists of a single activity—in this case, rolling a die once. We will describe the sample space and find the probability of obtaining a 4. This is the probability that a trial will give a result in the set $\{4\}$, whose only element is 4. This probability is written $P(\{4\})$.

The sample space consists of six elements—namely, the values of the faces of the die:

$$\Omega = \{1, 2, 3, 4, 5, 6\}$$

The probability of obtaining a 4 is the number of faces with 4's divided by the total number of faces on the die (that is, the number of points in the sample space):

$$P(\{4\}) = \frac{\text{Number of faces with a 4}}{\text{Total number of faces}} = \frac{1}{6}$$

EXAMPLE 4.2 ▪ **FINDING THE PROBABILITY OF DRAWING AN ACE OF HEARTS**

Now let's consider a standard deck of 52 playing cards. The number of possible draws is 52. The sample space, Ω, for this experiment has 52 elements. There are four aces. The probability of the event Drawing an ace is the ratio of the number of aces to the total number of cards in the deck:

$$P(\text{Ace}) = \frac{\text{Number of aces}}{\text{Total number of cards in deck}} = \frac{4}{52}$$

The ratio of the number of points corresponding to a particular event to the total number of points in the sample space is the probability of that event.

In the classical situation depicted in Example 4.2, each possible outcome has probability $1/N$, and the probability of an event that contains k simple events is the sum of the probabilities of those simple events or, simply put, k/N. This makes Ω into a **uniform probability space.** A uniform probability space is what casinos and dice makers strive for in their services and products. No outcome is favored over another; in that sense, the game is fair. Casino owners assure their profits by setting the sizes of the payoffs in their favor and then relying on the laws of chance rather than by rigging the wheels or loading the dice.

But we are not always dealing with a casino, which works to make every possible outcome equally likely. Let's consider the weather in Los Angeles. Our sample space will be $\Omega = \{\text{Rainy, Not rainy}\}$, where by "Rainy" we mean at least a hundredth of an inch of rain on a given day. Over the last 54 years, the average has been 35 rainy days per year. Slightly less than 1 day in 10 has been rainy. It is clear that the probability of rain on a random day in Los Angeles can't be figured by using the classical definition, based on equal probabilities. To do this problem, we need an inductive, or experimental, definition of probability based on observed frequencies of events—the **relative frequency definition of probability.**

Probability of a simple event: relative frequency approach

Relative Frequency Definition of Probability

The probability of an event A is the limiting frequency of the number of occurrences of A to the total number of trials, as the number of trials increases without limit. At any stage in the sequence of trials, we may use the ratio obtained so far as an approximation to the limiting ratio, which defines the actual probability.

$$\text{Probability of event } A,\ P(A) = \frac{\text{Number of times event } A \text{ has occurred}}{\text{Total number of trials}}$$

The **law of large numbers** says that as the number of trials grows, the approximation to $P(A)$ defined above converges almost surely to the correct value.

EXAMPLE 4.3 ■ FINDING THE PROBABILITY OF RAIN IN LOS ANGELES

If our sample space is $\Omega = \{\text{Rainy, Not rainy}\}$ and our experiment consists of checking the weather on a random day in Los Angeles, we can use the information from 54 years of observation to find $P(\{\text{Rainy}\})$. The *Statistical Abstract of the United States* for 1991 reports an average of 35 rainy days per year, giving a total of 35 days/year × 54 years = 1890 rainy days out of 365 days/year × 54 years = 19,710 days, ignoring leap years. Using our frequency definition, we have

$$P(\{\text{Rainy}\}) = \frac{\text{Number of rainy days}}{\text{Total number of days}} = \frac{1890}{19{,}710} = .096$$

Taking these 54 years together, we get an overall percentage of just under 10%. The probability of encountering a rainy day in Los Angeles varies from month to month. Over the same period, the average number of rainy days in January has been 6. There are 31 days in January, so the probability of a rainy day in January is 6/31 = .194, or roughly twice the probability for the year as a whole. By contrast, over this same period there has been an average of less than ½ rainy day per year in August, so the probability of a rainy day in August is less than ½/31 = .016, or about a sixth of the probability for the year as a whole.

EXAMPLE 4.4 ■ INTERPRETING THE CHANCE OF RAIN ON A GIVEN DAY

Suppose a weather forecaster tells us that there is a 75% chance of rain. What does this mean? This means that, over all previously recorded days on which conditions were similar, it rained three times out of four. Each day constitutes an observation. If our forecaster has records of 100 days like this one, then 75 of them were rainy.

From the relative frequency definition of probability, the probability of an event consists of the number of times that particular event occurred divided by the total number of possible events. The probability of rain is then

$$P(\text{Rain}) = \frac{\text{Total number of days it rained}}{\text{Total number of days considered}} = \frac{75}{100}$$

With either definition of probability, $P(A)$ is a ratio lying between 0 and 1, inclusive. If an event never occurs, its probability is 0. If an event occurs every time an experiment is performed, its probability is 1. Therefore, the probability, p, of any event must fall within this 0 to 1 range:

$$0 \leq p \leq 1$$

With respect to the weather forecast in Example 4.4, it cannot rain fewer than 0 days in any set of 100 days, and that makes the minimum value of the probability of rain 0. Nor can it rain more than 100 days in any set of 100 days, which makes the maximum value of the probability 1. Because it can rain any number of days from 0 to 100, the probability can assume values within the range from $0/100 = 0$ to $100/100 = 1$.

EXERCISES 4.1–4.11

Applying the Concepts

4.1 A child has a set of alphabet blocks. He selects one block at random. Describe the sample space, Ω, for his selection. What is the probability he selects a vowel? Explain your answer.

4.2 Suppose that, on her way to work each morning, a commuter may or may not buy a newspaper and may or may not buy a cup of coffee. If we consider each morning commute an experiment, describe the sample space, Ω, with respect to these purchases.

4.3 A Chinese restaurant offers seven different soups: three American cuisine soups and four Chinese style soups:

Column A (American)	Column B (Chinese)
Chicken noodle	Wonton
Chicken rice	Egg drop
Vegetable	Hot & sour
	Subgum wonton

a. Two quarts of soup are to be selected for a take-out order; one is to be from column A and one from column B. What is the sample space, Ω, for this event?

b. If two quarts are to be selected without regard to column, describe the sample space, Ω.

4.4 An urn contains three marbles: a red marble, a blue marble, and a green marble.

a. Two marbles are to be drawn at once. Describe the sample space, Ω, for this experiment.

b. When the representatives are all drawn at once, this is "sampling without replacement." Suppose, now, that two marbles are to be drawn with replacement. Describe the new sample space, Ω, for this experiment.

4.5 John will select one of three weekly news magazines—*Newsweek, Time,* or *U.S. News & World Report*—and one of two newspapers—the *New York Times* or *The Wall Street Journal.* Describe the sample space, Ω, for his selection.

4.6 A microchip manufacturer historically has produced an average of 5 defectives per 100 chips produced. What is the probability that a randomly selected component will be defective? What definition of probability are you using?

4.7 There are 30 bottles of iced tea and 20 bottles of cola in the display case in the school cafeteria. What is the probability that a bottle selected at random will be iced tea? What definition of probability are you using?

4.8 A spinner for a game has 30 equal-size slots. A total of 15 of these are red. What is the probability that, when the spinner is spun, it will land on a red slot? Explain how you arrived at your answer.

4.9 M&M's™ come in red, green, yellow, brown, and blue. Suppose your box of these candies contains 25 reds, 25 greens, 20 yellows, 10 browns, and 10 blues.
a. Describe the sample space for an experiment of drawing one piece of candy.
b. What is the probability that a blue piece will be drawn?

4.10 In the game of Yahtzee, five dice are tossed. A Yahtzee consists of obtaining five dice with the same face—for example, all 6s.
a. Describe the sample space for the experiment of tossing all five dice. (Do not attempt to list all points in the sample space.)
b. How many events are there in this sample space? Give examples of three of these points.
c. What is the probability that, on a single toss of the five dice, five 6s will show up?
d. What is the probability of throwing a Yahtzee?

4.11 In the experiment of flipping a thumbtack on a surface, each trial ends with the point sticking up (U) or down (D), but the probabilities of these outcomes depend on the tack and the surface. Perform this experiment with a given thumbtack 100 times. Record the resulting string of U's and D's.
a. For your series, what is the relative frequency of landing point up?
b. What is your best estimate of the probability, p, of your thumbtack's landing point up?
c. If a "fair" thumbtack is equally likely to land point up or point down, do you think your thumbtack is "fair"? Explain your reasoning.
d. Compare the values you and others got for p for different thumbtacks thrown on different surfaces—say, carpet versus hard floor. Are they close to each other? If not, can you think of differences in the experimental situations that can account for the different values of p?

SECTION 3 ■ EVENTS AND PROBABILITY

In probability, as in life, events occur in context. The extent to which knowledge about one event sheds light on the probability of other events is the quantitative expression of informal notions of related and independent events.

Drawing a red card from a standard well-shuffled deck is one event, getting a heart is another, and drawing a face card is a third. Intuitively, we know that our probability of drawing a face card is 3/13, whether or not we know the card is red. The probability of a face card is independent of the color of the card. We know that the probability of drawing a heart is 1/4 under normal circumstances. But the probability of a heart is 1/2 if the card is red, because half the red cards are hearts. So knowing that a red card has been drawn, we recalculate and get a different probability for the draw's being a heart. Understanding how events and their probabilities are related is useful in calculating probabilities and in understanding the situation.

With the example above, we can let F be the event of getting a face card and R the event of getting a red card. If both events occur on a given trial, our card is a red face card—that is, a card in the intersection of F and R, which is written F ∩ R. This is a joint event, as defined below.

Joint probability

> If A and B are events, their intersection, $A \cap B$, is the **joint event** of A and B. For $A \cap B$ to occur on a given trial, both event A and event B must occur on that trial.

Looking back at the Tool Kit picture for events in dice rolling, we see that the events Even and Top half define the joint event Even ∩ Top half = {4, 6}. Let's look at some probabilities for separate and joint events.

EXAMPLE 4.5 ■ FINDING JOINT PROBABILITY FOR A PENNY AND A NICKEL

Consider the experiment of tossing a penny and a nickel simultaneously. What is the sample space, Ω? What is the probability of the penny's coming up heads? Call this event HP. What is the probability of the nickel's coming up tails? Call this event TN. What is the joint probability of these two events?

The possible outcomes are shown in Figure 4.1, with the notation indicated.

Figure 4.1
Coin tossing

$$\Omega = \{H_pH_n, H_pT_n, T_pH_n, T_pT_n\}$$

The event of heads on a penny is HP $= \{H_pH_n, H_pT_n\}$, and its probability is $P(\text{HP}) = n(\text{HP})/N = 2/4 = 1/2$, since N is the number of outcomes in Ω, which is 4. The event of tails on the nickel is TN $= \{H_pT_n, T_pT_n\}$, and its probability is $P(\text{TN}) = n(\text{TN})/N = 2/4 = 1/2$. The probability of the joint event of heads on the penny and tails on the nickel is $P(\text{HP} \cap \text{TN}) = n(\text{HP} \cap \text{TN})/N = 1/4$ because HP ∩ TN $= \{H_pT_n\}$.

Dependence and Independence

Intuitively we know what the outcome for the penny has no effect on that for the nickel because they act independently. Computationally, we notice that

$$\frac{1}{4} = P(\text{HP} \cap \text{TN}) = P(\text{HP}) \times P(\text{TN}) = \frac{1}{2} \times \frac{1}{2}$$

Whenever separate and independent processes determine the two events, statistical independence exists; that is, the probability of the joint event is the product of the probabilities of the two separate events.

> Events A and B are **statistically independent** if the occurrence of one does not affect the likelihood of the occurrence of the other.

Statistical Independence

The probability of the joint event is the product of the probabilities of the individual events:

$$P(A \cap B) = P(A) \times P(B)$$

In our coin flipping example, the event of getting at least one head, $H = \{H_pH_n, H_pT_n, T_pH_n\}$, is not independent of the event of tossing matching coins, $M = \{H_pH_n, T_pT_n\}$, because $P(H \cap M) = 1/4$, while $P(H) \times P(M) = 3/4 \times 1/2 = 3/8$. Among those trials where the coins match, only half show heads; among all trials, three-quarters show at least one head. So once we know that the coins match, we know that getting a head is less probable. Thus, these events are not statistically independent; they are **statistically dependent.**

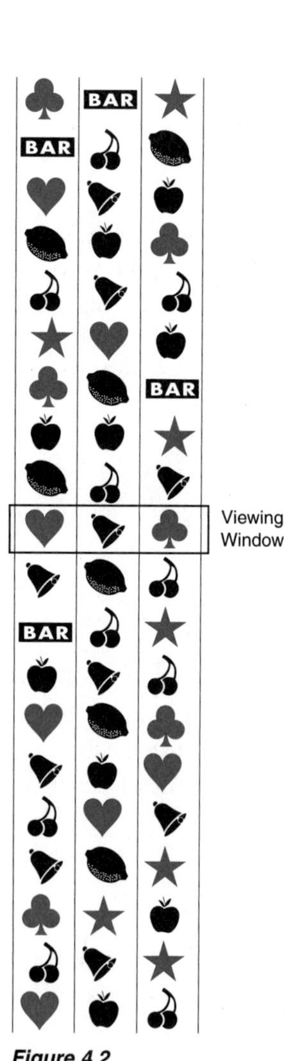

Figure 4.2
Mechanical slot machine

EXAMPLE 4.6 ▪ FINDING JOINT PROBABILITY FOR A SLOT MACHINE

Mechanical slot machines have three wheels that spin separately and lock up at random, starting with the left-most wheel. The payoff is determined by the symbols aligned in the viewing window, indicated by the box in Figure 4.2. Each wheel has a total of 20 symbols; a typical set is shown in the figure. What is the probability that three bars will appear in the window?

The wheels are independent and random in their stopping points. The classical uniform probabilities hold for each wheel. Let the separate events be B_L, bar on the left; B_C, bar in the center; and B_R, bar on the right wheel. Then $P(B_L) = 2/20 = 1/10$, and $P(B_C) = P(B_R) = 1/20$. All of these events are independent, so the joint probability is the product of $P(B_L)$, $P(B_C)$, and $P(B_R)$.

$$P(\text{Three bars}) = P(B_L \cap B_C \cap B_R) = \frac{1}{10} \times \frac{1}{20} \times \frac{1}{20} = \frac{1}{4000}$$

What sort of payoff do you think a casino might offer for getting three bars on a quarter slot machine? The casino has to know how to figure the probabilities in order to set its payoffs so as to assure its edge.

In daily life, events such as buying auto insurance and buying a car are clearly dependent. The next example shows how to test for dependence in a situation where the probabilities are found empirically.

EXAMPLE 4.7 ■ EXPLORING STATISTICAL DEPENDENCE OF EVENTS

Voters were surveyed about their opinions of a job creation program. A breakdown of the results with respect to party affiliation appears in the following table, called a **contingency table.** In such a table, the row and column headings are *always* taken to be the simple events for the analysis, and the table entries are considered joint events.

PARTY VS. OPINION OF JOBS PROGRAM

Opinion / Party	Favor	Oppose	Total
Democrat	50	10	60
Republican	5	35	40
Total	55	45	100

Are party affiliation and opinion of this jobs program independent?

To test for independence, we find the separate and joint probabilities.

Let $P(\text{D})$ be the probability of drawing a Democrat from the sample.

Let $P(\text{F})$ be the probability of drawing someone who favors the program from the sample.

Let $P(\text{D} \cap \text{F})$ be the joint probability of drawing a person who is a Democrat and is in favor of the program.

If $P(\text{D} \cap \text{F})$ equals the product of $P(\text{D})$ and $P(\text{F})$, we may conclude that party affiliation and opinion about the program are statistically independent. The required probabilities are

$$P(\text{D}) = .60 \qquad P(\text{F}) = .55 \qquad P(\text{D} \cap \text{F}) = .50$$

Since $P(\text{D} \cap \text{F}) = .50$ and $P(\text{D}) \times P(\text{F}) = .33$,

$$P(\text{D} \cap \text{F}) \neq P(\text{D}) \times P(\text{F})$$

Because the joint probability is not equal to the product of the simple probabilities, we may conclude that party affiliation and opinion about the program are not independent.

Mutually Exclusive and Exhaustive Events

Getting a head on the toss of a coin and getting a tail on the toss of a coin are mutually exclusive events. Day and night are mutually exclusive events. Being a registered Republican and being a registered Democrat are mutually exclusive events. For an item selected from a production line, being satisfactory and being defective are mutually exclusive events. All these sets of events have one common thread: the occurrence of the first precludes the occurrence of the second, and vice versa.

When events are mutually exclusive, their joint probability is 0. The probability of obtaining a head and a tail simultaneously is 0. The probability that day and night will occur simultaneously is 0. The probability of simultaneously passing and failing statistics is 0.

> Events A and B are **mutually exclusive** if they cannot occur simultaneously. In this case, the joint probability $P(A \cap B)$ equals 0.

EXAMPLE 4.8 ■ EXPLORING MUTUAL EXCLUSIVITY OF EVENTS

Consider the chance of obtaining a 3 on the toss of a single die (event Three) and the chance of obtaining an even value (event Even). Now consider the sample space for this experiment:

$$\Omega = \{1, 2, 3, 4, 5, 6\}$$

Our events are Three = $\{3\}$ and Even = $\{2, 4, 6\}$.

Events Three and Even have no common points. Therefore, they are mutually exclusive. If we represent the chance of obtaining a 3 as $P(\text{Three})$ and that of obtaining an even value as $P(\text{Even})$, then $P(\text{Three} \cap \text{Even})$ is 0, the defining characteristic of mutually exclusive events in probability.

For the roll of a single die, the events Odd = $\{1, 3, 5\}$ and Even are mutually exclusive, and every point in the sample space lies in one or the other of these sets. This tells us that the event that is their union (that is, the set of all points belonging to Odd or Even) is all of Ω. The union of two sets A and B is denoted $A \cup B$ and is the set of all points belonging either to A or to B. Here, the word *or* is used in its inclusive sense, as in "You will keep dry if you wear a raincoat or use an umbrella." If you do both, you stay all the drier. The events Odd and Even cover all the possibilities listed in Ω, so we say they *exhaust* the possibilities in Ω.

> A family of events A, B, and C is **exhaustive** for the sample space if their union equals Ω: $A \cup B \cup C = \Omega$. For an exhaustive family, the probability of its union is the probability of getting some point in Ω, which is 1.

In our example, $P(\text{Even} \cup \text{Odd}) = P(\Omega) = 1$, which is what we would expect, because every outcome must be either even or odd. (Although here we consider three events, the idea extends to more events.)

Even and Odd are mutually exclusive as well as exhaustive. Each point of Ω belongs to exactly one of this pair of sets. Two sets that are mutually exclusive and exhaustive are called **complementary,** and each of these sets is called the *complement* of the other. If A is an event, its complementary event is written $\overline{A}$ and consists of all the points of Ω that are not in A. Here $\overline{\text{Even}}$ = Odd and $\overline{\text{Odd}}$ = Even. Because each point of Ω belongs to exactly one of a pair of complementary sets, $n(A) + n(\overline{A}) = N$, as you can see in the example of Even and $\overline{\text{Even}}$ = Odd.

So,

$$P(A) + P(\overline{A}) = \frac{n(A)}{N} + \frac{n(\overline{A})}{N} = \frac{n(A) + n(\overline{A})}{N} = \frac{N}{N} = 1$$

Fat Chance

by Joseph J. Neuschatz

From birth to death, life depends on odds. This is probably the reason behind the world being what is today an elite group of professional statisticians, surrounded by a huge mess of amateurs, the rest of us.

Statistics decide one's chances of finding a mate in a particular age group, landing a job in a defined geographical area, winning the lottery or flying a plane containing a bomb. The odds are always good for casino owners, often bad for casino gamblers. Our artificial sweetener-du-jour depends on the latest statistical data from laboratory rats. Curious about chances for survival in the hospital of your choice? Check the federal government list of mortality statistics.

And what would a weather report be like without a "50 percent chance of showers by the weekend"?

I have an odd obsession with probabilities and odds. I don't mean a challenger's chances to win the Super Bowl. I am too selfish for that. All I care about are my own personal odds, and how to improve them.

Here is what I've found so far.

If I get sick, I stay home. More Americans die in a hospital bed than in any other location: motel, hotel, or ice cream parlor. I bought my bunk in the spring of 1958 and nobody's ever died in it. If I am brought to the hospital by ambulance, I refuse to go to ICU. More people die in Intensive Care Units than in regular hospital wards.

All my socks are of the same color, basic black. If I had gray, brown and black, I would have to grab four socks in the dark of the morning to make sure I have one pair. With two colors, the number of socks to grab goes down to three. But with black alone all I need are two grabs to make a pair. Who can beat those odds?

I never trust my watch. When it gains (or loses) one second a day, it shows the exact time once every 236.7 years. I always carry a second broken timepiece. This one at least shows the correct time twice in 24 hours.

When in Las Vegas, I employ the wait-and-watch attitude. I'll play a slot machine only after another guy has used his last coin without a win. It drives the losing gambler crazy, but odds are odds!

My most impressive victory is my statistical success over the fear of flying. Don't get me wrong. I'm not afraid of being up in the air—I'm scared of exploding while I'm up in the air. I calculated that being in a plane in which one crazy guy has a bomb is a one in a million—unacceptable—possibility. But I figured out that being in a plane in which two unrelated crazy guys carry one bomb each represents a one-in-a-trillion chance. My kind of odds! Now in my black-humor fantasy, when I fly I bring my own bomb.

From *Newsday Magazine,* June 24, 1990. Reprinted with permission of Joseph J. Neuschatz, M.D.

Figure 4.3
An amusing look at probabilities

The general rule that follows from this pattern is given below.

Probabilities for Complementary Events

The sum of the probabilities of complementary events is 1:

$$P(A) + P(\overline{A}) = 1$$

If a family of events is exhaustive and if each of these events is mutually exclusive, the sum of their probabilities is 1.

Probabilities play a role in almost every aspect of life. Dr. Joseph Neuschatz takes an amusing look at probabilities in his article "Fat Chance!", reproduced in Figure 4.3.

EXERCISES 4.12–4.21

Applying the Concepts

4.12 In *U.S. News & World Report* (November 26, 1990), Jo Ann Tooley wrote "43% of Christmas gifts are returned." What is the complement of this event? What is the probability of the complementary event?

4.13 The classic mortality tables indicate that, for a male of age 60, the probability of surviving for 1 year is .98. The repeatable experiment is to pick a man at random as he turns 60 and see what happens to him in the next year.

a. What outcomes make up the sample space for the chosen individual?

b. What is the probability that the randomly chosen 60-year-old male will not survive 1 year?

4.14 An article entitled "The Nicest Investor Research That Money Can Buy," in the *New York Times* (May 23, 1995), reported that in-house research analysts are nearly twice as likely to produce a favorable report on a company if that company is a client of the brokerage firm's corporate finance department. Assume that Moxie Brokers has produced a report on XYZ Corporation, a client of the finance department. What is the probability that the report is unfavorable? (Assume that Moxie never issues a neutral report.)

4.15 A traffic light is timed such that it is green for 30 seconds, yellow for 10 seconds, and red for 20 seconds. What is the probability that, if you are approaching the intersection, you will have to stop? Explain your reasoning.

4.16 A poll of chief executive officers (CEOs) appearing in *Fortune* (May 31, 1993) indicated that they did not have much confidence in President Clinton's economic plan. When asked whether they believed the plan would create jobs, 76% indicated that they did not believe it would. What is the probability that a CEO believed the plan would create jobs? Explain your reasoning.

4.17 In its "Vox Pop" section, *Time* magazine (May 31, 1993) asked 1000 adults the following question in a telephone poll: If Robert Kennedy had not been assassinated 25 years ago, do you think the country would be in better shape than it is today? A total of 400 respondents answered "yes." What is the probability that a particular respondent did not answer "yes"? Does this mean that he or she answered "no"?

4.18 Chinese cuisine is classified into two types: milder Cantonese and spicy Szechuan-Hunan. Two types of soups often top the menus: wonton (mild) and hot and sour (spicy). The following contingency table reflects the orders of 100 customers on a given day.

TYPE OF SOUP VS. TYPE OF CUISINE

Soup / Cuisine	Wonton	Hot & Sour
Cantonese	50	5
Szechuan-Hunan	25	20

a. What simple events are suggested by this table?

b. List the four joint events in this table.

c. What is the complement of ordering wonton soup?

d. Is the selection of type of cuisine independent of soup selection? Give your reasoning. (In the next exercise set, you will be asked to compute the probabilities.)

4.19 A survey in *U.S. News and World Report* (November 26, 1990) examined Christmas sales. One of the most interesting findings concerned rewrapping gifts and giving them to someone else. The table below is based on these findings.

DISPOSITION OF GIFT VS. INCOME

Disposition / Income	Rewrap	Keep
High Income	155	345
Low Income	115	385

a. What simple events should be considered in analyzing this table?
b. What are the joint events?
c. What is the complement of "rewrap"?
d. What is the complement of "high income"?
e. Are disposition and income independent? Explain your reasoning. (In the next exercise set, you will be asked to compute the probabilities.)

4.20 A baseball player's batting average is 300 (meaning that the probability that he gets a hit is .300).
a. What is the probability that, when he gets up to the plate, he will not get a hit?
b. Assuming that this batter's performances at bat are independent trials, answer the following question. If the player gets up to the plate three times, what is the probability that he will get a hit the first time but not the remaining times?
c. If the player gets up to the plate three times, what is the probability that he will get a hit all three times?

4.21 The field goal percentage for a basketball player is 48.6% (meaning that the probability that she scores a field goal is .486). Assuming each attempt is an independent trial, answer the following questions:
a. What is the probability that she will not score a field goal on the next opportunity?
b. What is the probability that she will be successful on both of the next two opportunities?
c. What is the probability that she will miss on both of the next two opportunities?
d. Are successive attempts independent? There is a lot of literature on the idea of the "hot streak."

SECTION 4 ▪ SOME LAWS OF PROBABILITY

In the study of probability, the words *and* and *or* take on very specific meanings. The law of addition examines the probabilistic meaning of *or,* which can be understood arithmetically or in terms of diagrams of sets.

The Law of Addition

The twelve signs of the zodiac, in chronological order, are Aries, Taurus, Gemini, Cancer, Leo, Virgo, Libra, Scorpio, Sagittarius, Capricorn, Aquarius, and

Pisces. The six from the vernal equinox to the autumnal equinox are called the ascending signs. Each sign is also assigned to a group based on characteristics, such as fire or water signs. Supposing that a random person is equally likely to be born under any of these twelve signs, we will look at the probability that a random person has either an ascending sign *or* a fire sign. The sample space, Ω, and the events Ascending and Fire are shown in Figure 4.4.

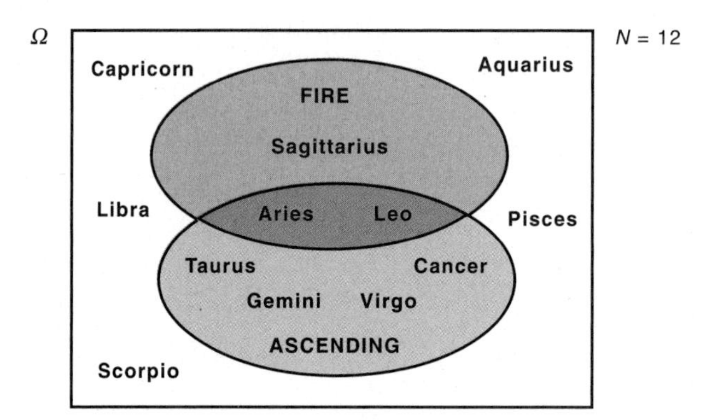

Figure 4.4
Signs of the zodiac

The probability of having a fire sign is $P(\text{Fire}) = 3/12$, and that of having an ascending sign is $P(\text{Ascending}) = 6/12$. The probability of the joint event of having a sign that is both fire *and* ascending is 2/12.

From this information, we can find the probability of having a fire sign or an ascending sign. This is the probability of the union of the events Fire and Ascending, $P(\text{Fire} \cup \text{Ascending})$. The probability is

$$\begin{aligned} P(\text{Fire} \cup \text{Ascending}) &= P(\text{Fire}) + P(\text{Ascending}) - P(\text{Fire} \cap \text{Ascending}) \\ &= \frac{3}{12} + \frac{6}{12} - \frac{2}{12} \\ &= \frac{7}{12} \end{aligned}$$

We can verify this result by simply looking at Figure 4.4 and counting the points in Fire and Ascending; there are 7.

Another approach is to add

$$\begin{aligned} n(\text{Fire} \cup \text{Ascending}) &= n(\text{Fire}) + n(\text{Ascending}) - n(\text{Fire} \cap \text{Ascending}) \\ &= 3 + 6 - 2 \\ &= 7 \end{aligned}$$

The last term is subtracted to compensate for the fact that the points in the intersection of the two events were counted twice when we added $n(\text{Fire})$ and $n(\text{Ascending})$.

This pattern always holds for probabilities and gives us a way to calculate $P(A \cup B)$, given $P(A)$, $P(B)$, and $P(A \cap B)$. This method is known as the **law of addition.**

Law of Addition

For any events A and B, the probability that A or B will occur is $P(A \cup B)$, which can be found by the formula

$$P(A \cup B) = P(A) + P(B) - P(A \cap B)$$

In words, the probability that A or B will occur is the sum of the probabilities of A and B minus the probability of the joint event $A \cap B$. This formula is especially useful when A and B are *mutually exclusive,* in which case $P(A \cap B) = 0$ and then $P(A \cup B) = P(A) + P(B)$. How would you use words to express this formula for the probability that one or the other of two mutually exclusive events will occur?

The law of addition can be seen at work in a contingency table. The simple events here are those defining the columns and rows, and the joint probabilities are those in the body of the table. In Example 4.9, we look at the event of being over 21 or being male.

EXAMPLE 4.9 ▪ USING THE LAW OF ADDITION

For the class depicted in the table, what is the probability of being male and over 21?

AGE CATEGORY VS. GENDER FOR A CLASS

Age Category / Gender	Age 21 and under	Over 21	Total
Male	5	10	15
Female	6	13	19
Total	11	23	34

A total of 15 individuals are male. Dividing by the total number of individuals in the table (34) yields the probability of being male:

$$P(\text{Male}) = \frac{15}{34}$$

A total of 23 individuals are over 21. Dividing by the total number of individuals in the table (34) yields the probability of being over 21:

$$P(\text{Over 21}) = \frac{23}{34}$$

Of the 34 individuals, there are 10 individuals who are male and over 21. Thus,

$$P(\text{Male} \cap \text{Over 21}) = \frac{10}{34}$$

and

$$P(\text{Male} \cup \text{Over 21}) = \frac{23}{34} + \frac{11}{34} - \frac{10}{34} = \frac{24}{34}$$

Note that the 10 people who give us the probability 10/34 were included in the "over 21" as well as the "male" category. We cannot count these 10 individuals twice, so we subtract 10/34.

The Law of Multiplication

Earlier, we found that the probability of a random person's having a fire sign is $P(\text{Fire}) = 3/12$. But what if we knew that the random person had an ascending sign? There are only six ascending signs, two of which are fire signs, so the probability of a person's having a fire sign, given that she has an ascending sign, is 2/6.

STATISTICS TOOL KIT

Using Conditional Probabilities to Solve Problems

The law of multiplication is useful because we can often figure out the conditional probability quite easily. Suppose we want to know the probability of dealing out two aces as the first two cards from a well-shuffled deck. The probability of the top card's being an ace is just the number of aces, 4, divided by the number of cards, 52. If the top card is an ace, the probability of the next card's being an ace is just the number of aces remaining, 3, divided by the number of cards remaining, 51. So

$$\begin{aligned} P(\text{Top two cards are aces}) &= P(\text{Second card is an ace} \mid \text{first card is an ace}) \\ &\quad \times P(\text{First card is an ace}) \\ &= \frac{4}{52} \times \frac{3}{51} = \frac{1}{221} \end{aligned}$$

or 1 chance in 221(.0045, or .45%).

We call this value the conditional probability of a fire sign, given that the sign is ascending. It is written $P(\text{Fire} \mid \text{Ascending})$. **Conditional probabilities** tell us how knowing that one event has occurred affects the probability of another event.

Conditional Probability

For events A and B, if $P(B)$ is not 0, the conditional probability of A given B, written $P(A \mid B)$, is given by

$$P(A \mid B) = \frac{P(A \cap B)}{P(B)}$$

This gives us the **law of multiplication.**

Law of Multiplication

The probability of A and B is the conditional probability of A, given B, times the probability of B:

$$P(A \cap B) = P(A \mid B) \times P(B)$$

In Example 4.10, the idea of conditional probability is applied to the probability of passing a course.

EXAMPLE 4.10 ▪ USING THE LAW OF MULTIPLICATION

Consider the following contingency table, which relates passing or failing to whether students attend class. What seems to be the best strategy for passing this course?

GRADE VS. ATTENDANCE

Attendance \ Grade	Pass	Fail	Total
Attends regularly	90	1	91
Attends sometimes	7	2	9
Total	97	3	100

As usual, the simple events are given by the column and row headings. The probability of passing is given by

$$P(\text{Pass}) = \frac{97}{100}$$

The probability of passing, given that the individual has attended class regularly, is a conditional probability and is calculated as follows:

$$P(\text{Pass} \mid \text{Attends regularly}) = \frac{P(\text{Pass} \cap \text{Attends regularly})}{P(\text{Attends regularly})} = \frac{90/100}{91/100} = \frac{90}{91} = .99$$

This last fraction is equal to the proportion of those attending regularly who passed the course. It is a conditional probability, where the condition is attending the class regularly.

$$P(\text{Pass} \mid \text{Attends sometimes}) = \frac{7/100}{9/100} = \frac{7}{9} = .77$$

We know that the statistical independence of events A and B should indicate that these events are unrelated. How does this fact tie up with conditional probability? If A and B are statistically independent, $P(A \cap B) = P(A) \times P(B)$. So

$$P(A \mid B) = \frac{P(A \cap B)}{P(B)} = \frac{P(A) \times P(B)}{P(B)} = P(A)$$

Knowing that B has occurred leaves the probability of A unchanged.

Conditional Probability and Independence

Events A and B, each having non-zero probability, are independent precisely when $P(A \mid B) = P(A)$ and $P(B \mid A) = P(B)$.

EXERCISES 4.22–4.26

Applying the Concepts

4.22 Consider, once again, the soup selections in a Chinese restaurant:

TYPE OF SOUP VS. TYPE OF CUISINE

Cuisine \ Soup	Wonton	Hot & Sour
Cantonese	50	5
Szechuan-Hunan	25	20

a. What is the probability that a customer will order Cantonese cuisine and wonton soup?

b. What is the probability that a customer will order Cantonese cuisine or wonton soup?

c. What is the probability that a customer who orders hot and sour soup will order Cantonese cuisine?

4.23 Recall the Christmas gift example:

DISPOSITION OF GIFT VS. INCOME

Income \ Disposition	Rewrap	Keep
High income	155	345
Low income	115	385

a. What is the probability that an individual has a high income and rewraps a gift?
b. What is the probability that a person has a low income or keeps a gift?
c. What is the probability that a person with a high income will rewrap a gift?

4.24 Consider the following data, which describe a sample of air travelers.

CHOICE OF AIRLINE VS. DESTINATION

Destination \ Airline	Alpha	Beta	Gamma
London	200	500	200
New York	200	100	0
Paris	600	400	800

a. How many simple events are described in this table? What are they?
b. Identify the joint events in this table.
c. What is the complement of the event Paris?
d. Are choice of airline and destination independent?
e. What is the probability of going to London and flying Gamma Air?
f. What is the probability of flying Beta Air or going to New York?
g. Given that a person is going to New York, what is the probability that she or he will fly Alpha Air?
h. If Gamma Air went out of business, how would the above answers change? (Ignore the 1000 Gamma passengers.)

4.25 Blood may be classified by type as A, B, AB, or O. Furthermore, blood may be classified according to whether it is Rh positive (Rh+) or Rh negative (Rh−). Consider a sample of 1000 individuals, with the distribution shown in the table.

BLOOD TYPE VS. RH FACTOR

Factor \ Type	A	B	AB	O
Rh+	357	85	34	374
Rh−	63	15	6	66

a. What are the simple events for this contingency table?
b. How many marginal probabilities are there? List them. Find the value of each one.
c. What is the probability that an individual will have blood type A and be Rh+?
d. What is the probability that an individual will be type O or Rh−?
e. Given that a person is Rh−, what is probability that she is type AB?
f. Are blood and Rh factor statistically independent? Explain your answer.

4.26 When a new drug is introduced, it is tested in comparison with a placebo (a harmless control). Patients trying a new drug for asthma experienced a number of "adverse events," as did those taking the placebo. A total of 4 patients taking the drug had stomachaches, while none taking the placebo had stomachaches. A total of 23 taking the placebo reported headaches; 28 taking the drug reported headaches. Overall, 187 people took the placebo and 184 took the drug.
a. Set up a contingency table for these data.

b. What is the probability of an "adverse event"?
c. What is the probability that an individual taking the drug will experience an "adverse event"?
d. What is the probability of a person's having a headache?
e. Do "adverse events" and treatment appear to be statistically independent? Explain your reasoning.

FORMULAS

Probability of an Event A

Classical definition (equal likelihood):

$$P(A) = \frac{n(A)}{N}$$

where $n(A)$ is the number of outcomes in event A and N is the number of outcomes in the sample space.

Relative frequency definition:

$$P(A) = \frac{\text{Number of times event } A \text{ has occurred}}{\text{Total number of trials}}$$

Joint Events

For statistically independent events A and B, the probability that both A and B will occur is given by

$$P(A \cap B) = P(A) \times P(B)$$

For dependent events A and B, the probability that both A and B will occur is given by the law of multiplication:

$$P(A \cap B) = P(A \mid B) \times P(B)$$

Complementary Events

If exactly one of A or $\overline{A}$ *must* occur, then

$$P(A) + P(\overline{A}) = 1$$

Law of Addition

The probability that A or B will occur is $P(A \cup B)$:

$$P(A \cup B) = P(A) + P(B) - P(A \cap B)$$

Conditional Probability

When $P(B) \neq 0$,

$$P(A \mid B) = \frac{P(A \cap B)}{P(B)}$$

NEW STATISTICAL TERMS

classical definition of probability
complement
complementary
conditional probability
contingency table
event
exhaustive
experiment
intersection
joint event
law of addition
law of large numbers
law of multiplication
mutually exclusive
relative frequency definition of probability
sample space
simple event
statistically dependent
statistically independent
trial
uniform probability space
union

EXERCISES 4.27–4.30

Supplementary Problems

4.27 The table below reflects results of a survey in *USA Today* (December 20, 1990) concerning the reasons people eat the way they do.

EATING CHOICES VS. GENDER

Diet \ Gender	Male	Female
Eat a healthy diet to live longer	3710	2880
Eat a healthy diet to look better	2800	4800
Do not eat a healthy diet	490	320

a. What is the probability that a person does not eat a healthy diet?
b. If the person selected eats a healthy diet, what is the probability that the person is female?
c. What is the probability that a female eats a healthy diet to look better?
d. Are gender and reasons for eating a healthy diet independent?

4.28 For a single woman who has not yet married and is now age 20–24, the probability of getting married sometime is .84. What is the probability that a single woman in that age cluster will not get married?

4.29 In 1993, the average total compensation of chief executive officers (CEOs) in the aerospace and defense industries was $1,533,000. The average value of sales, as reported by *Forbes* (May 24, 1993), was $5,592,000,000. Data for the 21 major firms are summarized in the table.

SALES VS. CEO COMPENSATION IN AEROSPACE CORPORATIONS

Compensation \ Sales	Above average	Below average
High compensation	8	3
Low compensation	4	6

a. What is the probability that CEO compensation is high?

b. What is the probability that CEO compensation is high and sales are above average?

c. Given that compensation is high, what is the probability that sales are above average?

d. Are level of sales and level of compensation independent?

4.30 The game of roulette involves bouncing a ball on a numbered spinning wheel. There are 38 slots on the wheel: 18 red, 18 black, and 2 green (numbered 0 and 00, considered neither even nor odd). A variety of bets may be made, all at different odds. A player buys chips and uses them to bet on individual numbers or any combination.

a. What is the probability of winning on an even or red bet? Both of these bets pay even money. What effect does this probability have on the casino in the long run?

b. A bet is made on the row with a 1 on the left and a 34 on the right. What is the probability of winning on this bet? (As a point of interest, the payoff odds on this bet are 2 to 1.)

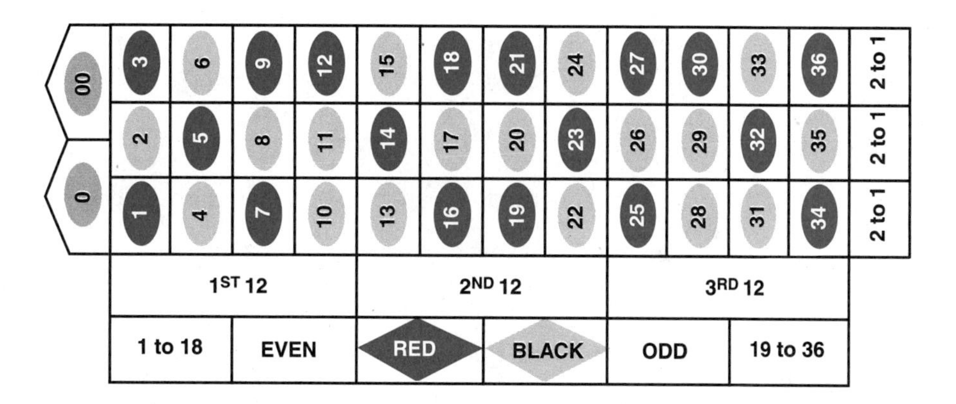

5

Probability Distributions

OUTLINE

LEARNING OBJECTIVES

1. Describe random variables and their probability distributions.
2. Calculate expected value and variance for a random variable.
3. Recognize the binomial distribution and its uses.
4. Discuss the Poisson distribution and its uses.
5. Identify the normal probability distribution and its uses.
6. Compute approximate binomial probabilities.

BUILD YOUR KNOWLEDGE *OF STATISTICS*

What is the probability that a gambler will win better than half of his next 10 bets on red or black at roulette? What is the probability that you will get through more than half the untimed traffic signals between your home and the airport? What are the short-term fluctuations in the number of customers trying to connect to an Internet service provider? How many cars go through an intersection near your house at 5 P.M.?

Numbers of wins at roulette, green lights encountered, calls per minute, and cars at an intersection are all random variables, as are finishing times, heights, and weights. They can be thought of as numbers produced by an experiment. To capture the probabilities of these outcomes in a table or formula for everyday scheduling or planning in business, we use the probability distribution for the random variable.

In Chapters 2 and 3, distributions were used to describe populations and reveal patterns in data. You saw how the area below a density curve gives the probability that a randomly chosen element from the population will fall in any given interval. Chapter 3 showed how a few numbers can give a good summary of a data set. In this chapter, you will learn how the mean and standard deviation determine the distribution of any one of these random variables exactly, providing a completely accurate summary with only two numbers. Here, you will study the binomial, Poisson, and normal distributions. These distributions can be applied to the questions mentioned above. The probability of winning at roulette or getting through untimed traffic signals can be determined using a binomial distribution. Variations in finishing times or heights and weights fit a normal distribution. Fluctuations in numbers of customers or cars follow a Poisson distribution.

In Chapter 6, you will see how a normal distribution quantitatively describes the distribution of the means of random samples. In Chapter 8, this description will provide tools for statistical estimation.

SECTION 1 ▪ RANDOM VARIABLES

When a contestant spins the Wheel of Fortune on the television show of the same name, the outcome is a random variable.

> A **random variable** is a variable that has a single numerical value (determined by chance) for each outcome of an experiment.

Each spin of the wheel is a trial, and each trial may have a different outcome. In every Wheel of Fortune game, there is a set of potential outcomes, the probabilities of which are determined by the wheel. The wheel, shown in Figure 5.1, is marked with several wedges. Some of the wedges indicate dollar amounts of as little as \$200. One wedge usually has a large dollar amount, sometimes \$10,000. There is a wedge labeled "Lose a Turn," and another labeled "Bankrupt." The random variable in this game puts a dollar value—a number—on the outcome of a spin of the wheel. In the third round of one game the outcomes were as follows: \$450, Lose a Turn, \$400, \$550, Bankrupt, \$200, \$250, \$250, \$200, \$1000, \$300, etc. In this experiment, the outcomes are the wedges, and the random variable is the value assigned to each wedge. We will assign a value of zero to the random variable X for the Bankrupt and Lose a Turn wedges. Indeed, these wedges may correspond to a large negative payoff.

Figure 5.1
The "Wheel of Fortune"

"Wheel of Fortune" photo, courtesy of Columbia TriStar Television.

The outcomes of an experiment make up the sample space for the experiment (see Chapter 4). The sample space for the Wheel of Fortune experiment is

$$\Omega = \{\text{Bankrupt, \$500, \$400, \$250, Lose a Turn, \$250, \$350, \$450, \$700, \$300, \$600, \$1000, \$250, etc.}\}$$

Each of these outcomes has a specific probability of occurring. So there is a probability for the outcomes on the wheel and for the values of the random variable X. The outcomes correspond to a probability distribution.

> The **probability distribution** for a random variable X is a table or formula for calculating the probability that the variable takes on a particular value or falls within a given range. Such tables or formulas provide models for various sorts of probabilistic processes.

Probability distributions generally have an expected value, which is analogous to the arithmetic mean and yields a measure of the center of the distribution. They also have a variance, which is the square of the standard deviation. (See Chapter 3 for a review of these descriptive statistics.)

The Wheel of Fortune has a discrete probability distribution; it takes on only a finite number of values (see Section 4 of Chapter 1). The **expected value** of a discrete probability distribution may be found by considering each value and its likelihood of occurring. The expected value is a weighted average of values, where each value is weighted by its respective probability.

Expected Value (Discrete Variable)

The expected value of a probability distribution is its mean value. It is obtained by taking the sum of the products of each outcome and its respective probability:

$$\text{Expected value, } E(X) = \sum x_i P(x_i)$$
$$= x_1 P(x_1) + x_2 P(x_2) + \cdots + x_n P(x_n)$$

where n is the number of values in the set of outcomes and $P(x_i)$ is the probability of the outcome x_i.

EXAMPLE 5.1 ▪ FINDING THE EXPECTED VALUE FOR A DISCRETE RANDOM VARIABLE

There are 23 wedges on the Wheel of Fortune. Remember that we are treating "Lose a Turn" and "Bankrupt" as representing outcomes of zero. We find the expected value of this game as follows:

$$E(X) = \$0\left(\frac{1}{23}\right) + \$800\left(\frac{1}{23}\right) + \cdots + \$250\left(\frac{1}{23}\right) = \$430.43$$

This means that, on average, a player can expect to "earn" \$430.43 per spin. Note that this is actually an overstatement of earnings, because an outcome of "Bankrupt" would cause the player to lose the whole of his or her earnings up to that spin!

Given the distribution of a random variable X, we can find its expected value and its variance. The expected value, $E(X)$, for a random variable is what we called the mean when we first looked at populations. The expected value of $[X - E(X)]^2$ is the mean square deviation from the mean of X. This value $E[(X - E(X))^2]$ is called the **variance** of X; it is a measure of the dispersion of the random variable

X. As with the parameters for populations, the square root of this average is called the standard deviation of X.

Variance and Standard Deviation (Discrete Variable)

$$\text{Variance, } V(X) = E[(X - E(X))^2]$$
$$= \sum [x_i - E(X_i)]^2 P(x_i)$$
$$\text{Standard deviation, } S(X) = \sqrt{V(X)}$$

Using the fact that $[X - E(X)]^2 = X^2 - 2E(X)X + [E(X)]^2$, we could show that the variance is also given by $V(X) = E(X^2) - [E(X)]^2$. The latter form is useful in hand calculations but less essential now that we have computers to do our number crunching for us.

EXAMPLE 5.2 ■ FINDING THE EXPECTED VALUE, VARIANCE, AND STANDARD DEVIATION FOR A DISCRETE RANDOM VARIABLE

Let's consider the experiment of tossing a pair of dice. Each toss is a trial, and the outcome is the sum of the numbers on the two faces. The probability data appear in Table 5.1.

Table 5.1

PROBABILITIES FOR THE SUM OF THE NUMBERS ON THE FACES OF TWO DICE

Value x	Probability $P(x)$	Value x	Probability $P(x)$
2	1/36	7	6/36
3	2/36	8	5/36
4	3/36	9	4/36
5	4/36	10	3/36
6	5/36	11	2/36
		12	1/36

To calculate the expected value, we apply the formula:

$$E(X) = \sum xP(x)$$
$$= 2\left(\frac{1}{36}\right) + 3\left(\frac{2}{36}\right) + 4\left(\frac{3}{36}\right) + 5\left(\frac{4}{36}\right)$$
$$+ 6\left(\frac{5}{36}\right) + 7\left(\frac{6}{36}\right) + 8\left(\frac{5}{36}\right) + 9\left(\frac{4}{36}\right)$$
$$+ 10\left(\frac{3}{36}\right) + 11\left(\frac{2}{36}\right) + 12\left(\frac{1}{36}\right)$$
$$= \frac{252}{36} = 7$$

Note that for each toss there is an x value, and associated with each x value is a probability.

To calculate the variance, we apply the formula:

$$V(X) = \sum [x - E(X)]^2 P(x)$$

$$= (2 - 7)^2 \frac{1}{36} + (3 - 7)^2 \frac{2}{36} + \cdots + (12 - 7)^2 \frac{1}{36}$$

$$= \frac{210}{36} = 5.83$$

To calculate the standard deviation, we apply the formula:

$$S(X) = \sqrt{V(X)} = \sqrt{5.83} = 2.42$$

For a nonnegative random variable, there is an easy geometric interpretation of the expected value in terms of the cumulative density function (CDF). Figure 5.2 shows the CDF for our dice-tossing experiment. The complete area of the rectangle is its width (13) times its height (a probability of 1); the portion of the area that is unshaded represents $E(X)$. The unshaded horizontal bar at mid-height has a width of 7 and a height proportional to $P(X = 7)$, so its area is part of the sum defining $E(X)$. The sum of the areas of all the unshaded rectangles is exactly equal to the sum $\Sigma\, xP(x)$, which defines $E(X)$.

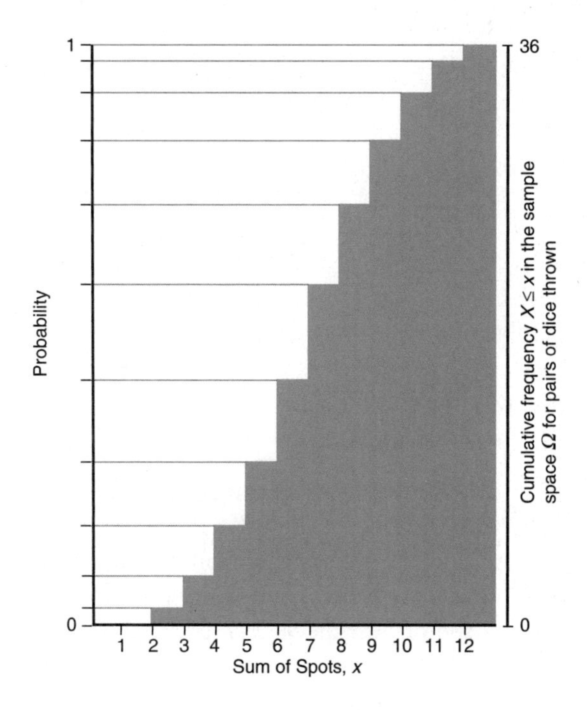

Figure 5.2
Cumulative density function for X

EXERCISES 5.1–5.10

Skill Builders

5.1 Consider the following set of random numbers with their respective probabilities:

X	P(X)
52	.50
4	.25
29	.12
6	.06
28	.03
79	.02
86	.02

a. Calculate the expected value of this distribution.
b. Compute the variance of this distribution.
c. Find the standard deviation.

5.2 Consider the two distributions below:

	A			B	
i	x_i	$P(x_i)$	*i*	y_i	$P(y_i)$
1	0	.10	1	0	.05
2	1	.20	2	1	.15
3	2	.30	3	2	.25
4	3	.30	4	3	.35
5	4	.10	5	4	.20

a. Compute the mean for X and Y.
b. Compute the standard deviation for X and Y.
c. Calculate the coefficient of variation for X and Y.
d. Which variable shows greater variation?

Applying the Concepts

5.3 A game involves rolling a 10-sided die. The values on the faces are the first 10 integers, listed here:

1 2 3 4 5 6 7 8 9 10

All faces have an equal probability of turning up. The random variable X is the number showing on top when the die stops rolling.
a. What is the expected value of X?
b. Compute the variance of X.
c. Calculate the standard deviation of X.

5.4 A poll of economists revealed that 25% believe that the economy will grow by only 2% in the next year, 70% believe the economy will sustain a 3% growth rate, and 5% believe the economy will sustain 5% growth. Our experiment consists of picking an economist at random and asking what growth rate she or he expects. The growth rate is the random variable X.
a. Calculate the expected value for X.
b. What is the variance for X?
c. Compute the standard deviation for X.

5.5 Review of the maintenance records on a certain machine yields the following information on breakdowns over a period of 50 working days.

Number of Breakdowns	Number of Days
0	40
1	7
2	2
3	1

a. What is the expected number of breakdowns per day?
b. Use these data to estimate the probability of having more than one breakdown on any given day.
c. If a service contract for the machine for a period of 50 working days costs $500 and a visit by the repairperson costs $50, would it pay to get a service contract? (*Hint:* Compare the probability of breakdown to the probability of no breakdown.)

5.6 A study entitled "Secure Detention for Juveniles in Suffolk County: An Analysis of the Population of the Children's Shelter" (September 1976) revealed that when persons in need of supervision (PINS) were charged with being runaways, they spent anywhere from 1 to 120 days in detention in the Suffolk County Juvenile Detention Facility. The data appear below:

Approximate Days in Detention	Number of Runaways
1	40
2	22
5	18
14	38
30	37
60	24
120	8
Total	187

a. Calculate the expected length of stay for PINS runaways.
b. Find the variance for this distribution.
c. Compute the standard deviation for this distribution.
d. Complete this paragraph: In 1975 there were 187 PINS runaways who were detained at the Suffolk County Juvenile Detention Facility. Their average number of days in detention was 22 days. The standard deviation was 27.7. Those held less than one week accounted for 43 % of all those detained.

5.7 A study of recidivism showed that the time between release and re-arrest varied in the pattern indicated below:

Time to Re-arrest (years)	Proportion of Those Re-arrested
1	.5
2	.3
3	.1
4	.05
5	.05

Source of Data: *Uniform Crime Report* (1966 statistics).

a. What is the mean time for re-arrest?
b. What is the variance?
c. Compute the standard deviation.
d. Complete this paragraph: Re-arrest occurs on average after _____ years. _____ percent of repeat offenders are re-arrested within three years of their release.

5.8 A study of tourists visiting Suffolk County, New York indicated that their lengths of stay varied as shown in the table:

	Visiting from	
Length of Visit	**New York City**	**Elsewhere**
Day trip (1 day)	36%	31%
Weekend (2 days)	35	49
1 week (7 days)	23	18
2 weeks (14 days)	6	2

Source of Data: *Tourism in Suffolk County*, 1979.

a. Compute the expected length of stay for someone from New York City.
b. Compute the expected length of stay for someone from outside the region.
c. Is the expected length of stay longer for someone from elsewhere?
d. Calculate the standard deviation for each group.
e. Find the coefficient of variation for each group.
f. Which group exhibits greater variability?

5.9 Response time by the emergency medical service in a particular city follows the distribution in the table below:

Response Time (minutes)	Percentage of Responses
5	50%
10	30
15	18
20	2

a. Compute the average response time.
b. What is the standard deviation of the distribution?
c. What is the probability that response time will be at most 10 minutes?

5.10 The number of pieces of mail received in the mailroom of a large company varies between 1000 and 1500 items each day. The distribution of the number of items received over a 100-day period is as follows. Treat this as the exact distribution of the random variable.

No. of Items	Frequency
1000	5
1100	20
1200	30
1300	20
1400	15
1500	10

a. Calculate the expected number of items received per day.
b. Find the standard deviation.

c. If the data follow the empirical rule (see Chapter 3), what range in number of items can be expected approximately 68% of the time?
d. Complete this paragraph, again assuming that the empirical rule holds: A study of the mailroom operations revealed that, on average, _____ items are received per day. The number of items falls between _____ and _____ about two-thirds of the time.

SECTION 2 ▪ THE BINOMIAL PROBABILITY DISTRIBUTION

Two discrete probability distributions play an important role in statistical analyses of events because they describe a variety of situations particularly well: the binomial probability distribution and the Poisson probability distribution. The **binomial probability distribution** deals with counts of successes and failures in a finite number of independent trials. In this section, we will consider the binomial distribution.

During 1994, it was reported that approximately 1 out of 13 flights departing from Newark Airport was delayed 15 minutes or more. Suppose these numbers still hold true. If 10 flights are about to take off, what is the probability that at most 2 flights will be delayed? Does that differ from the probability that the *first* two flights will be delayed? Is that different from the probability that *exactly* two flights will be delayed? The answers to these questions can be found with the aid of the binomial distribution.

The binomial distribution is the basis for much acceptance sampling (see Chapter 1). Based on the fraction of defective products in an incoming lot, decisions are made to accept or reject that lot. Because there are only two outcomes, this situation is clearly binomial. Many experiments may be characterized as binomial. A mortality table deals with the outcomes living and dead; a game may have the outcomes winning and losing.

Conditions Necessary for a Binomial Experiment

1. The experiment must have a fixed number of trials.
2. The trials must be independent.
3. Each trial must result in one of two mutually exclusive categories: success or failure.
4. The probabilities must remain constant for each trial.

Consider the following situation. A group of five 50-year-old men and women get together to form a life insurance club. The probability that an individual who is 50 years of age will survive to age 75 is approximately .53. Suppose that the club wishes to provide insurance to its surviving members 25 years hence, and only at that time. What premium should the group establish to realistically fund a payout to the surviving members? The premium is the payment made for the insurance coverage. Assume for simplicity's sake that a single premium payment will be made by those in the club at the start.

Insurance schemes in which benefits go to the surviving members are called *tontines,* after the 17th-century Neopolitan banker Lorenzo Tonti, who proposed such schemes to the French government. Tontines have figured in mystery movies (where one tontine member kills off the others). We shall assume, for the sake of our example, that no extraordinary events shorten the lives of the tontine members.[1]

The experiment of observing the status of each group member after 25 years is clearly binomial, because a participant in the club is either alive or dead at the end of the period. Several factors must be taken into account to determine the premium required. First, we consider the number of people in the group. Second, we decide what payout should be expected for those surviving at the end of the period. Putting aside considerations of the payout, let's look at the survival probabilities. The group consists of five people (realistically this is too small, but it works for the sake of the example). What are the possible outcomes after 25 years? At best, all might survive. At worst, none might survive. In between, we have the possibilities that 1, 2, 3, and 4 survive.

We consider the survival of each individual as an independent event. Then, from the law of multiplication for independent events, the probability of survival of all the individuals is $.53 \times .53 \times .53 \times .53 \times .53$. This is the probability of the first person's surviving times the probability of the second person's surviving, etc. (It's just like calculating the probability of getting five heads when a coin is flipped five times.) If p represents the probability of one person's surviving and $P(k)$ that of k people's surviving, then

$$P(5) = p^5 = .53^5 = .0418$$

For a group of five, this probability would also be represented by

$$P(5) = p^5q^0 = .53^5 \times .47^0 = .0418$$

where q represents the probability of an individual's dying in the 25-year period (.47).

Now let's look at the other extreme. What is the probability that none survive? It is represented by $P(0)$:

$$P(0) = p^0q^5 = .53^0 \times .47^5 = .47^5 = .0229$$

How would we represent the probability of just one survivor? Would it be like this?

p q q q q

This notation indicates that the first member survived and all the others died. But that is not the only way the outcome of one survivor could occur. Four other possibilities exist:

q p q q q

q q p q q

q q q p q

q q q q p

Each represents a different arrangement of outcomes, all with one survivor. By the law of multiplication for independent events, the probability of each of these simple events is $.53^1 \times .47^4$, so the probability of just one survivor is five times this product, or .1293. We will now see how these ideas can be extended to give a formula for the probability of exactly x successes in n independent trials.

---

[1]The idea of a tontine was the basis for the novel *The Tontine* by Thomas Costain, as well as the movie *The Wrong Box* (1966), a comedy with Peter Sellers, Dudley Moore, Michael Caine, and Ralph Richardson.

There are many possible patterns of success and failure that could give exactly x successes. The number of these patterns is called "n choose x" and is represented $\binom{n}{x}$. By the law of multiplication for independent events, each of these patterns has a chance $p^x q^{n-x}$ of occurring, where p is the probability of success and q is the probability of failure. We add these $\binom{n}{x}$ numbers, which are all equal to $p^x q^{n-x}$, to get the probability of achieving exactly x successes in n trials. The sum is $\binom{n}{x} p^x q^{n-x}$, which gives us the formula for the binomial distribution.

You may recognize $\binom{n}{x}$ from algebra, where it appears as the coefficient of $p^x q^{n-x}$ when the binomial power $(p + q)^n$ is expanded and like terms are collected. For this reason, $\binom{n}{x}$ is called a binomial coefficient. It is this link to binomials in algebra that gives binomial trials and the binomial distribution their names.

Binomial Probability

The probability of x successes in n trials is given by

$$P(x) = \binom{n}{x} p^x q^{n-x} = \frac{n!}{x!(n-x)!} p^x q^{n-x}$$

where p is the probability of success on a single trial and $\frac{n!}{x!(n-x)!}$ is a formula for $\binom{n}{x}$, expressed in terms of factorials. In general, $k! = k(k-1)(k-2) \cdots (2)(1)$ for an integer $k \geq 1$, while $0! = 1$.

To compute the probability of one death out of five, we apply this formula to obtain

$$\begin{aligned} P(1) &= \frac{5!}{1!4!} (.53)^1 (.47)^4 \\ &= \frac{5 \times 4 \times 3 \times 2 \times 1}{1 \times 4 \times 3 \times 2 \times 1} (.53)^1 (.47)^4 \\ &= 5(.53)(.0488) = .1293 \end{aligned}$$

When we consider two survivors, the formula takes into account the number of arrangements and the new number of survivors:

$$\begin{aligned} P(2) &= \frac{5!}{2!3!} (.53)^2 (.47)^3 \\ &= 10(.2809)(.1038) = .2916 \end{aligned}$$

Note that when $n!/k!\,(n-k)!$ is written out, there will be many common factors shared by the numerator and denominator. It is generally advisable to eliminate these in order to simplify the calculations.

After computing the probability of three and four survivors, we may complete Table 5.2 on page 158 for this binomial probability distribution. (The total does not equal exactly 1.0000 because of rounding errors.)

Table 5.2

DISTRIBUTION OF OUTCOMES FOR SURVIVAL OF LIFE INSURANCE GROUP

Number of Survivors	Probability of Survival
0	.0229
1	.1293
2	.2916
3	.3289
4	.1854
5	.0418
Total	.9999

Returning now to the question of the premium, we must look at the expected value of this distribution. Recall that $E(X) = \Sigma\ XP(X)$ for a random variable. Fortunately, for the binomial distribution, the expected value may be calculated with much less effort by using the ideas in the Tool Kit.

STATISTICS TOOL KIT

Independence, Sums, Expected Values, and Variances

You learned that the probability of getting x successes in n independent trials is given by $\binom{n}{x}p^x q^{n-x}$, where p is the probability of success on each trial and q is the probability of failure. If we let X be the number of successes in n trials, $X = X_1 + X_2 + \cdots + X_n$, where X_i is 1 if the ith trial yields a success and 0 otherwise. So the binomially distributed X is the sum of n identically distributed and independent variables, one for each trial in the series. As a result of this interpretation of X as a sum of independent identically distributed random variables, we can find the expected value and the variance for X very easily. The key facts follow:

The expected value of a sum is the sum of the expected values:

$$E(X_1 + X_2 + \cdots + X_n) = E(X_1) + E(X_2) + \cdots + E(X_n)$$

The variance of a sum of independent random variables is the sum of their variances:

$$V(X_1 + X_2 + \cdots + X_n) = V(X_1) + V(X_2) + \cdots + V(X_n)$$

Note that the *only* situation we will consider in which we can automatically apply the formula for variances is when the X_i's are independent. However, when the X_i's represent the outcomes of independently chosen random samples from our population, this formula is valid and thus it applies to the situations of greatest interest in statistics.

We will apply these ideas to the binomial distribution, and you will see them at work again in Chapter 6.

As explained in the Tool Kit, we can think of X, the number of successes in n independent trials, as the sum of the X_i's, each of which is either 1 or 0, depending on whether the ith trial yields a success or a failure. The probability that $X_i = 1$ is p and the probability that $X_i = 0$ is q, so $E(X_i) = 1p + 0q = p$. Thus,

$$E(X) = E(X_1) + E(X_2) + \cdots + E(X_n) = p + p + \cdots + p = np$$

Expected Value (Binomial Distribution)

$$E(X) = np$$

This result is not surprising. Because each of the n people has probability p of surviving, the number of survivors we expect out of the group of n people is np. In the case of the survival distribution, $n = 5$ and $p = .53$. Then, $E(X) = 5(.53) = 2.65$. In a large group of people, the average number of survivors per five people would be 2.65. Estimating conservatively, let's assume that three individuals in the group survive. If each survivor receives \$100,000, the total payout will be \$300,000. As this amount will be needed 25 years from the inception of the policy, each of the five participants should contribute \$60,000 (discounted to present value at an 8% rate of interest, compounded annually, the amount would be \$8761).

Direct calculation shows that $V(X_i) = pq$, so

$$V(X) = V(X_1 + X_2 + \cdots + X_n) = pq + pq + \cdots + pq = npq$$

because the X_i's are independent and the variance of their sum is the sum of their variances.

Variance and Standard Deviation (Binomial Distribution)

$$\text{Variance, } V(X) = npq$$
$$\text{Standard deviation, } S(X) = \sqrt{npq}$$

In the case of the five potential insurees, the variance is

$$V(X) = 5(.53)(.47) = 1.246$$

The standard deviation for the five insurees is calculated as the square root of the variance:

$$S(X) = \sqrt{1.246} = 1.116$$

Applying the empirical rule, we conclude that, for groups of five, we can expect between 1.53 and 3.77 survivors 68% of the time (2.65 ± 1.116). Of course, the number of survivors must be an integer, so this result corresponds to 2 or 3 survivors. From Table 5.2, the probability of 2 or 3 survivors is $.2916 + .3289 = .6285$. Not 68%, but close. As we will see, the empirical rule works better for a larger group. Although it is impractical to compute binomial probabilities for large groups, we can use the empirical rule to obtain good bounds for the expected number of survivors.

EXAMPLE 5.3 ▪ TESTING THE WATERS ON AN ASSAULT GUN BAN

A recent survey conducted by Prodigy explored people's feelings about an assault gun ban proposed to Congress. Even though the Prodigy poll did not involve a random sample, let's assume that its results reflect the general sentiments of the population. If the true proportion of individuals favoring an assault gun ban is the 61% found in the Prodigy poll, what is the probability that, in a sample of six people, a majority will favor this type of ban?

To solve this problem we must consider the probability that 4, 5, or 6 of the individuals sampled (representing a majority of the sample) will favor the ban.

$$\begin{aligned} P(4) &= \frac{6!}{4!2!}(.61)^4(.39)^2 \\ &= 15(.1385)(.1521) = .3160 \\ P(5) &= .1976 \\ P(6) &= .0515 \end{aligned}$$

The probability that a majority of the individuals in a group of six will favor an assault gun ban is the sum of $P(4)$, $P(5)$, and $P(6)$. We add these probabilities because the question is "What is the probability that either 4 or 5 or 6 individuals will favor an assault gun ban?" The question asks about the probability of the union of three mutually exclusive events. (See Chapter 4 for the law of addition.)

$$P(4 \text{ or } 5 \text{ or } 6) = P(4) + P(5) + P(6) = .3160 + .1976 + .0515 = .5651$$

EXAMPLE 5.4 ▪ ACCEPTANCE SAMPLING OF COMPUTER CHIPS

A producer of computers will accept lots of computer chips only if not more than 2 defectives are found in a random sample of 25 chips. The manufacturing process produces 15% defectives. What is the probability that testing a sample of 25 chips will lead to rejecting the whole lot? What are the mean, variance, and standard deviation for the distribution of the number of defectives found in a random sample of size 25?

If the true percentage of defectives is 15%, then $p = .15$. We will let $n = 25$ and compute the probability of $X = 0$, 1, and 2 (the probability of accepting the shipment).

$$\begin{aligned} P(0) &= \frac{25!}{25!0!}(.15)^0(.85)^{25} = .0172 \\ P(1) &= .0759 \\ P(2) &= .1607 \end{aligned}$$

The probability of acceptance is .0172 + .0759 + .1607, or .2538. The probability of rejection is 1 − .2538, or .7462.

The mean of this distribution is calculated as

$$E(X) = np = 25(.15) = 3.75$$

The variance of this distribution is calculated as

$$V(X) = npq = 25(.15)(.85) = 3.1875$$

The standard deviation of this distribution is computed as

$$S(X) = \sqrt{V(X)} = \sqrt{3.1875} = 1.7854$$

STATISTICS TOOL KIT

Using Formulas, Tables, or Software

Suppose you work for a microchip company that is experimenting with a new prototype on a small scale. The new chips are laid out 24 to a wafer. Yields have been running 80%, so the production manager expects to get about 19 good chips out of each wafer. She notices that a few wafers produced only 16 good chips, and she asks for your help in deciding whether this is a sign of trouble.

You look at this situation as a sequence of 24 binomial trials, with the probability of success on each trial equal to .8. You tell the production manager that the probability of getting 16 or fewer [$P(x \leq 16) = P(0) + P(1) + \cdots + P(16)$] good chips from a wafer is

$$^{24}C_0(.8)^0(.2)^{24} + {}^{24}C_1(.8)^1(.2)^{23} + \cdots + {}^{24}C_{16}(.8)^{16}(.2)^8$$

This formula is what you get by adding the binomial probabilities for 1, 2, . . ., 16 successes in 24 trials. The production manager politely says, "Yes, but what does this say about my problem?"

To answer her question, you must go further. At the very least, you must evaluate the formula. You could use paper and pencil, aided by a calculator, but this would be tedious. You could look up the answer in a table of the binomial distribution, as we do in Example 5.5. But the easiest and most powerful method would be to use computer software. The advantages of using a computer are that the calculations are done rapidly and accurately and the answers can be presented in numerical or graphic form. The catch is that you must learn how to get the answer you want from the software available. Either you must learn how to use a new program quickly or you must have ready access to one that you know.

In this book, we will use paper and pencil, tables, and software, and you should try all three methods yourself. For use in the workplace, however, the best bet is a standard software package, such as Excel or Minitab. Such a tool will lighten your load in the future, and you should feel free to use it in tackling the exercises in this book.

Returning to the production manager's questions, consider the following computer generated table of the probability that $X < x$, where X is the number of successes out of 24 independent trials when the probability of success on each trial is .8:

X=	16	17	18	19	20	21	22	23	24
P(X ≤ x)	0.0892	0.1889	0.3441	0.5401	0.736	0.8855	0.9669	0.9953	1

As you can see from this cumulative probability distribution for binomial trials, the probability of 16 or fewer successes is .0892, or about 1 chance in 11. The production manager should be concerned about the low yields she observed, but she shouldn't worry excessively, unless they occur quite frequently.

Most software will also produce graphics, and a picture is often the best way to present your results to others. You might well include a graph of this cumulative distribution, as shown on page 162, in your memo to the production manager. When we discuss quality control procedures in Chapters 8 and 9, we will develop systematic methods for answering questions such as the one raised here.

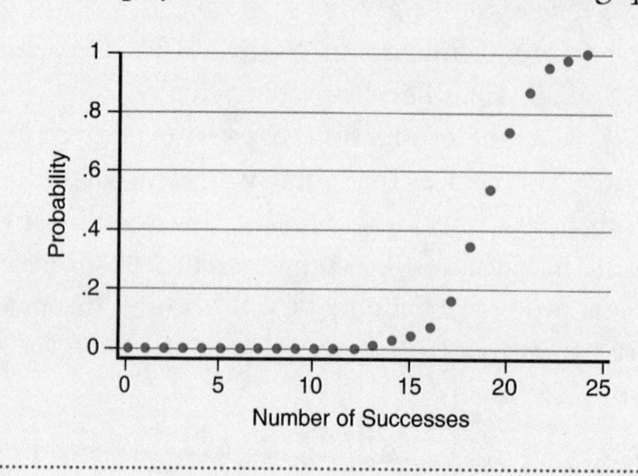

EXAMPLE 5.5 ▪ FINDING BINOMIAL PROBABILITIES FOR A NEW DRUG

When the antibiotic Zithromax™ was introduced through an advertisement in the *Journal of the American Medical Association* (December 11, 1996), Pfizer Pharmaceuticals claimed that 93% of a group of chronic bronchitis patients were cured or improved with a five-day dosing regimen of the drug. Using the table of binomial probabilities (Table II in the Appendix), let's calculate the following probabilities for a test group of 25 patients:

a. the probability that exactly 15 patients will experience a cure or improvement
b. the probability that at least 15 patients will experience a cure or improvement
c. the probability that no more than 20 patients will experience a cure or improvement

Using binomial tables

a. To find the probability that exactly 15 patients will experience a cure or improvement, we use the equation

$$P(15) = \frac{25!}{15!10!}(.93)^{15}(.07)^{10}$$

Because calculation of the factorial expression would be complicated, we turn to Table II. On the last page of the table, we find $n = 25$. Our next step is to find $x = 15$ in the x values in the second column from the left. The third step is to find the probability at the top of the table. Since .93 is not in the table, we take the closest value, .95. At the intersection of the $x = 15$ row and $p = .95$ column, we find .0000. It should be noted that this is only an approximation. Using the Excel Function Wizard, we can obtain a more precise value of .0000031089.

b. To find the probability that at least 15 patients will experience a cure or improvement, we use

$$P(X \geq 15) = P(15) + P(16) + \cdots + P(25)$$

$$= \sum_{i=15}^{25}\left(\frac{25!}{x!(n-x)!}p^x q^{n-x}\right), \quad \text{where } x \text{ varies from 15 to 25}$$

Clearly, performing this calculation would be tedious. Going to the binomial table or using software simplifies the task immensely. Again, we go to $n = 25$ and $p = .95$ (the value closest to .93). This still leaves the work of adding the values in the column from $x = 15$ down to $x = 25$:

$$.0000 + .0000 + .0001 + .0010 + \cdots + .2774 = 1.000$$

We get $P(X \geq 15) = 1.000$. (Note that in this case the values through $P(16)$ are all positive, but when rounded to four decimal places, they are shown as .0000.) Although the table is helpful, the Excel Function Wizard provides a more exact value of 0.999996542.

c. To determine the probability that no more than 20 patients have a satisfactory experience, we use

$$\begin{aligned} P(X \leq 20) &= P(0) + P(1) + \cdots + P(20) \\ &= 1 - [P(21) + \cdots + P(25)] \end{aligned}$$

From the table, the sum of the probabilities $P(21)$ through $P(25)$equals

$$.0269 + .0930 + .2305 + .3650 + .2774 = .9928$$

Subtracting .9928 from 1.00 yields .0072. Therefore, $P(X \leq 20) = .0072$.

The computer approach using Excel with $p = .95$ yields the more precise value of .007164948, rounded to nine decimal places, and it does so with far less work. Clearly, you will want to use a computer if you have to deal with problems like this often.

EXERCISES 5.11–5.22

Skill Builders

5.11 Consider a binomially distributed random variable X, with parameters $n = 7$ and $p = .4$.

a. Find $P(X = 3)$.
b. Find $P(X > 3)$.
c. Find $P(X \geq 3)$.
d. Find $P(X < 3)$.
e. Find $P(X \leq 3)$.

5.12 Suppose X is binomially distributed, with $n = 5$ and $p = .35$.

a. Find the probability that X is more than 4.
b. Find the probability that X is less than or equal to 4.
c. Find the probability that X equals 5.
d. Find the probability that X is greater than 0.

5.13 Let $n = 6$ and $p = .5$ for a binomially distributed random variable X.

a. Calculate $P(X \geq 4)$.
b. Calculate $P(X \leq 4)$.
c. Find the expected value of X.
d. Find the variance and standard deviation of X.

5.14 Let $n = 9$ and $p = .6$ for a binomially distributed random variable X.
a. Find the probability that X is at least 6.
b. Find the probability that X is at most 6.
c. Find the probability that X is greater than 6.
d. Find the probability that X is not equal to 6.

5.15 Consult the table of binomial probabilities (Table II in the Appendix) for the distribution of a random variable X with $p = .70$ and $n = 25$.
a. Find the probability that X is greater than 20.
b. Find the probability that X is less than 20.
c. Find the probability that X is more than 15.
d. Find the probability that X is at most 14.

Applying the Concepts

5.16 A growing number of companies are seeking the facilities to bring up-to-the-minute news to the workplace. *Broadcasting News* (October 31, 1994) reported that 20% of offices in the United States presently have cable television. These offices represent the potential market for CNN/Headline News–Intel Services.
a. If 10 offices at random are visited by representatives of CNN, what is the probability that exactly 5 will have cable television?
b. What is the expected number of offices with cable television per group of 10?
c. What is the likelihood that a salesperson visiting 10 offices will find that none of them have cable television?

5.17 A survey of large companies (*Working Woman*, March 1994) revealed that as many as 24% of all large companies offer their employees "corporate leaves." Consider a random sample of 30 large companies.
a. What would be the expected number offering such leave?
b. What is the probability that at most 2 companies offer such leave?
c. What is the probability that at least 2 companies offer such leave?

5.18 According to *The Great Divide: How Females and Males Really Differ* (Daniel Evan Weiss, 1991), 31% of new physicians are female.
a. What is the probability that, of 15 doctors selected at random, at least 8 are female?
b. What is the probability that no more than 4 are female?
c. What is the probability that fewer than 5 are female?

5.19 A computer consultant reports purchasing 50 monitors for a client and having to send 30 back. Use this observed percentage of defective monitors as the basis for your calculations.
a. What is the probability that, of 10 monitors purchased, more than 5 will be defective?
b. What is the probability that, if 10 monitors are purchased, fewer than 2 will be defective?
c. What is the expected number of defectives in a batch of 30?
d. Compute the standard deviation of the number of defectives in a batch of 30.

e. Based on the empirical rule, what range of numbers of defective monitors would one expect to find approximately 68% of the time in a batch of 30?

5.20 The record for highest career batting average is held by Ty Cobb, who had a career batting average of 366 (Les Krantz, *The Best and Worst of Everything*, 1991). Suppose he came up to bat six times during a game.

a. Compute the probability that Cobb would get six hits in six at-bats.
b. What is the probability that Cobb would be hitless in six at-bats?
c. What is the probability that Cobb would get three hits in six at-bats?

5.21 In the dice game called craps, a player wins on the first throw if he or she gets a 7 or an 11, called a natural. The probability of throwing a natural is 8/36, or about .222.

a. What is the expected value for the number of naturals in a series of 20 throws?
b. What is the probability of getting 5 naturals in 20 throws?
c. What is the probability of getting 5 or fewer naturals in 20 throws?

5.22 In the dice game of craps a player loses on the first throw if he or she gets a 2, 3, or 12. The probability of throwing one of these three numbers is 5/36, or about .1388.

a. What is the expected value for the number of games lost on the first throw in a series of 20 games?
b. What is the probability of losing 3 games out of a series of 20 on the first throw?
c. What is the probability of losing 5 or fewer games out of a series of 20 on the first throw?

SECTION 3 ▪ THE POISSON PROBABILITY DISTRIBUTION[2]

While the binomial distribution deals with the number of successes and failures in a fixed number of trials, the Poisson distribution describes the number of events that may be expected to occur in an interval of time or within a given area—for example, the number of earthquakes of magnitude 6 or greater expected to strike San Francisco per decade or the number of meteorites expected to strike the United States in a given century. Thus, both the binomial and the Poisson distribution deal with counts, but in very different ways.

In this section, we will explore the attributes of the Poisson distribution. Pioneering uses of the Poisson distribution were made in business and industry by the Swedish actuary Lundberg, in 1903, and the Danish engineer Erlang, in 1909. Erlang was a director of the telephone company in Copenhagen at the time.

Suppose an Internet service provider knows that, on an average weekday afternoon, 60 customers per hour will call in for a connection. How probable is it that 10 or more customers will call within a 5-minute period and exceed the supplier's service capacity? Users may call in at any instant of the 5-minute interval. What is constant in this process is the average number of calls per unit time over a reasonably long stretch of time.

[2]This is an optional section.

Rush hour does not occur only on streets and highways. In the sky around La Guardia Airport at 5 P.M., you may see a plane coming in for a landing every 30 seconds. What is the probability that three planes will enter the landing queue in a single minute? If this probability is high enough, the Port Authority might need to upgrade its traffic-handling ability. If the probability is low, its present facilities are sufficient to handle the existing air traffic.

A baker mixes chocolate chips into cookie batter, with enough chips to average six per cookie. What is the probability that a chocolate chip cookie will end up chipless? Should the baker increase the number of chips in the batter? Can the baker reduce the number of chips and still have chocolate chip cookies?

In all these examples, the random variables are counts (see Chapter 1). A model commonly used for such random variables is the Poisson distribution. In contrast to the binomial distribution, which deals with the probability of x successes out of n trials, the Poisson distribution deals with the probability of x successes within a certain period of time or x successes within a certain physical area or space. The average over a large number of equal intervals is known, but the counts fluctuate from interval to interval. In the air traffic example, these are intervals of time, whereas in the baking example, they are intervals of space (that is, area or volume).

> The **Poisson distribution** is the probability distribution of a random variable *X* representing the number of random, independent occurrences over time or space.

Poisson Probability

$$P(X = x) = P(x) = \frac{\lambda^x}{x!} e^{-\lambda}$$

where x is the observed number of occurrences.

Variance and Standard Deviation (Poisson Distribution)

$$\text{Variance, } V(X) = \lambda$$
$$\text{Standard deviation, } S(X) = \sqrt{\lambda}$$

The Poisson distribution for the Internet service provider appears in Figures 5.3 and 5.4. Figure 5.3 is a graph of the cumulative density function, or CDF. Figure 5.4 shows the probability density function, or PDF. Recall that this distribution has a mean of 5; that is, $\lambda = 5$.

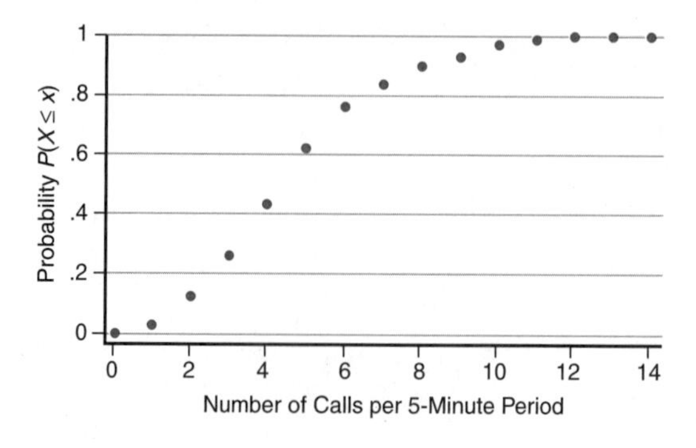

Figure 5.3
The cumulative distribution function for the Internet provider, with $\lambda = 5$

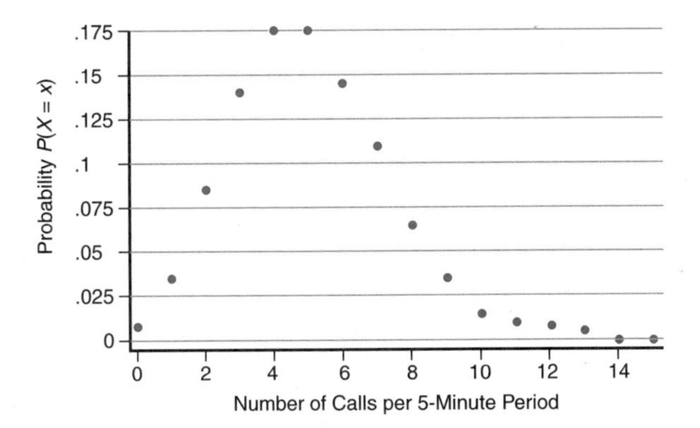

Figure 5.4
The probability density function for the Internet provider, with $\lambda = 5$

To answer the question "How probable is it that 10 or more customers will call within a 5-minute period and exceed the supplier's service capacity?" we look at Figure 5.4. We locate 10 calls per minute at the bottom of the graph and examine the probability density function to that point. The probability of $X \leq 10$ is $1 - .96817$, or about 3.2%. Since the probability that this event will occur is relatively low, the ISP

Figure 5.5
A change in probabilities wreaks havoc

Source: CyberTimes (The New York Times on the Web), January 12, 1997. Reprinted with permission by Associated Press.

Access Providers Rethinking Flat-Rate Pricing for Internet

by Elizabeth Weise

San Francisco (AP)—Time is running out for Internet users who meander for hours through cyberspace without worrying about the bill. All-you-can-eat pricing plans allow online lingering that leaves others with long waits for connections. For some, particularly businesses that rely on e-mail, the tie-ups are becoming intolerable. In some corners of the Internet flat rates have already come and gone. Instead, Internet access companies are charging more, even by the hour, promising reliable connections.

Netcom On-line Communications Service Inc. of San Jose, Calif., a pioneer of the flat-rate price, plans to announce a return to hourly rates next month. Others, while still allowing unlimited usage, are charging double the going rate to keep lines open.....

America Online, with 6.5 million members, just a month ago started offering a flat-rate, a step some blame for the current logjam. The company says the pricing has been popular with its customers and that the company is upgrading its systems to handle the crush....

Patsy Northcutt, who runs Northcutt Productions, a video and multimedia production company in Sausalito, Calif., likes using America Online for her business. But she says she has gotten so frustrated with delays that she is planning to open a second account with an Internet-only provider.

"Sometimes I'll actually go into the set and change the number I'm dialing in to get a better line. I can always get on, but it can take three or four tries," she said. "So far it's been inconvenient but never disastrous—but at the wrong time it could be horrible."

Netcom is looking to convert business owners like Northcutt. "We want to service small to mid-sized business users who can't afford not to get their e-mail, who can't afford not to get on the Net—and they're willing to pay a premium price to ensure access to their accounts," said a Netcom spokesman, Curtis Kundred.

Zilker Internet Park, an Austin, Tex.–based Internet service provider, offers its customers a flat-rate monthly fee—but at $39.95 instead of $19.95. Expensive, yes, but at that price the company can ensure that customers get more than a busy signal. "It's similar to those specialty airlines that only have first class seats," said Zlker's president Smoot Carl-Mitchell. "You'll always have a few who will pay more for leg room."

Flat-rate can easily be a money-loser for online companies. At a cost of between 90 cents and $1.80 per hour to connect a user, Internet providers who charge no more than $19.95 per month start losing money after as little as 11 hours, said Eric Paulak, an analyst at Gartner Group. Paulak said the average user is online about 16 to 18 hours per month.

AOL and MSN add to their revenue with advertising, merchandising and transaction-handling fees. For example, in fiscal 1996 AOL made $1.1 billion, 91 percent of which came from online service and 9 percent from ads and transactions. For fiscal 1997 the company has projected 85 percent of its earnings coming from the online service and 15 percent from secondary sources.

AOL subscribers are bearing the worst of the crush. Busy signals and long waits may be the price customers pay for cheap Net access.

"Like anything that's a great value, consumers are flocking to it," said David Gang, AOL's vice-president of product marketing. "Sometimes that means they have to wait a bit, but it's worth it. Think about Disney. People will wait in line for hours to get into Disneyland."

decides not to expand. Yet, as Steve Case of America OnLine found out, a slight change in marketing conditions can change all that, and havoc can ensue. (See "Access Providers Rethinking Flat-Rate Pricing for Internet" in Figure 5.5 on page 167.)

EXAMPLE 5.6 ▪ STUDYING AIRPORT ARRIVAL TIMES

If planes arrive at La Guardia Airport at an average rate of 1 arrival per 30 seconds, what is the probability that no planes will arrive during a 60-second interval? 1 plane? 3 planes? We will assume that the average rate of 1 arrival per 30 seconds holds over a 15- to 30-minute period. Under this assumption, the mean number of arrivals during a 60-second interval is $\lambda = 60/30 = 2$. Therefore,

$$P(X = 0) = \frac{2^0}{0!} e^{-2} = e^{-2} = .1353$$

Poisson probabilities

The probability of at least 1 arrival during 60 seconds is given by

$$P(X \geq 1) = 1 - P(0) = 1 - .1353 = .8647$$

and the probability of exactly 3 arrivals during that time is

$$P(X = 3) = \frac{2^3}{3!} e^{-2} = .1804$$

Because calculations with the Poisson distribution may be cumbersome, tables have been constructed to make such calculations easier. Minitab, Excel, and other software packages will also calculate Poisson probabilities.

EXAMPLE 5.7 ▪ SEEING WHERE THE CHIPS MAY FALL

Returning to our chocolate chip cookies, let's suppose we have an average of 3 chips per cookie.

a. What is the probability that there are no chips in a cookie?
b. What is the probability that there are no more than 3 chips in a cookie?
c. What is the probability that there are at least 3 chips in a cookie?
d. What is the standard deviation of this distribution?

Using Poisson tables

a. To calculate the probability that there are no chips in a cookie, we may use the formula for the Poisson distribution or we may turn to the table of Poisson probabilities (Table III in the Appendix). In the table of Poisson probabilities, we find the value $\lambda = 3.0$, which is followed by a column of probabilities. On the left side of the same page, we observe that x goes from 0 to 12. To find the probability that the random variable X is 0, we look up $P(0) = .0498$ in the table. The probability that there are no chips in a cookie is nearly .05.
b. In part a, calculating the answer would have been just as efficient as looking it up in the table. In answering the second question, however, use of the table yields significant economies. No more than 3 chips in a cookie translates to 0, 1, 2, or 3 chips in a cookie. Finding the probability would require calculating $P(0)$, $P(1)$, $P(2)$, and $P(3)$. It is much easier to simply look down the $\lambda = 3.0$ column and add the probabilities.

$$P(X \leq 3.0) = P(0) + P(1) + P(2) + P(3)$$
$$= .0498 + .1494 + .2240 + .2240$$
$$= .6472$$

c. If using the table made life easier in part b, it is even more useful in finding the probability that a cookie has at least 3 chips. "At least 3 chips" includes 3, 4, 5, 6, ..., etc. To calculate this value, we would begin with $P(3)$, then go on to $P(4)$, and continue until we calculated a probability of .0000. Using the table, we just add the results, beginning with $P(3)$:

$$P(X \geqslant 3) = P(3) + P(4) + \cdots + P(12)$$
$$= .2240 + .1680 + \cdots + .0001$$
$$= .5768$$

Looking at it another way, we have

$$P(X \geq 3) = 1 - P(X \leq 2) = 1 - (.0498 + .1494 + .2240) = .5768$$

The empirical rule does not apply to Poisson distributions

d. To find the standard deviation of the distribution, we simply take the square root of the mean, λ. As $\lambda = 3.0$, the standard deviation is approximately 1.73. In this case, the standard deviation is simply a measure of variation. We should not apply the empirical rule to these data because they depart sharply from the normal distribution.

EXERCISES 5.23–5.32

Skill Builders

5.23 Consider a Poisson distribution with a mean of 1.3.
a. Find $P(X = 3)$.
b. Find $P(X \leq 3)$.
c. Find $P(X \geq 3)$.
d. Find $P(X > 3)$.
e. What is the variance of this distribution?

5.24 Consider a Poisson distribution with $\lambda = 2.0$.
a. Find the probability that X equals 3.
b. Find the probability that X is less than 3.
c. Find the probability that X is at most 3.
d. Find the probability that X is greater than 3.
e. Find the standard deviation of this distribution.

5.25 Let $\lambda = .7$ for a Poisson distribution.
a. Calculate $P(X = 1)$.
b. Calculate $P(X \geq 1)$.
c. Calculate $P(X < 2)$.
d. Calculate $P(X > 2)$.
e. What is the variance of this distribution?

5.26 Suppose $\lambda = 2.2$ for a Poisson distribution.
a. Find the probability that X is greater than 1.

b. Find the probability that X is at most 1.
c. Find the probability that X is at least 1.
d. What is the probability that X is less than 2?
e. What is the standard deviation of this distribution?

Applying the Concepts

5.27 The Third Engine Company of Phoenix responds to an average of 2.1 emergency calls per hour (Les Krantz, *The Best and Worst of Everything,* 1991). Using this average, answer the following questions:
a. What is the probability that the engine company receives 0 calls in a given hour?
b. What is the probability that the engine company receives at most 1 call in a given hour?
c. What is the probability that the engine company receives more than 3 calls in a given hour?

5.28 It has been calculated that an average of 3.3 motor vehicle thefts occur every minute in the United States (*Places Rated Almanac,* 1993). Using this number, estimate the following:
a. What is the likelihood that exactly 2 motor vehicle thefts occur in a given minute?
b. Calculate the probability of fewer than 3 motor vehicle thefts in a given minute.
c. What is the probability that more than 6 motor vehicle thefts occur in a given minute?

5.29 Phone calls come into a switchboard at the rate of 5 per minute. One operator can handle 3 per minute. There are two operators at the switchboard. To evaluate the need for a third operator, it is necessary to estimate the probability of having more calls arrive than can be handled simultaneously by two operators. Draw the PDF and CDF for this distribution. Use your results to answer the following:
a. What is the probability of at least 6 calls during one minute?
b. What is the probability of between 7 and 9 calls during one minute?
c. What is the probability of more than 9 calls during one minute?
d. Find the probability of fewer than 3 calls during one minute.
e. What is the standard deviation of the number of calls coming into the switchboard during one minute?

5.30 A recent survey of traffic on the Garden State Parkway revealed that, on average, 2.7 cars per minute go through a toll plaza lane.
a. Calculate the probability that more than 3 cars will pass through the toll plaza lane in a given minute.
b. What is the probability that no more than 1 car will pass through in a given minute?
c. What is the probability that no cars will pass through during a 2-minute interval?

5.31 During World War II, London was bombarded by V1 flying bombs. The British studied the number of hits per area for 576 areas of 25 hectares each. There were 537 hits, or .9323 per area, on average. The pattern was as follows:

Number of Hits	Number of Areas
0	229
1	211
2	93
3	35
4	7
≥5	1

Source of Data: R. D. Clarke, *The 1946 J. Actuaries.*

a. Use these data to make a table for $P(k)$, the probability of an area's getting k hits.

b. For these same k's, make a table for the Poisson possibilities $P(k)$ when $\lambda = .9323$.

c. Compare the probabilities found in parts a and b, and decide whether the data fit the Poisson model with $\lambda = .9323$.

5.32 You are in charge of communications for a company that must choose between two levels of service for a data line between its offices. One level is fairly noisy, with a probability of 1/10 of introducing one error during transmission of a single letter or number. The other level is less noisy, with a probability of 1/100 of introducing a single error. Your company uses a coding system that will catch and correct single errors in sending a letter or number, so the text will be complete unless two or more errors occur while a single character is being sent.

a. Using the Poisson distribution, find the probability of a missent letter or number on the noisy line ($\lambda = .1$).

b. Find the same probability on the better quality line ($\lambda = .01$).

c. Suppose that the monthly cost for leasing the higher quality line is five times that for the lower quality line and that the line chosen will be used mostly for informal memos. Which line would you recommend choosing and why? What suggestions might you make for handling critical transmissions?

SECTION 4 ▪ THE NORMAL PROBABILITY DISTRIBUTION

While the binomial and Poisson probability distribution describe count or discrete data, the **normal distribution** describes random variables that are continuous, such as weights, measures, monetary amounts, and time. The normal distribution is of particular interest to statisticians because many populations and their data sets follow its bell-shaped pattern. The normal probability distribution model is widely applied in the inferential methods used in the remainder of this text. It is, for example, essential for understanding the sampling models in Chapter 6 and the methods of estimation and hypothesis testing of sampled data in Chapters 7 and 8. The empirical rule for analyzing distribution models described in Chapter 3 is derived from the normal probability distribution model.

Normal Probability

Normal probabilities

$$Y = \frac{1}{\sqrt{2\pi}\sigma_X} e^{-\frac{1}{2}\left(\frac{x-\mu_X}{\sigma_X}\right)^2}$$

where Y is the height of the curve, π is approximately 3.14159, σ_X is the standard deviation of the distribution, μ_X is the mean of the distribution, and e is the base of the natural logarithm system. The normal curve is bell-shaped, as shown in Figure 5.6.

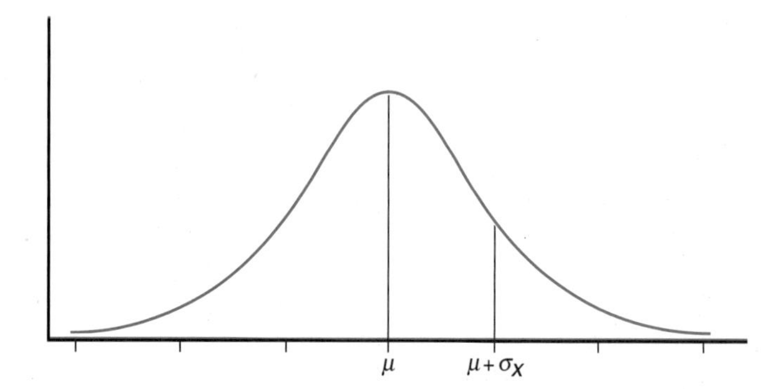

Figure 5.6
Normal curve

Like the binomial distribution, the normal distribution is determined by two parameters: the mean (μ) and the standard deviation (σ). The mean and the standard deviation for a normal curve can be identified geometrically. The mean lies on the axis directly below the peak of the curve. The normal curve is symmetrical around the mean—that is, one side may be folded exactly over the other. The points $\mu + \sigma$ and $\mu - \sigma$ lie on the axis below the "knees" of the curve. These are the points where the curve changes its concavity—that is, from arching upward to arching downward or vice versa (see Figure 5.6).

Back in the 1840s, Adolphe Quetelet made extensive studies of measurements, and he found that they commonly assume a bell-shaped pattern. Figure 5.7 shows the distribution of chest measurements, in inches, for 5738 Scottish soldiers. The mean measurement is 39.83 inches, and the standard deviation is 2.04 inches. A normal curve at the same scale and with the same mean and standard deviation has been superimposed to show how close the fit is. The data are from Quetelet's 1846 book, *Lettres à S.A.R. le Duc Régnant de Saxe-Cobourg et Gotha.*

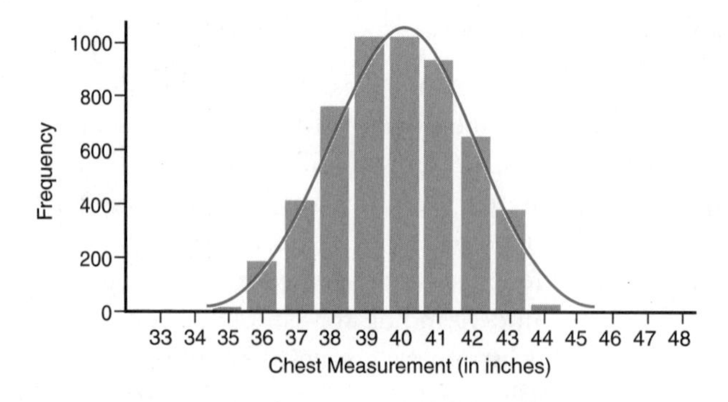

Figure 5.7
Chest measurements for 5738 Scottish soldiers

Now let's consider a more recent illustration of the normal distribution: the distribution of times to completion for the 30,000 runners in the 1994 New York City

Marathon. Just as data can be continuous or discrete (see Chapter 1), random variables can be either continuous or discrete. Because finishing times can vary continuously within a given range, they constitute an example of a continuous random variable.

> A **continuous random variable** may take on any value within some given range. Such a variable usually represents measurements on a continuous scale, with no gaps or interruptions.

Of the 30,000 marathon runners, we want to focus on those who finished the race within the first 8 hours 40 minutes. The average time to completion for those runners was 4 hours 22 minutes. The standard deviation was approximately 51 minutes. Figure 5.8 contains an illustration of the distribution. The curve is a smooth approximation; dots show the original data.

Figure 5.8
The 1994 New York Marathon: times to completion

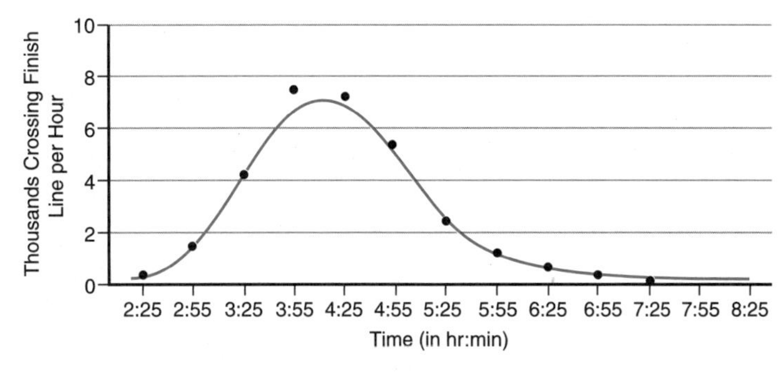

Source of Data: *New York Times,* November 7, 1994.

According to the empirical rule (introduced in Chapter 3), if the data follow a normal distribution, approximately 68.26% of the data will fall between the mean and ±1 standard deviation. In this instance, that would be between 3 hours 31 minutes and 5 hours 13 minutes. The original data for the marathon indicate that a total of 21,643 times (approximately 72% of the data) fell within this interval. About 95.44% of the data are expected to fall within 2 standard deviations of the mean—that is, between 2 hours 40 minutes and 6 hours 4 minutes. Actually, 95.99% of the data fell in this interval. Lastly, when data follow a normal distribution, it is expected that 99.73% of the data will fall within 3 standard deviations of the mean. See Table 5.3 for a comparison of actual percentages and those predicted by the empirical rule.

Table 5.3

COMPARISON OF DATA TO PREDICTIONS FOR MARATHON COMPLETION TIMES

Range	Actual Number	Actual Percentage	Percentage Predicted by Empirical Rule
Within 1 standard deviation			
3 hr 31 min–5 hr 13 min	21,643	72.14%	68.26%
Within 2 standard deviations			
2 hr 40 min–6 hr 4 min	28,797	95.99	95.44
Within 3 standard deviations			
1 hr 49 min–6 hr 55 min	29,916	99.58	99.73

Do the marathon data fit the empirical rule? The mean less 3 standard deviations is 1 hour 49 minutes. Did anyone run the marathon in a faster time? Not yet! German Silva, a 26-year-old man, made it in a record 2 hours 11 minutes. Did anyone take longer than 3 standard deviations beyond the mean? You bet! One determined individual took 27 hours, but she did it! Of the 30,000 runners in the marathon, 99.58% came within 3 standard deviations of the mean. The point is that the normal curve may be useful for describing large data sets even if the distribution is not precisely normal. Although the long right tail of this distribution (see Figure 5.8) shows that it is not really normal, the empirical rule gives a good prediction of the finishing times of the vast majority of runners. Many individual elements cluster around the mean, while fewer cluster away from the mean.

The predicted percentages of runners in different time slots in Table 5.3 are based on the probability that a normal random variable X will fall within a certain range. This probability is proportional to the area beneath the normal curve that lies within the range in question. Tables or computer programs can be used to find this probability. These tables are based on the concept of a ***z* score**. The z score is a new random variable calculated from values of X and its mean (μ) and standard deviation (σ).

z Score

For a normal random variable X with a mean μ and a standard deviation σ, the corresponding z score is defined by

$$Z = \frac{X - \mu}{\sigma}$$

Figures 5.9 and 5.10 illustrate the distributions of normal random variables and their z scores.

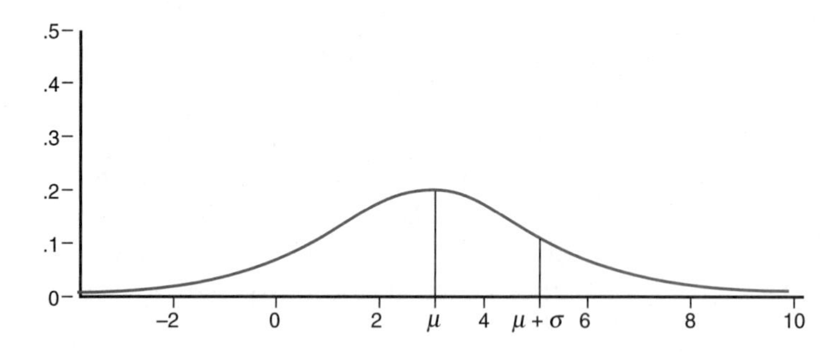

Figure 5.9
Distribution of a normal variable X with a mean of $\mu = 3$ and standard deviation of $\sigma = 2$

The z score for X is given by

$$Z = \frac{X - \mu}{\sigma}$$

Its distribution is shown in Figure 5.10.

Z has three very convenient properties:

1. It is a normal random variable.
2. It has a mean of 0.
3. It has a standard deviation of 1.

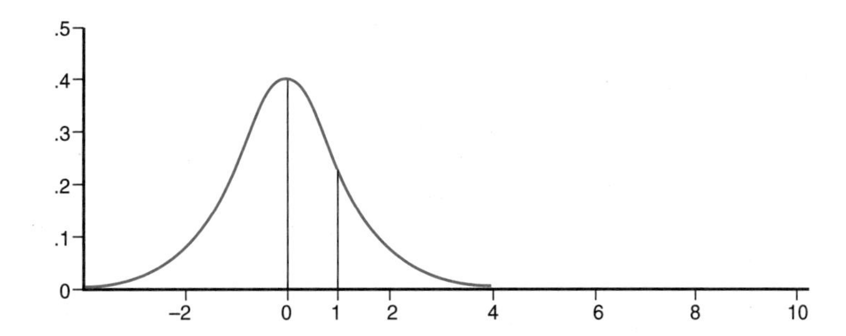

Figure 5.10
Distribution of the z score for X

In sum, Z is a **standard normal variable.** Mathematicians sometimes summarize this fact in the following manner:

$$Z \sim N(0, 1)$$

Given z scores, we can look up the corresponding area beneath the standard normal curve in Table IV in the Appendix. Figures 5.11 and 5.12 illustrate how this works. First consider Figure 5.11. The probability that the values of a normal random variable X with a mean of $\mu = 3$ and standard deviation of $\sigma = 2$ will lie between 3 and 7 is proportional to the area shown in color in Figure 5.11.

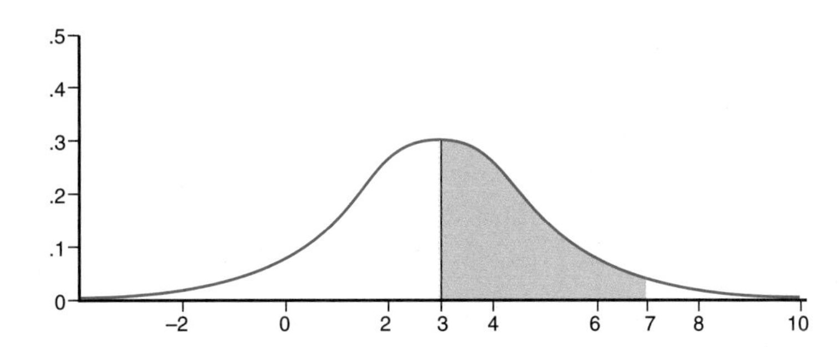

Figure 5.11
Probability and area for a normal random variable

Now turn to Figure 5.12. The probability that $3 \leq X \leq 7$ is the same as the probability that X's z score, given by $Z = (X - \mu)/\sigma$, lies between the corresponding z values of $(3 - 3)/2 = 0$ and $(7 - 3)/2 = 2$, in the area shown in color in Figure 5.12. This probability can be found from a standard normal table, such as Table IV in the Appendix. Table IV gives the area below the curve, bounded on the left by 0 and on the right by the z value you look under. In this case, this area is exactly the area we are looking for.

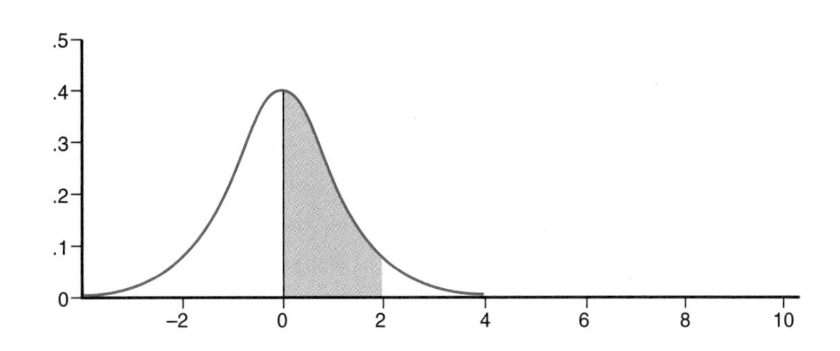

Figure 5.12
Probability and area for a z score

Examples 5.8–5.12 show how to work with z scores. In these examples, we will calculate area beneath the normal curve as the boundary of the area moves from right to left.

EXAMPLE 5.8 ▪ FINDING THE PERCENTAGE OF RUNNERS WHO COMPLETED THE MARATHON IN 3 HOURS OR LESS

We will use the normal distribution model to analyze the distribution of completion times of the marathon runners. The model does not fit exactly, but it can be a useful approximation *as long as we are clear about our assumptions.* The mean running time was 4 hours 22 minutes, and the standard deviation was 51 minutes. To find the percentage of runners who completed the marathon in less than 3 hours (less than 180 minutes), we follow a four-step procedure.

Step 1. We draw a picture of the situation in Figure 5.13.

Figure 5.13
Times under 3 hours

Normal probability: left extreme

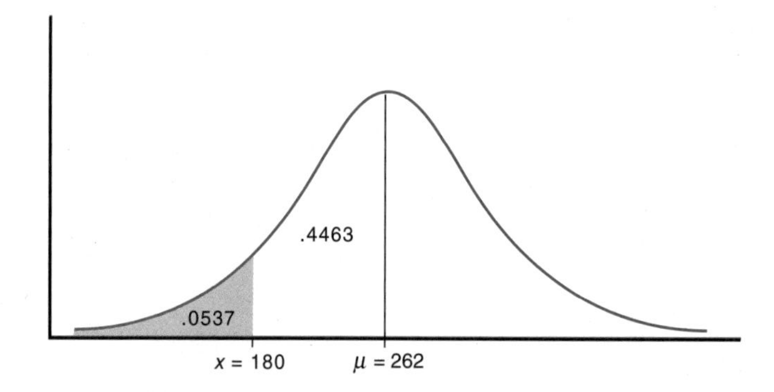

Step 2. We convert 180 minutes into a z score:

$$Z(180) = \frac{180 - 262}{51} = -1.61$$

Step 3. We look up the z score in Table IV in the Appendix. Disregarding the negative sign, we look for the row that has 1.6 on the left side. At the intersection of that row with the column headed .01 (as $1.6 + .01 = 1.61$), we read the value .4463. This entry gives us the proportion of the area under the curve that falls between the mean and the point in question. (The negative sign in the original z score means that the area falls on the left side of the distribution.)

Step 4. To find the desired area, we subtract .4463 from .5000, which is the total area beneath the curve to the left of μ. We may conclude that only .0537, or 5.37%, of the runners finished in 3 hours or less.

EXAMPLE 5.9 ■ FINDING THE PERCENTAGE OF RUNNERS WITH TIMES BETWEEN 3½ HOURS AND THE MEAN

To find the percentage of runners with completion times between 3½ hours and the mean of 4 hours 22 minutes, we again follow a four-step process.

Step 1. We draw a picture of the situation in Figure 5.14.

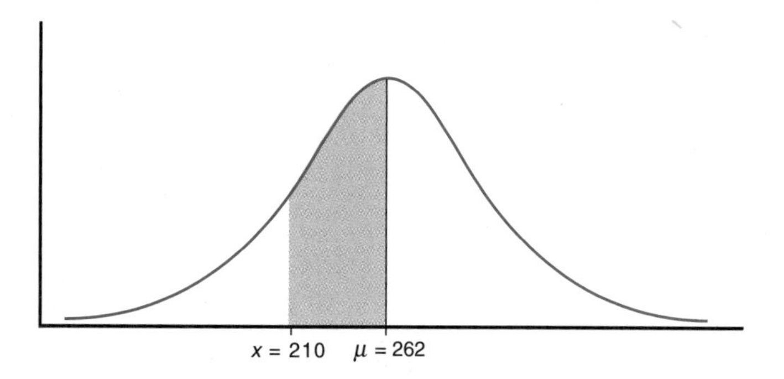

Figure 5.14
Times between 3½ hours and the mean

Normal probability: just left of mean

Step 2. We convert 3½ hours into a z score, remembering to convert the time to minutes:

$$Z(210) = \frac{210 - 262}{51} = -1.02$$

Step 3. We look up -1.02 in the z table, disregarding the sign. (A negative sign simply means that the area in question falls to the left of the mean.) At the intersection of the row for 1.0 and the column headed .02, the area is .3461.

Step 4. Since .3461 corresponds exactly to the area we seek, we may conclude that 34.61% of the marathon completion times were between 3½ hours and the average time of 4 hours 22 minutes.

EXAMPLE 5.10 ■ FINDING THE PERCENTAGE OF RUNNERS WITH TIMES BETWEEN 4 AND 5 HOURS

Let's now find the percentage of runners with times between 4 and 5 hours.

Step 1. We draw a picture of the situation in Figure 5.15.

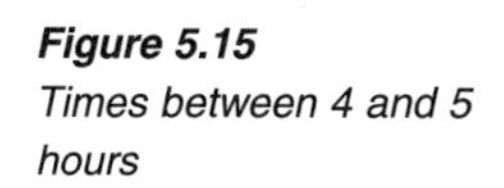

Figure 5.15
Times between 4 and 5 hours

Normal probability: area straddling mean

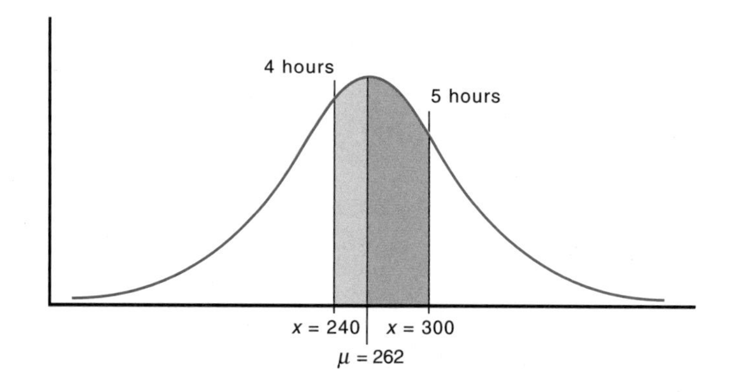

Step 2. We compute a z score for the area on *each* side of the mean:

$$Z(240) = \frac{240 - 262}{51} = -.43$$

$$Z(300) = \frac{300 - 262}{51} = .75$$

Step. 3 We look up the z scores in Table IV. A z score of .43 corresponds to an area of .1664, and a z score of .75 corresponds to an area of .2734.

Step 4. We add the two areas together to compute the probability of a completion time between 4 and 5 hours. The conclusion is that 43.98% of the runners will finish in between 4 and 5 hours.

EXAMPLE 5.11 ▪ FINDING THE PERCENTAGE OF RUNNERS WITH TIMES BETWEEN 5 AND 6 HOURS

Finding the percentage of runners with times between 5 and 6 hours requires a process similar to the one followed in Example 5.10.

Step 1. We draw a picture of the situation in Figure 5.16.

Figure 5.16
Times between 5 and 6 hours

Normal probability: a slice

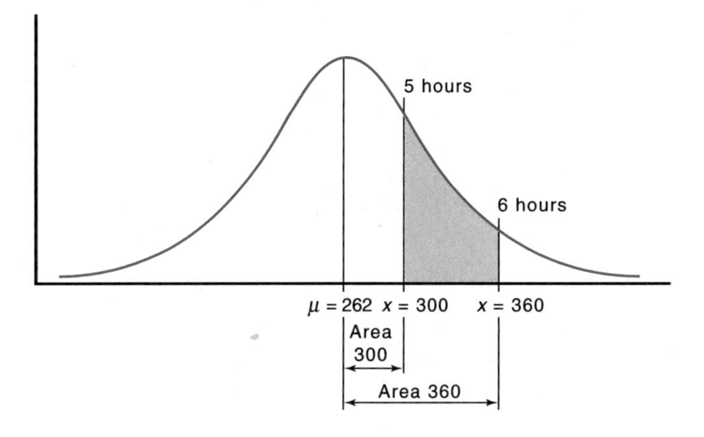

Step 2. We compute the z scores for 5 hours (300 minutes) and 6 hours (360 minutes):

$$Z(300) = \frac{300 - 262}{51} = .75$$

$$Z(360) = \frac{360 - 262}{51} = 1.92$$

Step 3. We find the areas corresponding to $Z(300)$ and $Z(360)$. $Z(300)$ corresponds to an area of .2734, and $Z(360)$ corresponds to an area of .4726.

Step 4. Because the area for $Z(360)$ includes the area for $Z(300)$, we must subtract the smaller area from the larger to arrive at the area of the shaded slice. The difference in the areas is .1992. The percentage of runners with marathon completion times between 5 and 6 hours is 19.92%.

STATISTICS TOOL KIT

Using Probability Models

The probability calculations discussed in this chapter are a means to better describing of various situations and reaching decisions. For example, when we looked at the 1994 New York Marathon data, we decided that it looked approximately normal. So the mean (4 hours 22 minutes) and standard deviation (51 minutes) give us a reasonable summary of the data. Suppose your neighbor, who just took up running, mentions that she ran in a race the weekend of the marathon, finishing in 2 hours 33 minutes. Because you have been working on the marathon example, you can't help thinking that this would be a good finishing time for the marathon: 1 hour 49 minutes under the mean, giving a z score of about -2.14.

The z score gives the number of standard deviations between an observation and the mean, if the data are normally distributed, with a known mean and standard deviation. The table of normal probabilities allows us to go from z scores to the probability that a random observation will be as far from the mean as the one we observed. This correspondence between z scores and probabilities is shown in the figure below. The top ruler gives the probability that a normal random variable X yields a z score between 0 and z. The lower ruler gives the z score, z.

Probability

0 .1 .2 .3 .4 .49 .495 .499

0 1 2 3

z score, z

The relationship between z scores and probabilities reflected by these scales is the same one tabulated in the back of this book. For example, from the rulers, the probability that a normally distributed X gives a z score between 0 and 1 is just a bit larger than .34. The table provides the more precise value of .3413. Although the rulers provide less detail, they make the idea of the correspondence clear.

You calculated that your neighbor's running time corresponds to a z score of about -2.14 for New York Marathon times. The probability of a result this extreme is the same as that of getting a z score of 2.14 or above, something we will now explore, using the rulers above.

From the ruler, the probability of a normal variable's falling between the mean and the z score of 2.14 is about .4838, and the probability of its falling below the mean is .5. So the probability of having a z score of 2.14 or less is $.5 + .4838 = .9938$. Thus, the probability of achieving a z score of 2.14 or above is $1 - .9938 = .0062$, or a bit less than 1 chance in 150.

Here is where statistical reasoning comes into play. You should always question a hypothesis that makes the observed data seem highly unlikely. Your neighbor's finishing time is improbable for a randomly chosen finisher in the New York Marathon. As you will see in Chapters 8 and 9 on hypothesis testing, this is reason to reexamine and possibly reject assumptions—in this case, the assumption that your neighbor ran a marathon. You might ask your neighbor what race she ran in.

The neighbor's race turns out to have been a half-marathon on Long Island. Her time puts her about in the middle of the pack, with a z score that raises no troubling questions. But you couldn't have resolved the question her time raised without additional information; she might simply have been a very fast runner in the New York Marathon. Uncertainty is an inescapable feature of statistical thinking; be careful that it doesn't lead you astray.

The important thing to remember about the distributions and equations described in this chapter is how they can be used to model various real-life situations. In practice, the calculations are often done by a computer which determines the probability of an observation's being as extreme as the one you have observed, given the model that you have assumed. This process will be discussed further in Chapters 8 and 9. What is important now is that you understand the principle behind this type of probability and its implications for the problem you are working on.

We have calculated the probabilities that runners will fall within various ranges. In Example 5.12, we confront a different type of situation—determining the maximum time for the fastest 5% of the runners. The solution to this type of problem has implications for the areas of statistical estimation (Chapter 7) and hypothesis testing (Chapter 8).

EXAMPLE 5.12 ▪ FINDING THE MAXIMUM TIME FOR THE FASTEST 5% OF RUNNERS

To find the maximum time for the fastest 5% of runners, we basically reverse the procedure used in Examples 5.8–5.11.

Step 1. We draw a picture of the situation in Figure 5.17, shading the area of interest.

Figure 5.17
Times of fastest 5%

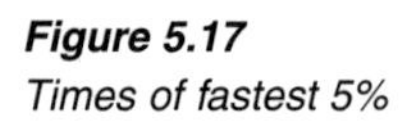

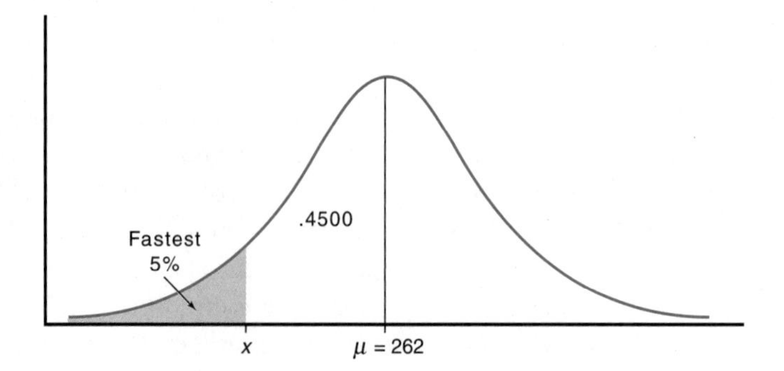

Step 2. In the body of Table IV, we look up .4500, the area between the cutoff and the mean. We find that .4500 is between .4495 and .4505 (which

correspond to 1.64 and 1.65, respectively). Splitting the difference, we let $z = -1.645$. Note that z is negative.

Step 3. We solve for x:

$$-1.645 = \frac{x - 262}{51}$$

$$x = 178 \text{ minutes}$$

Step 4. We state the conclusion: The fastest 5% of the runners made the 26.2-mile run in less than 2 hours 58 minutes.

EXERCISES 5.33–5.42

Skill Builders

5.33 Consider the following z scores:

1.18	−2.03	.47	−.99	3.09

a. Compute the area between each z score and the mean of the curve.
b. What is the area to the right of each z score?
c. What is the area to the left of each z score?

5.34 Find the indicated probabilities for a standard normal random variable Z.
a. $P(Z \leq 1.00)$
b. $P(-1.00 \leq Z \leq 1.00)$
c. $P(Z \geq 1.64)$
d. $P(.18 \leq Z \leq .23)$
e. $P(-1.94 \leq Z \leq 2.05)$

5.35 Assume a normally distributed random variable X with a mean of 100 and a standard deviation of 25.
a. Find the z score corresponding to a value of 125 for X.
b. What is the area under the distribution curve for X to the right of $X = 125$? What is the area to the left of $X = 125$?
c. Find the z score corresponding to a value of 87.
d. What is the area between 87 and the mean of the distribution for X?
e. Calculate the probability that X will fall between 87 and 125.

5.36 Suppose a normally distributed random variable X has a mean of 357 and a standard deviation of 43.
a. Find the cutoff for X for the lowest 10% of the distribution.
b. Find the range for X for the middle 50% of the distribution.
c. Above what value of X does 14% of the distribution fall?
d. What is the probability that X falls between 368 and 379?
e. What is the probability that X falls below 257?

5.37 Let $\mu = 22$ and $\sigma = 3$ for a normally distributed random variable X.
a. Compute the probability that X exceeds 25.
b. What is the probability that X falls between 19 and 27?

c. Find the likelihood that X falls between 23 and 29.
d. What is the cutoff for X for the highest 10% of the values in this distribution?
e. Find the range for X for the central 95% of the distribution.

Applying the Concepts

5.38 A "Car Buyer's Guide" (*Money,* March 1995) listed over 300 models of cars, with prices ranging from under $10,000 to nearly $92,000. The average suggested retail price for these cars was $23,454. The standard deviation was $13,875. Assume the data follow a normal probability distribution.
a. What is the proportion of models costing less than $15,000?
b. What proportion of the models cost between $18,000 and the average price?
c. What proportion of the models cost between $20,000 and $30,000?
d. What proportion of the models cost at least $25,000 but less than $35,000?
e. Above what price are the most expensive 10% of the cars?

5.39 A study of lease rates for a selection of 1995 cars revealed that the average monthly rate for a vehicle was $220.67. The standard deviation was $59.63. Assume the rates follow a normal distribution.
a. What is the probability that a random car will lease for less than $150?
b. Find the probability that a random car will lease for more than $350.
c. Calculate the range of monthly lease costs for the central 50% of the distribution.
d. What is the cutoff cost for the most expensive 5% of the leases?
e. What percentage of leases fall between $200 and $300 per month?

5.40 An important measure of the performance of CD-ROM drives is access time. *PC Magazine* (March 28, 1995) reported "seek-and-read" times for a total of 30 drives. (Seek-and-read time is the time it takes a CD-ROM head to position itself to read the data you seek.) The average time for the sample of quadruple-, triple-, and double-speed drives was 430 milliseconds (ms). The standard deviation was 142 milliseconds. Assume the data follow a normal distribution.
a. What is the probability that a randomly chosen drive takes over 500 ms to seek and read?
b. Calculate the probability that such a drive will take between 400 ms and 600 ms to seek and read.
c. Above what seek-and-read time are the slowest 25% of the drives?
d. If you wanted to purchase a CD-ROM drive that was faster than 95% of the drives available, what maximum seek-and-read times would you look for?
e. Complete this paragraph: At present, approximately _____% of CD-ROM drives require at least 500 ms to seek and read a CD-ROM. About _____% take between 400 ms and 600 ms. The slowest 25% take at least _____ ms to do their job. However, for those who are willing to pay the money, the fastest 5% are characterized by a seek-and-read time of under _____ ms.

5.41 A popular measure of the effectiveness of mutual funds is their five-year average annual total return. The Fidelity Organization offers a wide variety of funds, including 16 in the tax-free money market category (*Fidelity Focus,* Spring 1995). Assume that these represent a much larger population of such funds, whose five-year average annual return has a mean value of 3.27% and a standard deviation of .33%.

a. Calculate the 90th percentile value for these funds.

b. Calculate the 10th percentile value for these funds.

c. If you randomly selected a fund from this population, what is the likelihood that it would return at least 3%?

d. If you randomly selected a fund from this population, what is the likelihood that it would return over 4%?

e. Complete this paragraph: The top 10% of tax-free money market funds provided an annual average return of at least _____%. The worst performing 10% yielded no more than _____%. The probability that a fund yielded at least 3% is _____. The probability that a fund returned over 4% is _____.

5.42 Statistics published by the Wine Institute of America and the U.S. Department of Commerce reveal that the highest per capita consumption of wine in the United States is in Washington, D.C.: 6.95 gallons per capita per annum. The lowest is in the state of Mississippi: .63 gallon per capita per annum. Assume that the mean is 3.79 gallons and the standard deviation is 1.58 gallons and that the data follow a normal distribution.

a. Calculate the proportion of states with annual per capita wine consumption over 5 gallons.

b. Compute the probability of a random state's having annual per capita consumption below 1 gallon.

c. What is the probability that a random state will fall in the 3- to 4-gallon per capita consumption category?

d. What number of gallons per capita separates the driest 10% of the states from the rest?

e. Clearly, the best market for wine should be the states at the top of the distribution. Find a resource in which to verify this assumption. Are the data approximately normal?

SECTION 5 ■ APPROXIMATING THE BINOMIAL DISTRIBUTION

Using the Normal Approximation

Because the normal distribution makes calculating statistics for large numbers of events relatively easy, it is often used to approximate other distributions. In particular, when the total number of trials, n, is large, the normal distribution serves as a good approximation to the binomial distribution, and we can easily compute z scores from the mean and standard deviation of the distribution. The following rule of thumb tells when to use this approximation.

Rule of Thumb for Normal Approximation to the Binomial Distribution

Use the normal approximation to the binomial distribution if $np \geq 5$ and $nq \geq 5$. This approximation uses the mean and standard deviation of the binomial distribution, where $(1 - p) = q$.

$$\text{Mean, } \mu = np$$
$$\text{Standard deviation, } \sigma = \sqrt{np(1 - p)} = \sqrt{npq}$$

Example 5.13 illustrates the application of the normal distribution to approximate the binomial distribution.

EXAMPLE 5.13 ▪ ASSESSING THE PUBLIC'S PREFERENCE FOR A FLAT TAX

Suppose the public generally favors a flat tax instead of the present income tax system. Suppose, too, that the true percentage of the population favoring the flat tax is 80%. Let's find the probability that, in a survey of 500 individuals, at least 390 people favor the flat tax.

Step 1. We calculate the value of the mean and standard deviation for the distribution:

$$\mu = np = 500(.80) = 400$$

Since $np > 5$ and $nq = 500(.20) = 100$, the rule of thumb allows us to use the normal approximation:

$$\sigma = \sqrt{np(1 - p)} = \sqrt{500(.80)(.20)} = 8.94$$

In the next step, we will use the normal curve with this μ and σ as an approximation.

Step 2. We draw a picture of the situation in Figure 5.18.

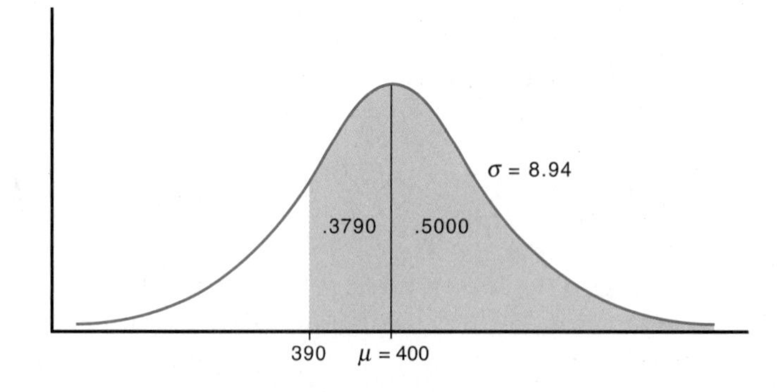

Figure 5.18
Probability that at least 390 people favor the flat tax

Approximating binomial probabilities: normal distribution

Step 3. We calculate the z score associated with 390. To make the approximation as accurate as possible, we subtract ½ unit from 390 and calculate the z score for 389.5. (You'll see why soon, in Figures 5.19 and 5.20.) This adjustment improves accuracy when we are using a continuous distribution to approximate a discrete distribution.

$$z = \frac{x - \mu}{\sigma} = \frac{389.5 - 400}{8.94} = -1.17$$

Step 4. We find the corresponding area under the normal curve. A z score of -1.17 corresponds to an area of .3790. Note that this is the area between the point in question and the mean; it is *not* the area desired.

Step 5. We state our conclusions. The complete area falling to the right of 390 consists of the area just calculated between 390 and the mean (.3790) *and* the complete area to the right of the mean (.5000). Thus, the probability that at least 390 individuals out of the 500 surveyed favor the flat tax is 87.90%.

Figures 5.19 and 5.20 show how well the **continuity correction factor** works. The dots represent the cumulative distribution function for the binomial distribution in Example 5.13 (n = 500, p = .80). The smooth curve in Figure 5.19 is the ogive (CDF) for the uncorrected normal distribution that approximates the binomial distribution in this example (μ = 400, σ = 8.94). Notice that the binomial counts are just a bit ahead of the cumulative normal distribution. If we move the normal distribution ½ step to the left, we get the almost perfect match shown in Figure 5.20, with the continuity correction.

Figure 5.19
Ogive for the normal approximation to the binomial distribution without the CCF

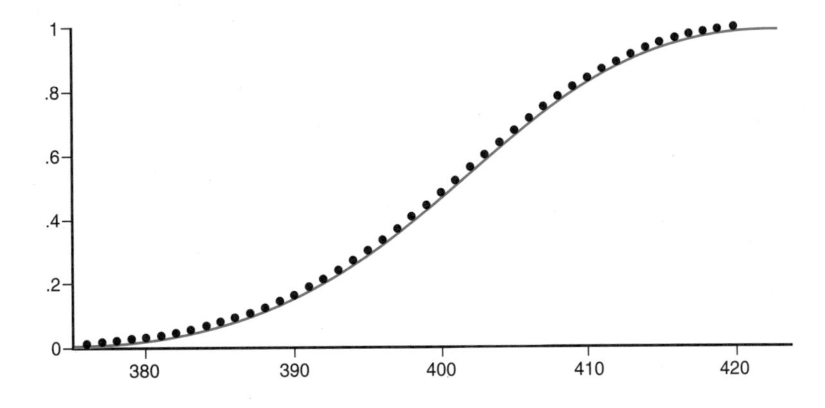

Figure 5.20
Ogive for the normal approximation to the binomial distribution with the CCF

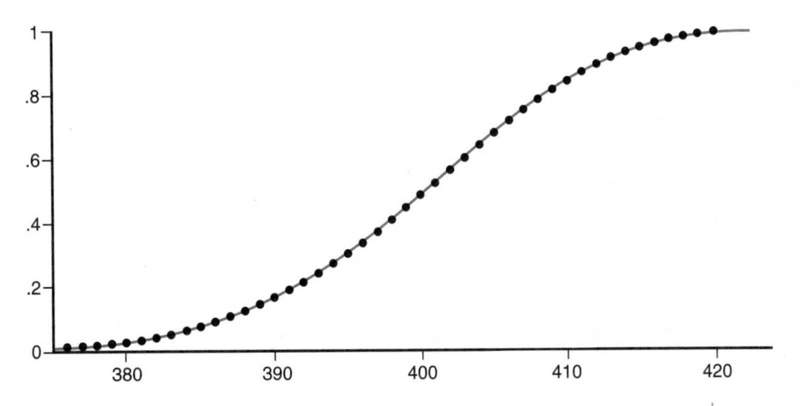

Using the Poisson Approximation

The Poisson distribution may be used to approximate the binomial distribution when the number of trials is large enough and the probability of success is low enough. When the number of trials is very large, calculations using the binomial approximation become nearly impossible. Consider finding the probability of 900 successes in a situation with 1000 trials. With $n = 1000$ and $x = 900$, the number of combinations of successes and failures alone would exceed the capacity of many calculators. The probability would be represented by the binomial coefficient

Approximating binomial probabilities: Poisson distribution

$$\binom{1000}{900} = \frac{1000!}{(900!)(100!)}$$
$$= \frac{(1000 \times 999 \times 998 \times \cdots \times 1)}{(900 \times 899 \times 898 \times \cdots \times 1)(100 \times 99 \times 98 \times \cdots \times 1)}$$

This expression would be too cumbersome for most calculators, even after cancellations.

Rule of Thumb for Poisson Approximation to the Binomial Distribution

Use the Poisson approximation to the binomial distribution if n is large (≥ 20) and p is very small ($\leq .05$).

By interchanging the roles of p and q, we can use the Poisson approximation to determine the probability of k failures out of n trials when n is large ($n \geq 20$) and q is very small ($\leq .05$).

EXAMPLE 5.14 ■ FINDING THE PROBABILITY OF A BAD REACTION TO A VACCINE

Consider the problem of calculating the probability of a bad reaction to a vaccine. Suppose that 1000 vaccinations are to be administered by a medical team from the local health department. The probability of an allergic reaction is .001. A bad reaction can be counteracted with drugs. What is the probability that there will be at most 4 bad reactions if 1000 shots are given?

This is a binomial problem, as there can be only two outcomes for any individual: no bad reaction or a bad reaction. To solve this problem using the binomial formula, we compute

$$P(X \leq 4) = P(0) + P(1) + P(2) + P(3) + P(4)$$

where $P(0)$ is calculated as

$$P(0) = \frac{1000!}{0!1000!}(.001)^0(.999)^{1000}$$
$$= .999^{1000}$$

$P(1)$ is calculated as

$$P(1) = \frac{1000!}{1!999!}(.001)^1(.999)^{999}$$

and so forth.

Although $P(0)$ and $P(1)$ can be readily calculated since so many factors cancel out, calculating $P(4)$ becomes a major undertaking. The values are beyond the capacity of some pocket calculators, and even some computers! Since n is very large and p is very small, we can use the Poisson approximation. To use the Poisson, we need the mean or expected value λ, which is the expected number of reactions. We find it using the simple expression for the expected value of a binomial random variable: $E(X) = np$. So

$$\lambda = np = 1000(.001) = 1.00$$

To calculate the probability of at most four bad reactions, we can use a calculator or simply turn to the Poisson table, using $\lambda = 1$:

$$P(X \leq 4) = P(0) + P(1) + P(2) + P(3) + P(4)$$

$$P(X \leq 4) = .3679 + .3679 + .1839 + .0613 + .0153$$
$$= .9963$$

If the health team carries four kits to counteract bad reactions, that should be more than sufficient for a day in the field.

EXERCISES 5.43–5.50

Skill Builders

5.43 Let $n = 500$ and $p = .32$ for a binomially distributed random variable X.
- **a.** Find the mean of X.
- **b.** Compute the standard deviation of X.
- **c.** Which approximation should be used in any probability calculations? Explain your reasoning.
- **d.** What is the approximate probability that X exceeds 172?
- **e.** Calculate the approximate probability that X falls between 150 and the mean.

5.44 Consider a binomially distributed random variable X with $n = 1000$ and $p = .03$.
- **a.** Find the mean of X.
- **b.** What is the standard deviation of X?
- **c.** What is the approximate probability that $X = 3$? (Hint: $2\frac{1}{2} \leq X \leq 3\frac{1}{2}$.)
- **d.** What is the approximate probability that X is greater than 3? (See the hint above.)

5.45 Under what conditions can the normal distribution be used to approximate the binomial distribution? If a sample is very large and the probability of a success very small, what distribution can be used to approximate the binomial distribution?

Applying the Concepts

5.46 Match the variables described below with the sort of distribution you would expect them to follow. Give your reason for choosing a binomial, Poisson, or normal distribution.
- **a.** the number of people killed by lightning in the United States from year to year

b. the heights of army recruits
c. the number of spades in a random seven-card hand
d. the average weights of a dozen oranges chosen at random from a truckload
e. the number of no-shows on a given New York to Chicago flight, when the plane is booked to carry 140 passengers

5.47 According to *The American Consumer* (R. H. Bruskin Associates Market Research, 1988), 44% of males would buy a very expensive car if they won a million-dollar lottery. Consider a group of 1000 males.
a. Find the expected value for the number in the group who would buy a very expensive car if they won a million-dollar lottery.
b. Calculate the standard deviation.
c. What is the approximate probability that more than 450 of these 1000 males would purchase an expensive car if they won a million-dollar lottery?
d. What is the probability that fewer than 430 of these 1000 males would purchase an expensive car if they won a million-dollar lottery?

5.48 The *Places Rated Almanac* (1993) indicates that Tucson, Arizona experiences an average of 198 clear days over the course of a year; that is, the probability of a clear day is .54. A sample of 30 days is randomly selected.
a. What is the probability that at least 20 out of the 30 days will be clear?
b. What is the probability that there will be no more than 10 clear days in the 30 days?
c. Find the probability that there will be between 20 and 25 days of clear weather in 30 randomly selected days.
d. Complete this statement: In a random sample of 30 days, between _____ and _____ clear days can be expected 90% of the time.

5.49 It has been reported that a drug called Norvasc, used in treating hypertension and angina, has a discontinuation rate of 1.5%. (Drugs are discontinued when the patients exhibit adverse effects.)
a. If a cardiologist prescribes this drug to 100 patients, what is the probability that more than 5 will suffer adverse effects?
b. What is the likelihood that, in a group of 100 patients, no more than 3 will suffer adverse effects?
c. Find the standard deviation of this distribution.
d. Complete this paragraph: Norvasc has been found to be effective in treating hypertension and angina. The likelihood that more than 5 persons in 100 will have an adverse effect is _____%. In fact, there is a _____ probability that no more than 3 will require a change in treatment. The average number of failures for the drug in a group of 100 is _____. The standard deviation is _____.

5.50 An analysis of a survey on industrial migration (the reasons firms move from one area to another) revealed that occupancy costs were rated as a very important factor by 57% of manufacturers. (Assume that this is a population percentage.)

a. If a random sample of 100 manufacturers is taken, what is the probability that over 60% will rate occupancy costs as a very important factor?
b. What is the probability that, in a random sample of 100 manufacturers, less than 50% will rate occupancy costs as a very important factor?
c. Calculate the likelihood that, in a random sample of 100 manufacturers, between 55% and 60% will rate ocupancy costs as a very important factor.
d. Is it realistic to expect that over 75% in a sample of 100 will rate occupancy costs as a very important factor?

FORMULAS

Discrete Random Variable

Expected value:

$$E(X) = \sum x_i P(x_i)$$
$$= x_1 P(x_1) + x_2 P(x_2) + \cdots + x_n P(x_n)$$

Variance:

$$V(X) = E[(X - E(X))^2] = \sum [x_i - E(X_i)]^2 P(x_i)$$

Standard deviation:

$$S(X) = \sqrt{V(X)}$$

Binomial Probability

The probability of x successes in n trials is given by

$$P(x) = \frac{n!}{x!(n - x)!} p^x q^{n-x}$$

where p is the probability of success on a single trial and ! indicates the factorial.

Expected value:

$$E(X) = np$$

Variance:

$$V(X) = npq$$

Standard deviation:

$$S(X) = \sqrt{npq}$$

Poisson Distribution

$$P(X = x) = P(x) = \frac{\lambda^x}{x!} e^{-\lambda}$$

Expected value:

$$E(X) = \lambda$$

Variance:

$$V(X) = \lambda$$

Standard deviation:

$$S(X) = \sqrt{\lambda}$$

z Scores

For a normal random variable X with a mean μ and standard deviation σ,

$$Z = \frac{X - \mu}{\sigma}$$

Normal Approximation to the Binomial Distribution

Conditions: Use if $np \geq 5$ and $nq \geq 5$.
Mean:

$$\mu = np$$

Standard deviation:

$$\sigma = \sqrt{np(1 - p)} = \sqrt{npq}$$

Remember that for any value, normal approximation begins ½ unit below the value and ends ½ unit above the value. Boundary calculations begin ½ unit beyond the stipulated boundary. This adjustment is known as the *continuity correction factor.*

Poisson Approximation to the Binomial Distribution

Conditions: Use if $n \geq 20$ and $p \leq .05$.
Mean:

$$\lambda = np$$

NEW STATISTICAL TERMS

binomial probability distribution
continuity correction factor
continuous random variable
expected value
normal probability distribution
Poisson probability distribution
probability distribution
random variable
standard normal variable
variance
z score

EXERCISES 5.51–5.62

Supplementary Problems

5.51 In a 1990 Medicare beneficiary survey (*Journal of the American Medical Association,* May 18, 1994), it was reported that, of 2243 HMO members who had chest pain, a total of 69.8% own their homes.

a. If a random sample of 20 HMO members with chest pain is taken and the random variable X is the number that own their homes, what is the probability that at least 15 own their homes?

b. What is the probability that $X \leq 10$ in a sample of 20?

c. Find the expected value of X.

d. Complete this paragraph: If a doctor cares for a group of 20 HMO patients with chest pain, it can be expected that _____ own their homes. There is a _____% chance that no more than 10 own their homes.

e. Do you think home ownership is a risk factor for chest pain? Why or why not?

5.52 In September 1994, the probability of winning on a single spin of the 25-cent slot machine at Caesar's Palace in Atlantic City was .888. Consider a person who plays the slots for 100 spins. Let the random variable X be the number of winning spins.

a. What type of distribution does X have?

b. Find the expected value for X.

c. What is the probability of no more than 25 wins?

d. What is the probability of at least 50 wins?

5.53 In the late 1970s, annual life insurance premium payments for people under the age of 30 were distributed approximately as shown below (*American Demographics,* March 1980). Treat the premiums as a random variable X with this distribution.

Premium Value	Percent Purchased
x	$P(x)$
$ 60	21%
195	40
365	25
500	14

a. Calculate the expected value of X.
b. Find the variance of X.
c. Compute the standard deviation of X.
d. What is the probability that X exceeds $195?
e. Complete this paragraph: Annual life insurance premiums for individuals below the age of 30 averaged $_____ in the late 1970s. The standard deviation for policy holders was $_____. While the majority of individuals under 30 years of age spent $195 or less on premiums, _____% spent at least $195.

5.54 Company ABC guarantees a 99.5% win factor with its "proven roulette method." Assume their claim is valid, and use the method on 10 trials.
a. What is the probability that you will win in exactly 8 spins?
b. What is the probability that you will win in no more than 5 spins?
c. What is the expected value of your number of wins?
d. Calculate the standard deviation for your number of wins.
e. Complete this paragraph: A test of the ABC Company's "proven roulette method" should yield an expected win in _____ out of 10 trials. The standard deviation for this variable is _____. If the claims are valid, there is a _____ probability that you will win in exactly 8 spins. There is a _____% chance that you will win in no more than 5 spins.

5.55 In 1977, according the the New York State Division of Motor Vehicles, there was .68 motor vehicle accident per minute in New York. Use this information to answer the following questions:
a. Calculate the probability that in a given minute there are no accidents.
b. What is the probability that there is exactly 1 accident?
c. What is the probability that there is more than 1 accident?

5.56 There were 113,867,000 voters in the 1992 presidential election. Their age distribution looked like this:

Age	Proportion
19	.03
23	.06
30	.20
40	.22
50	.17
60	.13
70	.12
80	.07

Source of Data: *Information Please Almanac*, 1994.

a. What is the expected value of voter age for this distribution?
b. What is the variance of voter age for this distribution?
c. Calculate the standard deviation of voter age.
d. What is the probability that a voter is at least age 50? Estimate how many voters are over 50.

5.57 It was reported that larcenies occurred an average of 1 every 13 minutes in Suffolk County, New York in 1976 ("1978 Suffolk County Criminal Justice Plan: Crime Analysis"). This value converts to an average of 4.6 larcenies per hour. Using this average, estimate the following:
a. What is the standard deviation of the number of larcenies per hour?
b. Find the probability that no more than 2 larcenies will occur in a given hour.
c. What is the likelihood that more than 5 larcenies will occur in a given hour?
d. What is the probability that no larcenies will occur in a given hour?

5.58 According to a study published in the September 1, 1992 issue of the *Journal of the American Medical Association,* 60% of all medical school programs have an ethics curriculum. Consider a survey of 15 randomly selected programs.
a. What is the expected number of schools in the survey with an ethics curriculum?
b. What is the probability that no school in the survey will have a medical ethics program?
c. What is the probability that all schools in the survey will have medical ethics programs?
d. What is the probability that a majority of the schools in the survey will have such a program?

5.59 Sales and use taxes vary throughout the United States, from zero in Alaska to as high as 7% in Mississippi. The following data set summarizes the sales and use tax situation as reported in *Information Please Almanac,* 1994. Use this distribution as a basis for answering the following questions.

Tax Rate	Percentage of States
0%	10%
3	4
4	24
5	30
6	28
7	4
Total	100%

a. What is the expected value for a randomly selected state tax rate?
b. What is the standard deviation for state tax rates?
c. What is the probability that the sales tax rate for a randomly chosen state is less than 5%?
d. What is the probability that the sales tax rate for a randomly chosen state exceeds 6%?
e. Complete this paragraph: The average state sales tax rate in the United States is _____%. Only _____ % of the states have rates below 5%. A total of _____% have rates of 7%.

5.60 There are many drip coffee makers on the market. Their average price is $33.53, and the standard deviation is $13.10 (*Consumer Reports,* October 1994). Assume that the price has an approximately normal distribution.
a. What is the probability that a random consumer will pay less than $25 for a drip coffee maker?
b. What is the likelihood that a drip coffee maker costs between $35 and $45?
c. Calculate the probability that a drip coffee maker costs at least $30.
d. Find the cutoff price for the most expensive 10% of drip coffee makers.

5.61 A survey of forensic psychiatrists revealed that the average daily rate for court testimony is $2500; the standard deviation is $750 (*The Guide to Expert's Fees, 1992–1993,* National Forensic Center). Assume that these figures are for the population of forensic psychiatrists and that the daily rate has an approximately normal distribution.
a. What is the probability that the income of a randomly selected forensic psychiatrist falls between $2300 and $2700 per day?
b. What is the probability that a random forensic psychiatrist earns less than $1000 for a day's work?

5.62 In a study entitled *Older Workers in the New York City Labor Market* (Albert C. Ovedovitz, 1983), it was found that 29.3% of unemployed workers over age 55 were unemployed 27 weeks or more. Suppose samples of 300 older workers are randomly selected.

a. What is the probability that at least 35% of the workers in such a sample will have been unemployed 27 weeks or more?

b. In what proportion of such samples will between 20% and 30% of the workers have been unemployed for 27 weeks or more?

c. Is it conceivable that more than 40% of the workers in such a sample would have been unemployed for 27 weeks or more?

6

Sampling Distributions and Sampling

OUTLINE

LEARNING OBJECTIVES

1. Understand the central limit theorem and its implications.
2. Apply the central limit theorem to practical problems of sampling.
3. Describe the sampling distribution of the mean.
4. Describe the sampling distribution of the proportion.
5. Distinguish between nonrandom samples and random or probability samples.
6. Evaluate a variety of sample designs.
7. Select simple, systematic, stratified, and cluster random samples.

BUILD YOUR KNOWLEDGE OF *STATISTICS*

The Shell Poll, conducted for Shell Oil Company in March 1999, examined the values of Americans with respect to morality and ethics. A total of 1277 randomly selected adults, ages 18 and over, were asked whether they thought there had been a general let-down in honesty among the American people as a whole. Seventy-five percent responded that they believed such a let-down had occurred (Peter D. Hart Research Associates, Inc., Shell Poll, Study No. 5404, March 1999, p. 8). If other samples were drawn, we would expect similar—but not identical—percentages. Another sample might yield 70%, and yet another 83%. More of the sample results would lie near the population parameter and fewer would fall further away, yielding a sampling distribution.

The purpose of all statistical analysis is decision-making. Samples of data are drawn from populations, and descriptive measures are taken. A political pollster draws a sample of the voting public, an auditor draws a sample of accounts, and a quality control inspector draws a sample of output from a production line. In each of these instances, sample data are used because it would be too costly or impractical to take a complete survey or because a carefully taken sample may yield greater accuracy by avoiding errors that might enter into a large-scale census. However, if sample statistics are to be used, it is imperative to know the characteristics of the sampling distributions.

Utilizing your knowledge of the normal distribution and descriptive measures, we will explore the nature of sampling distributions. You will learn about an all-important description of the characteristics of sampling distributions—the central limit theorem. We will also examine the types of random samples that can be selected from a population. The knowledge you gain in this chapter will serve as a foundation as you explore the topics of estimation and hypothesis testing in the following chapters.

SECTION 1 ■ THE SAMPLING DISTRIBUTION OF THE MEAN

In this section, you will see that the distribution of the sample means, $\bar{x}$, for simple random samples of size n, taken with replacement from a given population, can be usefully approximated by a normal distribution whose mean is the mean, μ, of the random variable X being sampled and whose standard deviation, $\sigma_{\bar{x}}$, is $\sigma_x/\sqrt{n}$, where σ_x is the standard deviation of the underlying X. Although we do not always know the exact distribution of the sample means, $\bar{x}$, we do know that this distribution will be approximately normal; its mean, $\mu_{\bar{x}}$, is the mean, μ_x, for the underlying population variable; and the standard deviation of $\bar{x}$, $\sigma_{\bar{x}}$, is given by $\sigma_x/\sqrt{n}$, where n is the sample size. When the underlying population variable is obvious from the context, we can suppress it in our notation and simply write $E(\bar{x}) = \mu$ and $\sigma_{\bar{x}} = \sigma/\sqrt{n}$.

If the underlying population variable X that is being sampled is normally distributed or close to normal in its distribution, the normal approximation for the distribution of $\bar{x}$ will be exact even for small values of n. For practical purposes, the normal approximation may be used when $n \geq 30$, although you should always take further samples if results run counter to your expectations. An adjustment should be made for finite population size when the population is small (the total population is not larger than 10 times the sample size) and when the sample is made without returning the sampled individuals to the population between random draws (called *sampling without replacement*).

Here the behavior of sample means, $\bar{x}$, will be illustrated by experiments in which random samples are generated by a computer. Such experiments can be run automatically with a software package like Minitab. You can also work out a sampling distribution by hand, as shown in Example 6.1, though calculations would be cumbersome for samples of any significant size.

The validity of the normal approximation to the distribution of the sample means, $\bar{x}$, when $n \geq 30$ is confirmed by evidence from computer simulations and backed by the central limit theorem. This theorem spells out in detail just how good the normal approximation to the sampling distribution is, given the nature of the underlying population and the size of n. Here, we are concerned with the general applicability of the normal approximation—not its technical specifics—so the important point is that for large n the normal approximation is excellent.

Figure 6.1 shows the distribution of prices of sports utility vehicles, as reported in *Money* (March 1995). For the 69 vehicles considered, prices ranged from a low of \$11,670 for a Geo Tracker to a high of \$52,500 for a Land Rover. The distribution is roughly normal, with the price of a Land Rover as an outlier. The mean price is \$22,409.

In Figure 6.2, we see the distribution of means of 50 random samples of size 20, selected from the basic population of sports utility vehicles. If we were to redo this sampling many times, results would vary, as these are random samples selected by a computer using Minitab. But the overwhelming probability is that the sample means will show the same general behavior as these particular 50 means: (1) They fall within a narrow range concentrated near the mean of the underlying population; these particular means range over the narrow interval bracketed in Figure 6.1.

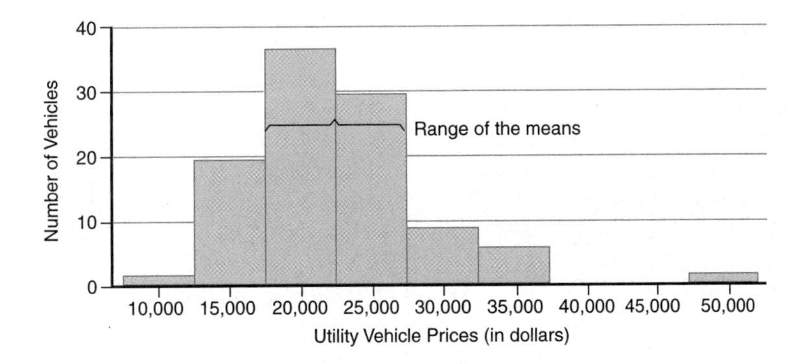

Figure 6.1
Distribution of 1995 manufacturers' suggested retail prices for sports utility vehicles

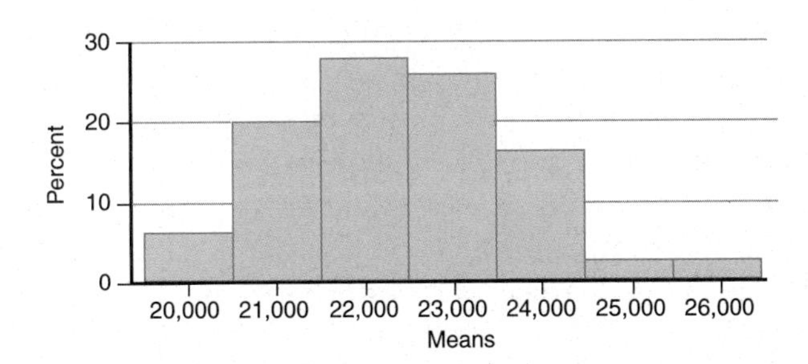

Figure 6.2
Distribution of means for 50 SUV samples of size n = 20

(2) Their distribution is more symmetrical and closer to the bell shape of a normal curve than was the distribution of the original population. In short, the distribution of the sample means is more concentrated about the population mean and more normal than is the distribution of the underlying population. The larger the sample size n is, the more obvious these effects are.

The underlying data distribution does not have to be normal—or nearly so—for the distribution of the means to follow a normal curve. Consider the distribution of the assets of billionaires (*Fortune,* September 9, 1991) shown in Figure 6.3. Again using the simulation features of Minitab, in Figure 6.4 on page 200 we look at the distribution of the means of 50 random samples, $n = 20$, from this population. Once again, the software tossed out the tails of the sampling distribution; this can be done with the sampling function of Excel as well.

The distribution of the population is very skewed; the distribution of the means from the sample data is much closer to a bell-shaped pattern. Figure 6.4 contains the distribution of means from 50 samples of 20 observations each. The distribution

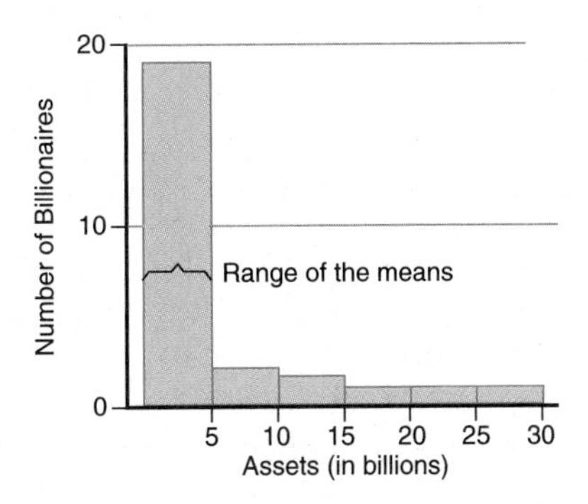

Figure 6.3
Distribution of billionaires by assets, 1991

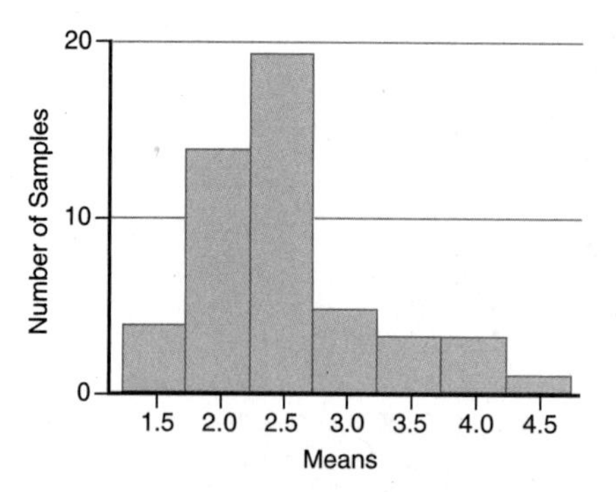

Figure 6.4
Distribution of means for 50 billionaire samples of size n = 20

is concentrated about the population mean of \$2.637 billion. The range for these samples is indicated by the brace in Figure 6.3. Whereas the initial distribution began at \$1 billion and extended to over \$30 billion, the means for the 50 random samples range from about \$1 billion to \$5 billion. It should be noted that when population histograms are heavily skewed, a larger sample size is required in order to yield a relatively normal distribution for the sample means.

Let's consider an example in which the distribution can be worked out in detail for a sample of size 100. Our population consists of 10 balls, 6 of which have 0's on them and 4 of which have 1's on them. We pick a ball at random and record the number on it. This draw is a binomial trial for which the probability of success (getting a 1) is $p = .4$. The population mean, μ, is given by $1 \times p + 0 \times q = 1 \times .4 = .4$, and the standard deviation, σ, is equal to $\sqrt{pq} = \sqrt{.4 \times .6} = \sqrt{.24} = .49$.

The data points in Figure 6.5 show the distribution of means of 1000 random samples of size 100 drawn from this population. The smooth curve is a normal curve with a mean of .4, the same as that of the original population. The standard deviation of this normal curve is $\sigma_x/\sqrt{100} = \sigma_x/10 \approx .049$, where σ_x is the standard deviation of the original population.

The data fall close to the normal curve, and the scatter of data points about that curve seems patternless. This is the match we would expect between a random process and its underlying distribution. There are samples of size 100 with means of .8 or .9, but in the 1000 trials performed, none of these turned up. Nothing rules these means out, but the greater number of samples with means in the range from .3 to .5 causes them to appear more frequently. In Figure 6.5, 3 standard deviations from the mean of .4 does not let us quite reach .25 on the low side or .55 on the high side. In the long run, for normally distributed random variables, 99.7% of the

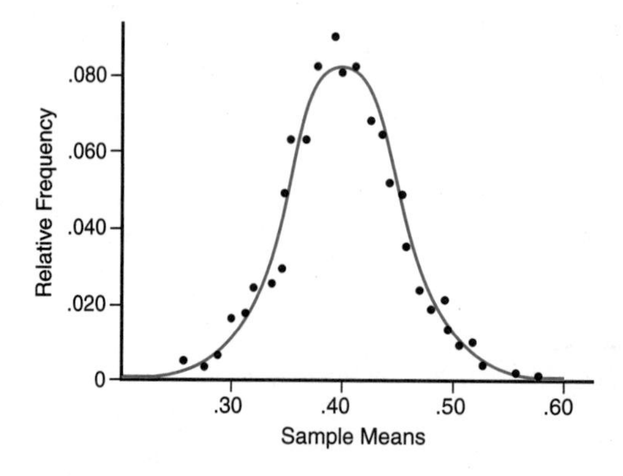

Figure 6.5
Comparison of the normal approximation with a relative frequency plot for the means of 1000 samples of size n = 100

data will fall in this range. It is not surprising that in 1000 trials the lowest observed mean was no smaller than .24 and the highest was no larger than .58. Typically, for samples that are even moderately large, sample means are close to the population mean. This is why sampling is so effective.

The distribution of sample means tends to approximate a normal distribution, and so the match between the sample means and the population mean improves as the sample size grows. Furthermore, the distribution of sample means becomes more concentrated about the population mean as the sample size grows. These facts are summarized in the central limit theorem.

Central Limit Theorem:
Sampling Distribution of the Mean

The sample means, $\bar{x}$, for simple random samples of size n, drawn from a population with mean μ and standard deviation σ, may be treated as approximately normally distributed. The distribution has a mean equal to the population mean:

$$\mu_{\bar{x}} = \mu_x$$

The distribution has a standard deviation equal to the population standard deviation divided by the square root of the sample size:

$$\sigma_{\bar{x}} = \frac{\sigma_x}{\sqrt{n}}$$

The larger n is or the closer the original distribution is to normal, the better the normal distribution approximates the distribution of $\bar{x}$. It is generally quite good if $n \geq 30$. The value $\sigma_{\bar{x}}$ is called the standard error of the mean.

Note that the means tend to have a normal distribution *whether or not* the underlying population is normally distributed.

Consider a very small population, consisting of just the five elements A, B, C, D, and E. Table 6.1 shows all 25 samples of size 2, where A is 45, B is 18, C is 23, D is 12, and E is 52. The expected value of the sample means, $E(\bar{x})$, equals the

Table 6.1

SAMPLING DISTRIBUTION FOR A POPULATION WITH A MEAN OF 30 AND STANDARD DEVIATION OF 15.66

Sample	Data	Mean	Sample	Data	Mean
AA	45, 45	45.0	DB	12, 18	15.0
AB	45, 18	31.5	BE	18, 52	35.0
BA	18, 45	31.5	EB	52, 18	35.0
AC	45, 23	34.0	CC	23, 23	23.0
CA	23, 45	34.0	CD	23, 12	17.5
AD	45, 12	38.5	DC	12, 23	17.5
DA	12, 45	38.5	CE	23, 52	37.5
AE	45, 52	48.5	EC	52, 23	37.5
EA	52, 45	48.5	DD	12, 12	12.0
BB	18, 18	18.0	DE	12, 52	32.0
BC	18, 23	20.5	ED	52, 12	32.0
CB	23, 18	20.5	EE	52, 52	52.0
BD	18, 12	15.0			

$E(\bar{x}) = 30 \qquad \sigma_{\bar{x}} = 15.66/\sqrt{2} = 11.07$

population mean. The standard error of the mean, $\sigma_{\bar{x}}$, equals the standard deviation divided by the square root of the sample size, $\sigma_x/\sqrt{n}$.

With a larger population and larger sample size than in Table 6.1, listing all the samples would be impractical or impossible. They are listed here simply to show that the values predicted by the formulas given in the central limit theorem are exactly what we observe. The values assigned to A, B, C, D, and E are completely arbitrary in order to emphasize that the tendency toward a normal distribution for the means holds even without special assumptions on the original finite distribution.

Examples 6.1 and 6.2 apply the central limit theorem to the population of completion times of the 1994 New York City marathoners.

EXAMPLE 6.1 ▪ FINDING THE PROBABILITY THAT THE SAMPLE MEAN OF RUNNING TIMES IS LESS THAN 4 HOURS

Suppose samples of size 50 are drawn from the population of completion times of 1994 New York City Marathon runners. What is the probability that the sample mean, $\bar{x}$, is less than 4 hours? The mean running time for the 1994 marathon was 4 hours 22 minutes; the standard deviation was 51 minutes.

Step 1. We calculate the standard error of the estimate for the sampling distribution of the means:

$$\sigma_{\bar{x}} = \frac{\sigma_x}{\sqrt{n}} = \frac{51}{\sqrt{50}} = 7.21$$

Step 2. We draw a picture of the situation in Figure 6.6.

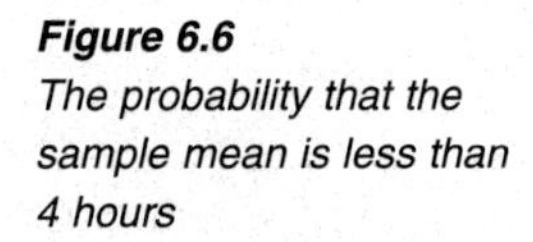

Figure 6.6
The probability that the sample mean is less than 4 hours

Calculating the probability that $X \leq x$

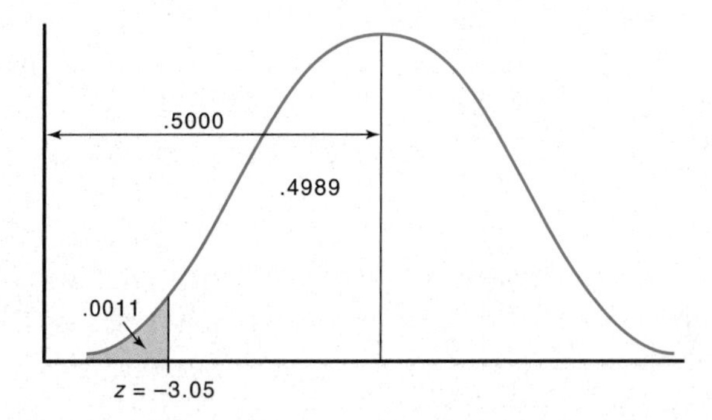

Step 3. We compute the z score of the cutoff value. Converting hours to minutes, we get 4 hours = 240 minutes and 4 hours 22 minutes = 262 minutes.

$$z = \frac{240 - 262}{\sigma_{\bar{x}}} = \frac{-22}{7.21} = -3.05$$

Step 4. We look up the z score in Table IV in Appendix B and calculate the area in question. The area falling between the mean and the cutoff point is .4989. Since we wish to calculate the area to the left of the cutoff, we subtract this value from .5000. The resulting area is .0011.

Step 5. We state the conclusion: The probability that the sample mean falls below 4 hours is .0011. In other words, there is a .11% chance that a sample mean will fall below 4 hours—that is, 11 out of 10,000 sample means will fall below 4 hours.

As you can see, the procedure for calculating the probability that a sample mean, $\bar{x}$, falls within a certain range differs from that for calculating the probability that a single observation falls within a certain range, but only in the first step—calculating the standard error of the mean. In Example 6.2, the objective is to calculate the cutoff for a given percentage of the distribution of samples drawn from the marathon population.

EXAMPLE 6.2 ■ FINDING THE CUTOFF FOR THE CENTRAL 90% OF THE DISTRIBUTION OF SAMPLE MEANS

For the marathon data, the mean was 4 hours 22 minutes and the standard deviation was 51 minutes. The original distribution was close to normal, so the normal distribution with the same mean and standard deviation is a good approximation. By the central limit theorem, the agreement will be even better for sample means. Suppose samples of size 30 are drawn from the population. What are the cutoffs for the central 90% of the distribution of sample means?

Step 1. We calculate the standard error of the mean:

$$\sigma_{\bar{x}} = \frac{\sigma_x}{\sqrt{n}} = \frac{51}{\sqrt{30}} = 9.31$$

Step 2. We draw a picture of the situation in Figure 6.7.

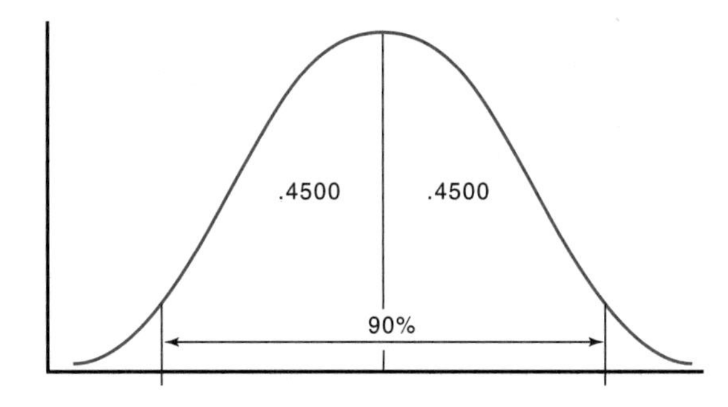

Figure 6.7
The central 90% of the distribution of sample means

Finding the limits on the center of a sampling distribution

Step 3. We look up .4500, which is one-half of the central value of 90% (or .9000), in the body of the z table.

Since .4500 falls between .4495 and .4505, we split the difference between corresponding values of 1.64 and 1.65, yielding a z score of 1.645.

Step 4. We solve for the upper value of the mean, recalling that $\sigma_{\bar{x}} = 9.31$:

$$1.645 = \frac{\bar{x} - 262}{9.31}$$

$$\bar{x} = 277 \text{ minutes (or 4 hours 37 minutes)}$$

Step 5. We solve for the lower value of the mean:

$$-1.645 = \frac{\bar{x} - 262}{9.31}$$

$$\bar{x} = 247 \text{ minutes (or 4 hours 7 minutes)}$$

Step 6. We state the conclusion: The central 90% of the means fall between 4 hours 7 minutes and 4 hours 37 minutes.

We have seen that sampling distributions tend to be sharply concentrated about the population mean. Figure 6.8 shows the probability density functions for the distribution of X for four values of n: $n = 1$ (the basic population), $n = 10$, $n = 30$, and $n = 100$. The lowest curve is for sampling with $n = 1$—that is, the population distribution. As n increases to 10, 30, and 100, the standard deviations decrease to .155, .094, and then .049 and the sample means $\bar{x}$ cluster more tightly about the population mean $\mu = .4$. This tighter clustering as sample size increases will be called *consistency* in Chapter 7. The population mean here is similar to that in Figure 6.5 in that $\mu = .4$ and $\sigma = \sqrt{.4 \times .6} = .49$. These are the values associated with the binomial distribution with $p = .4$. The most concentrated distribution, for $n = 100$, has $\sigma_{\bar{x}} = .49/\sqrt{100} = .049$. These X's fall within 3 standard deviations, or .15, of the population mean better than 99.7% of the time. For large samples, the sample means are very likely to be relatively close to the population mean—this is the key to sampling. We will soon quantify this observation more precisely.

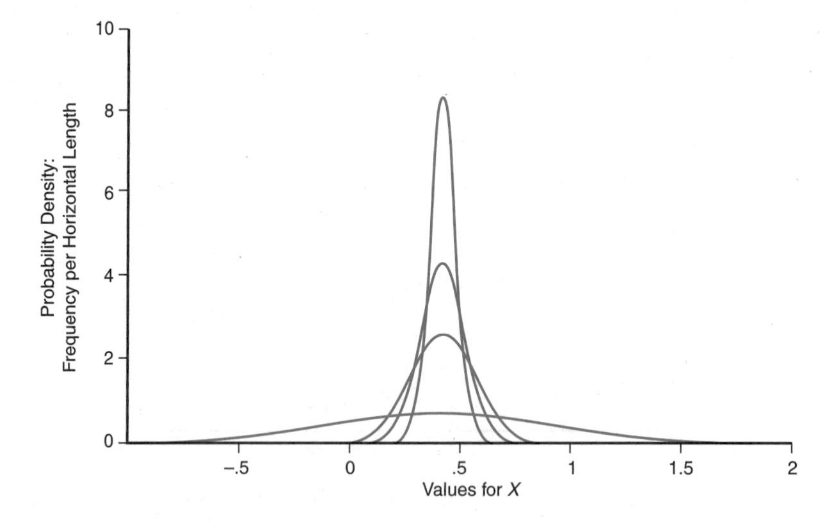

Figure 6.8
Sampling distributions for n = 1, n = 10, n = 30, and n = 100 for a normal population with μ = .4 and σ = .49

EXERCISES 6.1–6.10

Skill Builders

6.1 If the population standard deviation, σ, is 100, calculate $\sigma_{\bar{x}}$ for samples of the following sizes:

a. $n = 4$ **b.** $n = 9$ **c.** $n = 16$ **d.** $n = 25$

6.2 Assuming that the underlying distribution is normal, find the z score for a sampling distribution of the mean if $n = 100$, the population mean, μ, is 50, the population standard deviation, σ, is 10, and $\bar{x}$ is as follows:

a. $\bar{x} = 48$ **b.** $\bar{x} = 53$ **c.** $\bar{x} = 49$ **d.** $\bar{x} = 51.3$

6.3 Find an upper bound for the sample mean, $\bar{x}$, such that the shaded area in the figure equals .2500. The true mean for this distribution is 52, the population standard deviation is 13, and the basic population has a normal distribution. The sample sizes are as follows:

a. $n = 4$ **b.** $n = 9$ **c.** $n = 16$ **d.** $n = 25$

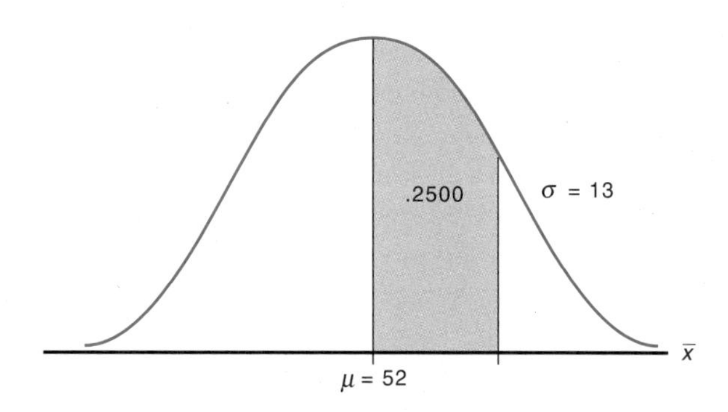

6.4 Given a normal distribution with $\mu = 451$ and $\sigma = 57$, consider random samples of size $n = 25$.

a. Find the probability that $\bar{x} \geq 462$.
b. Find the probability that $\bar{x}$ falls between 440 and 460.
c. What percentage of sample means fall below 436?
d. Calculate the cutoff for the top 10% of the distribution of $\bar{x}$.
e. Find the limits of the central 95% of the distribution of $\bar{x}$.

6.5 If the population mean of a distribution is 1000 and the population standard deviation is 300, calculate the probability that a sample mean does each of the following:

a. exceeds 950 for a sample of size $n = 100$.
b. is less than 1057 for a sample of size $n = 81$.
c. falls between 990 and 1040 for a sample of size 36.

Applying the Concepts

Note: Given the sample sizes in the following exercises, you may assume that the distribution of sample means is normal even if the distribution of the original population is not.

6.6 Suppose the population average income for family physicians in May 1993 was \$109,900 and the population standard deviation was \$16,115 (based on data from the American Academy of Family Physicians, *Computer Usage and Community Information and Current Fees,* May 1993). Consider samples of size $n = 100$.

a. What is the probability that the average income of a random sample of family physicians will exceed \$115,000?

b. What is the probability that the average income of a random sample of family physicians will fall between \$100,000 and \$112,000?
c. What is the probability that an individual family physician earns over \$130,000 per year?
d. What is the probability that the average income of a random sample of 25 family physicians will exceed \$130,000 per year?

6.7 It was reported by the National Forensic Center in the *Guide to Experts' Fees, 1992–1993* that the average daily fee of forensic accountants for court testimony was \$1600. Take this figure as the population mean, and assume that the population standard deviation is \$320.
a. What is the probability that, in a random sample of 400 forensic accountants, the average daily fee is less than \$1550 per day?
b. Find the probability that, in a random sample of 400 forensic accountants, the average fee falls between \$1500 and \$1700 per day.
c. What is the probability that an individual forensic accountant earns at least \$1800 per day?
d. What is the probability that, in a random sample of 400 forensic accountants, the average fee exceeds \$1800 per day?
e. Calculate the range for the central 95% of sample means for samples of size 400.

6.8 The fastest Amtrak Metroliners average 90.79 mph; the standard deviation of speeds on high-speed runs is 4.44 mph. Consider samples of speeds of 40 trains on such runs.
a. What is the probability that the sample mean speed will be less than 85 mph?
b. What is the probability that the sample mean speed will exceed 100 mph?
c. What percentage of sample means fall between 90 and 95 mph?
d. Find the cutoff of the top 5% of this distribution of means.

6.9 It is estimated that average household television usage per day in 1991–1992 (Nielsen Media Research, as reported in the *Information Please Almanac,* 1994) was 7 hours 4 minutes. Assume that the population standard deviation is 15 minutes. Consider a sample of 100 households.
a. Calculate the probability that mean television usage of a random sample will exceed 7 hours 20 minutes.
b. What is the likelihood that the sample mean usage will fall between 6 hours 58 minutes and 7 hours 5 minutes?
c. What percentage of random samples will have a mean usage exceeding 6 hours 45 minutes?
d. The highest 25% of the sample means fall above what cutoff?

6.10 An examination of the "50 Hottest Jobs" (*Money,* March 1995) indicated that the mean estimated growth rate in these fields would be nearly 45% for the 5-year period extending through the year 2000. The standard deviation for these growth rates was 18%. Suppose a sample of 30 jobs is randomly selected with replacement from this list.

a. What is the probability that the sample mean estimated growth rate exceeds 50%?
b. Find the likelihood that the sample mean estimated growth rate is under 40%.
c. What is the range for the middle 50% of the sample means?
d. What is the cutoff growth rate for the slowest 10% of the "50 hottest jobs"?

SECTION 2 ▪ THE SAMPLING DISTRIBUTION OF THE PROPORTION

Like sample means, sample proportions follow a normal distribution centered around the population parameter. For a given population, the population proportion, π, is defined as the number of times, X, a particular condition is found, divided by the size, N, of the total population. This definition indicates that we are looking at binomial trials and that the population proportion, π, is the same as the probability of a success, p, for these trials.

$$\text{Population proportion, } \pi = \frac{\text{Number of favorable cases}}{\text{Total population}} = \frac{X}{N}$$

Similarly, a sample proportion, p, is equal to the number of times, x, a condition is found in a sample, divided by the total sample size, n. (See Figure 6.9.)

$$\text{Sample proportion, } p = \frac{\text{Number of favorable cases}}{\text{Size of sample}} = \frac{x}{n}$$

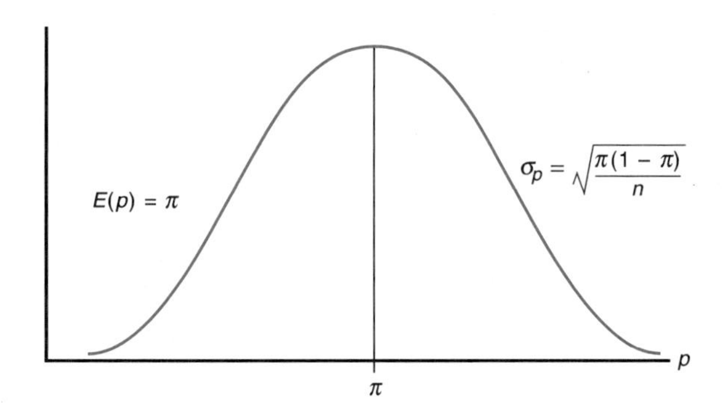

Figure 6.9 *The sampling distribution of a proportion*

Central Limit Theorem:
Sampling Distribution of the Proportion

For $n \geq 30$, the sampling distribution of a proportion, defined as x/n, may be treated as approximately normally distributed. The distribution has an expected value equal to the population proportion, π:

$$E(p) = \pi$$

The distribution has a standard deviation equal to the square root of the product of the population proportion and its complement divided by the sample size:

$$\sigma_p = \sqrt{\frac{\pi(1 - \pi)}{n}}$$

The value σ_p is known as the standard error of the sample proportion.

EXAMPLE 6.3 ■ FINDING THE PROBABILITY OF A FALSE NEGATIVE RESULT

Consider a situation in which the percentage of the population favoring a candidate is 52%. What is the probability that a sample of 1000 individuals will yield a percentage of 49% or less in favor of the candidate?

Step 1. We draw a picture of the situation in Figure 6.10.

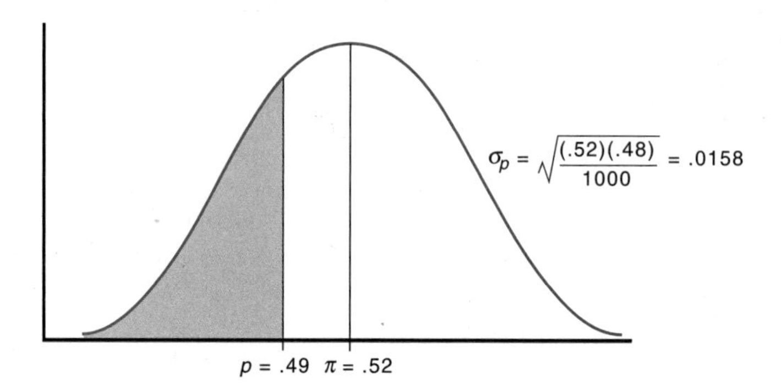

Figure 6.10
Probability of a false negative

Finding an area with the distribution of sample proportions

Step 2. We calculate the z score corresponding to the sample proportion $p = .49$:

$$z = \frac{p - \pi}{\sigma_p} = \frac{.49 - .52}{.0158} = -1.90$$

Step 3. We find the corresponding area, which is the area between the mean and the point in question. From the z table, the area between the mean and a z value of 1.90 (we can ignore the negative sign) is .4713. This is not the area of interest. The area of interest is the shaded region lying to the left of the cutoff. The shaded area is .5000 − .4713 = .0287.

Step 4. We state the conclusion: The probability that the sample proportion will be 49% or less when the true proportion is 52% is .0287. In other words, there is a very small chance that, when a candidate has a majority of the vote, a random sample will yield the false conclusion that he or she will not win.

Note that the calculations follow the same pattern as in other problems involving the normal curve. The mean of the distribution is noted, the standard error is calculated, the z score is determined, and its associated area is found in the table.

SECTION 3 ▪ THE FINITE POPULATION CORRECTION FACTOR

One additional factor that we must take into account when considering the distribution of sample means or proportions is the finite population correction factor (fpcf). The central limit theorem is based on the assumption that samples were selected with replacement from a population. However, in survey research, samples are usually drawn without replacement from a finite population. Most people regard being asked to respond to a poll as something of an imposition and would be unlikely to cooperate if approached twice by the same pollster with the same questions. When the population size is finite, the relative sizes of the population and the sample have an effect on the variance and the standard error of the mean and proportion. This effect may be ignored when the sample size, n, is small relative to the population size, N. However, when the sample size is 5% or more of the population size, the finite population correction factor (fpcf) should be incorporated:

Finite Population Correction Factor

$$\text{fpcf} = \sqrt{\frac{N - n}{N - 1}}$$

With the fpcf, the formula for the standard error of the proportion becomes

$$\sigma_p = \sqrt{\frac{\pi(1 - \pi)}{n}}\sqrt{\frac{N - n}{N - 1}}$$

When n is less than 5% of N, this correction is so small that we can neglect it. But when n is a large fraction of N, the sample is close to being a census for the population and p is closer to π than we would expect if we were sampling with replacement. So we use the fpcf to reduce the standard deviation for p. To apply this correction, we must know the size of the total population. In situations such as a market survey of potential car buyers, the target population may not be defined sharply enough for its size to be known. This is a practical consideration in the use of the fpcf. An illustration of the use of the finite population correction factor appears in Example 6.4.

EXAMPLE 6.4 ▪ USING THE fpfc IN EXAMINING FACULTY ATTITUDES

To examine faculty attitudes about a proposal that all students at a business school be required to take a comprehensive examination before graduation, a survey was conducted. The survey revealed that 45% of the faculty favored such an examination. What is the probability that, in a random sample of 35 members of the 200-person faculty at the college, at least 60% would favor such an idea?

Step 1. We draw a picture of the situation in Figure 6.11 and calculate the standard error of the proportion.

Using the finite population correction factor

$$\sigma_p = \sqrt{\frac{\pi(1 - \pi)}{n}}\sqrt{\frac{N - n}{N - 1}} = \sqrt{\frac{(.45)(.55)}{35}}\sqrt{\frac{200 - 35}{199}} = .0766$$

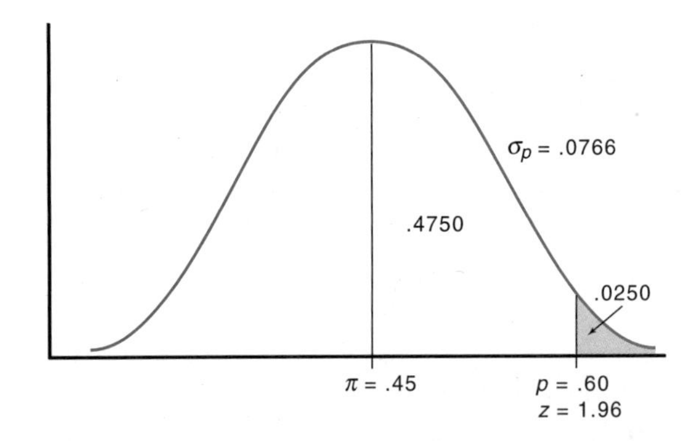

Figure 6.11
Probability that at least 60% of faculty favor a comprehensive examination

Step 2. We calculate the z score for the sample proportion, $p = .60$:

$$z = \frac{p - \pi}{\sigma_p} = \frac{.60 - .45}{.0766} = 1.96$$

Step 3. We find the corresponding area, which is the area between the mean and the z value calculated. Then the shaded region is $.5000 - .4750 = .0250$.

Step 4. We state the conclusion: From the sample drawn, we can infer that the probability of finding a sample in which at least 60% of the faculty favor the examination is .025.

EXERCISES 6.11–6.19

Skill Builders

6.11 Given a true proportion, π, of .57 and a sample size of 100 from a very large population, do the following.

a. Find the probability that the sample proportion will exceed .62.

b. Compute the probability that the sample proportion will fall between .54 and .59.

c. What is the probability that the sample proportion will be less than .63?

d. What is the cutoff for the highest 10% of the distribution of sample proportions?

6.12 Let $N = 50$, $n = 30$, and $\pi = .78$.

a. What are the values of the fpcf and σ_p?

b. Compute the probability that a sample proportion will exceed .80.

c. What is the probability that a sample proportion will fall between .72 and .76?

d. Find the cutoff for the highest 5% of the distribution of sample proportions.

6.13 Let $n = 500$ and x = number of successes in these 500 trials, with a probability of success $p = \pi = .32$. Assume that the sampling is done with replacement or, alternatively, that N is very large relative to $n = 500$.

a. Find $E(x)$.

b. Compute $S(x)$.

c. What is the probability that x exceeds 172?
d. Calculate the probability that x falls between 150 and $E(x)$.

6.14 A sample of 50 is drawn from a population of only 100 items. The population proportion is .49.
a. What is the probability that the sample proportion will fall below .48?
b. Find the probability that the sample proportion will exceed .51.
c. Compute the probability that the sample proportion will fall between .53 and .55.
d. Find the value of p that corresponds to a z score of 1.73.

6.15 Let $N = 1000$, $n = 200$, and $\pi = .67$.
a. What are the values of the fpcf and σ_p?
b. Calculate the probability that the sample proportion will not be greater than .65.
c. What is the probability that the sample proportion will be at least .71?
d. Compute the likelihood that $p \leq .50$.
e. Below what value will 30% of the distribution of sample proportions be found?

Applying the Concepts

6.16 Insurance rates for cars priced between $15,000 and $19,999 are classified from "very low" for the Ford Taurus GL to "very high" for the Pontiac Firebird. The 1995 population of this category of cars consists of 90 models. A total of 20% of these models fall into the "high" or "very high" insurance rate category (*Money,* March 1995). Suppose samples of 20 vehicles are selected at random with replacement from cars in the $15,000 to $19,999 range for more intensive study. We are interested here in the proportion of cars in the sample that fall into the "high" or "very high" insurance rate class.
a. What is the standard error of the sample proportion for such samples?
b. What is the probability that at least 25% of the cars will fall into the "high" or "very high" insurance category?
c. Calculate the probability that less than 10% of the cars in a 20-car sample will fall into the "high" or "very high" insurance category.
d. Find the likelihood that between 15% and 25% of the models will fall into this insurance classification.

6.17 Suppose a survey of 100 telecommunications firms reveals that 50% of them have price/earnings multiples (P/E ratios) of 20 or more, perhaps suggesting overpriced stock. A random sample of 35 of these telecommunications firms is selected without replacement for further study.
a. Find the fpcf for this study.
b. What is the probability that no more than five of these firms will have a P/E ratio of 20 or more?
c. What is the probability that more than eight of these firms will have a P/E ratio of 20 or more?
d. What is the probability that at most three of these firms will have a P/E ratio of 20 or more?

6.18 According to *The American Consumer* (R. H. Bruskin Associates Market Research, 1988), 44% of males say that they would buy a very expensive car if they won a million-dollar lottery. Consider a randomized survey of 1000 males.

a. Find the expected value for the number of males in this survey who would purchase such a car.
b. Calculate the standard deviation.
c. What is the probability that more than 450 of the males will say that they would purchase an expensive car if they won a million-dollar lottery?
d. What is the probability that fewer than 430 of the males will say that they would purchase an expensive car if they won a million-dollar lottery?

6.19 *Places Rated Almanac* (1993) reports that Tucson, Arizona experiences an average of 198 clear days over the course of a year; that is, the probability of a clear day is .54. Samples of 30 days are randomly selected.

a. What is the probability that at least 20 out of the 30 days will be clear?
b. What is the probability that there will be no more than 10 clear days in any sample of 30 randomly selected days?
c. Complete this statement: When random samples of 30 days are selected, between _____ and ____ clear days can be expected 90% of the time.

SECTION 4 ▪ TYPES OF SAMPLES

Now that you are familiar with sampling statistics, let's take a closer look at the practical problems of sampling. Circumstances often determine what methods are used. Each sampling procedure yields a different type of sample, with its own set of advantages, costs, and limitations. Simple random sampling was introduced in Section 1 of Chapter 1, and systematic random sampling in Section 1 of Chapter 2. Here, we shall examine six types of samples: convenience samples, judgment samples, simple random samples, systematic random samples, stratified random samples, and cluster samples.

Convenience Samples

When data collection is determined by circumstance—for example, interviewing the first willing passers-by at a mall or walk-ins to a retail store—the sample is known as a **convenience,** or **chunk, sample.** This is the sort of sample that led the *Literary Digest* to the error discussed in the introduction to Chapter 1. That magazine then compounded its error by relying on voluntary responses from those polled. Such an approach puts undue emphasis on responses from those with strong feelings about the survey topic and almost surely will give a biased result. Convenience samples are haphazard rather than random, and thus they lack the characteristics of random samples.

All convenience samples have one advantage in common: The data are readily available. Convenience samples may be good for pilot testing a survey, but not for drawing sound conclusions.

Judgment Samples

In **judgment sampling,** a sample is selected by an expert, based on her or his subjective judgment. Although judgment sampling is a quick and convenient method for getting needed numbers, there is no systematic way to assess the error in the results.

Every month, the Bureau of Labor Statistics compiles the wholesale price index. First published in 1902 to help a Senate subcommittee investigate the effects of tariff laws, it has been published regularly ever since. This monthly index is based on a judgment sample of commodities, chosen on the basis of their shipment values. Those with the largest shipment values are included in the index (*BLS Handbook of Methods,* April 1997).

Simple Random Samples

Unlike judgment sampling, **simple random sampling (SRS)** allows us to obtain a measure of sampling error—that is, a measure of the variation in results from one sample to the next. We saw in Figure 6.5 that sample means for samples of size 100 closely follow a normal distribution. We know from experimenting and from the central limit theorem that the distribution of sample means from SRS is approximately normal. We have seen how to quantify the estimates of the variation in sample means using the normal distribution. Simple random samples avoid all bias; they are affected only by the smaller and more controllable sampling error.

> **Sampling error** is the absolute difference between sample results and the parameter being estimated. Sampling error results from the sample-to-sample variation that is found in the random sampling process.

Recall from Section 1 of Chapter 1 that a simple random sample of size n is selected in such a way that every subset of size n from the population has an equal chance of being selected. How can this be done? Consider a hat containing 100 equal-sized slips of paper, each of which has someone's name and age on it. If the hat full of slips is the universe (or population), random samples may be selected by blindly drawing slips from the hat. If slips are put back in the hat after each draw, the process is sampling with replacement; if slips are put aside, the process is sampling without replacement. The resulting sample mean provides an estimate of the average age of the group. We can also make a statement about the margin of error for that estimate—for example, "With 95% certainty, the average age of the group is 21 years, plus or minus 2 years." Such a quantified estimate of the margin of error cannot be provided for a judgment sample.

Selecting a simple random sample

How can you carry out the selection process without the hat and the slips of paper? You can use a table of random numbers to make your selection. Table 6.2 on page 214 is a table of random numbers generated by Excel. To select a sample from the population of 100, you use any two adjacent columns in the table of random numbers. If each population element is assigned a number (for example, from 00 to 99), then selecting the first two elements in column 1 for each of rows 1 to 30 will generate a random sample of size 30. (Leading spaces should be treated as zeros; Excel does not print leading zeros.)

Table 6.2

RANDOM NUMBERS FOR SIMPLE RANDOM SAMPLING

Row	Column 1	Column 2	Column 3	Column 4	Column 5
1	776543	784386	170507	120426	433576
2	42085	962827	291543	543961	37110
3	407361	444227	300638	101199	277199
4	455214	986235	148778	42207	774742
5	170202	210059	895565	472273	197516
6	90670	399334	424512	249061	314737
7	280556	52339	756034	7172	408917
8	512924	448713	872127	487258	752860
9	821618	922665	486281	329325	295510
10	119449	354594	404706	858424	573412
11	35524	821466	465865	469008	321665
12	663197	982634	364909	578630	208594
13	946378	27467	869258	683767	935025
14	453749	509262	719992	818902	240455
15	586077	922726	678670	954588	471633
16	458021	670888	662220	860438	63021
17	621631	272194	364513	386455	630359
18	849970	196295	65218	464461	558671
19	970182	443922	26215	383465	155034
20	342967	673146	295205	208075	781944
21	673268	9735	921628	64028	576372
22	999389	14008	128391	782463	912350
23	496078	576006	88595	56917	843927
24	275734	495956	106021	509903	956937
25	57375	678975	802697	694234	47487
26	956266	667500	691396	236213	983977
27	398236	270302	547257	823755	812493
28	471053	882167	204748	920407	169835
29	392193	833978	619250	120517	978361
30	20600	444624	826410	301217	771690

Source of Data: Excel.

Where simple random sampling is practical, it offers the surest way of avoiding bias in the way the sample is selected. From theory and from practical examples, we know that simple random sampling works well and gives the sharpest bounds on errors for sample means. Its main disadvantage is that it is sometimes cumbersome to carry out, as the chosen sample may be widely scattered geographically or otherwise hard to assess. Thus, although systematic random sampling or stratified random sampling may be used for convenience or to ensure that special groups are adequately represented in the sample, simple random sampling is generally the tool of choice.

Systematic Random Samples

Though simple random samples give reliable population estimates, matching the list of random numbers with population elements requires you to jump back and forth from 31 to 24, to 75, and so on. If you can't computerize some of this work, you may want to use the more convenient **systematic random sampling.** Systematic random sampling was applied to a list of stock funds in Section 1 of

Chapter 2. The process begins with a list of the population to be sampled. Such a list of population elements is known as a **sampling frame.** Once you have a sampling frame, systematic random sampling is easy. The starting point is selected at random in the sampling frame, and then every *k*th item thereafter is selected, where *k* is chosen to give the right sample size. The process is illustrated in Example 6.5.

EXAMPLE 6.5 ▪ USING SYSTEMATIC RANDOM SAMPLING FOR VOTERS

Selecting a systematic random sample

How might we obtain a sample of the voting public? Registration rolls are the natural sampling frame. In this instance, we might select a value K from the table of random numbers as a starting point, and then another value N as the interval between selected values.

Suppose we threw a dart at Table 6.3 and it landed on the number 3 on line 14 in column 9. We would then let $K = 3$. Then a second dart landed on the number 4 on line 25 in column 9. So, we let $N = 4$. Our random selections would then begin with the third item and include every fourth item until the appropriate sample size was generated, yielding a systematic random sample.

Table 6.3

PORTION OF TABLE OF RANDOM NUMBERS

Row	Column 6	Column 7	Column 8	Column 9		Column 10
12	276065	174269	847693	735534		476546
13	393420	798757	645733	90175		337244
14	119267	228835	561292	672783	K	290998
15	593001	179609	61182	657581		292834
16	794465	27067	290628	753416		574723
17	300964	706500	275852	597943		280449
18	357403	812043	88206	60444		189680
19	322319	132246	647518	271478		425131
20	868943	797787	789862	469458		234157
21	35330	350707	817047	130193		515195
22	276150	364712	137724	885907		409470
23	238394	669881	496736	548996		303033
24	724	481088	300797	87345		105035
25	263027	169679	118752	175714	N	531094
26	451452	600216	855571	437916		78657
27	650183	551963	847544	748617		687323
28	845913	509354	660677	546840		733663
29	347115	51981	60697	207221		285526
30	599896	549999	361040	389307		474011

Source of Data: Minitab.

The advantage of systematic random sampling is that it is easy to select the sample by hand. To understand the disadvantages of systematic random sampling, we need to compare it with simple random sampling.

The number of simple random samples of size 100 that can be drawn from a population of 10,000 is staggering. But for at least 95% of these samples, the sample proportions will fall within .1 of the true ratio.

Now consider systematic random samples from the same population. If we use the sampling method and sampling frame given in Example 6.5, in which the sample is determined by two single-digit numbers, there will be only 100 different systematic random samples. Will these 100 samples be as representative as *all* simple random samples? They may be, but we can't be sure, given that there are over 100^{100} times more simple random samples of size 100 than there are systematic random samples that can be taken from a population of 10,000.

You can vary the method of taking systematic random samples, as noted in Section 1 of Chapter 2, but there are still far fewer systematic random samples than simple random samples. Because systematic random samples are limited in number, the possibility of bias always exists in such sampling. For this reason, simple random samples are generally preferred over systematic random samples, despite the fact that systematic random samples are easier to construct, especially when working by hand.

Stratified Random Samples

Neither simple random sampling nor systematic random sampling is as efficient as stratified random sampling or cluster sampling. These two techniques modify the nature of the selection process without destroying its randomness.

Stratified random sampling utilizes additional information to divide the population into subgroups. These subgroups are called strata (stratum in the singular). A study of appliance sales in a particular area might divide the stores selling the appliances by type: discount outlets, department stores, and stores specializing in appliances. Each of these categories would be a stratum. If sales were to be analyzed by store type, it would be important to ensure that enough data were gathered from each type of store (stratum). The plan for the study would include deciding how large a sample to take from each stratum.

> **Strata** are subgroups of homogeneous observations used in stratified sampling procedures.

After a random sample was drawn from each stratum, the sample results for the various strata would be combined, based on their numerical importance, to arrive at an overall estimate. Greatest efficiency is attained if there is general homogeneity within strata and larger differences between strata.

EXAMPLE 6.6 ■ USING STRATIFIED RANDOM SAMPLING TO SURVEY TOURISM

Selecting a stratified random sample

In June 1979, the Suffolk County Department of Economic Development conducted a survey of tourism in the county. The aim of the survey was to find out where the tourists came from, what they liked to do, and whether they found the "Discover Suffolk" guide useful.

A stratified random sample was used to gather the data. Using the mailing list of people requesting "Discover Suffolk" as a sampling frame, the researchers divided the respondents into groups based on geography. Recipients of "Discover Suffolk" who lived outside of Suffolk County were divided into three strata:

Nassau County (the neighboring county)
New York City (just west of Nassau County)
Other (including respondents from as far away as California)

By stratifying geographically, the Department of Economic Development was able to choose a random sample from each group. The results of the random samples were combined, according to the percentage of requests for "Discover Suffolk," to generate realistic estimates for several variables. For example, it was found that 43% of the tourists came to Suffolk County to visit restaurants. However, analysis by strata revealed that 50% of the tourists from Nassau County cited restaurants, whereas only 38% of those from New York City did so. This suggests that Nassau County may represent a fertile market in which to develop an interest in dining in Suffolk County, while other features of Suffolk might be highlighted in an advertising campaign in New York City. It also suggests where Suffolk County merchants might direct their promotional efforts.

To use stratified random sampling, then, you set up strata in your population, reflecting portions of the population whose opinions you wish to assess, and use random sampling on these strata. With this method, you are assured adequate sampling of strata important to you, so that you can estimate the population parameters within these strata. However, stratified random sampling will not give as precise an overall estimate of the population parameters for a given sample size as will simple random sampling.

Cluster Samples

When the universe under study is spread out over a large geographical area, **cluster sampling** may help to reduce costs, though simple random sampling remains the gold standard for statistics. Cluster sampling is particularly useful when it is important to conserve resources, such as gasoline, interviewers' time on the road, and so forth. The Bureau of the Census uses cluster sampling in the Current Population Survey, which is the basis for United States employment statistics. Cluster sampling is frequently used in housing surveys. City blocks are randomly selected, and a complete enumeration is made of these blocks, or clusters.

Using cluster sampling

Another application of cluster sampling is in evaluating the size of an agricultural crop. A systematic sample of plots might be drawn, and a cluster of trees selected in each plot. The number of trees in the cluster will depend on what is workable for the field staff; for example, a cluster might consist of six trees. All the fruit on each of these trees is then counted, and the results are used to estimate the entire crop.

The advantages of cluster sampling include greater convenience and reduced costs in cases in which travel is involved or a bundled unit must be broken up to examine any of the elements in the bundle. However, because the units in a cluster may be more alike than randomly selected units from the entire population, the accuracy for a given sample size will not be as great as with simple random sampling.

EXERCISES 6.20–6.28

Applying the Concepts

6.20 Identify the sampling method used in each of the following situations, and describe briefly its advantages and limitations.

a. Windsor Shirt Company leaves service evaluation cards on its register counter for customers to fill out if they wish to do so.

b. To conduct a crop survey, the Department of Agriculture randomly selects plots and makes a count of the crop from each plot.

c. To obtain a sample of registered voters, a polling agency chooses a random voter from a list of voters and then selects every 30th voter thereafter on the list.

d. To find out more about its policyholders, an insurance company categorizes its policyholders by age group and surveys a random sample from each group.

e. An insurance company uses its computerized files to select 750 customers at random for a survey of customer satisfaction.

6.21 University Food Service wants feedback on its plans for a lounge and food service for faculty but does not want to poll the entire faculty. Develop a sampling design that would properly reflect the faculty's feelings. Indicate why you chose this particular design.

6.22 A local bookshop has a questionnaire that it wishes to distribute to its customers. The shop's mailing list is relatively current. Propose a sampling method for this survey in a brief draft memo to the bookstore owner.

6.23 The National Association of Purchasing Management conducts a survey of purchasing executives in over 300 industrial companies across a broad spectrum of industries. Design a sampling procedure that will provide a representative sample of industrial companies, and justify the approach you chose.

6.24 A local manufacturer of nuts and bolts wishes to set up quality control procedures to ensure uniform high-quality output. Clearly, every nut and every bolt coming off the production line cannot be tested. Write a memo proposing a sampling technique to monitor quality.

6.25 A real estate agency wants to obtain a representative value for houses in a particular community. Which sampling method should it employ? Explain your reasoning.

6.26 A university's athletic department wants to use a new ticket distribution method in the coming year. In addition to setting aside a block of tickets for past season ticket holders, the department wants to set aside a block of tickets for alumni. The department must conduct a survey to determine the opinions of the university community.

a. Design a sampling procedure that reflects the opinions of all the groups in the community.

b. Make up a brief questionnaire that could be used.

6.27 The secretary of an investment club of some 250 members wants to ascertain the objectives of its members.
a. Design a questionnaire for this survey.
b. What type of sampling method should be employed? Justify your choice.

6.28 An accountant auditing the books of a firm wishes to check accounts receivable. Write a memo proposing a sampling plan to check accounts receivable.

FORMULAS

Expected Value of Sample Means

$$E(\bar{x}) = \mu = E(X)$$

where the random variable X is being sampled.

Standard Error of the Mean

$$\sigma_{\bar{x}} = \frac{\sigma_x}{\sqrt{n}}$$

z Score for Sample Mean

$$z = \frac{\bar{x} - \mu_x}{\sigma_{\bar{x}}}$$

Population Proportion

$$\pi = \frac{\text{Number of favorable cases}}{\text{Total population}} = \frac{X}{N}$$

Identify π with p for binomial trials.

Sample Proportion

$$p = \frac{\text{Number of favorable cases}}{\text{Size of sample}} = \frac{x}{n}$$

Expected Value of the Proportion

$$E(p) = p = \pi$$

Standard Error of the Proportion

$$\sigma_p = \sqrt{\frac{\pi(1 - \pi)}{n}}$$

z Score for Sample Proportion

$$z = \frac{p - \pi}{\sigma_p}$$

Finite Population Correction Factor

$$\text{fpcf} = \sqrt{\frac{N - n}{N - 1}}$$

NEW STATISTICAL TERMS

central limit theorem
chunk sampling
cluster sampling
convenience sample
finite population correction factor (fpcf)
judgment sampling
sampling distribution of the mean
sampling distribution of the proportion
sampling error
sampling frame
simple random sampling (SRS)
standard error of the mean
standard error of the sample proportion
strata
stratified random sampling
systematic random sampling

EXERCISES 6.29–6.32

Supplementary Problems

6.29 There are many drip coffee makers on the market. In 1994, the average price of those models on the market was $33.53, and the standard deviation was $13.10 (*Consumer Reports,* October 1994). Suppose a random sample of 30 coffee maker models is selected.

a. What is the probability that the average price will be less than $25?

b. What is the likelihood that the average cost of a drip coffee maker will be between $35 and $45?

c. Calculate the probability that the average cost of the drip coffee makers will be at least $30.

6.30 A survey of forensic psychiatrists revealed that the average daily rate for court testimony is $2500; the standard deviation is $750 (National Forensic Center, *Guide to Experts' Fees, 1992–1993*). Assume these figures are for the population of forensic psychiatrists. Consider a random sample of 50 forensic psychiatrists.

a. What is the probability that the average remuneration of forensic psychiatrists falls between $2300 and $2700 per day?

b. What is the probability that an individual forensic psychiatrist earns less than $1000 for a day's work?

c. What is the probability that the average daily remuneration for a random sample of 50 forensic psychiatrists is less than $1000?

d. Within what range may we expect to find the central 95% of daily rates for samples of size 50?

6.31 An analysis of a survey concerning industrial migration (that is, the reasons firms move from one area to another) revealed that 57% of manufacturers considered occupancy costs to be a very important factor. (Assume that this is the population percentage.)

a. In a random sample of 100 manufacturers, what is the probability that over 60 will rate this as a very important factor?

b. What is the probability that in a sample of 100 manufacturers, fewer than 50 will rate this as a very important factor?

c. Calculate the likelihood that, in a random sample of 100 manufacturers, between 55% and 60% will rate occupancy costs as a very important factor.

d. Is it realistic to expect that over 75% in a random sample of 100 will rate occupancy costs as a very important factor?

6.32 A study entitled *Older Workers in the New York City Labor Market* (Albert C. Ovedovitz, March 1983) found that 29.3% of unemployed workers over age 55 had been unemployed for 27 weeks or more. Samples of 300 older workers were randomly selected. Using the reported percentage as the population percentage, answer the following questions.

a. What is the probability that a sample will be found in which at least 35% of the workers have been unemployed for 27 weeks or more?

b. In what proportion of samples have between 20% and 30% of the workers been unemployed for 27 weeks or more?

c. Is it conceivable that a sample will have more than 40% of the workers unemployed for 27 weeks or more?

7

Estimation

OUTLINE

LEARNING OBJECTIVES

1. Define a point estimate.
2. Enumerate the characteristics of a good point estimate.
3. Describe a confidence interval.
4. Understand the relationship between sample size and confidence levels.
5. Estimate a mean for a large sample with a given level of confidence.
6. Estimate a proportion with a given level of confidence.
7. Distinguish between a statistically large and a statistically small sample.
8. Use the *t* distribution to estimate a population mean from a small sample.
9. Determine sample size for a survey.

BUILD YOUR KNOWLEDGE OF *STATISTICS*

Polls provide sample statistics that are then used to reach conclusions about the public at large. As a reflection of the public's views on health care reform, the *New York Times* published responses from a random telephone poll, in which respondents were asked to complete the phrase "That every American receives health care coverage is . . . ." Approximately 82% said "very important," 16% said "somewhat important," and the remainder said "not very important" or had no answer (March 15, 1994).

We can't ask everyone what she or he thinks, but a random survey of 1000 to 2000 people can give us a general picture. As you will see in this chapter, such surveys can pin down the percentage of the public in favor of or against a given proposal within ±3%, with 95% confidence of being correct.

Survey results such as those reported in the *New York Times* are part of the foundation for business decisions. Other surveys have more direct business effects. For example, the Nielsen Media Research survey of audience size is used to determine advertising rates. Based on Nielsen ratings, the cost of a 30-second commercial during the 1994 Superbowl was $600,000. To attract advertisers, newspapers conduct surveys of their readership (see "How Newspapers Use Demographic Data," *American Demographics,* February 1980). Each year, the Department of Agriculture estimates expected crop yields, and these estimates drive the commodity markets. Surveys and sampling affect manufacturers' estimates of the subsequent year's sales, which in turn determine staffing and purchasing decisions.

In Chapter 6, we worked with the sampling distributions of the mean and the proportion. Given the parameters of a particular population, we found the probability that the sample statistics would fall within particular ranges. But generally we do not know the values of the parameters of sampling distributions. What we measure are sample statistics drawn from those distributions. In this chapter, you will see how knowing a sample statistic and the sampling distribution allows us to estimate the population parameter using the central limit theorem.

SECTION 1 ■ POINT ESTIMATES

Based on a sample, statisticians often derive a single figure that is then used to characterize the population. Such a figure is known as a point estimate.

The thought of going to graduate school to get an MBA, a medical degree, or a law degree is typically followed by the question "How much will it cost?" In 1993, the average out-of-state tuition for the "Top 25" MBA programs was $16,996 per year. For one of the top-rated medical schools, it was $19,465 per year. Out-of-state tuition at the most highly ranked law schools was $16,195 annually (*U.S. News & World Report,* March 21, 1994). These numbers are all point estimates—specific values, derived from a sample, that allow the reader to estimate what his or her out-of-state tuition costs might be.

> A **point estimate** is a value used in place of an unknown population characteristic (parameter). Often, point estimates are based on sample statistics such as the mean or standard deviation.

A point estimate is a single figure that is representative of the larger population. We are concerned here with *scientific* point estimates—that is, point estimates obtained through random and representative sampling techniques. There is no way of telling whether other types of point estimates, such as hunches or guesses, are good or bad. Just as businesses are evaluated with respect to profits and price/earnings multiples, statistics used as estimators are evaluated on the basis of numerical standards. Good estimators are *unbiased* and *consistent.* These criteria can be systematically applied only when the estimators are determined by specific procedures applied to a known type of data, such as random samples drawn from an underlying population.

Unbiased Estimators

The sample mean is an unbiased estimator of the population mean in the following sense: If we repeatedly draw random samples from a population, the average of the means of those samples will tend toward the population average, μ. Suppose you wanted to know the average age of your classmates. A census of your classmates' ages would yield the population mean for your class. If you then drew every possible subsample of a fixed size from your class and found the grand mean of all these averages, it would be the same as the population mean. This is one conclusion of the central limit theorem, and it is supported by the experiments and analysis in Section 1 of Chapter 6.

> A sample statistic is an **unbiased estimator** of a population parameter if the expected value of the sample measures (statistics) equals the population value (parameter).

EXAMPLE 7.1 ■ **EVALUATING THE SAMPLE MEAN AS AN UNBIASED ESTIMATOR**

Unbiased estimators

Table 7.1 gives commuting times for the five areas with the longest workers' commutes in the United States.

Table 7.1

THE LONGEST COMMUTES

Area	Commuting Time
A. New York, NY	75.8 minutes
B. Long Island, NY	64.6 minutes
C. Washington, DC vicinity	63.0 minutes
D. Chicago, IL	61.5 minutes
E. Riverdale–San Bernadino, CA	59.3 minutes

Source of Data: *Places Rated Almanac,* 1993.

In Table 7.2, we list all 10 possible samples of size $n = 3$ that can be drawn from this population. Is the sample mean an unbiased estimator of the population mean?

Table 7.2

DISTRIBUTION FOR SAMPLES OF SIZE $n = 3$

Sample	Data	Mean	Sample	Data	Mean
ABC	75.8, 64.6, 63.0	67.80	ADE	75.8, 61.5, 59.3	65.53
ABD	75.8, 64.6, 61.5	67.30	BCD	64.6, 63.0, 61.5	63.03
ABE	75.8, 64.6, 59.3	66.57	BCE	64.6, 63.0, 59.3	62.30
ACD	75.8, 63.0, 61.5	66.77	BDE	64.6, 61.5, 59.3	61.80
ACE	75.8, 63.0, 59.3	66.23	CDE	63.0, 61.5, 59.3	61.27

The mean for the population of $N = 5$ is 64.84 minutes. We compute the expected value for the 10 sample means as follows:

$$E(\overline{x}) = \frac{67.80 + 67.30 + \cdots + 61.27}{10} = 64.86$$

We find that $E(\overline{x}) = \mu_x$.

In Example 7.1, we found that the sample mean was an unbiased estimator of the population mean. As discussed in Chapter 6, the first part of the central limit theorem states that the expected value for the sample mean is always equal to the population mean; in short, sample means are unbiased estimators of the population mean.

Consistent Estimators

An important characteristic of any statistic is consistency. Intuitively, we know that a statistic is consistent if it becomes more reliable as an estimator as the sample size increases. The pattern of consistency for sample means was apparent in Figure 6.8, where the distribution of sample means became more concentrated about the population mean as the sample size became larger.

A statistic is a **consistent estimator** of a parameter if the probability that the estimator falls within a given range about the parameter increases to 1 as the sample size grows.

The limit of 1 will be achieved when the sample size equals the population size, as illustrated in Example 7.2.

EXAMPLE 7.2 ■ EVALUATING THE SAMPLE MEAN AS A CONSISTENT ESTIMATOR

Consistency

From the central limit theorem, we know that the sampling distribution of the mean has an expected value equal to the population mean, μ, and we know that the standard deviation is $\sigma/\sqrt{n}$, where σ is the population standard deviation and n is the sample size. We have seen that as n increases, the sample means are more and more concentrated about the population mean. Thus, in sampling with replacement, the sample means are consistent estimators of the population mean. How consistent are the sample means when sampling is done without replacement? Consider the data in Table 7.3 on yearly expenditures on deodorant in 5 of the top 10 cities in the United States. The mean of this population is \$6.10.

Table 7.3

YEARLY EXPENDITURES ON DEODORANT

City	Per Capita Expenditure on Deodorant
A. Pittsburgh	\$8.04
B. Salt Lake City	6.70
C. Seattle	5.60
D. Dallas	5.24
E. Denver	4.93

Source of Data: Les Krantz, *The Best & Worst of Everything*, 1991, p. 95.

In Table 7.4, we consider all samples of size $n = 3$. The standard error of the means is \$.49, and the range of the means is \$1.52.

Table 7.4

DISTRIBUTION FOR SAMPLES OF SIZE $n = 3$

Sample	Data	Mean	Sample	Data	Mean
ABC	\$8.04, 6.70, 5.60	\$6.78	ADE	\$8.04, 5.24, 4.93	\$6.07
ABD	8.04, 6.70, 5.24	6.66	BCD	6.70, 5.60, 5.24	5.85
ABE	8.04, 6.70, 4.93	6.56	BCE	6.70, 5.60, 4.93	5.74
ACD	8.04, 5.60, 5.24	6.29	BDE	6.70, 5.24, 4.93	5.62
ACE	8.04, 5.60, 4.93	6.19	CDE	5.60, 5.24, 4.93	5.26

In Table 7.5, we consider samples of size $n = 4$. The standard error of the means is \$.32, and the range of the means is \$.78. What we see in this case is a pattern indicating the consistency of the sample means as estimators of the population mean. The larger the samples, the less variation around the true population

Table 7.5

DISTRIBUTION FOR SAMPLES OF SIZE $n = 4$

Sample	Data	Mean
ABCD	$8.04, 6.70, 5.60, 5.24	$6.40
ABCE	8.04, 6.70, 5.60, 4.93	6.32
ABDE	8.04, 6.70, 5.24, 4.93	6.23
ACDE	8.04, 5.60, 5.24, 4.93	5.95
BCDE	6.70, 5.60, 5.24, 4.93	5.62

mean of $6.10. In the smaller samples, the means range as far as $.84 from the true mean. In the larger samples, the means range no further than $.48 from the population mean.

Our definition of consistency requires that the estimator eventually fall within any given range about the population parameter as larger samples are taken. In this case, to estimate the population mean using the means of samples taken without replacement, we need only increase our sample size to 5 for the sample mean to equal the population mean exactly. Thus, for samples of size 5, we hit the population mean dead center, and the sample means fall within any given range about the population mean with a probability of 1. So the estimator not only looks consistent but is provably consistent.

EXERCISES 7.1–7.5

Skill Builders

7.1 Five respondents to a survey answered as follows:

A. Yes
B. Yes
C. Yes
D. No
E. No

a. Find the population proportion responding positively.
b. Draw all samples of size $n = 4$ from this universe without replacement. List them.
c. Show that the sample proportions provide an unbiased estimate of the true proportion of respondents answering positively.

7.2 The following four measurements were taken of a particular machine part:

3.4 mm 3.6 mm 3.5 mm 3.5 mm

a. Find the population mean for this data set.
b. Draw all samples of size $n = 3$ from this population without replacement. List them.
c. Calculate the sample means.
d. Repeat parts b and c for samples of size $n = 2$.
e. Demonstrate that the means exhibit consistency for this data set.

7.3 According to the 1990 Survey of Buying Power in Sales and Marketing Management, the five metropolitan areas with the highest percentages of households earning over $50,000 per annum were as follows:

A. Nassau/Suffolk Counties, NY	44.8%
B. Stamford/Norwalk/Danbury, CT	43.4
C. Middlebury/Somerset, NJ	43.2
D. San Jose, CA	40.8
E. Washington, DC	40.4

Source of Data: Les Krantz, *The Best & Worst of Everything,* 1991, p. 78.

a. Draw samples of size 3 and then of size 4 without replacement to illustrate the concept of consistency of the sample means as estimators of the population mean.

b. Show that the sample means are unbiased estimators of the population mean.

7.4 Wholesale sales of legal medicines run into the hundreds of millions of dollars annually. Data for the top nonprescription medicines are shown below:

A. Tylenol	$585 million
B. Advil	235 million
C. Anacin	150 million
D. Metamucil	125 million
E. Bayer Aspirin	120 million

Source of Data: Les Krantz, *The Best & Worst of Everything,* 1991, p. 263.

a. Calculate the mean for this data set.

b. Draw all samples of size of $n = 2$ without replacement. Calculate the means of the samples.

c. Show that the sample means are unbiased estimators of the population mean.

7.5 The editors of *Total Baseball* developed a rating scale of pitcher excellence by combining over two dozen measures. Using this scale, they rated the following pitchers as the greatest of all time:

Pitcher	Rating
A. Walter Johnson	81.5
B. Cy Young	81.0
C. Kid Nichol	65.3
D. Pete Alexander	64.8
E. Lefty Grove	59.9

Source of Data: Les Krantz, *The Best & Worst of Everything,* 1991, p. 11.

a. Generate all samples of size $n = 4$ from this population without replacement.

b. Calculate the sample means and medians.

c. Calculate the population mean and median for all samples of size $n = 4$.

d. Generate all samples of size $n = 3$ from this population without replacement, and calculate the sample means and medians.

e. Evaluate these means and medians with respect to bias and consistency.

SECTION 2 ▪ CONFIDENCE INTERVALS

We have seen that sample means give point estimates tightly clustered about the population mean. However, because points have no breadth, the chance that one of these point estimates will hit the population mean dead center is essentially zero. To achieve a 95% chance of hitting the population mean, we need to use something with breadth—a confidence interval, computed from information in the random sample in such a way that for 95% of the random samples of size n it will contain the population mean, μ. Other procedures will give confidence intervals for other population parameters.

Consider Figure 7.1. The basis for the construction of this figure is the observation, made in Chapter 6, that sample means follow a normal distribution (see especially Figure 6.8). At the top of Figure 7.1 is the sampling distribution for samples of size 20, drawn at random from a normal population with a known mean, μ. The thick bar below the distribution is centered on μ, and its width is such that 95% of the sample means, $\bar{x}$, will fall in this range. The width of this bar determines the widths of the 100 95% confidence intervals for μ shown below it. These confidence intervals are centered on the values of $\bar{x}$ obtained from the 100 random samples.

Here, μ is known, and we can see that 96 of the 100 confidence intervals contain μ. (The four confidence intervals that miss μ have been highlighted in color in Figure 7.1.) This is about what we would expect for 95% confidence intervals. Generally, μ is unknown; we can be sure that, on average, 95 out of 100 95% confidence intervals will contain μ, but we can't tell which ones these will be.

Estimates of the population mean obtained from the averages of random samples will follow the sampling distribution described by the central limit theorem. These sample estimates of the mean will cluster around the true population mean. A confidence interval takes into account the inherent variability of sample statistics about the population parameter, providing a range that has a specified probability of capturing the population parameter.

> A **confidence interval** is a range that has a specific probability of containing the population parameter. Such intervals are calculated from data obtained by random sampling. If the procedure gives a 90% confidence interval based on a sample of size *n*, the intervals for 90% of such samples will contain the population parameter.

The process of estimating the true mean of a population is similar to that of drilling for oil. An oil company will drill a well where its geologists believe it is highly probable that oil will be found. The greater the number of positive signs generally associated with oil deposits, the greater the likelihood of finding oil. If an oil company wants to increase the likelihood that its leases will cover the oil pool, it can either (1) lease a larger area based on the existing amount of geophysical evidence or (2) gather more data so that the location of the oil pool can be better pinpointed. This situation is analogous to that of the survey researcher selecting a confidence interval based on survey data. The researcher will use a wider confidence interval when data are scarce and a narrower interval when data are plentiful. You can see the contrast by comparing Figure 7.1 with Figure 7.2.

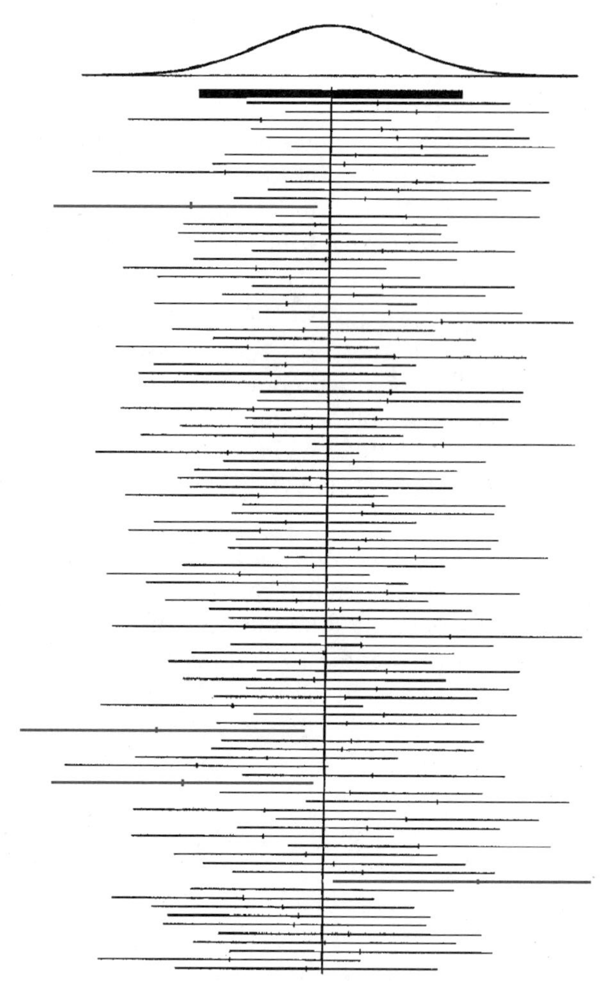

Figure 7.1
Sampling distributions and 95% confidence intervals

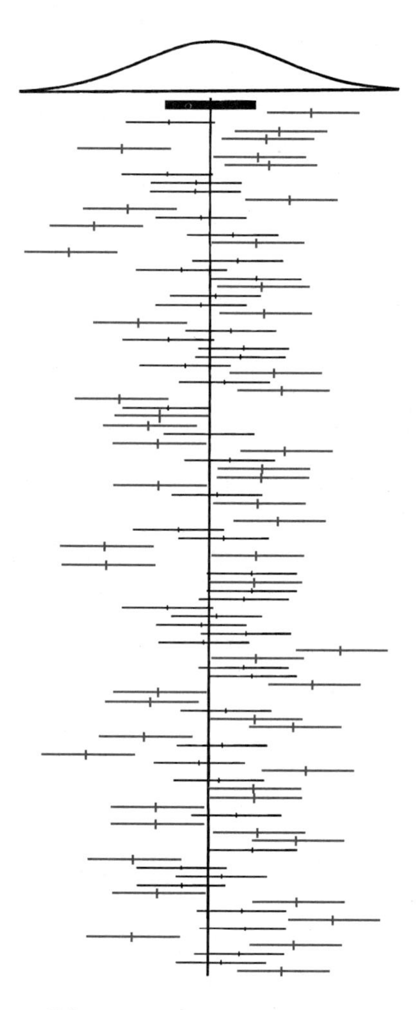

Figure 7.2
Sampling distributions and 50% confidence intervals

Working out confidence intervals requires the same sort of reasoning geologists use when they are deciding whether to drill within a broad tract or try to pin down the location of an oil pool to a smaller area. How much data to gather depends on business considerations—the costs of gathering data, the payoffs, and the costs of discovering that the sought-after target is not in the predicted range. The critical issues are how to find confidence intervals at various levels of confidence and precision and how to determine what size sample is needed to achieve the right combination of a probability of correctness and precision. To see the effect on confidence intervals of increasing the sample size from 20 to 80, compare Figure 7.1 to Figure 7.3. Once you understand the principles behind these tradeoffs, you can decide how much data should be gathered in order to make the business decisions at hand.

STATISTICS TOOL KIT

How Confidence Intervals Depend on Sample Size and Confidence Levels

You can extend your understanding of confidence intervals by comparing Figure 7.1 with Figures 7.2 and 7.3, which show how the behavior of confidence intervals changes when confidence level or sample size changes. All three figures were constructed using computer-generated random sampling.

As you look at Figures 7.2 and 7.3, observe the following:

- The population mean, μ, stays fixed, but the confidence intervals vary depending on the particular random sample chosen. Repeated sampling will produce varied results, even though the procedure and confidence level are fixed.
- The higher the confidence level for a given sample size and population, the wider the confidence intervals are and the more likely they are to include the value of μ.
- The shorter the confidence intervals, the more likely they are to miss μ.
- If you keep the same confidence level but increase the sample size, the distribution of $\bar{x}$ becomes more concentrated and the confidence intervals become shorter. Thus, your estimate becomes more precise.

When you use confidence intervals, all you know is where the interval is and the *probability* that such a randomly constructed interval will contain μ. Generally, you can't tell whether or not a *particular* confidence interval contains μ, because in most situations you don't have any independent way of determining μ.

Figure 7.2 shows the same sampling distribution as Figure 7.1, but the thick bar below the graph extends just $.674\sigma$ to either side of the population mean. The value .674 is the critical value of z for a 50% confidence level. It determines the length of the 50% confidence interval for μ.

Below the bar are 100 50% confidence intervals for μ, obtained with the same data used for the 95% confidence intervals in Figure 7.1. Note that the 50% confidence intervals are shorter, and only 47 of them overlap μ, whose value is marked by the vertical line.

Figure 7.3 is the sampling distribution for $\bar{x}$ with $n = 80$. Compare this distribution with the distribution for $n = 20$ in Figure 7.1; both have the same underlying population. Of the observed values of $\bar{x}$, 95% will fall within the interval indicated by the thick bar, which is the width of a 95% confidence interval. Since the sample size is four times the earlier sample size, the confidence intervals are now $1/\sqrt{4} = \frac{1}{2}$ the earlier width, yielding a corresponding increase in the precision of our estimate. All but 3 of the 100 95% confidence intervals below the graph contain μ.

The confidence intervals shown in Figures 7.1, 7.2, and 7.3 are for samples from normal populations, so the sample means are also normally distributed. But the same methods can be used whenever sample sizes are sufficiently large that the central limit theorem holds. The sample means will then be approximately normal.

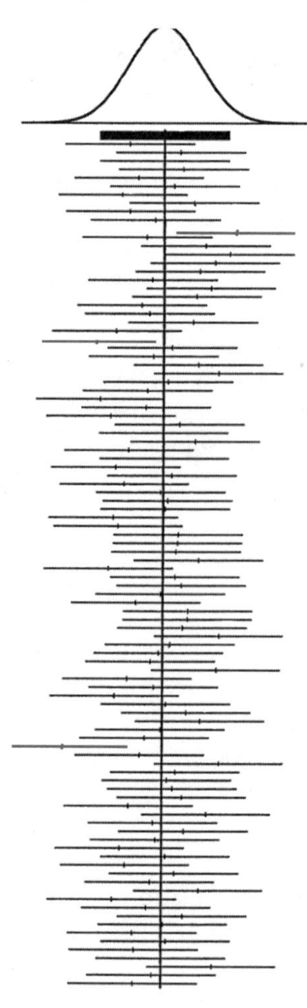

Figure 7.3
Sampling distributions and 95% confidence intervals for a sample size of 80

Large Samples

In Figure 7.4, we consider the widths of intervals about μ in terms of σ for a normal random variable. These widths are the widths of confidence intervals used to pin down the population mean, μ, using the sample mean, $\bar{x}$.

We know that

50% of the values of $\bar{x}$ lie within the top interval from $\mu - .674\sigma_{\bar{x}}$ to $\mu + .674\sigma_{\bar{x}}$.

90% of the values of $\bar{x}$ lie within the second interval from $\mu - 1.645\sigma_{\bar{x}}$ to $\mu + 1.645\sigma_{\bar{x}}$.

95% of the values of $\bar{x}$ lie within the lowest interval from $\mu - 1.960\sigma_{\bar{x}}$ to $\mu + 1.960\sigma_{\bar{x}}$.

Thus, to get a 95% confidence interval for μ, we use a range of $\bar{x} - 1.960\sigma_{\bar{x}}$ to $\bar{x} + 1.960\sigma_{\bar{x}}$.

The figure confirms that confidence intervals work the same way as an oil company's efforts to locate an oil pool: the wider the interval, the higher the confidence level; the narrower the interval, the lower the confidence level. There is always a tradeoff between accuracy and precision. If we wish to be more accurate (correct more often), then we cannot be as precise in our attempt to pin down the mean.

This tradeoff between accuracy and precision is a common feature of ordinary experience. Suppose I ask you at what time you arrived home last night. If you respond "At 2:23 A.M.," your answer is quite precise, but it may not be correct. Your watch may have been fast or slow. If you reply "Between 2 and 3 in the morning," your answer is far less precise but more likely to be correct. Just how far off could your watch be? In choosing how to respond to this question, you might ask yourself whether the exact minute you arrived home is important. Probably not. The fact that you arrived home in the early morning hours is the important information. Statistical inference works in the same manner. The requirements of the situation determine the proper tradeoff between greater precision and a higher probability of being right.

Confidence intervals for large samples are constructed using the central limit theorem. We know that the sample means fall in approximately a normal distribution when the sample size is sufficiently large. For practical purposes, a large sample is defined as $n \geq 30$. The sample means have a standard deviation equal

Figure 7.4
How the widths of confidence intervals for μ vary with the confidence level

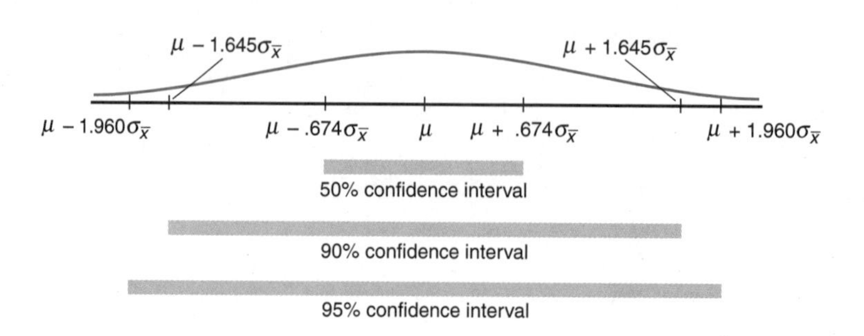

to the standard deviation of the population divided by the square root of the sample size:

$$\sigma_{\bar{x}} = \frac{\sigma_x}{\sqrt{n}}$$

Confidence intervals are constructed at varying **confidence levels,** such as 90%, 95%, and 99%. Their general form is determined by finding a critical value for z that satisfies the formula

$$P(\mu - z\sigma_{\bar{x}} \leq \bar{x} \leq \mu + z\sigma_{\bar{x}}) = \text{Confidence level}$$

The process is the same as finding z so that $P(-z \leq X \leq z)$ equals the confidence level, where X is a standard normal random variable; therefore, z can be found from a table of the standard normal distribution, such as Table IV in Appendix B.

As shown in Figure 7.5, the confidence interval itself stretches from $\bar{x} - z\sigma_{\bar{x}}$ to $\bar{x} + z\sigma_{\bar{x}}$, where the values of z are as listed in Table 7.6. The higher the confidence level, the larger z must be and the wider the confidence interval will be. To achieve a confidence level of 90%, we choose z so that the probability that a standard normal random variable will fall between $-z$ and $+z$ is 90%. The tables show that a standard normal variable has a probability of 45% of lying in the interval from

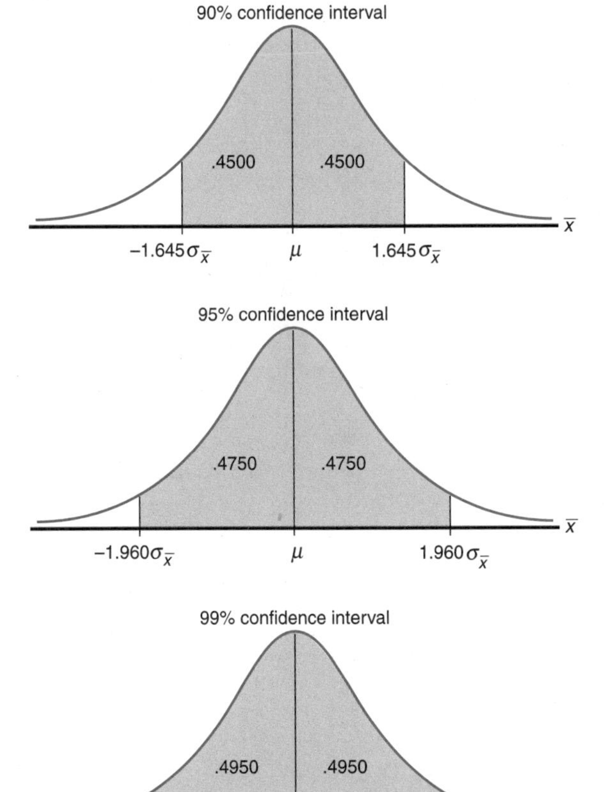

Figure 7.5
Confidence levels and width of confidence intervals for the arithmetic mean figured from a random sample of fixed size

0 to 1.645, so such a variable has a probability of 90% of lying between −1.645 and +1.645. This is how we found z for a 90% confidence level in Table 7.6. The other values were found in a similar fashion.

Table 7.6

z STATISTICS AND CONFIDENCE LEVELS

z	Confidence Level
1.645	90%
1.960	95
2.575	99

We know how to find z for a given confidence level, so we can find the confidence interval itself, given $\bar{x}$ and $\sigma_{\bar{x}}$. We get $\bar{x}$ from the data and either calculate $\sigma_{\bar{x}}$ from the population variance or estimate it using $s_{\bar{x}}$, as in the following examples.

EXAMPLE 7.3 ■ ESTIMATING A MEAN WITH 95% CONFIDENCE

Suppose an investor wants to determine the true average yield on a selection of investments from the *Business Week* 1000. Suppose the investor draws a random sample of three dozen firms and finds an average yield on investment of 2.37% and a sample standard deviation of 2.13% (*Business Week,* March 28, 1994). Construct a 95% confidence interval for the true average yield, using this value of s as an estimate for σ for the population.

Confidence intervals for large samples: the mean

Step 1. We write the confidence level formula:

$$P(\bar{x} \pm z s_{\bar{x}}) = \text{Confidence level}$$

Step 2. We select the z statistic for the 95% confidence level:

$$z = \pm 1.96 \text{ for 95\% confidence level}$$

Step 3. We calculate an estimate for the standard error of the mean, $s_{\bar{x}}$, using s_x as a point estimate for the population standard deviation, σ_x:

$$s_{\bar{x}} = \frac{s_x}{\sqrt{n}} = \frac{2.13}{\sqrt{36}} = .36$$

With a large sample, this approximation is highly likely to be very close to what we would get using the unknown $\sigma_x/\sqrt{n}$ for $s_{\bar{x}}$.

Step 4. We substitute the values in the confidence level formula to find the interval with endpoints at 2.37 ± 1.96(.36), or 2.37 ± .71. The interval is from 1.66 to 3.08. As we saw in Figures 7.1 and 7.2, we can't be sure whether μ lies in this interval. What we do know is that if we construct a large number of intervals in this same way, based on random sampling, on average 95% will contain μ.

Step 5. Interpreting the results, we conclude, with a 95% confidence level, that the average yield among the top 1000 firms lies between 1.66% and 3.08%.

EXAMPLE 7.4 ■ ESTIMATING A PROPORTION WITH 95% CONFIDENCE

When a random sample of 50 students taking a required business statistics course at a large urban university were asked about their major, 10 responded that their major was accounting. Estimate with a 95% confidence level the true percentage of students choosing accounting as their major. To estimate the standard error of the proportion, use an approximation analogous to that used to approximate the standard error of the mean in Example 7.3. From Section 2 of Chapter 6, the standard error of the sample proportion is

$$\sigma_p = \sqrt{\frac{\pi(1 - \pi)}{n}}$$

Replacing the population proportion, π, by the sample proportion, p, and replacing $1 - \pi$ by $q = 1 - p$, we get

$$s_p = \sqrt{\frac{pq}{n}} = \text{Estimated standard error of the proportion}$$

Step 1. We write the confidence level formula:

$$P(p \pm zs_p) = \text{Confidence level}$$

Step 2. We select the z statistic for the 95% confidence level:

$$z = \pm 1.960 \text{ for 95\% confidence level}$$

Step 3. We calculate the estimated standard error of the proportion, s_p. The formula involves p and q, which are the unknown population parameters we seek. Using the observed sample values of .20 and .80 as point estimates for p and q, respectively, we have

Confidence intervals for large samples: a proportion

$$s_p = \sqrt{\frac{pq}{n}} = \sqrt{\frac{.20(.80)}{50}} = .057$$

(Alternatively, we could simply elect to be conservative. The largest value is obtained for s_p is when $p = q = .5$, so we could use $\sqrt{.5^2/n}$ as our estimate. This might give us a larger than necessary confidence interval, but the interval will contain the population proportion at least 95% of the time.)

Step 4. We substitute the values in the confidence level formula to find the interval with endpoints at $.20 \pm 1.960(.057)$, or $.20 \pm .11$. The interval goes from .09 to .31. We don't know whether this *particular* confidence interval contains π, but we do know that the procedure used produces random intervals depending on the data and that these have a 95% chance of containing π.

Step 5. Interpreting the results, we are 95% certain that the population proportion is between .09 and .31. In other words, the true percentage of students desiring to major in accounting would appear to lie between 9 and 31%. Our level of confidence in our estimate is 95%. If we wished to narrow the range, we could increase the sample size. Alternatively, we could arrive at a more precise estimate by reducing our level of confidence to, say, 90%. At a confidence level of 90%, the proportion would range from .11 to .29, inclusive.

Small Samples

If $n \leq 30$, the large-sample estimates just discussed are not reliable. Small-sample methods should be used, which work when only limited data are available. If the sampling distribution for a small sample is approximately normal, all the conclusions of the central limit theorem hold. But, as with confidence intervals based on large samples, we usually do not know σ and must use the sample standard deviation s instead of σ in the formula to determine the interval. And when n is less than 30, s is not a reliable point estimate for σ. So, to find the confidence interval, we make use of the t distribution instead of the normal distribution. The t distribution was first studied by W. S. Gosset while he was working for Guinness Breweries. Guinness had a policy of not allowing publications to be associated with the brewery, so Gosset published under the pen name "Student." This is why the distribution is often referred to as *Student's t distribution.*

The ***t* distribution** is defined by

$$t = \frac{\bar{x} - \mu_x}{s_x/\sqrt{n}}$$

If the underlying population is normal, with a mean of μ and the sample size of n, then t follows a *t distribution with $n - 1$ degrees of freedom.* The various t distributions have been studied, graphed, and tabulated just as the standard normal distribution has been. Note the difference in the density functions for t distributions with 2 degrees of freedom and 30 degrees of freedom, which are plotted on the same graph in Figure 7.6. The fewer the degrees of freedom, the more spread out the t distribution is. Thus, the confidence intervals giving a desired level of confidence also spread out more when the number of degrees of freedom is small. The intervals on the t axis corresponding to a 90% confidence interval are shown below the graphs in Figure 7.6. The shorter interval is for the more concentrated distribution with 30 degrees of freedom. It runs from $t = -1.07$ to $t = +1.07$. The larger interval is for the t distribution with 2 degrees of freedom and runs from $t = -2.9$ to $t = +2.9$.

The number of degrees of freedom for a distribution depends on the number of independent quantities needed to determine it. With the t distribution, there are n independent data values; but in the denominator, we encounter s, which is the square root of $1/n[(x_1 - \bar{x})^2 + (x_2 - \bar{x})^2 + \cdots + (x_n - \bar{x})^2]$. These n numbers that are squared are not independent. Because they sum to 0 (see Example 2.9 in Chapter 2), only $n - 1$ of them can be independently specified and those $n - 1$ numbers determine the values of all n numbers. Hence, for the t distribution we have $n - 1$ degrees of freedom. (We will also encounter degrees of freedom in connection with the chi-square distribution, where they have the same sort of meaning.) As you can see, it may take some mathematical detective work to figure out how degrees of freedom should be determined for a distribution.

The t distribution with 30 degrees of freedom (shown in Figure 7.6) is indistinguishable from a standard normal distribution shown at the same scale. This is why large-sample methods work when $n \geq 30$, even though they use the point estimate $s/\sqrt{n}$ in place of the unknown $\sigma/\sqrt{n}$. Using the t distribution for samples larger than size 30 is complex and yields no gain in accuracy.

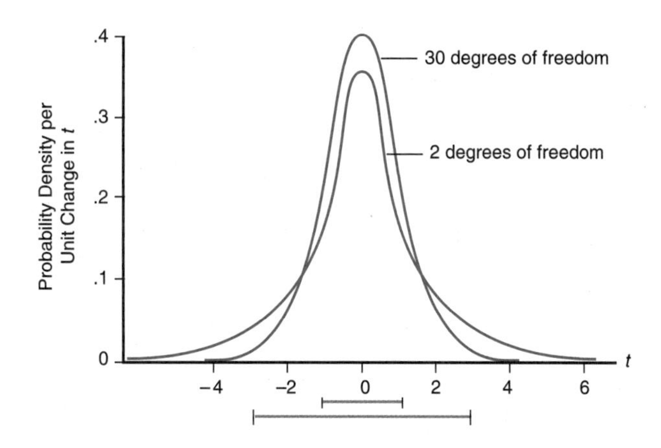

Figure 7.6
Probability density functions for t *distributions with 2 and 30 degrees of freedom*

What we are most interested in is the range of values of t that will account for a given percentage of the values—say, 80% or 90%. These ranges are often given in terms of the *tail probabilities* or *tail areas,* as shown in Example 7.5.

EXAMPLE 7.5 ■ USING THE t DISTRIBUTION TO ESTIMATE A POPULATION MEAN FROM A SMALL SAMPLE

A random sample of 20 university research libraries had an average staff of 240.90 individuals, with a standard deviation of 124.59 individuals (*Chronicles of Higher Education,* March 10, 1993). Calculate the 95% confidence interval for the true mean staff at university libraries. Assume that the staff sizes are approximately normally distributed.

Step 1. We write the confidence interval formula:

Confidence intervals for small samples: the mean

$$P(\bar{x} \pm ts_{\bar{x}}) = \text{Confidence level}$$

Step 2. We are looking for an interval like those shown in Figure 7.6, except that we want it to span 95% of the area below the graph. We need the probability density curve for a t distribution with $20 - 1 = 19$ degrees of freedom. (Keep in mind that the number of degrees of freedom is the sample size minus one.) The interval we want spans 95% of the area, so it leaves tail probabilities of 2.5%, or .025, that the variable will fall above or below this interval. The t values corresponding to such tail probabilities are given in Table V in Appendix B. A portion of Table V is shown in Table 7.7, where the values for different degrees of freedom are given in separate rows. The *critical value* for t for a tail probability of .025 is 2.093, so the range is from $t = -2.093$ to $t = +2.093$. Tail probabilities are often represented by the Greek alpha, α. For a t distribution with n degrees of freedom, t_α is the value whose tail probability, $P(t \geq t_\alpha)$, is α.

Table 7.7

PORTION OF THE *t* DISTRIBUTION TABLE

Area of probability

Degrees of Freedom (df)	RIGHT-TAIL AREAS .10	.05	.025	.01	.005
15	1.341	1.753	2.131	2.602	2.947
16	1.337	1.746	2.120	2.583	2.921
17	1.333	1.740	2.110	2.567	2.898
18	1.330	1.734	2.101	2.552	2.878
19	1.328	1.729	2.093	2.539	2.861
20	1.325	1.725	2.086	2.528	2.845
21	1.323	1.721	2.080	2.518	2.831

Step 3. We calculate the standard error of the mean, $s_{\bar{x}}$:

$$s_{\bar{x}} = \frac{s_x}{\sqrt{n}} = \frac{124.59}{\sqrt{20}} = 27.86$$

Step 4. We substitute the values in the confidence level formula to find the interval with endpoints $240.90 \pm 2.0930(27.86)$, or 240.90 ± 58.31. For a 95% confidence level, endpoints are 182.59 and 299.21.

Step 5. Interpreting the results, we say with a 95% confidence level, based on the sample drawn, that the true mean staff at university research libraries falls between 183 and 299 individuals. We do not know whether the population mean falls in this particular confidence interval, but we do know that if we use repeated random samplings to obtain a sequence of such confidence intervals, in the long run we can expect 95% to contain the population mean.

EXERCISES 7.6–7.20

Skill Builders

7.6 Consider the following results:

$$\bar{x} = 52.9 \qquad s_{\bar{x}} = 27.6 \qquad n = 44$$

a. Based on sample size, what distribution would you use to find a confidence interval?

b. Calculate the standard error of the mean.

c. Calculate the 90% confidence interval for the population mean.

d. Interpret your results.

7.7 If $n = 295$ and $p = .041$, find the 90% confidence interval for π, the population proportion. Clearly describe your reasoning

7.8 Consider a sample whose mean is 285 and whose standard deviation is 79. There are 9 observations, and you may assume that the underlying population is roughly normal in its distribution.

a. What distribution should you use to find a confidence interval for these data?

b. Find the standard error of the mean.

c. Construct a 99% confidence interval for the mean.

d. What does the result in part b tell you?

7.9 The following data were collected in an experiment:

.156	.002
.202	.722
.562	.912
.715	.518
.316	.306

a. Compute the sample mean.

b. Find the standard deviation.

c. Calculate the standard error of the mean.

d. Construct a 90% confidence interval for the population mean, justifying the method you chose.

e. Interpret your result in part d.

7.10 Let $n = 554$, $\bar{x} = 699$, and $s_{\bar{x}} = 64$. Construct a 99% confidence interval for the mean. Give the reasoning behind your choice of methods.

Applying the Concepts

7.11 For a sample of five aerospace and defense firms, the CEOs' mean salary (excluding bonuses and stock gains) was $728.6 thousand, with a standard deviation of $247.1 thousand (*Forbes,* 1993). Assume that the population is approximately normal.

a. Calculate the standard error of the estimate.

b. Using your result from part a, construct a 95% confidence interval for the true mean salary of top aerospace and defense CEOs.

7.12 When the Alaskan Department of Fish and Game conducted a survey of Kodiak Island king crabs in 1973, the average number of male crabs per pot was 335.5. The standard deviation was 259.1. A total of 40 crab pots were examined.

a. Calculate the standard error of the estimate.

b. Construct a 95% confidence interval for the true average, justifying your choice of method.

c. Interpret your finding in part b.

7.13 According to the National Household Survey on Drug Abuse (U.S. Department of Health and Human Services, 1988), 35% of males drink alcohol at least once a week. A total of 14% of females drink alcohol at least once a week.

a. Suppose a sample of 1200 males was surveyed. Estimate the standard error of the estimate for the proportion of males drinking alcohol at least once a week.
b. Construct a 90% confidence interval for the true percentage of males drinking alcohol at least once a week.
c. Repeat parts a and b for females. Assume a sample size of $n = 1200$.
d. Compare the results for males and females. What do you conclude?

7.14 Mercedes-Benz sponsors a marathon each year. Qualifiers are held at the Central Park Delacorte Oval. A total of 62 men completed the qualifier held on August 8, 1990, finishing in an average of 5 minutes 20 seconds, with a standard deviation of 58 seconds. Construct a 99% confidence interval for the true average time.

7.15 A sample of 20 GMC trucks averaged 16.2 mpg (city driving), with a standard deviation of 1.5 mpg. Construct a 99% confidence interval for the true average city mileage. Assume that gas mileages for these trucks are approximately normally distributed.

7.16 The price/earnings multiples of a random sample of the "Best Small Companies" (*Business Week,* May 23, 1994) appear below. A total of 36 companies were selected. Construct a 90% confidence interval for the true average P/E ratio. How could you use this information?

47	57	14	30	16	48	39	42	31	31
18	17	18	24	11	21	40	17	19	15
11	13	25	19	22	16	8	15	40	34
11	21	12	13	26	14				

7.17 The *Sourcebook of Criminal Justice Statistics* (U.S. Department of Justice, 1987) states that 43% of females think having a gun in the home makes it a more dangerous place. Assume that the estimate was made using a random sample of 1000 females.

a. Construct a 90% confidence interval for the true proportion of females who feel this way.
b. What is your margin of error?
c. The percentage of males responding similarly was 26%. Assume that this estimate was made using a similar sample of 1000 males. Construct a 90% confidence interval for this estimate.
d. What is your margin of error?
e. Compare your results for males and females. On the basis of this evidence, does it appear that males and females differ significantly in their understanding of the risks of guns?

7.18 Data compiled by the Department of Transportation in September 1990 indicated that 83.8% of United Airlines' flights were on time. American Airlines had an 84.6% record. "On time" means within 15 minutes of the scheduled time, and the reported percentages are the observed proportions of on-time flights.

a. Calculate the standard error of the estimate for the United Airlines proportion, assuming that the data were drawn from a sample of 600 flights.
b. Estimate a 95% confidence interval for the true on-time percentage for United.
c. Repeat parts a and b for American Airlines.
d. Using your results from parts a through c, write a paragraph comparing United's and American's on-time performances.

7.19 Suppose that 42% of a random sample of 600 investors believe that the market will go up in the next month. Estimate the true population proportion with 99% confidence.

7.20 A *Consumer Reports* survey of soups (November 1993) listed 25 brands of chicken soup. The data for these brands appear below:

Brand	Price per Container
Campbell's Homestyle	$ .93
Progresso	1.56
Campbell's (regular)	.48
Nissin Cup Noodles, chicken flavor	.58
Progresso Healthy Classics	1.54
Lipton Soup Mix	1.25
Campbell's Ramen Soup, chicken flavor	.18
Oodles of Noodles, chicken flavor	.22
Campbell's Soup Mix	1.28
Pathmark	.45
Shoprite	.50
Maruchan Ramen, chicken flavor	.18
Lady Lee (Lucky)	.51
Weight Watchers	1.00
Knorr Chicken Flavor Noodle Soupmix	1.63
Campbell's Home Cookin'	.99
Hain	2.29
Mrs. Grass Noodle Soup with Real Chicken Broth	.93
Campbell's Cup Instant Microwavable	.36
Lipton Cup-a-Soup	1.08
Campbell's Chunky Classic	.99
Campbell's Healthy Request Hearty	1.40
Pritikin Chicken Soup with Ribbon Pasta	1.75
Campbell's Low Sodium Chicken with Noodles	1.08
Healthy Choice Old-Fashioned	1.47

a. Find the mean and standard deviation for this data set.
b. Compute the standard error of the mean.
c. Assuming that the population follows a normal distribution, construct a 99% confidence interval for the true mean price.
d. Interpret the results of part c in English.

SECTION 3 ▪ DETERMINATION OF SAMPLE SIZE

When a researcher sets out to conduct a survey—whether to determine the characteristics of the readers of a magazine, the size of this year's wheat crop, the durability of a product, or the level of customer satisfaction at an HMO—the first question that comes to mind is "How large a sample will I need?" The statistical tools needed to answer such a question will be explained in this section.

Mathematics determines the relationships among the sample size, the precision of the statistical estimate, and the confidence level of that estimate. But business considerations determine how precise an estimate must be to be useful, as well as the level of uncertainty and costs that will be acceptable. Before moving to the mathematics, let's take a brief look at these business factors.

Suppose you are working in a group that is planning future online banking services for Bank of America customers. Your group needs to know what proportion of the bank's customers regularly use online services of any sort. A survey would give the answer, but your bank has millions of customers. How large a survey would be needed? In Example 7.7, you will see that a sample of 1068 will suffice to pin the proportion down to within 3% with a confidence level of 95%. Is this good enough? Yes, because you are concerned with making initial plans, not fine-tuning some huge, well-defined operation.

Would a larger sample be helpful? Probably not, because cutting the error by one-third, to $\pm 1\%$, would require increasing the sample size by a factor of $3^2 = 9$ and would be much more expensive. Would a smaller sample be helpful? Quite possibly. Suppose the planning group, in consultation with management, had already decided to proceed if a survey indicated that more than 60% of the bank's customers were Internet users. A pilot survey of 300 customers might be a reasonable first step. If enough customers in the survey turned out to be Internet users, the estimate could be confirmed at the confidence level set by the group. The money budgeted for the 1068-customer survey might be set aside for later use, in testing more detailed plans.

A cautionary word about costs for the type of survey outlined above: Depending on how many questions are asked and what overhead costs are, the cost per person surveyed might be \$10 to \$20. Given what is under consideration—the digital future of a major bank—a total cost of \$10,000 to \$20,000 is quite reasonable, but substantial enough to make it worthwhile to plan the survey well, so that the group's purposes will be achieved.

If your group were monitoring the manufacturing process for computer chips and considering some change aimed at increasing the percentage yield of chips, you would probably be looking at small changes in a carefully controlled process and correspondingly small changes in percentage yields. However, because thousands of chips are produced and tested each day in the ordinary business of manufacturing, you would be able to accumulate enough data to see whether yields could be pushed up by as little as a few tenths of a percent. This would be worth doing, because over the long run the payoffs from even small increases in yield could be large and the cost of gathering the data is low. This is a situation where precise estimates are needed and a high level of confidence is desired.

Suppose your job is to choose items to be included in a direct mail catalogue. You have detailed information on orders drawn from past catalogues and can use these draw rates to guide your choices. But what about trying new items that have no track record with your customers? You may decide to do this without any market research at all, based on your judgment and that of a few colleagues. However, relying on the comments of a few colleagues makes you dependent on a very small sample; worse, your sample is an inherently biased group, consisting of people who would like to believe that *their* product has wide appeal. You can't even specify a level of precision or a confidence level for your estimate of the draw of an untested item. But you will learn something by trying the item out on your customers and the cost is reasonable, so in this case, where the margins of error are huge and the level of confidence undeterminable, you might sensibly base a business decision on a small or nonexistent sample.

Once you are clear about your purposes for gathering and analyzing data, about your budget, and about how precise an answer you need, you can find the right sample size fairly easily. Four basic factors affect sample size:

Variance. The larger the variance of the variable being estimated, the larger the sample size needed to pin it down with a given precision at a given confidence level.

Confidence Level and Precision. The down side of a higher confidence level (a higher reliability) is a loss of precision—that is, a larger confidence interval. If we were willing to settle for a confidence interval that stretched from $-\infty$ to $+\infty$, we could be 100% certain the population parameter would lie within it—and we could make this prediction based on a sample of size 0. Reliability is 100%, but we have lost all precision. The cost of achieving high reliability and great precision is that of taking a large random sample.

Costs. Survey work, acceptance testing, and quality control all involve costs. In a business setting, these costs must be kept in line with the benefits to be gained. Some costs are direct expenses and can be easily measured; others, in the form of time and human resources, are less clearly defined. To achieve a high level of precision and reliability, we want larger samples. To economize, we would like smaller ones. A balance must be struck, and that balance must be worked out in each business situation. There are also logistic considerations. The larger and more complex any undertaking becomes, the greater the probability that processing errors will occur. Documents get misplaced. Clerks, far removed from the supervision of the study and not necessarily attuned to the needs of the statistician, may not conscientiously collect data. Data may be fudged just to submit a report that looks appealing.

Variability. It is possible that every individual in a population represents the population mean, just as every electron has the same charge. But even in such a situation, a sample of some reasonable size should be looked at, if only to confirm this lack of variability. In a business situation, variability often represents opportunity. A magazine is not interested just in the average income of its subscribers. It wants to give its advertisers an idea of the *distribution* of subscribers' incomes. For this reason, samples of size $n = 1$, though cheap, are not really useful. When values are limited to a small range, it usually isn't necessary to draw too large a sample. For example, in a study of daily lunch expenditures for students at a university, lunch expenses might range from \$0 to \$5.00, so a sample of about 12 would provide an estimate accurate to within about \$.50 of the true mean. In contrast, figures for the household income of readers of a magazine might range between \$20,000 and \$100,000. In that case, a sample of nearly 700 would be needed to provide an accurate estimate of average readership income to within \$1000.

Sample Sizes for Estimating an Average

It is often convenient to think of the endpoints of confidence intervals as being determined by

Point estimate ± Margin of error

With regard to determining sample size, margin of error refers to the acceptable range of variation about the true mean. Political polls are frequently reported with a ±3% margin of error. If we were looking at the household income of magazine readers, we might wish to measure error within, say, \$5000. The more precise we wish to be, the smaller the margin of error must be. The smaller the margin of error, the larger the sample size needed.

> The **margin of error** consists of the z statistic multiplied by the standard error of the statistic being estimated—for example, $\pm zs_{\bar{x}}$ or $\pm zs_p$.

Given the concept of a margin of error, the technical factors for determining sample size can be reduced to the following formula:

Sample size: estimating the mean

$$n = \frac{z^2 s^2}{E^2}$$

where n is the sample size, z corresponds to the level of confidence we wish to associate with our estimate, s is the estimated standard deviation of the variable being studied, and E is the desired margin of error.

It should be noted that the z values here correspond to those discussed under confidence intervals. If we wanted an estimate with a 95% confidence level, we would choose a z value of 1.96. We need an estimate of s. If we're doing a survey, it is because we do not know population values. What can we use as a reasonable estimate for s? There are a number of approaches. Prior studies may provide estimates of s, but if there are no prior studies, a pilot survey may be conducted and the resulting s used as a point estimate.

Alternatively, a pilot study can provide an estimate of the range, or the difference between the high and low values. If history provides observations, the range can be estimated from the data. It is a rule of thumb that *the standard deviation may be estimated from the range by dividing the range by 6.*

Quick Formula for Estimating the Standard Deviation

$$s = \frac{r}{6}$$

where s is the standard deviation and r is the range.

If you have a few observations (say, 20 to 40), this estimate of s will probably be low, but with a sample of size 100 to 200 it serves as a good guide. Using this method with 10 random samples of size 160, taken from a normally distributed population with $\sigma = 1$, yielded the following estimates for s:

.87	.89	.90	1.06	.90	.89	.83	.81	.86	1.01

These results show a small tendency to underestimate the standard deviation. Doing the same experiment with samples of size 100 yielded the following estimates:

1.09	1.06	0.93	1.20	1.00	1.11	1.20	1.24	1.05	1.12

Here we see a small tendency to overestimate the population parameter. Taken together, these experiments suggest that the rule of thumb works fairly well over a wide range of sample sizes when the data are approximately normal.

EXAMPLE 7.6 ▪ DETERMINING SAMPLE SIZE TO ESTIMATE THE AVERAGE STARTING SALARY OF LAW SCHOOL GRADUATES

Each year, *U.S. News & World Report* ranks graduate schools in various disciplines. Law schools are ranked in five tiers. The top 50 schools are in the first two tiers; approximately 120 other law schools are in tiers 3, 4, and 5. Needless to say, average starting salary of graduates is a primary concern.

Average starting salaries range from about \$26,000 for a graduate of North Carolina Central University to \$64,000 for a graduate of Rutgers at Camden. What size sample should be selected to estimate average starting salary with a \$2000 margin of error with 95% confidence?

Step 1. We write the formula:

$$n = \frac{z^2 s^2}{E^2}$$

Step 2. Then we fill in known data:

$$z = 1.96 \qquad E = 2000$$

Step 3. We approximate the standard deviation, s, based on the range, r. Since $r = 68{,}000 - 26{,}000 = 42{,}000$,

$$s = \frac{r}{6} = \frac{42{,}000}{6} = 7000$$

Step 4. We solve for n:

$$n = \frac{1.96^2(7000)^2}{2000^2} = 47.06 \approx 48$$

Since $n > 30$, our large-sample approximation, with $z = 1.96$, works well here.

Step 5. We interpret the result: A sample of at least 48 schools would be necessary to determine the true average starting salary for graduates of schools in tiers 3, 4, and 5 with 95% confidence and a margin of error equal to $2000.

Sample Sizes for Estimating a Proportion

The procedure for estimating the sample size for a proportion or percentage is much the same as that for estimating the sample size for an average. The confidence level (the z statistic), the margin of error (E), and the variance or standard error of the proportion (s_p) must all be taken into account. As with the sample mean, the greater the variability in the data, the larger the sample needed. You may recall that the standard error of the proportion is calculated by the formula

$$s_p = \sqrt{\frac{pq}{n}}$$

where p is the proportion with a given attribute, q is the proportion without a given attribute, and n is the sample size.

The formula for the sample size needed to estimate a proportion is analogous to the one used for averages, with s^2 replaced by $(\sqrt{pq})^2$, or pq:

$$n = \frac{z^2 pq}{E^2}$$

where n is the sample size, z corresponds to the level of confidence we wish to associate with our estimate, pq is the numerator of the estimated standard deviation of the variable being studied, and E is the desired margin of error.

Because we do not know prior to a survey what variability to expect in our results, it is safest to assume that the proportion will exhibit the greatest variation possible. You can see by graphing or experiment that this occurs when $p = .5$. Thus, the largest sample size is required, all other things being equal, when $p = .5$ and $q = .5$. When we know nothing about the population under study, the conservative course is to take a sample large enough to deal with this worst-case scenario. This is called the "know nothing assumption" or "equally likely assumption."

EXAMPLE 7.7 ■ DETERMINING SAMPLE SIZE TO ESTIMATE THE PROPORTION FAVORING CLINTON'S HEALTH CARE PLAN

Suppose we want to examine the public's view of President Clinton's health care plan—an issue of public concern in 1994. We wish to have a margin of error of no more than 3% and a confidence level of 95% or higher. How large a sample size do we need?

Step 1. We write the formula:

$$n = \frac{z^2pq}{E^2}$$

Step 2. Then we fill in known data:

$$z = 1.96 \qquad E = 3\% = .03$$

Sample size: estimating a proportion

Step 3. We estimate p and q. Not knowing anything about the population up front, we should assume a worst-case scenario of the know nothing assumption—that is, a 50–50 split. Therefore, we assume that $p = .5$ and $q = .5$.

Step 4. We solve for n:

$$n = \frac{1.96^2(.5)(.5)}{.03^2} = 1067.11 \approx 1068$$

Step 5. We interpret the result: A sample size of 1068 will yield a sample with no more than a 3% margin of error at a 95% confidence level. (If you are familiar with national surveys of public opinion, you are aware that the usual margin of error is $\pm 3\%$ and the poll size is usually 1000 to 1500, which is in line with our estimate.)

If it is known that the population proportion is close to either 0 or 1, the values of p and q in the formula for estimating sample size may be replaced by more reasonable figures. Such a replacement would be appropriate in estimation of the unemployment rate, for example. Instead of $p = .5$, it would be reasonable to use $p = .1$. The resulting sample size would be considerably smaller, permitting significant cost savings over the course of the survey.

One last point about sample size: The sample size is equal to the number of individuals who respond to a survey. It may be necessary to poll many more people to arrive at a specific sample size. Response rates will depend on the nature of the survey, the types of questions asked, and the population being polled. If the decision to respond is related to the response itself, then serious bias can result, as we saw in Chapter 1 with the *Literary Digest* poll. Any responsible survey will include a second stage in which nonrespondents are sampled to determine if such a relationship exists. Incentives may increase the response rate, but they have the potential to bias the response.

A Poll on Health Care

From April 15 to 29, 1994, the *New York Times*/CBS News Poll of business executives was conducted with senior executives at companies throughout the United States by means of telephone interviews. The sample of companies and executives was randomly drawn by Dun and Bradstreet Information Services from its database of privately held and publicly owned companies. As with previous polls in this series, companies with \$5 million or more in annual revenues were eligible for the survey. However, because the subject of this particular poll—health care—was especially relevant to small business, an additional sample of companies with less than \$5 million in revenues was also drawn for comparative purposes.

At each company, one senior executive was interviewed. Individuals eligible for the poll included those with the title of owner, partner, chief executive, president, executive vice president, or senior vice president, with first opportunity for an interview given to the most senior executive.

The completed sample included 115 interviews in companies with annual revenues of \$500 million or more, 121 interviews in companies with \$100 million to \$499 million in annual revenue, and 248 interviews in companies with \$5 million to \$99 million in annual revenue. This combined sample of 484 was then weighted by revenue size to reflect the actual distribution of all listed companies in the country, because smaller companies predominate. The separate sample of very small businesses—those with revenues of less than \$5 million—had an additional 191 interviews.

In theory, in 19 cases out of 20, results based on such samples will differ by no more than 6 percentage points in either direction from what would have been obtained by seeking senior executives at all listed companies in the country with revenues exceeding \$5 million. The potential sampling error for groups with fewer cases is larger. For example, for the group of companies with less than \$5 million in revenue, the potential error is ± 7 percentage points. To provide a more comprehensive picture, pollsters then compared the responses of the business executives with those of the general public. The results depicted in Figure 7.7 are based on a total sample size of 1215.

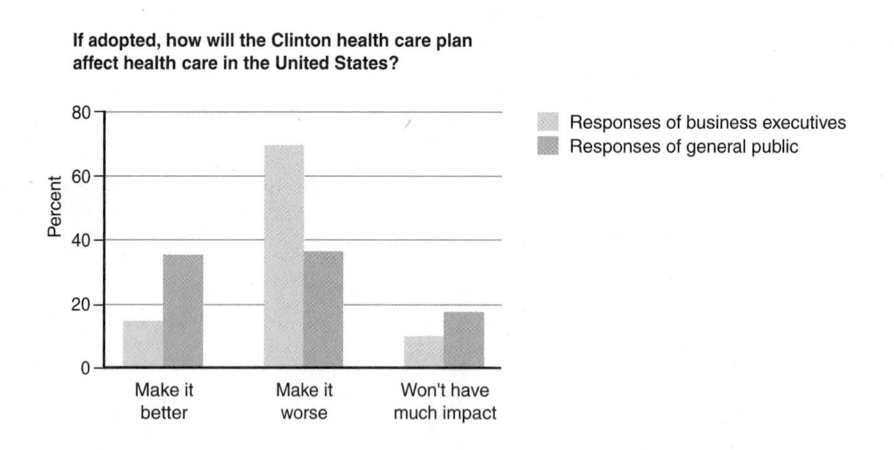

Figure 7.7
Results of a New York Times*/CBS News poll on President Clinton's health care plan*

EXERCISES 7.21–7.34

Skill Builders

7.21 If $z = 1.645$, $s = 100$, and $E = 25$, what sample size n is indicated?.

7.22 Suppose $p = .4$, $z = 2.575$, and $E = .05$. What is the right sample size?

7.23 Suppose we are working with a sample size of 600, a 95% confidence level, and a variance of 900. What is the margin of error for this survey?

7.24 Let $z = 1.96$, $s = 25$, and $E = 10$. What sample size is in order?

7.25 Suppose we want to conduct an opinion poll. We wish to be 99% confident of the result, and a margin of error no greater than 4% is desired. Employing the "know nothing assumption," determine what size sample we need.

Applying the Concepts

7.26 A survey in *The Great Divide—How Females and Males Really Differ* (Daniel Evan Weiss, 1991) reported that, as of the late 1980s, 34% of females age 25 and over finished at least one year of college, compared with 41% of males. Suppose we wished to confirm these percentages in today's world.

a. For 95% confidence and a 5% margin of error, what size samples should be drawn? Assume the old percentages hold. Redo your estimate based on the "know nothing assumption."

b. What are the percentages of males and females over 25 who have finished at least one year of college today? (*Hint:* Look at the Web site for the National Center for Educational Statistics at http://nces.ed.gov/.)

7.27 In a study entitled "Effects of Alternative Family Structures on Managerial Career Paths" (*Academy of Management Journal,* August 1993), Joy Schneer and Frieda Reitman examined how family structure was related to income and career satisfaction. They sent a survey to over 1300 individuals. Suppose they desire to estimate the average age of their respondents to within 5 years with 90% confidence. Assume the ages of respondents ranged from 25 to 45 years. What size sample would they need? What method did you use for estimating s?

7.28 Each month, the National Association of Purchasing Managers publishes the NAPM Index. Purchasing executives are asked "Do you think the economy is expanding or contracting?" The association compiles responses from purchasing executives in over 300 industrial companies. Assuming that their sample consists of exactly 300 companies and that survey results are presented with 95% confidence, estimate the margin of error. Indicate what assumptions and methods you used.

7.29 In February, *Business Week* publishes the Mutual Fund Scoreboard. This scoreboard includes a rundown of all the major mutual funds. Suppose you wished to estimate the average price/earnings multiple to within 3 points (that is, a 3-point margin of error). How large a sample would you need if you wished to make your estimate with 99% confidence? Assume the P/E ratios exhibit a standard deviation of 5.

7.30 Suppose the tenures of CEOs of electric utility companies range from 1 year to 37 years. To estimate the average tenure of electric utility CEOs with a 5-year margin of error with 95% confidence, what size sample is needed? (See *Forbes*, May 24, 1993, for the data.)

7.31 According to *Street and Smith's Baseball,* National League batting averages in 1993 ranged from a high of about 370 to a low of about 140. Suppose you wish to estimate the average for the league within ±20 points. What size sample should you draw in order to make an estimate with 90% confidence?

7.32 Suppose you wished to take a poll regarding the anti-smoking bill introduced by the Clinton administration in 1994. In an unscientific survey conducted by the Prodigy Computer Service, 66% of the respondents supported the legislation. If you were to conduct a scientific survey, carefully selecting a random sample, what size sample would you select? What population would you survey? Assume that the true percentage is 66%, a margin of error of 4% is allowable, and a 99% level of confidence is desired.

7.33 Child daycare can eat up as much as 20% of a family's annual income. In *The Best & Worst of Everything* (Les Krantz, 1991), the most and least expensive cities for daycare are listed. New York, at $104 per week, was the most expensive; Ogden, Utah, at $39 per week, was the least expensive. If you wished to draw a sample to estimate average daycare expenses for cities in the United States, what size sample would you need? Assume a $10 margin of error is wanted and a 95% level of confidence is desired.

7.34 Graduating seniors from the business school of a major eastern university had grade point indexes ranging from 2.000 to 3.789 during the fall semester of 1993. If we wished to estimate the average to within .100 point with 99% accuracy, what size sample should be drawn?

SECTION 4 ▪ USING THE COMPUTER TO ESTIMATE

Using the computer is essential when handling large data sets. Computer programs perform the needed calculations accurately and rapidly, and they can calculate confidence intervals for large samples as well as small samples.

Making Large-Sample Estimates by Computer

The executives of a large firm are considering relocating one of its facilities, and one area under consideration is Parsippany, New Jersey. An important factor in the decision is the cost of housing. To get a quick estimate of housing costs in Parsippany, researchers gathered sample data from a real estate booklet. The data were entered into an Excel worksheet, and the column containing the price data (column A) was labeled PRICES.

Under the Tools menu, the Data Analysis Add-In was selected. In Excel, once a particular statistical tool has been selected, a series of instructions appears in the dialog boxes. Confidence intervals for large samples can be calculated under "Descriptive Statistics."

The results of the Excel analysis appear below.

Using Excel to obtain a confidence interval for a large sample

PRICES	
Mean	179300.9697
Standard Error	8941.962884
Median	169000
Mode	124900
Standard Deviation	51367.66597
Sample Variance	2638637107
Kurtosis	−0.592945096
Skewness	0.628927805
Range	190000
Minimum	94900
Maximum	284900
Sum	5916932
Count	33
Confidence Level (95.0%)	18214.16693

Looking at the printout, we can see several things. First, the average price of a house in Parsippany is approximately \$179,301. Although the result reads \$179,300.9697 because most of the data are given to the nearest hundred, the ".9697" creates a false impression of precision. Attributing more precision to the result than the data warrant is counterproductive. Second, we may state with 95% confidence that the true mean lies between \$161,100 and \$197,500. Although calculations show that the mean lies within 18,214 of \$179,300—that is, between \$161,086 and \$197,514—these numbers are rounded to reflect the fact that the original data were rounded to the nearest hundred dollars.

The same task can be performed using Minitab. Once the data have been entered into a Minitab worksheet, a confidence interval can be calculated by taking the following steps.

Figure 7.8
Histogram of home prices, produced using Minitab

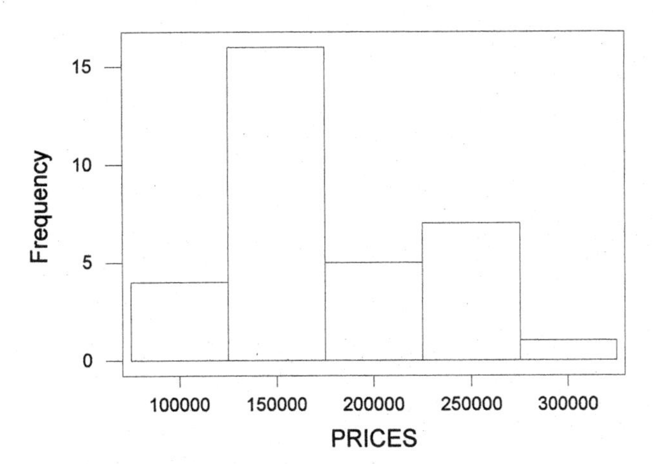

Using Minitab to obtain a confidence interval with the *z* statistic

1. Select "File."
2. Open PARSIP.MTW.
3. Select "Stat."
4. Go to "Basic Statistics."
5. Select "Sample z...."
6. Highlight "Prices" in the dialog box.
7. Enter sigma (say, 35000).
8. Select "OK."

The results appear as follows:

Variable	N	Mean	Stdev	SE Mean	95% C.I.
Prices	33	179301	51368	6093	(167355, 191247)

Minitab also allows us to draw a histogram of the data, as shown in Figure 7.8.

Getting a Small-Sample Confidence Interval by Computer

Consumer Reports publishes car ratings each year. In its April 1994 issue, annual fuel costs for 17 models were presented. Data were included on models from the Geo Prizm to the Hyundai Electra (the top and bottom entries on the list, respectively). Suppose a firm is planning to buy a fleet of 24 cars for its sales force, and the comptroller would like us to estimate annual fuel expenses for this fleet.

There are only 17 observations. Assuming the population histogram is approximately normal, the *t* statistic can be used to estimate the confidence interval for the average. When we wish to estimate total cost, we begin by estimating the average and then use n times x to find the total. The following procedure will give us not only a point estimate, but also high-side and low-side estimates.

Using Excel to obtain a confidence interval with the *t* statistic

1. Double-click on "Sheet 1" at the bottom of the screen and type CI on Mean.
2. Type Data in cell A1.
3. Fill in the data in column A.
4. In cells C1 to C9 of column C, type in n, mean, std dev, std err, conf level, t, samp err, lower bnd, upper bnd. (Leaving column B empty makes the printout easier to read.)
5. Go to cell D1. Using the function wizard f_x (a button on the tool bar), go to statistical functions and select COUNT. On the first "values" line, enter A:A. This function will count all the data elements in column A. When it finishes, n (in this case, 17) will appear in cell D1.
6. Repeat the process in Step 5 for the mean and standard deviation. The statistical functions are, respectively, AVERAGE and STDDEV.
7. Using the formula capability of Excel, go to cell D4 and enter =D3/sqrt(D1).
8. In cell D5, enter 95%.
9. Go to cell D6. Using the function wizard f_x, go to statistical functions and select TINV. Enter .05 and the appropriate degrees of freedom. The program will yield the $t_{critical}$ value for a 95% confidence interval.
10. Enter =D6*D4 in cell D7.
11. Enter =D2−D7 in cell D8.
12. Enter =D2+D7 in cell D9.

The results appear below.

Data		
550	n	17
510	mean	622.9
670	std dev	82.1
530	std err	19.91217
620	conf level	95%
530	t	2.1199
740	samp err	42.21192
530	lower bnd	580.7
740	upper bnd	665.2
530		
740		
710		
610		
620		
650		
680		
630		

The printout reveals several interesting points. The mean fuel cost for this size vehicle is about $623 per annum. The data exhibit a standard deviation of approximately $82. Based on the empirical rule, we may expect that, if the population is normal, costs for any individual vehicle will fall in the interval $623 ± $82 two-thirds of the time. The confidence interval for average annual fuel costs runs from $581 to $665.

We can calculate the answer to the comptroller's question regarding the annual cost of fuel for a fleet of two dozen cars. Looking at the printout, we find that the mean is 622.9. This is the average, $\bar{x}$. Multiplying $\bar{x}$ by 24 yields $14,949.60. A midrange estimate of the fleet's annual fuel cost is $14,950. At the lower end of the spectrum is 580.7. Multiplying by 24 yields $13,936.80. In other words, the company might spend as little as $13,937. A high range estimate is derived by multiplying 665.2 by 24. The high end figure is $15,964.80, which becomes $15,965.

To sum it all up for the comptroller, we expect the fleet of cars to incur fuel costs (rounded to the nearest $100) of between $14,000 and $16,000 per year. If we had to be pinned down to a single figure, it would be approximately $15,000 per year. Assuming that the population histogram of fuel costs is approximately normal, there is a 5% risk that the actual costs will be outside this interval.

EXERCISES 7.35–7.38

Applying the Concepts

Use a statistical calculator or computer for these exercises.

7.35 To establish that it produces truly domestic cars, the Ford Motor Company publishes a list giving the percentage of domestic content in each of its vehicles. Using the data below, calculate the average domestic content of Ford's motor vehicles. Use a 95% confidence level. Assume that the underlying population data follow a normal distribution and that all of the vehicles should be given equal weight in your calculations. Can Ford make the claim that, on average, domestic content is over 75%? How could it improve its percentage?

Vehicle	% Domestic	Vehicle	% Domestic	Vehicle	% Domestic
Taurus	91	F-Series	96	Town Car	95
Tempo	94	Explorer	85	Continental	97
Festiva	22	Ranger	89	Mark VIII	95
Crown V	72	Aerostar	91	Grand Marquis	72
Escort	85	Econoline	97	Sable	94
Mustang	92	Bronco	98	Capri	19
Probe	80	Villager	80	Cougar	94
Thunderbird	94	Tracer	81	Topaz	94

7.36 In its "Money Guide" (1994), *Money* magazine provided a list of 1958 major funds. Below are the one-year percentage returns for a random sample of these funds. Based on this sample, calculate the 99% confidence interval for the population average return. Assume that the population data are approximately normal.

Fund	Return	Fund	Return
AARP Capital Growth	1.8	API Capital Income	.0
Arch Growth & Income Equity	32.0	Colonial Global Equity	19.6
Columbia Balanced	4.2	Dreyfus Third Century	1.1
Eagle Growth Shares	−8.2	Fidelity Select Automotive	15.1
Fiduciary Capital Growth	5.6	Founders Growth	17.9
Fountain Square Balanced	−2.7	Hancock Global Technology	27.1
Harbor Capital Appreciation	8.7	Ivy International A	29.4
Jackson National Growth	.8	Mackenzie North America A	2.8
Mainstay Capital Appreciation	9.7	Mutual Qualified	11.8
National Industries	−.1	Parkstone Equity Income	4.3
Parnassus	15.8	Putnam Diversified Income A	7.5
Quantitative Boston	28.7	Seligman Communications & Information A	39.7
Sentinel Aggressive Growth	5.8	Templeton American	8.7
Third Avenue Value	5.0	USAA Investment Balanced	5.3
UST Master Equity	6.5	WPG Tudor	9.1
Yacktman	−5.0		

7.37 In its May 23, 1994 issue, *Forbes* published information on the ages of CEOs in the major firms in each industry. Using the data below, determine the average age of CEOs in the health care industry. Provide a 95% confidence interval. Assume that the data come from an approximately normal distribution, and describe the methods you used.

Firm	Age	Firm	Age
Amgen	58	Beverly	57
Columbia	41	Health Trust	60
Hillenbrand	53	Mallinckrodt	51
Merck	64	PacifiCare	46
St. Jude's	51	Upjohn	54

7.38 Graduation rates for athletes have long been a subject of controversy at major universities. Data for a sample of 19 universities follow.

School	Rate	School	Rate
Appalachian State	58	Butler University	80
Colgate University	91	Duquesne University	72
Georgetown University	95	Jacksonville University	56
Marshall University	51	Niagara University	82
Penn State	78	St. John's University	79
Southern Methodist	71	Texas Southern	24
University of California, Santa Barbara	67	University of Illinois, Chicago	55
University of Montana	38	University of Oregon	66
University of S. Mississippi	43	University of Virginia	88
Washington State	49		

Source of Data: *Chronicles of Higher Education,* July 7, 1993.

a. Estimate the average graduation rate with 90% confidence, assuming that the underlying distribution is approximately normal

b. In your school library, use the *Chronicles of Higher Education* or data from the National Center for Educational Statistics to estimate with 90% confidence the average graduation rate for non-athletes at these schools. How do these rates compare?

FORMULAS

Standard Error of the Mean

$$\sigma_{\bar{x}} = \frac{\sigma_x}{\sqrt{n}}$$

where σ_x is the population standard deviation and n is the sample size.

Estimated Standard Error of the Mean

$$s_{\bar{x}} = \frac{s_x}{\sqrt{n}}$$

where s_x is the estimated sample standard deviation.

Estimated Standard Error of the Proportion

$$s_p = \sqrt{\frac{pq}{n}}$$

where p is the sample proportion and $q = 1 - p$.

Confidence Interval for the Mean

For a large sample:

1. Choose a confidence level.
2. Choose z so that a standard normal variable will have a probability c of containing the population mean, μ.
3. Find the confidence interval, which is from $\bar{x} - z\sigma_{\bar{x}}$ to $\bar{x} + z\sigma_{\bar{x}}$.

If the x's are obtained by random sampling, this normal random variable falls between $-z$ and $+z$.

For a small sample, when the population histogram is approximately normal and n = sample size:

1. Choose a confidence level.
2. Choose t so that a t-distributed variable with $n - 1$ degrees of freedom will have a probability c of falling between $-t$ and $+t$.
3. Find the confidence interval, which is from $\bar{x} - ts_{\bar{x}}$ to $\bar{x} + ts_{\bar{x}}$.

If the x's are obtained by random sampling, this t-distributed random variable falls between $-t$ and $+t$.

Confidence Interval for a Proportion

A large sample size is required.

1. Choose a confidence level.
2. Choose z so that a standard normal variable will have a probability c of containing the population proportion, π.
3. Find the confidence interval, which is from $p - zs_p$ to $p + zs_p$, where p is the sample proportion and s_p is the estimated standard error for the sample proportion.

If the observations are obtained by random sampling, this normal random variable falls between $-z$ and $+z$.

Rule of Thumb for Estimating the Standard Deviation

$$s = \frac{r}{6}$$

where r is the range.

Sample Size

For estimating an average:

$$n = \frac{z^2 s^2}{E^2}$$

For estimating a proportion:

$$n = \frac{z^2 pq}{E^2}$$

NEW STATISTICAL TERMS

confidence interval
confidence level
consistent estimators
large sample
margin of error
point estimate
small sample
t distribution
unbiased estimators

EXERCISES 7.39–7.48

Supplementary Problems

7.39 A popular motor inn chain wants to know what percentage of its guests are staying at one of its motor inns for the first time. What size sample should be drawn to find the true percentage with a 5% margin of error? The results should be reported with 95% confidence. What problems do you foresee if the chain decides to rely on a self-selected sample based on a mail survey?

7.40 In *Money* magazine's 1994 "Money Guide," five mutual funds were listed as "top picks" for the future:

Fund	One-Year Return
Crabbe Huson Special	24.2
Legg Mason Special Investment	21.3
Babson Value	16.8
Vista Capital Growth A	10.5
Tweedy Browne Global Value	22.6

a. Generate all samples of size $n = 3$ without replacement from this population.
b. Calculate the sample means.
c. Show that the sample means are unbiased estimators of the population mean.

7.41 Each quarter, *Business Week* publishes a "Corporate Scoreboard," which includes data on sales, profits, and returns on equity. Data for 40 banks revealed average profits of 9.85% of gross, with a standard deviation of 3.58% (assume that this is the population standard deviation).
a. Calculate the standard error of the mean.
b. Construct a 95% confidence interval for the mean.
c. Complete the following paragraph: Average profits for the banking industry were about _____ as of the third quarter of 1993. They varied considerably, as evidenced by a coefficient of variation of _____. It is expected that the true mean for the industry falls between _____ and _____. Sales, as measured by an all-industry composite, increased by 4%. The banking industry could be said to have performed (better than/about the same as/worse than) the rest.

7.42 The Belmont Stakes has been run since 1867. The track is 1½ miles long. Data for the 1981–1990 period follow.

Year	Horse	Jockey	Time
1981	Summing	G. Martens	2:29.0
1982	Conquistador Cielo	L. Pincay	2:28.1
1983	Caveat	L. Pincay	2:27.4
1984	Swale	L. Pincay	2:27.1
1985	Crème Fraiche	E. Maple	2:27.0
1986	Danzig Connection	C. McCarron	2:29.4
1987	Best Twice	C. Perret	2:28.1
1988	Risen Star	F. Delahoussaye	2:26.2
1989	Easy Goer	P. Day	2:26.0
1990	Go and Go	M. Kinane	2:27.5

Source of Data: *Universal Almanac*, 1991.

a. Calculate the standard error of the estimate.
b. Construct a 90% confidence interval for the true average time it takes a winner to complete the race, assuming that winning times follow an approximately normal distribution.
c. Complete the following paragraph: The average time it took a horse and jockey to win the Belmont Stakes in the decade of the 1980s was _____. The standard deviation was _____. We can expect that 90% of the time the true average will fall in the _____ to _____ range.

7.43 *Smart Money* (June 1994) did a study of frequent flyer miles. One part of their study concerned "blackout days" on which frequent flyer miles cannot be used. Data for five of the major airlines appear below:

Airline	Number of Blackout Days
American	14
Delta	0
TWA	34
United	32
US Air	7

Treat these as a sample drawn from a larger population of airlines whose numbers of blackout days are roughly normally distributed.
a. Calculate the average number of blackout days.
b. Find the standard error of the mean.
c. Construct a 99% confidence interval for the true average number of blackout days.
d. Complete the following paragraph: A practice of most of the major airlines is the use of blackout days, when frequent flyer miles may not be used. On average, there are _____ blackout days. We may state with 99% confidence that the true average number of blackout days falls between _____ and _____.

7.44 A random survey of 48 accommodations in the Greater Boston area (*Getaway Guide, Massachusetts,* 1994) revealed that 29 offered handicap-accessible rooms.
a. Calculate the standard error of the proportion.
b. Construct a 95% confidence interval for the true proportion of locations offering handicap-accessible accommodations.
c. Complete the following paragraph: Approximately _____ % of the hotels and motels in the Greater Boston area have handicap-accessible facilities. We can say with 95% confidence that the true percentage falls in the _____ to _____ range.

7.45 Homes shown in the June 1994 issue of *Las Vegas Real Estate Showcase* by Bar-K Realty ranged in price from \$67,900 to \$528,000. What size sample would be needed to estimate the average value of real estate sold by Bar-K with a \$10,000 margin of error and 95% confidence?

7.46 According to *Fortune* (April 1994), profits as a percentage of sales in the oil industry during 1993 ranged from −5% to 7%. If we wish to estimate average profits as a percentage of sales with 99% confidence, what size sample should be drawn? Assume a margin of error of 2%.

7.47 Using the Mercury Lease Vehicle database found in Appendix A and on the CD in the back of this book, draw a random sample of responses to estimate the average age of lease holders. Draw a sample large enough to estimate the average age within three years and with 90% confidence. How could a marketing manager use this information?

7.48 Use the Tammy's Restaurants data set found in Appendix A and on the CD in the back of this book.

a. Estimate the average amount spent by visitors to Tammy's Restaurants. Draw a sample large enough to estimate the average with a $2 margin of error. Construct a 99% confidence interval.

b. Using the the same sample as in part a, estimate the percentage of customers who visit once a week. What is your margin of error?

c. Write a brief report on your findings in parts a and b. What revenue implications does your report suggest?

8

Hypothesis Testing and Decision Making Using Means

OUTLINE

LEARNING OBJECTIVES

1. Understand the nature of hypothesis testing.
2. Distinguish between Type I and Type II errors.
3. Identify a test statistic.
4. Explain a decision rule.
5. Perform tests about a single mean for a large sample.
6. Compute and discuss *p* values.
7. Perform tests about a single mean for a small sample.
8. Compare means from independent samples.
9. Use hypothesis testing in quality control.
10. Employ the computer for hypothesis testing.

BUILD YOUR KNOWLEDGE *OF STATISTICS*

In 1996, Boston's Inspectional Services Department made a randomized study of sticker prices in 20 large retail stores. The department looked at 1899 items and found that 26%, or 490 in all, were mismarked. Was this mismarking random and uncontrollable, or did it reveal a systematic pattern?

To answer this question, the department utilized the techniques of hypothesis testing. It noted that 277 of the mismarked items were overpriced, while 213 were underpriced. In the absence of a systematic pattern, overpriced and underpriced items would occur with the same probability as heads and tails on the flip of a fair coin. If the number of overpriced items was a binomial random variable with $n = 490$ and $p = .5$, the expected value would be $np = 490 \times .5 = 245$ and the standard deviation would be $\sqrt{npq} = \sqrt{490 \times .5 \times .5} = 11.07$. The observed value was 277, nearly 3 standard deviations (3σ) above the expected value. Using the normal curve as a guide, we find that the probability of such an observation is about one in five hundred. Following the general practice of hypothesis testing, we reject the null hypothesis—the hypothesis that there is no systematic pattern—because the values actually observed are extremely improbable under the null hypothesis. Thus, Boston's Inspectional Services Department had found a systematic and correctable pattern.

The department worked with merchants to improve accuracy in pricing and reduce overpricing errors. A year later, the survey of 20 stores was repeated, this time checking 5290 items. Only 4.7% were mismarked; of these, 91 items were overpriced and 151 were underpriced. The ratio still appears lopsided, but this time in the consumer's favor. Boston officials are working to reduce the overall error rate to 2% or less. They want to see the positive and negative errors more nearly in balance, as one would expect from random or sampling errors.

Chapter 7 explored the techniques of estimation, one of the statistical tools applied to sample information to draw conclusions about a largely unexamined universe. This chapter focuses on another element in the statistics tool kit: hypothesis testing. The procedures of hypothesis testing constitute the foundation of the scientific method. You will see what a test statistic is and why all error is not the same. You will learn how to test and evaluate statistical claims. You will look at quality control as an application of hypothesis-testing techniques, and at the role of the computer in hypothesis testing.

This chapter concerns the testing of quantitative variables. Chapter 9 will explore the testing of qualitative variables.

SECTION 1 ▪ THE NATURE OF HYPOTHESIS TESTING

Hypothesis testing involves testing two propositions against each other. For example, the hypothesis that fluoride treatments reduce tooth decay (the alternative hypothesis) might be tested against the null hypothesis that fluoride treatments leave the rate of tooth decay unaffected. The null hypothesis is denoted by H_0; the alternative hypothesis is denoted by H_a.

The **null hypothesis** (H_0) claims that the observed differences are due to random variation rather than systematic differences between groups in a hypothesis-testing situation.

The **alternative hypothesis** (H_a) is the competing claim to the null hypothesis. It is associated with the idea that there are real systematic differences between groups or treatments studied.

Early clinical studies of fluorides and tooth decay produced evidence that the rate of tooth decay was lower in areas where the water contained moderate levels of fluorides and higher in areas where fluorides were absent. This evidence suggested that H_a might be true, but it did not constitute proof. Experiments needed to be performed and comparisons made. In this case, comparisons of the average number of cavities developed per person were carried out in a randomized study, with half the sampled subjects receiving fluoride treatments and the other half receiving a placebo. Large-scale tests of this sort were conducted in the 1950s and 1960s by the U.S. Public Health Service. Those people who received the fluoride treatments had significantly fewer cavities, on average.

At this stage, we merely had further evidence that fluorides can reduce the incidence of cavities, but on the tube of fluoride toothpaste you will see words to the effect that they have been *proven* effective. This is where quantitative hypothesis testing comes in. The null hypothesis, H_0, would state that fluorides do not reduce tooth decay and the fact that we observed fewer cavities in the groups treated with fluorides was due to sampling error—the luck of the draw. However, we know how to calculate the probability of such chance variations. If that probability, or *p* value, is less than 1%, we can reject the null hypothesis, H_0, at the 1% significance level and regard H_a—the hypothesis that fluorides are effective in reducing the incidence of cavities—as statistically proven at this level. Note that while our calculations show that we may be wrong in our conclusion, they set a bound of 1% or less, on average, for the probability that this will happen by chance alone. The possible outcomes of hypothesis testing are given in Figure 8.1.

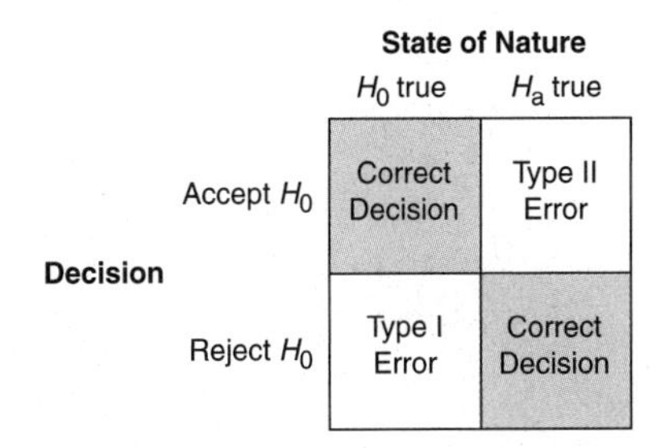

Figure 8.1
Sample-based decisions and their consequences

Type I and Type II errors

If you were running a dairy, your null hypothesis, H_0, might be that pasteurization was up to snuff and bacteria levels in the milk were low enough to meet state standards. So a Type I error, which consists of rejecting the null hypothesis (H_0) when it is actually true, might be to panic about the quality of the pasteurization when it's actually OK and dump a batch of good milk. The level of Type I error we are willing to accept depends on the cost of such an error. If the cost is low, we may be willing to take a larger risk of making a Type I error and suffering the consequences. Type I error risk is often referred to in business as *producers' risk* because it measures the probability of stopping a production process when, in fact, nothing is wrong. The probability associated with producers' risk is called the alpha level, or α.

Type II error is typically an error of caution, as it involves not rejecting the null hypothesis when it is false. Type II error risk is often referred to as *consumers' risk*. The probability associated with consumers' risk is called the beta level, or β. In the case of the dairy, a Type II error would cause consumers to buy milk that had bacteria levels above the standard. In general, β measures the probability of a consumer's accepting the claim that a product is OK when actually it is not. Consumers' risk arises from a problem that goes undetected, leaving the consumer to deal with a defective product.

A **Type I error** consists of rejecting a true null hypothesis. It involves accepting the existence of real systematic differences when none exist.

A **Type II error** consists of accepting a false null hypothesis. It involves wrongly assuming that there are no systematic differences between treatments when such differences exist.

Setting acceptable levels for Type I and Type II errors is outside the realm of statistics; it depends on business and social factors. A plant that manufactures latex goods may take a relaxed attitude toward punctures in its balloons, because the consequences of shipping lots of balloons in which 5 or 10 balloons per 100 have holes in them are slight. But if the plant is making surgical gloves, it must achieve a very high level of certainty that none of its gloves have punctures.

If you need to use hypothesis-testing methods in your work, you should determine whether your employer, the industry, or the government has established standards for acceptable levels of Type I and Type II errors. If they have, you should follow them. If not, you and your coworkers will have to consider costs, benefits, risks, and consequences and then use your own judgment to determine to what lengths you must go to avoid errors. Statistics can quantify the tradeoffs, but setting the level of risk is a business decision.

In any hypothesis-testing situation, a test statistic is used. A test statistic is a number based on sampling data and computed according to a formula suited to the purposes at hand. The resulting value of the test statistic is compared with a pre-established critical value obtained from a table. This critical value is chosen so that, under the null hypothesis, the probability that the test statistic will be as extreme as this critical value is α, the value that you have set as your limit for the probability of a Type I error.

Alternatively, your software may simply give you a p value for the test. The p value is the probability, under the null hypothesis, of observing as extreme a value as you have gotten. Then you may decide, on the basis of this p value and the potential consequences of a Type I error, whether or not to reject the null hypothesis.

In hypothesis testing, the **test statistic** is the measure used in the evaluation of sample data. A calculated value is compared with a predetermined critical value or used in computer calculation of a p value.

Fortunately, most test statistics used in statistical hypothesis testing have a standard form, so we can consider the *generic test statistic.* The values referred to in the following formula may be the values of the z statistic or the t statistic, depending on the type of test used.

The Generic Test Statistic

$$\text{Test statistic} = \frac{\text{Observed value} - \text{Expected value}}{\text{Standard error of test statistic}}$$

where the expected value is the value expected under the null hypothesis.

Before you can use statistical hypothesis testing to guide business decisions, you and your coworkers must have a clear understanding of your purposes and the uncertainties involved. Probabilities of error can be calculated in all hypothesis testing, but you must decide what risks you are willing to take. Usually, the decision comes down to specifying the **significance level** for the test, which is α, the probability of making a Type I error. Once you have done this, you can carry out the test using standard techniques.

There are six steps in formal hypothesis testing:

1. State the null hypothesis.
2. State the alternative hypothesis.
3. Choose the test statistic.
4. State a decision rule.
5. Calculate the test statistic.
6. Decide whether to reject or provisionally accept the null hypothesis.

Step 1 is the statement of the null hypothesis. The null hypothesis might be that a manufacturer's product has no benefits and all variation is due to random fluctuation in the population. In Step 2, the alternative hypothesis is stated. For example, in the case of the manufacturer of a gasoline additive, the alternative hypothesis might be that the product improves gas mileage. Certain evidence will be supplied supporting this claim, and we will use this evidence to compare the alternative hypothesis to the null hypothesis that there is no improvement in gas mileage.

The choice of a test statistic (Step 3) depends on the sampling distribution for the parameter in question. To test claims about means, the z or t statistic is appropriate. To test claims about variances, the F statistic, discussed in Chapter 11, is appropriate. The purpose of the test statistic is to produce a sample value that can be compared with a preset criterion determined by the significance level, α, that we established beforehand. The selection of the test statistic depends on the parameter being tested. In the first part of this chapter, it is the population mean, μ. (In Chapter 9, we will consider the population proportion, π.) The parameter might be the difference between population means, $\mu_1 - \mu_2$, or the difference between population proportions, $\pi_1 - \pi_2$. These differences come into play when two populations are being compared. In testing the value of the population mean, we have a choice between two test statistics: z and t. The decision as to whether to use z or t depends on the nature of the distribution and sample size.

To get a sense of the distribution of the data, it is important to use histograms or other graphic methods. If the underlying distribution of data from which the sample is drawn is perfectly normal, then the z statistic is appropriate. If the underlying distribution is only approximately normal, the t statistic is the proper choice.

Samples may be classified as either large or small. *Large* means a sample size of 30 or more. For large samples, the z statistic provides an acceptable result (roughly equivalent to that of the more accurate t statistic, used when the data do not follow a strictly normal distribution but approximate it). For small samples where the underlying distribution is roughly normal, the t statistic is the appropriate choice.

In Step 4, a decision rule is stated. A decision rule sets the dividing line between rejecting the null hypothesis and accepting it: If the **calculated value** of the sample test statistic falls beyond a certain **critical value,** the null hypothesis, H_0, is rejected and the alternative hypothesis, H_a, is accepted. Large values of the test statistic are unlikely if the null hypothesis is true.

A **decision rule** is the basis for concluding that a claim is or is not reasonable. The rule requires that (1) a comparison be made between the calculated and the critical value of the test statistic and (2) a course of action be specified, depending on the value of the test statistic.

Typically, hypothesis testing is done at one of three levels of confidence (see Chapter 7): 90%, 95%, or 99%. Depending on the level of confidence selected (and the associated probability, α, of Type I error the decision maker is willing to accept), critical z scores are chosen to serve as cutoffs (Step 5). If a test statistic falls beyond the range defined by these cutoffs, the null hypothesis is rejected and the alternative hypothesis is considered proven. If the test statistic falls within the range defined by the cutoffs, the null hypothesis is considered plausible.

Frequently, the decision maker is willing to risk a 5% chance (1 in 20) of making a Type I error. The decision rule should then be that if the test statistic falls more than 1.96 standard deviations away from the claimed mean, the decision maker will reject the null hypothesis. It is at $\pm 1.96\sigma$ that there is exactly a 5% chance that the test statistic will fall in the rejection region when the null hypothesis is true. This probability of a Type I error is exactly the α level chosen for the test. Figure 8.2 shows a normal curve with cutoffs at the 95% confidence level.

Figure 8.2
Normal curve with cutoffs at the 95% confidence level

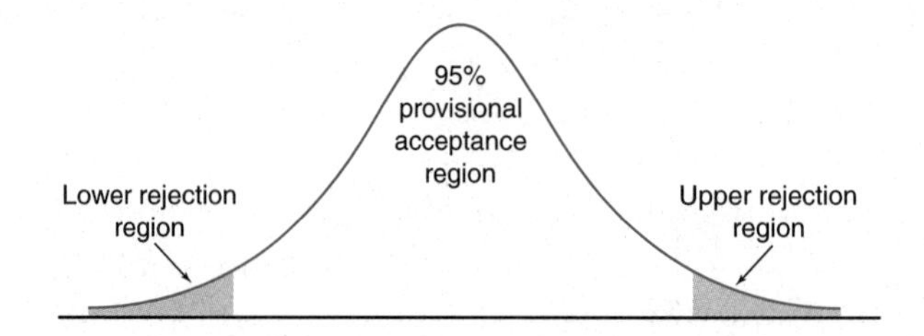

The z value that corresponds to 95% of the area beneath the curve is 1.96. This z value, $z_{critical}$, is the critical value for a two-tailed test at the 5% level, because the chance that the statistic will fall outside the acceptance region by chance alone when H_0 is true is 5%. The sample results will be compared with $z_{critical}$. The probability of a Type I error, α, is proportional to the area of the shaded tails. In this example, $\alpha = .05$.

SECTION 2 ■ TESTS ABOUT SINGLE MEANS

Large Samples

Hypothesis testing uses sample data to evaluate a claim statistically. The choice of a test statistic depends, in part, on sample size. When the sample size is 30 observations or larger, the z statistic may be used.

EXAMPLE 8.1 ■ USING THE z STATISTIC TO TEST A COMPANY'S CLAIM

Del Monte Foods packages Blue Lake Cut Green Beans in 16-ounce cans, which it claims contain 8.75 ounces of cut green beans. Suppose that, as quality control officers for Del Monte, we want to make sure that these cans are not being systematically overfilled or underfilled. We plan to draw a sample of 36 cans of cut green beans.

Step 1. We state the null hypothesis—that the average weight of cut green beans per can is 8.75 ounces:

The mean for a large sample, two-tailed test

$$H_0: \mu = 8.75 \text{ oz}$$

The null hypothesis presupposes that observed variations from this value are due to random variation about this mean.

Step 2. We state the alternative hypothesis. The alternative hypothesis and the null hypothesis together must be mutually exclusive and exhaustive. That is, all possible outcomes must be covered by either the null hypothesis or the alternative hypothesis. Thus, if the null hypothesis is that the average is 8.75 ounces, the alternative must be that the weight is *not* 8.75 ounces:

$$H_a: \mu \neq 8.75 \text{ oz}$$

Note: The statements of both the null and the alternative hypothesis are in terms of the value of the parameter, μ. *All hypothesis statements concern the value of the population measure—that is, the parameter.*

Step 3. We choose the test statistic. Since we are drawing a sample of 36 cans of cut green beans, it is appropriate to use the ***z* statistic,** which is calculated as follows:

$$z = \frac{\bar{x} - \mu_x}{s_{\bar{x}}}$$

where $s_{\bar{x}} = s_x/\sqrt{n}$.

Step 4. We state a decision rule. The test statistic provides a calculated value that can be compared with a predetermined cutoff value so that a decision can be made. The decision rule directly reflects the alternative hypothesis. In this instance, the alternative hypothesis is

$$H_a: \mu \neq 8.75 \text{ oz}$$

The conclusion, if the alternative hypothesis is true, is that the population mean of all cans of Blue Lake Cut Green Beans is not equal to 8.75 ounces. However, this conclusion does not imply that if the sample mean is, indeed, unequal to that exact value, the original claim (H_0) should be rejected. Means for samples will vary; variability is a fact of statistical life. One sample of cans may have a mean of 8.63 ounces, while another has a mean of 8.90 ounces. We need to determine the average weight that should cause us to reject the initial claim, H_0. If we draw a sample of 36 cans and the average weight is 8.50 ounces, should we assume that consumers are being cheated? If the weight is 9.00 ounces, should we conclude that Del Monte is cheating itself by putting too much in the cans? Given our alternative hypothesis, a value for the test statistic that is either too small or too large will be considered sufficient to reject the claim that there are 8.75 ounces of cut green beans per can on average.

Step 5. We calculate the test statistic. Suppose we find that the estimated average weight of our sample of 36 cans is 9.00 ounces and the standard deviation is 3.06 ounces. Does this mean that Del Monte is cheating itself? Or is this simply the result of chance variations?

$$z_{\text{test}} = \frac{9.00 - 8.75}{3.06/\sqrt{36}} = .49$$

Step 6. We decide whether to reject or provisionally accept the null hypothesis. Since .49 falls between −1.96 and +1.96, we provisionally accept the null hypothesis. We conclude that the company's claim of 8.75 ounces per can is plausible, realizing that a claim of 9.00 ounces per can would be equally plausible. What the test told us is that H_0 *could be true.* The data do not depart enough from the expected value to suggest that the mean differs from 8.75.

One Tail or Two? In Example 8.1, the null hypothesis—that the mean equals the weight claimed by the company—was tested against the alternative hypothesis—that the mean does not equal the weight claimed. This was a **two-tailed test** because the decision maker assumed that uncontrolled variability in the weights of the cans would allow those weights to fall on either side of the advertised 8.75 ounces. What we wanted to test was whether or not this variability systematically carried the average weight of the sample significantly above or below the advertised 8.75 ounces.

Often, a **one-tailed test** is more appropriate than a two-tailed test. Suppose a manufacturer wants to test the claim that its light bulbs will last at least 500 hours. In this instance, rejection of the claim involves proving that the bulbs lasted *less than* 500 hours. The hypotheses appear as follows:

$$H_0: \mu \geq 500$$

$$H_a: \mu < 500$$

The rejection region falls to the left of the distribution, in the direction in which the inequality symbol is pointing. The probability of a Type I error, α, is proportional to the shaded area in the left tail of the distribution in Figure 8.3.

Figure 8.3
Lower-tail rejection region

The mean for a large sample, lower-tail rejection region

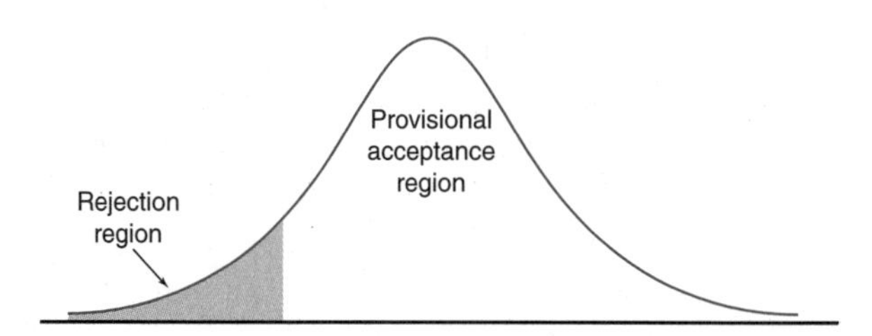

Your purposes determine how you set up H_0 and H_a. For example, if a light bulb company wants to support the claim that its bulbs have an average lifetime *greater than* 500 hours, the hypotheses must be H_0: $\mu \leq 500$ and H_a: $\mu > 500$. To support the claim of average lifetimes greater than 500 hours, you need data that will allow you to reject H_0. As shown in Figure 8.4, the rejection area will be the upper tail region, not the lower tail region. The company is claiming a longer lifetime for its bulbs, and it needs evidence in the upper tail to support this claim.

Figure 8.4
Upper-tail rejection region

The mean for a large sample, upper-tail rejection region

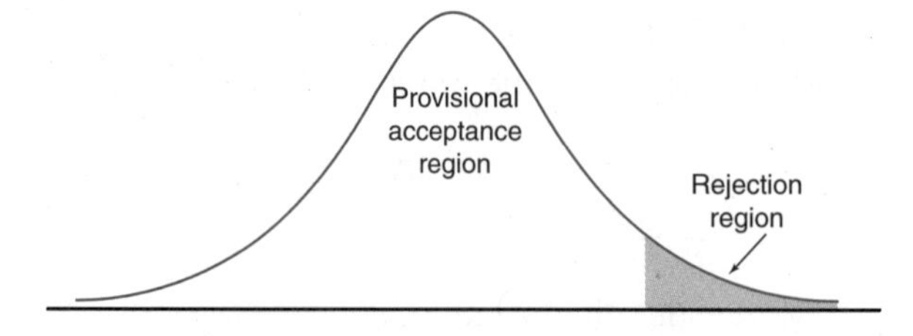

Here is another situation in which a one-tailed test is appropriate. Morningstar Farms claims its Breakfast Links have no more than 5 grams of fat. Suppose Consumers Union wants to shoot down this claim. What sort of test should the group's statistician use? The test should show an upper-tail region. Why? Because the Consumers Union statistician has the burden of proof to show that Breakfast Links have more fat than is claimed. This statistician must ask "What will disprove the claim?" The competing hypotheses follow:

$$H_0: \mu \leq 5 \text{ g}$$
$$H_a: \mu > 5 \text{ g}$$

The rejection region falls to the right of the distribution.

The quality control statistician at Morningstar Farms, on the other hand, wants evidence supporting her company's low fat claims. She wants to reject H_0: $\mu \geq 5$ grams, so her rejection region will be a lower-tail region. Examples 8.2 and 8.3 illustrate applications with an upper-tail and a lower-tail rejection region, respectively.

> The **rejection region** is the area beneath the probability distribution where the null hypothesis is to be rejected.

EXAMPLE 8.2 ■ USING A ONE-TAILED TEST TO CHECK TIRE DURABILITY

A tire manufacturer guarantees that its tires will last for at least 50,000 miles. If the tire fails earlier, the manufacturer offers a prorated replacement. Too many such replacements are bad for business, so the manufacturer tests tires at random to see whether corrective action is needed on the production line. A sample of 35 tires is examined. The average durability is 48,257 miles, with a standard deviation of 512 miles. At the .05 level, should the manufacturer be satisfied that its claim holds?

Step 1. We state the null hypothesis:

$$H_0: \mu \geq 50{,}000 \text{ mi}$$

Step 2. We state the alternative hypothesis:

$$H_a: \mu < 50{,}000 \text{ mi}$$

Step 3. We choose the test statistic. Because the sample is large ($n = 35$), the z statistic provides an acceptable result, regardless of the underlying nature of the distribution.

Step 4. We state a decision rule. We need to ask ourselves, "What would make us believe that the 50,000-mile claim was wrong?" Since only a significantly lower value would jeopardize this claim, the rejection region lies entirely within the lower tail of the distribution and a one-tailed test is appropriate. Rejecting the lowest 5% of the distribution corresponds to a critical z value of -1.645. If the test statistic falls below -1.645, that will force us to reject the null hypothesis and accept that $\mu < 50{,}000$ miles.

Step 5. We calculate the test statistic:

$$z_{\text{test}} = \frac{48{,}257 - 50{,}000}{512/\sqrt{35}} = -20.14$$

Step 6. We decide whether to reject or provisionally accept the null hypothesis. Since -20.14 is far below -1.645, we may safely reject the manufacturer's claim. The average life span of the tires is less than 50,000 miles.

p values

***p* Values.** When you use software such as Excel or a statistical calculator, you get not only the z value for a test, but also the corresponding p value. In the case of Example 8.2, with a z value of -20.14, the software will give a p value of .0000. What this p value means is that the probability that a standard normal random variable will have a value of less than -20.14 is 0 to the accuracy shown. (The actual probability is positive, though smaller than 7.7×10^{-175}, and can be worked out from formulas given in National Bureau of Standards tables.)

This p value of .0000 is our estimated bound on the probability of observing data as extreme as the data we have found, assuming that the null hypothesis is true. For Example 8.2, the null hypothesis is H_0: $\mu \geq 50{,}000$ miles. Our observed $\bar{x}$ is less than 50,000 miles, and we are conducting a lower-tail test. If H_0 is true and $\mu \geq 50{,}000$ miles, then the chance of observing a value of x that is less than 50,000 miles is maximized when μ is as small as it can be while still satisfying H_0. Thus, if H_0 is true, the probability of making an observation as small as or smaller than x is greatest when μ is 50,000 miles.

Now, if μ is 50,000 miles, if the distribution of tire lives is normal, and if $s_{\bar{x}}$ is a good estimate for the standard deviation of $\bar{x}$, then z_{test} is approximately normally distributed, with a mean of 0 and a standard deviation of 1. Hence, the probability of observing $\bar{x} = 48{,}257$ with an $s_{\bar{x}}$ of $512/\sqrt{35}$ is approximately the same as that of observing a value of -20.14 for a standard normal random variable. This is the p value reported for the z test, and it is 0 to more than 170 decimal places.

> The ***p* value** is the probability of encountering the calculated value of the mean when the null hypothesis is true.

A z value of -20.14 is so extreme that we can be certain that some of our assumptions are wrong; in particular, we can reject the idea that H_0: $\mu \geq 50{,}000$ miles. In addition, it is likely that $s_{\bar{x}}$ differs from $\sigma_{\bar{x}}$, so we should not have too much confidence in the estimate of 7.7×10^{-175} supplied for the p value. The answer of .0000, given to a few decimal places by standard software, tells us to reject H_0, and it suggests the limits in precision inherent in these calculations.

EXAMPLE 8.3 ■ USING A ONE-TAILED TEST TO CHECK STUDENTS' AVERAGE CUT RATE

A university announced that the number of cuts per course per semester averaged 3. A dean, skeptical of this figure, believes that the average number of cuts is greater. A random sample of 50 student records yields an average of 4.2 cuts per student per course, with a standard deviation of .8. Is the dean correct in his belief? Test at the .10 level.

Step 1. We state the null hypothesis:

$$H_0: \mu \leq 3$$

Step 2. We state the alternative hypothesis:

$$H_a: \mu > 3$$

Step 3. We choose the test statistic. Because the sample is relatively large ($n = 50$), the z statistic will provide an acceptable result, regardless of the underlying nature of the distribution.

Step 4. We state a decision rule. The dean's concern is that the cut rate is greater than stated. As a result, only a significant change in the upward direction, in the upper tail of the distribution, is of interest. Thus, a one-tailed test is employed. As shown in Figure 8.5, the rejection region consists of the area in the upper 10% of the distribution, so the critical value will be the z statistic that yields the 90% cutoff. z_{critical} is 1.28. If z_{test} falls above 1.28, it can be concluded that the cut rate is greater than 3.

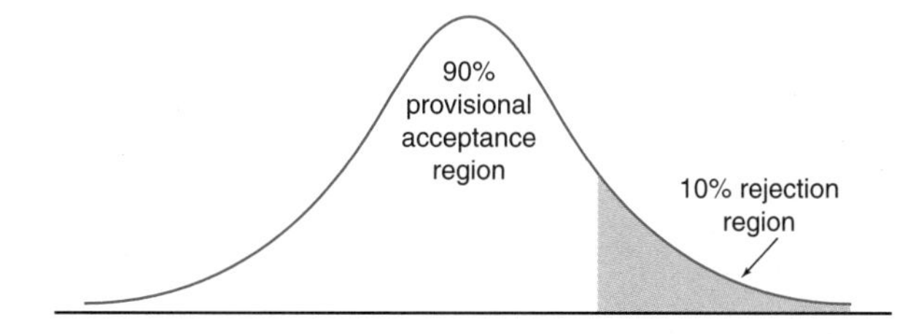

Figure 8.5
Upper-tail rejection region with a 90% cutoff

Step 5. We calculate the test statistic:

$$z_{\text{test}} = \frac{\bar{x} - \mu_x}{s_{\bar{x}}}$$

$$= \frac{4.2 - 3.0}{.8/\sqrt{50}} = 10.61$$

Step 6. We decide whether to reject or provisionally accept the null hypothesis. Since 10.61, the calculated value, is well beyond the critical value of 1.28, we can confidently conclude that the cut rate is greater than 3.

Small Samples

It would be ideal if, every time a claim were made, sufficient data could be collected to have a large, perfectly normal sample. Unfortunately, researchers must often settle for samples that are smaller than 30 observations and/or not perfectly normal. As long as the sample consists of at least 30 observations, the z statistic provides an accurate assessment, even if the data are not perfectly normal. The z statistic is a **robust statistic;** that is, it is relatively insensitive to minor variations in the assumptions about the distribution of the underlying population. On the other hand, when the sample is small—for example, 6 observations—the t statistic is the appropriate statistic for evaluating claims about the value of a mean.

The t statistic has a distribution quite similar to that of the z statistic for samples of 30 or more. However, the t statistic is a function of sample size, as well as the mean and standard deviation of the underlying population. For samples of less than 30, the t statistic is a more appropriate choice than the z statistic.

EXAMPLE 8.4 ▪ USING THE t STATISTIC TO EVALUATE A CLAIM ABOUT TRADING SOFTWARE

The magazine *Technical Analysis of Stocks and Commodities* is filled with advertisements for trading software. Each software package promises "untold profits" and "money-making opportunities." Suppose one package promises 35% profits on options trading. Following are four observations of the trading gains that would have resulted had the software been used:

53% 28% 34% 19%

Can we accept the claim that use of the software will result in an average return of 35% per annum or better? Test at the .05 level. Assume that the rate of return is approximately normally distributed.

Step 1. We state the null hypothesis:

$$H_0: \mu \geq 35$$

Step 2. We state the alternative hypothesis:

$$H_a: \mu < 35$$

Step 3. We choose the test statistic. Because the sample is small ($n = 4$), the t statistic must be used.

The mean for a small sample, lower-tail rejection region

The ***t* statistic** is calculated in basically the same way as the z statistic is calculated:

$$t_{\text{test}} = \frac{\bar{x} - \mu_x}{s_{\bar{x}}}$$

Step 4. We state a decision rule. Our concern is that the rate of return might be too low. Thus, this is a one-tailed test. A value falling in the lowest 5% of the t distribution will cause us to reject the software's claim.

Examining the t statistic tables, we see that it is not enough to specify the level at which we wish to draw the cutoff line; we also need to specify **degrees of freedom.** Critical values from the t distribution are a function of both the level at which the test is done and the sample size. If the sample size is n, the number of degrees of freedom (df) is $n - 1$. Thus, for a sample size of $n = 4$, df $= n - 1 = 3$. What we seek here is a critical value for the t statistic with 3 df at the .05 level, denoted by $t_{.05,3}$. Turning to Table V in Appendix B, a small portion of which is reproduced in Table 8.1, we find that $t_{.05,3}$ falls in the second column, third row. The value for t_{critical} is -2.353. Note that, as with z, a value to the left of the mean of the distribution is considered negative.

Step 5. We calculate the test statistic. In this case, we begin with the raw data, so the mean and standard deviation must be computed. Using a calculator or spreadsheet, we arrive at the following values:

$$\bar{x} = 33.50\% \qquad s = 14.39\%$$

Table 8.1

PORTION OF THE *t* DISTRIBUTION TABLE

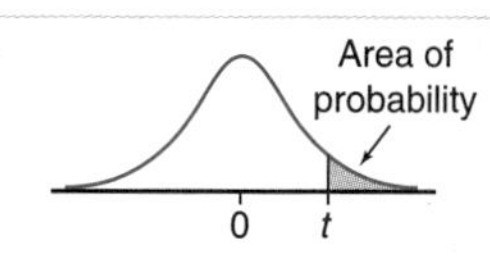

Degrees of Freedom	**RIGHT-TAIL AREAS**					
(df)	**.10**	**.05**	**.025**	**.01**	**.005**	**.001**
1	3.078	6.314	12.706	31.821	63.657	318.309
2	1.886	2.920	4.303	6.965	9.925	22.327
3	1.638	2.353	3.182	4.541	5.841	10.215
4	1.533	2.132	2.776	3.747	4.604	7.173
5	1.476	2.015	2.571	3.365	4.032	5.893

The t statistic is calculated as follows:

$$t_{\text{test}} = \frac{33.50 - 35.00}{14.39/\sqrt{4}} = -.21$$

Step 6. We decide whether to reject or provisionally accept the null hypothesis. As $-.21$ does not fall to the left of -2.353, the null hypothesis should be accepted. Although we may be somewhat skeptical about the software's claim, we have insufficient data to refute it. Perhaps, with more data, we might be able to do so. Our next action might be to compare these results to those achieved with other software packages.

EXERCISES 8.1–8.15

Skill Builders

8.1 You are given the following information:

$$H_0: \mu = 263.01 \qquad H_a: \mu \neq 263.01$$
$$\bar{x} = 251.05 \qquad s = 15.17 \qquad n = 23$$

a. What test should you use? Why?
b. Test at the 95% level.

8.2 You are given the following information:

$$H_0: \mu = 111 \qquad H_a: \mu > 111$$
$$\bar{x} = 120 \qquad s = 10 \qquad n = 100$$

a. What test should you use? Why?
b. Test at the .01 level.

8.3 You are given the following information:

$$H_0: \mu = 17 \qquad H_a: \mu < 17$$
$$\bar{x} = 16.52 \qquad s = .43 \qquad n = 63$$

a. What test should you use? Why?
b. Test with a Type I error of 10%.

8.4 If the mean of a sample is 25.6 and the standard deviation is 2.7, test the claim that the population mean is 25 ($n = 16$).
a. What test should you use? Explain your choice.
b. Test with 99% confidence ($\alpha = .01$).

8.5 It is claimed that the mean of a process is at least 100. A sample of 47 is drawn, with a mean of 95 and a standard deviation of 10. Is the claim provisionally acceptable at the .05 level? Indicate why you chose the test you did.

Applying the Concepts

For each of the following exercises, indicate the null and alternative hypotheses, the test statistic, your reason for choosing the test statistic, the critical value, and your conclusion.

8.6 In order to gather data to be used in the management of the King Crab Fisheries, the Alaskan Department of Fish and Game conducted a survey about Kodiak Island king crabs. In 1973, the average number of male crabs per pot was 335.5. The standard deviation was 259.1. In 1979, the mean number of male crabs per pot was 266.4 from a sample of 40 crab pots (Statlibd@stat.cmu.edu). Assuming the data follow a normal distribution, test (at the .05 level) the contention that there was a decline in the male crab population.

8.7 In 1979, the average labor force participation rate (LFPR) in New York State was 60.3%. The LFPR is the percentage of the population in the labor force. Data for 1992 appear below:

61.1	60.9	61.2	61.3	61.6	63.2
63.0	62.4	61.6	61.0	61.6	61.8

Source of Data: U.S. Department of Labor, Bureau of Labor Statistics.

Test (at the .01 level) the claim that more difficult economic times forced more people into the labor force in 1992. Assume that the data for the year follow a normal distribution.

8.8 It is claimed that a particular motor vehicle will stop in an average of 194 feet when traveling at 60 miles per hour. A total of 50 observations yield a mean of 203 feet and a standard deviation of 12. Test the claim at the .05 level.

8.9 A study entitled "Effects of Alternative Family Structures on Managerial Career Paths," by Joy Schneer and Frieda Reitman (*Academy of Management Journal,* August 1993), suggests that single women experience greater career satisfaction than do single men. Suppose that the mean career satisfaction score of the population of single men is 3.46. A sample of 100 single women is randomly selected, and their career satisfaction scores are obtained. The sample mean is 3.64, with a standard deviation of .86. Do the data show, at the 1% level, that the career satisfaction level of single women is greater than that of single men?

8.10 A major food processor produces ready-to-serve soups. Its machinery is set to produce cans of soup with an average weight of 19 ounces. A random sample of 10 cans is drawn, and the average is found to be 18.6 ounces, with a standard deviation of .5 ounce. Is this finding consistent with the hypothesis that the machinery is operating properly ("in control") and producing cans whose average weight is 19 ounces? Assume that the weights are approximately normally distributed. Test at the .05 level.

8.11 A bottler of iced tea wishes to ensure that an average of 16 ounces of tea are in each bottle. A sample of three dozen bottles reveals a mean fill of 15.9 ounces, with a standard deviation of .1 ounce. Should the bottler readjust the bottling process? Test at the .01 level.

8.12 The owner of a stationery store claims that he sells an average of at least 200 copies of the *New York Times,* Sunday edition, each weekend. Observations over eight weeks yield an average of 193 newspapers sold, with a standard deviation of 11 newspapers. Treat these eight weeks as though they represented a random sample, and assume that the number of newspapers sold has an approximately normal distribution. Test his claim at the .05 level.

8.13 It is claimed that retrieval time for a new secondary storage system is under 5 milliseconds. After 50 trials, the average is 5.1 milliseconds, with a standard deviation of .2 millisecond. Is the claim justified at the .10 level?

8.14 A survey of college students indicated that they spend an average of $251.73 on books each semester. The standard deviation is $25.12. Can it be claimed that the average expenditure is $275? Test at the .01 level. Let $n = 100$.

8.15 A pharmaceutical firm states that each capsule of its pain killer contains 200 milligrams of ibuprofen, the active ingredient. A test of a random sample of 40 pills shows the average dosage to be 197 milligrams. The standard deviation is 2 milligrams. Is the firm's claim justified? Test at the .05 level.

SECTION 3 ▪ COMPARISON OF MEANS FROM INDEPENDENT SAMPLES

Often, an analyst is asked to compare one group with another group. The process of comparing two means is similar to the hypothesis-testing procedures we've already examined. The basic steps are the same; the differences lie in the choice of the test statistic and the determination of the critical values.

Comparison of means might be used to investigate the relative performance of students in different classes, the durability of different brands of light bulbs, the efficiency of different machines, or whether one product is better than another. Questions of comparison and choice are the basis for serious decision making. Which supplier will you purchase from? Should you consider contracting with someone else to supply a specific product or service? As a consumer, where should you spend your dollars?

Comparisons can be made with either large samples or small ones. As always, samples should be as random as possible to avoid systematic biases that would throw off error estimates or affect significance levels.

Large Samples

Example 8.5 illustrates the comparison of two large samples.

EXAMPLE 8.5 ▪ USING THE z STATISTIC TO COMPARE COSTS OF MOTEL ROOMS

A company wishing to establish a travel policy for its employees decided to compare rates for single rooms at Howard Johnson's and the Hampton Inn. Independent random sampling from the 1993 directories for both chains yielded the results in Table 8.2 when room rates for a single were compared on a city-by-city basis.

Two means for large samples, two-tailed test

Table 8.2

COMPARISON OF MOTEL ROOM RATES

	Howard Johnson's	Hampton Inn
Average rate	$74.48	$60.83
Standard deviation	27.69	9.64
Number of observations	54	54

Is there a significant difference in room rates? Company management wants an α of .05 or less, so we will test at the .05 level.

Step 1. We state the null hypothesis:

$$H_0: \mu_{\text{HoJo}} = \mu_{\text{Hampton}}$$

(For the test statistic, think of this hypothesis as $H_0: \mu_{\text{HoJo}} - \mu_{\text{Hampton}} = 0$, because it suggests the form of the test statistic.)

Step 2. We state the alternative hypothesis:

$$H_a: \mu_{\text{HoJo}} \neq \mu_{\text{Hampton}}$$

Step 3. We choose the test statistic. Because we are dealing with large samples ($n > 30$), we may use the z statistic. The distribution of differences for large samples is approximately normal. The sum or difference of independent normal variables is again normal, so $\bar{x}_{\text{HoJo}} - \bar{x}_{\text{Hampton}}$ is approximately normally distributed. Here, the z statistic is calculated as follows:

$$z_{\text{test}} = \frac{(\bar{x}_{\text{HoJo}} - \bar{x}_{\text{Hampton}}) - (\mu_{\text{HoJo}} - \mu_{\text{Hampton}})}{\sqrt{\dfrac{s^2}{n_{\text{HoJo}}} + \dfrac{s^2}{n_{\text{Hampton}}}}}$$

The denominator is the standard deviation for $\bar{x}_{\text{HoJo}} - \bar{x}_{\text{Hampton}}$. Because $\bar{x}_{\text{HoJo}}$ and $\bar{x}_{\text{Hampton}}$ are from independent samples, they are independent random variables, and the variance of the difference of independent random variables is the sum of their variances. These variances are s^2/n_{HoJo} and s^2/n_{Hampton}, respectively. Because $\mu_{\text{HoJo}} - \mu_{\text{Hampton}} = 0$ under the null hypothesis, the formula can be simplified to

$$z_{\text{test}} = \frac{\bar{x}_{\text{HoJo}} - \bar{x}_{\text{Hampton}}}{\sqrt{\dfrac{s^2}{n_{\text{HoJo}}} + \dfrac{s^2}{n_{\text{Hampton}}}}}$$

Step 4. We state a decision rule. Our goal is to determine whether there is a difference in the average price, and initially we did not know that one chain would exhibit an average that appears significantly lower than the other. For these reasons, our test is done with an upper and lower rejection region. A sizable difference in either direction would be considered sufficient to reject the null hypothesis that there is no real difference between the average prices for these chains. Figure 8.6 illustrates where the cutoffs lie.

Figure 8.6
Testing for significant differences

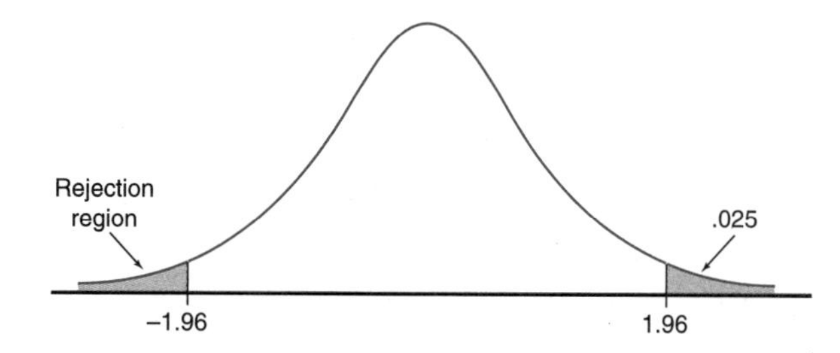

Testing at the .05 level means that rejection regions of 2.5% will appear at the far right and far left of the distribution. The critical values for the z statistic will equal 1.96 and −1.96, respectively.

Step 5. We calculate the test statistic:

$$z_{\text{test}} = \frac{74.48 - 60.83}{\sqrt{\dfrac{766.74}{54} + \dfrac{92.93}{54}}} = \frac{13.65}{3.99} = 3.42$$

Step 6. We decide whether to reject or provisionally accept the null hypothesis. We see that z_{test} exceeds the critical value of 1.96. We can conclude that there exists a difference between the rates offered at Howard Johnson's and the Hampton Inn.

Small Samples

As in estimation, there are often insufficient observations to justify the use of large-sample techniques to draw statistical inferences. We can use the t distribution if the population histograms are approximately normal. Example 8.6 illustrates this testing procedure.

EXAMPLE 8.6 ■ USING THE TWO-SAMPLE t STATISTIC TO COMPARE LIFE EXPECTANCIES

The *Journal of the American Medical Association* regularly publishes obituaries of its members. Included in these obituaries is age at death. *Information Please Almanac* publishes similar data on celebrities. A small random sample of physicians' obituaries

($n = 15$) was drawn, and a random sample of male celebrities' obituaries was also selected. The data for both are summarized in Table 8.3. We will assume that the age distributions at death of both these populations are approximately normal. Do physicians live longer? Test at the .10 level.

Two means for small samples, unequal variances

Table 8.3

COMPARISON OF AVERAGE AGES AT DEATH FOR TWO POPULATIONS

	Physicians	Celebrities
Average age at death	78.53	78.13
Variance	24.01	140.66
Sample size	15	15

Looking at the information in Table 8.3, we see that although the average ages of death are close, the variances are quite different. This difference is allowed for in the hypothesis test described below, called the *two-sample t procedure.* (Example 8.7 offers a second procedure that can be used when the variances are equal.)

Step 1. We state the null hypothesis. Since $\bar{x}_{\text{dr}}$ is slightly larger than $\bar{x}_{\text{celeb}}$, our null hypothesis should be

$$H_0: \mu_{\text{dr}} \leq \mu_{\text{celeb}}$$

This hypothesis states that what we have observed is due simply to sampling error.

Step 2. We state the alternative hypothesis:

$$H_a = \mu_{\text{dr}} > \mu_{\text{celeb}}$$

Step 3. We choose the test statistic. We are assuming that the distributions are normal, so $(\bar{x}_{\text{dr}} - \mu_{\text{dr}}) - (\bar{x}_{\text{celeb}} - \mu_{\text{celeb}})$ is normal. The null hypothesis holds that $\mu_{\text{dr}} \leq \mu_{\text{celeb}}$. Given our observation $\bar{x}_{\text{dr}} - \bar{x}_{\text{celeb}}$, the conservative course, which minimizes the probability of a Type I error, is to find the critical point when μ_{dr} is as large as it can be and still be consistent with H_0—namely, $\mu_{\text{dr}} = \mu_{\text{celeb}}$. So we look at

$$(\bar{x}_{\text{dr}} - \mu_{\text{dr}}) - (\bar{x}_{\text{celeb}} - \mu_{\text{celeb}}) = \bar{x}_{\text{dr}} - \bar{x}_{\text{celeb}}$$

Because the samples are small, we must use something like the t test. Here, as in Example 8.5, the variance of the difference is $s_{\text{dr}}^2/n_{\text{dr}} + s_{\text{celeb}}^2/n_{\text{celeb}}$, so the analog of the regular t statistic is

$$t_{\text{test}} = \frac{\bar{x}_{\text{dr}} - \bar{x}_{\text{celeb}}}{\sqrt{\dfrac{s_{\text{dr}}^2}{n_{\text{dr}}} + \dfrac{s_{\text{celeb}}^2}{n_{\text{celeb}}}}}$$

Step 4. We state a decision rule. Our alternative hypothesis, H_a, is that physicians live significantly longer than celebrities. This hypothesis is tested against

H_0, that physicians do not live as long as or live equally as long as celebrities. Therefore, we place the rejection region to the right of the distribution. Figure 8.7 shows where the cutoff value will lie.

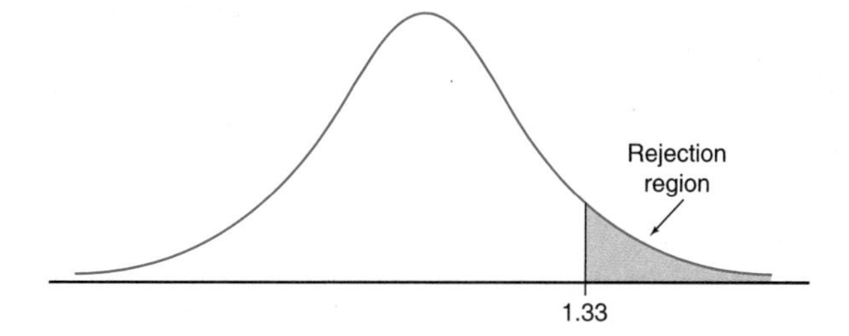

Figure 8.7
The t distribution: testing for the difference between two means

But two difficulties arise. First, to find a critical value for a t test, we need to know the number of degrees of freedom. Second, this statistic is not precisely t distributed, even when the underlying populations are normal. However, if the following formula is used to find the degrees of freedom, the test statistic will closely match the t distribution found in tables:

$$\text{df} = \frac{(v_1 + v_2)^2}{\dfrac{v_1^2}{n_1 - 1} + \dfrac{v_2^2}{n_2 - 1}}$$

where

$$v_1 = \frac{s_1^2}{n_1} \quad \text{and} \quad v_2 = \frac{s_2^2}{n_2}$$

if the sample sizes are n_1 and n_2 and the sample variances are s_1^2 and s_2^2. Then round down to an integer and use that value for df.

Here, we have

$$v_1 = \frac{24.01}{15} = 1.601 \quad \text{and} \quad v_2 = \frac{140.66}{15} = 9.377$$

$$\text{df} = \frac{(1.601 + 9.377)^2}{\dfrac{1.601^2}{14} + \dfrac{9.377^2}{14}} = 18.645$$

Thus, df = 18. (Remember to always round down when using this formula.) From Table V, $t_{\text{critical}} = t_{.10,18} = 1.33$.

Step 5. We calculate the test statistic:

$$t_{\text{test}} = \frac{78.53 - 78.13}{\sqrt{1.601 + 9.377}} = .12$$

Step 6. We decide whether to reject or provisionally accept the null hypothesis. We accept the null hypothesis because $t_{\text{test}} = .12$, which is less than 1.33. We conclude that physicians live, on average, no longer than celebrities.

Example 8.6 showed the use of the regular t test, a method that is applicable even when the sample variances are quite different. Example 8.7 shows how to use a simpler method, the *pooled variance t test,* which applies when samples are small and variances are approximately equal. For the pooled variance t statistic, the degrees of freedom is given by df $= n_1 + n_2 - 2$.

The pooled variance t test for the difference of means is a simpler alternative to the t test; it can be used when the data suggest that the variances are equal for the two populations sampled. Computer software will generally provide p values computed using both the pooled and the unpooled method. You will have to look at the data to see whether the assumption of equal variances can be justified. The conservative course is to use the ordinary t test when there is doubt about the variances.

EXAMPLE 8.7 ■ USING THE POOLED VARIANCE t TEST TO COMPARE SALARIES OF LAW SCHOOL GRADUATES

Law schools are ranked by tiers. Suppose data from a random sample of 10 third-tier law schools reveal that, upon graduation, their students earned an average salary of \$48.2 thousand, with a standard deviation of \$11.9 thousand. Suppose data from a random sample of 10 fourth-tier law schools indicate an average salary of \$36.5 thousand, with a standard deviation of \$9.7 thousand. Assume that the data come from approximately normal populations. Can we conclude that third-tier graduates do significantly better financially? Test at the .05 level.

Step 1. We state the null hypothesis:

$$H_0: \mu_t \leq \mu_f$$

Step 2. We state the alternative hypothesis:

$$H_a: \mu_t > \mu_f$$

Step 3. We choose the test statistic. Because the sample is small, the t statistic is appropriate. Since variances seem to be approximately equal, the pooled t statistic is the statistic of choice.

Remember that the generic test statistic is of the form observed value minus expected value, divided by the standard error term. The observed value here is $\bar{x}_t - \bar{x}_f$. Under H_0, we expect no difference, so the expected value is 0. The denominator of the test statistic, which gives the standard error of the estimate, is more complicated (as you see below), but its form suggests the pooling of the variances of the two samples.

$$\text{Pooled } t_{\text{test}} = \frac{\bar{x}_t - \bar{x}_f}{\sqrt{s_p^2\left(\frac{1}{n_t} + \frac{1}{n_f}\right)}}$$

where

$$s_p^2 = \frac{(n_t - 1)s_t^2 + (n_f - 1)s_f^2}{n_t + n_f - 2}$$

Step 4. We state a decision rule. Because we wish to see whether third-tier graduates do better financially than fourth-tier graduates, we place the rejection region to the right of the distribution, as we did in Figure 8.7. If t_{test} exceeds $t_{critical}$, then we reject H_0.

Two means for small samples, approximately equal variances

As in Example 8.6, the numerator of the formula for t is normally distributed, and the denominator gives the degrees of freedom: $df = n_t + n_f - 2$. In this case, $df = 10 + 10 - 2 = 18$. Thus, $t_{critical}$ is $t_{.05,18} = 1.734$, and the standard error of the estimate is $s_p^2 = [9(11.9) + 9(9.7)]/18 = 117.85$.

Step 5. We calculate the test statistic:

$$t_{test} = \frac{48.2 - 36.5}{\sqrt{117.85(.1 + .1)}} = 2.410$$

Step 6. We decide whether to reject or provisionally accept the null hypothesis. We reject the null hypothesis, because the t_{test} of 2.410 exceeds the $t_{critical}$ of 1.734. So we conclude that third-tier graduates appear to do better financially than fourth-tier graduates.

There are two basic steps in comparing means. The first step is choosing the test statistic. The second step is deciding whether the test should be one-tailed or two-tailed. The criterion for choosing the test statistic is sample size. If the sample is large (30 or more), use the z statistic. If the sample is small (less than 30), use the t statistic. For the t test, you must choose between the regular (unpooled) and pooled form, using the pooled form only when the assumption of equal variances is justified.

The choice between a one-tailed and a two-tailed test depends on the nature of the underlying question. If we are interested only in whether there is a difference between two samples—*not* whether one sample is greater than or less than the other—the test should be two-tailed. After all, a sizable difference in *either* direction will give us the answer to such a question. If the question involves direction (for example, whether the mean of population A is greater or less than the mean of population B), the test should be one-tailed.

Finally, a suggestion: It is often easier to set up the alternative hypothesis first, before setting up the null hypothesis. Take note of the inequality sign in the alternative hypothesis; it indicates where the rejection regions should be. If, for example, the alternative hypothesis reads $A < B$, the rejection region will fall to the left on the distribution. If the alternative hypothesis reads $A \neq B$, the rejection region must fall in the upper and lower tails. Lastly, if the alternative reads $A > B$, the rejection region must fall to the right on the distribution.

EXERCISES 8.16–8.30

Skill Builders

In Exercises 8.16 through 8.20, describe which test statistic you used and why, which test is appropriate, how you found the critical value, and what conclusions you drew.

8.16 Let H_0 be $\mu_1 = \mu_2$. Set up H_a. If $\bar{x}_1 - \bar{x}_2 = 20$, $s_1 = 15$, $s_2 = 25$, $n_1 = 35$, and $n_2 = 30$, test at the .01 level.

8.17 For data set 1, $\bar{x}_1 = 5$, $s_1 = .3$, and $n_1 = 15$. For data set 2, $\bar{x}_2 = 7$, $s_2 = 1.2$, and $n_2 = 10$. Assume the populations are normally distributed. Is the mean of set 2 significantly greater than that of set 1? Test at the .10 level.

8.18 Let H_a be $\mu_1 < \mu_2$. Set up H_0. The difference between the sample means is 6. The pooled sample estimate of variance is 3, with $n_1 = 13$ and $n_2 = 7$. Test at the .05 level. Assume the populations are normally distributed.

8.19 Consider the data sets below. Test for a significant difference at the .05 level. Assume that the populations are normally distributed.

Set 1			Set 2		
.5292	.0436	.2949	.0407	.8691	.1561
.0411	.2848	.7869	.5616	.3158	.7223

8.20 If $\bar{x}_1 = 37{,}500$, $\bar{x}_2 = 41{,}000$, $s_1 = 3000$, $s_2 = 4125$, $n_1 = 100$, and $n_2 = 300$, is μ_2 significantly greater than μ_1 at the .01 level?

Applying the Concepts

8.21 In a study in the *Academy of Management Journal* ("Effects of Alternative Family Structures on Managerial Career Paths," August 1993), Joy Schneer and Frieda Reitman examined two groups of managers who had children and an employed spouse. One group was male, and the other was female. The following statistics were compiled on total income (in thousands of dollars) for these groups:

	Male	Female
Mean	64.15	56.03
Standard deviation	23.46	24.04

If the male sample had 100 respondents and the female sample had 150, do the data show that males earn significantly more than females? Test at the .05 level. Describe which test statistic you used and why, which test is appropriate, how you found the critical value, and what conclusions you drew.

8.22 Do athletes have lower graduation rates than all students combined? Random samples of graduation rates for all students and athletes were selected for 50 schools. The summary statistics are listed below.

	All Students	Athletes
Average rate	56.02%	59.62%
Standard deviation	20.38%	18.24%

Is the claim that athletes have lower graduation rates than all students combined supported by the data? Test at the .05 level. This question can be answered without any calculations. Can you explain?

8.23 *Places Rated Almanac* for 1993 provides a variety of statistics, including a housing cost index, on many locations throughout the United States. The Mega Corporation has decided to locate in either Texas or California. A look at the housing cost indexes yielded the following data:

	Texas	California
n	27	21
$\bar{x}$	7087	11,632
Standard deviation	1180	3194

Is there a significant difference between the cost of housing in Texas and in California? Assume that the housing index is approximately normally distributed. Test at the .10 level.

8.24 Each year, *Fortune* publishes a list of the best cities for business. These cities are ranked with respect to percentage of labor force that is skilled, daily international flights, manufacturing competitive index, and pro-business attitude. Below are data on the variable percent skilled labor force for the 10 best cities and 10 other randomly selected cities. Do the 10 best cities have a significantly higher percentage of skilled labor? Assume that these populations are normally distributed. Test at the .10 level.

10 Best				10 Rest			
47.0	49.2	44.7	44.9	43.5	38.5	39.7	44.9
46.3	42.0	43.4	49.0	42.7	42.6	39.3	42.9
42.6	40.7			41.1	40.3		

Source of Data: *Fortune*, November 2, 1992.

8.25 In the April 1993 issue, *Consumer Reports* rated new cars. The cost of a year's fuel for sports/sporty cars ranged from $535 for the Toyota Paseo to $1230 for the Chevrolet Corvette. The average for 17 such models is $820; the standard deviation is $228. The cost of a year's gasoline for eight models of small cars ranged from $510 for the Toyota Tercel to $715 for the Hyundai Electra. The average for this group is $639; the standard deviation is $60. Assume that the gas mileage is approximately normally distributed. Is there a significant difference at the .01 level?

8.26 Each year, students in New York City schools take a citywide reading test. In 1993, the average score of students in 32 districts was 45.4; the standard deviation was 12.7. In 1992, the average was 45.3, with a standard deviation of 12.1 (*New York Times*, June 18, 1993). Is the headline "Public Schools' Scores Show Modest Increase" justified? Test at the .05 level.

8.27 In May of 1993, *Forbes* compiled a list of CEOs' compensation. It had long been contended that electric utility CEOs were overpaid. To test this contention, a random sample of utility executives was compared with a sample of CEOs from the entertainment industry. The data for each sample follow:

	Entertainment	Utilities
Sample size	10	10
Average	$2,400,100	$578,800
Standard deviation	1,877,909	123,029

Use a level of .01. Assume that these populations are approximately normally distributed. Do you think this comparison is a fair test of whether utility companies' CEOs were overpaid? Explain.

8.28 A recent study of agricultural households in Botswana compared cash outflows for households headed by males and those headed by females (Hilary Sims Feldstein and Susan V. Poats, *Working Together: Gender Analysis in Agriculture*, Kumarian Press, 1989). It was reported that households headed by males spent 105.99 pula (the currency unit of Botswana), while female-headed households spent, on average, 98.48 pula. There were 17 male-headed households and 10 female-headed households in the study. Suppose the standard deviation is 15 pula for male-headed households and 10 pula for female-headed households. Is there a statistically significant difference in spending between male- and female-headed households? Test at the .05 level. Assume that these populations are approximately normally distributed.

8.29 Each year, *Fortune* publishes a list of "America's Most Admired Corporations." The top 10 firms in each industry are scored on a scale from 0 to 10. Data for 1993 and 1994 for the telecommunications industry are found below. Has the rating of telecommunications corporations fallen significantly? Assume that these data are approximately normally distributed. Test at the .10 level.

	1993	1994
n	10	10
Mean	6.85	6.73
Standard deviation	.38	.65

8.30 Modems are used with computers to transmit data over telephone lines. They vary in transmission rates. In the early 1990s, slower modems operated at 1200 baud, while faster ones operated at 2400 baud or more. Data for the prices of 1200- and 2400-baud modems (*PC Today,* June 1990) appear below:

	1200 baud	2400 baud
n	19	44
Mean	$108.70	$171.60
Standard deviation	70.40	87.90

a. Were the faster modems significantly more expensive? Test at the .01 level.
b. Perform this analysis for modems available today.

SECTION 4 ▪ INFERENCE, DECISION MAKING, AND STATISTICAL PROCESS CONTROL

Total quality management (TQM) is an important application of hypothesis testing. TQM is the legacy of W. Edwards Deming, whose development and promotion of the concepts of quality control had a revolutionary impact on industry. In this section, you will see how quality control works.

Quality control (also known as **statistical process control**) begins with the recognition that variation occurs in all production processes. Drawing conclusions from data affected by random variations is a fundamental task of statistics. The range, standard deviation, interquartile range, and coefficient of variation are all tools that help us see underlying patterns in data affected by such variation (see Chapter 3). When anything is manufactured, no matter how careful or precise the production process, we find variations. Despite these variations, we must be able to see what is happening and also control the production process. The question is "How much and what kind of variation is acceptable?" Deming's genius was to recognize that variations take two forms: unavoidable random variation and assignable cause variation. Quality control methods can identify assignable cause variation, which may merit further investigation and correction.

© Catherine Karnow/Corbis

Figure 8.8
Quality control pioneer W. Edwards Deming, 1900–1993

Statistical process control

Understanding variation is important in many situations. Suppose that, as a packager of coffee, you want to ensure that the amount of fill matches the weight stated on your labels. Too little and you could get in trouble with Consumer Affairs. Too much and you could go broke! If the average fill is steadily increasing, the production process probably needs readjustment; but if it fluctuates around the stated weight, the process may be acceptable. Deming's tools for quality control are used to monitor such operations.

Quality control is really automated hypothesis testing. When we consider a production process, the null hypothesis is "The process is in control." The alternative hypothesis is "The process is out of control." Control may be measured in a number of ways, but these measurements can all be understood with charts that record the output as it varies over time. Because the sample averages are plotted, the term *Xbar* is often used. This type of chart was introduced in the 1920s by Walter Shewhart at Bell Labs. The center line on the chart represents the process mean. For the coffee packager, the process mean would be the 1-pound weight per bag that is the ideal.

> The **process mean** is the average value for a production process. It is a preset level that serves as the goal to be met in the course of production.

The upper and lower control limits are typically at the three-sigma level (3σ). If the data follow a normal distribution, we know from the empirical rule that, on average, 99.7% of the data will fall between $\mu - 3\sigma$ and $\mu + 3\sigma$. The **three-sigma limit** is popular because, even without assumptions concerning the population histogram, statistical theory tells us that the probability that a value will fall three sigmas from the mean is less than $\frac{3}{10}$ of 1%. Experience has shown that these limits minimize the likelihood of false alarms. Figure 8.9 depicts a process in control; in Figure 8.10 on page 286, we see a process going out of control.

Figure 8.9
A process in control

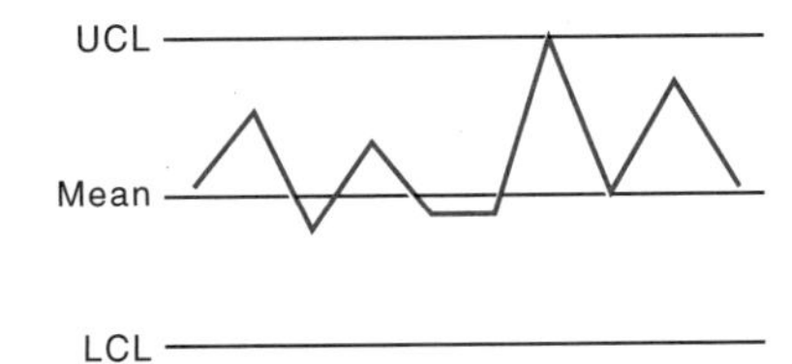

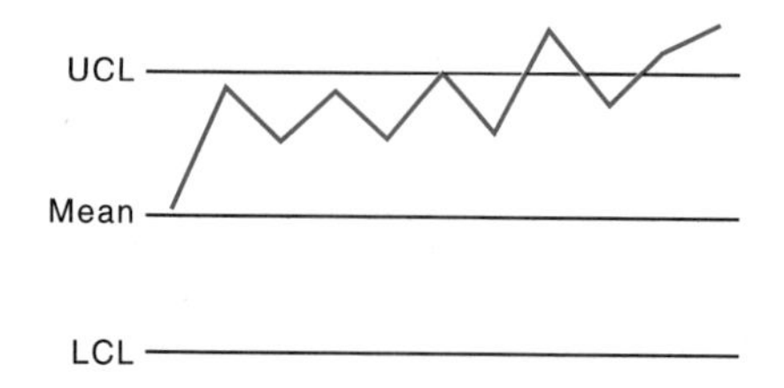

Figure 8.10
A process out of control

EXAMPLE 8.8 ▪ SETTING UP AND INTERPRETING A CONTROL CHART FOR CEREAL FILL

Data were collected over a 10-day period on the net weight of boxes of a certain dry cereal. The filling process was set to have a mean of 453 grams and a standard deviation of 6 grams. Each day, five samples were collected; the results are shown in Table 8.4.

Table 8.4

WEIGHT OF BOXES OF DRY CEREAL (in grams)

Day	9 A.M.	11 A.M.	1 P.M.	3 P.M.	5 P.M.
1	459	461	465	463	458
2	453	452	462	466	459
3	451	453	445	445	443
4	439	440	448	441	442
5	447	448	452	455	461
6	448	450	453	459	460
7	445	441	452	461	445
8	441	453	467	462	445
9	459	452	445	441	441
10	441	446	465	455	459

Considering the samples from each day as a subgroup, we will set up a control chart for the arithmetic mean and then decide whether or not the process is in control.

Our first task is to draw a chart with a center control line (CCL) at 453 grams, which is the process mean, μ. Next, we need to establish the upper control and lower control limits. The upper control limit (UCL) will be located at the mean $+ 3\sigma$ level. The lower control limit (LCL) will be located at the mean $- 3\sigma$ level. To determine these limits, we need to know sigma. If we had values of sigma from past history, when the filling machine was properly set and in good working order, we would use these values. Here, we will simply estimate sigma by finding s_x for the full data set. If we did not know the target μ (here, 453 grams), we would estimate it using the mean of all the data. Sigma for the daily average of n samples will be $s_x/\sqrt{n}$. Here, $n = 5$, so sigma for $\bar{x}$ equals $s_x/\sqrt{5}$.

The daily average is calculated by adding the observations for each day and dividing by 5. These daily averages are the $\bar{x}$ after which the Xbar charts were named. The UCL is calculated by use of this formula:

$$\text{UCL} = \mu + 3\sigma$$

Here,

$$\sigma = \frac{s_x}{\sqrt{n}} = \frac{6}{\sqrt{5}} = 2.68$$

where n is the sample size on a given day. The LCL is calculated similarly:

$$\text{LCL} = \mu - 3\sigma = 453 - 3(2.68) = 444.96$$

In this instance, the UCL is at 461 grams, the LCL is at 445 grams, and the CCL is at 453 grams.

The next step in this process is calculating each day's mean, or $\bar{x}$. These averages are given below:

Day	$\bar{x}$	Day	$\bar{x}$
1	461	6	454
2	458	7	449
3	447	8	454
4	442	9	448
5	453	10	453

Once the averages have been calculated, they are plotted, yielding the Xbar control chart in Figure 8.11. Is the process in control? It would appear so, with one exception: Day 4. On that day, the observed $\bar{x}$ fell below the lower control limit. This finding suggests that cereal boxes produced on that day should be examined and replaced with cereal boxes falling within the three-sigma limits. Moreover, the operating conditions on that day should be reviewed in order to prevent recurrences of such shortfalls.

Figure 8.11
Control limits for cereal fill

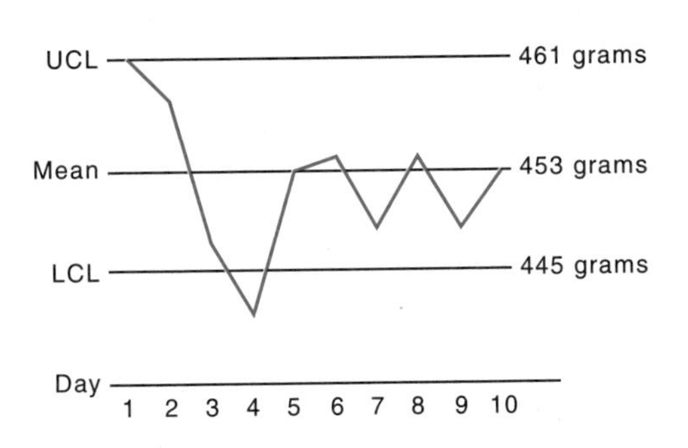

We have just examined one simple application of a quality control technique, in which the three-sigma control limit is used as a signal that a process is out of control. Other signals are used as well—for example, a run of nine or more values on the same side of the mean or a run of three values, two of which deviate from the mean by more than two sigmas in the same direction. More detailed information can be found in works like *Statistical Quality Control* (Eugene L. Grant and Richard S. Leavenworth, 1988). The importance of quality control as a tool for total quality management cannot be overemphasized. The article by Glenn Rifkin in Figure 8.12 relates Motorola's experience with implementing an improved quality control process.

Pursuing Zero Defects Under the Six Sigma Banner

By Glenn Rifkin

When Motorola Inc.'s fiscal year ended in December, the payoff from improving the quality of the company's manufacturing operations suddenly became evident: $500 million saved during the year from reducing defects.

The Digital Equipment Corporation benefited from improving its quality as well. Its storage division cut its product-defect rates by 80 percent in one year, which meant a 25 percent reduction in overall product costs.

What these success stories have in common is that they were accomplished under the banner of "Six Sigma," a three-year-old defect-reduction strategy. Six Sigma was first used by Motorola and is now being embraced by a growing number of American companies seeking to improve quality.

"If you control your processes very tightly, you don't produce defects and therefore you don't have to refine and repair defects," said Richard Buetow, senior vice president and director of quality at Motorola. "You don't need the extra people and equipment."

Six Sigma is a statistical term attached to the concept of achieving approximately zero defects—actually 3.4—per million opportunities. Sigma—the 18th letter of the Greek alphabet—refers to the number of standard deviations from the mean in any given statistically measurable process. The mean is the arithmetic average; the standard deviation is a measure of variation from that average.

Six Sigma requires no new manufacturing equipment, but it does require rethinking the manufacturing process. It is, therefore, very difficult to take an existing product line and bring it all the way up to Six Sigma. "You can't take an old product and move it to Six Sigma," said Mr. Buetow. "You need to start with a clean sheet of paper." . . .

If Six Sigma were used at home in the kitchen for baking bread—which would never be the case—the process might work like this: First the recipe would be analyzed to see if there were ways to reduce the number of ingredients. Next, the measuring process would be assessed to see which were the more difficult tasks. Goals would be established for the number of acceptable measuring errors. Technologists would assess whether the rising time of the dough could be shortened and they would refit the oven controls to make them more accurate. In the end, the bread would be sampled by the group that wrote the recipe to see if it conformed to expectations.

The quality of most manufacturing processes in the United States is plus or minus three sigma, which means that there are [2,700] defects per million opportunities for error. Motorola saw a clear potential to achieve better results.

"There's no great secret to all this," Mr. Buetow said. "You simply look at a process, center the mean and reduce the variation; do it the same way every time." In other words, with Six Sigma, the company sets a target for the number of permissible errors for each series of operations and holds to that goal.

Six Sigma was conceived at Motorola in early 1987. Motorola's chairman, Robert Galvin, tired of hearing customer complaints about quality, sent out a companywide directive. In it, he urged all employees to be relentless in improving quality. . . .

Motorola's 5-year quality plan, which began with Mr. Galvin's 1987 memo, called for the company to achieve Six Sigma capability by 1992. That meant a tenfold reduction in defects by 1989 and a hundredfold reduction by 1991 from the 1987 rate.

Mr. Buetow says Motorola is well on its way to Six Sigma. "We will be close," he said. "The evolutionary process is already paying off. We realized a savings of $500 million in manufacturing costs this year. The target is to reach $1 billion in savings in 1992." . . .

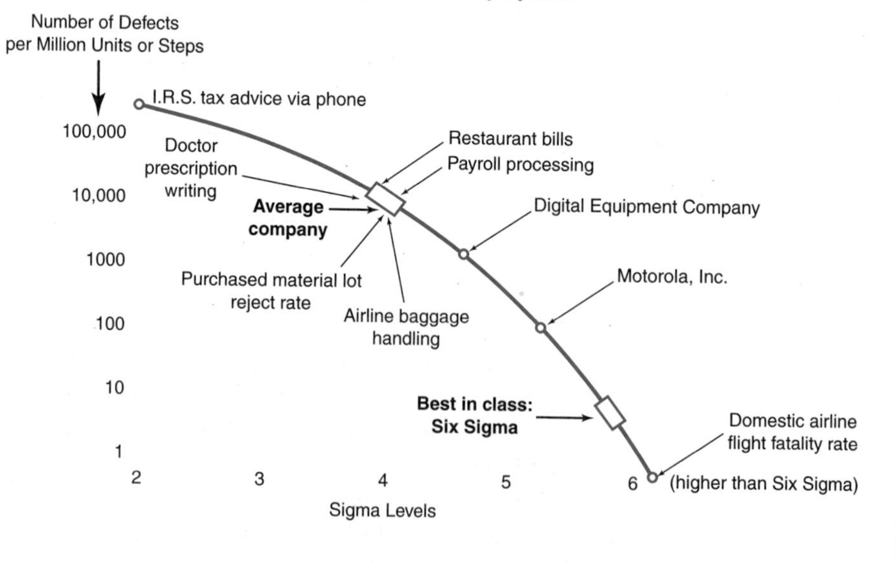

Figure 8.12
Motorola's experience with quality control

EXERCISES 8.31–8.35

Skill Builders

8.31 Consider a process in which the mean is 1.500 inches and the standard deviation is is .001 inch.

a. Set up an Xbar control chart for this process for samples of size 4.
b. Set up an Xbar control chart for this process for samples of size 8.
c. Compare the results of parts a and b.

Applying the Concepts

8.32 A production process produces a metal shaft. The mean for this process is set at 3.000 inches. The standard deviation is .005 inch. If the shaft is too wide, it will not satisfy buyer specifications; nor will it do so if it is too narrow. The shaft must fall within three sigmas of specifications in order to be acceptable. Samples of size 3 are to be drawn for control purposes. Set up the Xbar control chart.

8.33 In order to maintain reasonable delivery schedules, the ABC Company records the number of orders it receives each day. When the number of orders on a given day exceeds control limits, management knows it must operate overtime. In the past, the average number of orders per day has been 100. The standard deviation is 5. Do the data below suggest that operations should go into overtime? Consider each week as a sample, and assume that the numbers of orders per day are approximately normally distributed.

Sample	Orders per Day
Week 1	101, 103, 104, 99, 97
Week 2	81, 88 ,93, 101, 100
Week 3	102, 105, 110, 117, 125
Week 4	88, 88, 73, 91, 94
Week 5	125, 120, 110, 113, 99
Week 6	111, 99, 90, 100, 105
Week 7	87, 94, 101, 99, 84
Week 8	113, 114, 107, 103, 123

8.34 Soda bottles are filled with an average of 2.00 liters of cola. Inspection of past records reveals a standard deviation of .02 liter. In order to control the bottling process, random samples of five bottles are drawn from the production line.

Sample Number	Fillings Observed
1	1.99, 2.00, 2.01, 2.02, 2.00
2	1.98, 1.97, 1.98, 1.97, 2.01
3	2.01, 2.03, 2.02, 2.00, 2.04
4	2.02, 2.03, 2.03, 2.04, 2.02
5	1.99, 1.98, 1.99, 1.99, 2.00

a. Create an Xbar chart for this process, showing the UCL, LCL, and CCL.
b. Using the data above, can we conclude that the process is in control?

8.35 A process produces aluminum cans. It is important that these cans have a certain tensile strength. Random samples of four cans each are drawn from an assembly line, where the average tensile strength is set to be 40.00 pounds per square inch, with a standard deviation of .50 pound per square inch.

Sample Number	Tensile Strength Measured
1	39.35, 40.11, 40.85, 40.02
2	40.10, 39.36, 39.50, 38.75
3	40.12, 39.46, 38.96, 40.00
4	40.03, 40.74, 40.11, 41.00
5	40.50, 39.10, 41.02, 40.00
6	40.17, 39.87, 38.97, 39.75
7	40.54, 40.88, 39.80, 39.00
8	39.99, 39.54, 39.21, 39.10
9	38.77, 40.39, 39.37, 39.00
10	38.99, 40.02, 40.78, 39.34

a. Create an Xbar chart for this process, showing the UCL, LCL, and CCL.
b. Using the data above, determine if the process is in control.

SECTION 5 ■ COMPUTERS AND *p* VALUES

A computer is usually used to perform the calculations of hypothesis testing. A computer is essential in dealing with large data sets, and many statisticians use it routinely for all analyses. In the simplest situation, the computer is used to test a claim about an individual mean.

Large Samples

The z statistic is chosen when the sample is large. In Example 8.9, we will use Minitab to find the z statistic.

EXAMPLE 8.9 ■ USING MINITAB TO FIND THE z STATISTIC FOR A LARGE SAMPLE

Will the rate of return on a portfolio of randomly selected stocks be better than the money market rate (the rate of interest for certificates of deposit)? The relevant population parameter is μ, the average rate of return for all stocks. For our test, we will use a sample of 75 randomly selected stocks. (Data on these stocks, from *Business Week*, December 27, 1993, appear in the printout on page 291.) The money market rate at the time was 2.00%.

Minitab z-statistic calculations

Step 1. We state the null hypothesis:

$$H_0: \mu \leq 2.00\%$$

Step 2. We state the alternative hypothesis:

$$H_a: \mu > 2.00\%$$

Step 3. We choose the test statistic. Because we are dealing with a large sample ($n > 30$), the z statistic is the right choice.

Turning to the Minitab menu, under "Stat" we select "Basic Statistics," followed by "Descriptive Statistics." Then, we indicate that we want to describe the variable "YIELD."

The formula for the z statistic is

$$z = \frac{\bar{x} - \mu_x}{\text{SEMEAN}}$$

where SEMEAN = STDEV/$\sqrt{n}$. Thus, we need to provide the standard deviation, which is 1.495.

Step 4. We state a decision rule. We are interested in finding out whether a randomly selected portfolio of stocks will earn more than 2%, so the test should be one-tailed. We will reject the null hypothesis only if the z statistic is significantly greater than a critical cutoff. When no details are provided about the level of Type I error desired, this error should be set at .05 (or 5%).

Step 5. We calculate the test statistic. Given the standard deviation, the software will do the rest. The printout is shown below.

```
MTB> Describe 'Yield'
         N      MEAN      MEDIAN     TRMEAN     STDEV     SEMEAN
YIELD   75      1.909      1.960      1.862     1.495      0.175

         MIN        MAX        Q1        Q3
YIELD   0.000      5.300     0.070     3.240

MTB> ZTEST 2.00 1.495 'YIELD';
SUBC> ALTERNATIVE 1.

TEST OF MU = 2.000 VS MU G.T. 2.000
THE ASSUMED SIGMA = 1.50

             N      MEAN      STDEV     SEMEAN        Z     P VALUE
YIELD       75     1.909      1.495      0.173     -0.53      0.70

MTB> Save 'PORTFOLI.MTW'

Worksheet saved into file: PORTFOLI.MTW

MTB> STOP.
*** Minitab Release 8 *** Minitab, Inc. ***
Storage available 3500
```

Step 6. We decide whether to reject or provisionally accept the null hypothesis. Since our z statistic is $-.53$, which is certainly not greater than 1.645, we accept H_0. The evidence indicates that a randomly selected portfolio does no better than a 2% CD.

The associated graph is shown in Figure 8.13. The critical value of z, which separates the rejection region from the rest of the distribution,

Figure 8.13
One-tailed test with the z statistic

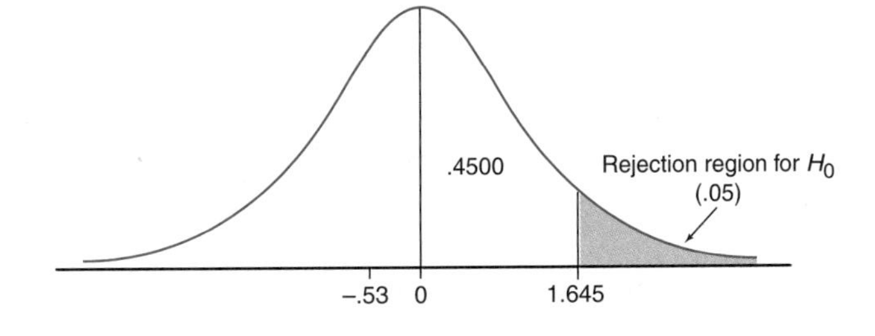

is 1.645, as that value leaves .05 of the area to the right. Thus, any value above 1.645 would be cause to reject the null hypothesis, but the observed value of −.53 is well to the left.

A software program like Minitab increases reliability and accuracy and saves time, as a computer can crunch numbers faster than we can manually with calculators. But it saves time in other ways as well. On the printout, it gives the very useful *p* value, highlighted in color in the portion of the printout below.

	N	MEAN	STDEV	SEMEAN	Z	P VALUE
YIELD	75	1.909	1.495	0.173	−0.53	0.70

The printout shows a *p* value of .70 for the one-tailed test of the alternative hypothesis $\mu > 2.00$ against the null hypothesis $\mu \leq 2.00$. As shown in Figure 8.14, this *p* value is the probability that, if $\mu = 2.00$, the mean of a sample of the chosen size will fall at or above the observed value. It is the greatest possible probability of observing such a sample, assuming that the null hypothesis $\mu \leq 2.00$ is true. The *p* value of .70 indicates that an observation at or above the observed value of 1.909 is quite likely under the null hypothesis. Therefore, the *p* value allows us to accept the null hypothesis and reject the alternative, H_a, that $\mu > 2.00$.

Figure 8.14
Interpreting the p value

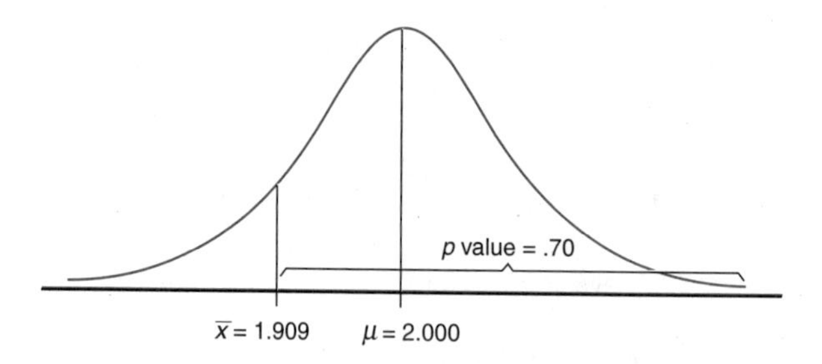

A *p* value of .0001 would mean that an observation as extreme as the one observed would occur by chance alone only 1 time out of 10,000 under the null hypothesis. The decision rule requires that we reject any hypothesis that makes what we actually observe highly unlikely. So, *when the* p *value is small, we reject the null hypothesis.* If we reject the null hypothesis when *p* is less than .05, we are rejecting the null hypothesis at the 5% level.

The *p* value given by hypothesis-testing software provides the probability that, under the null hypothesis, a random sample will yield a value as extreme as the one for the sample. Thus, if you were to make the critical value equal to the observed value, the *p* value would give the probability of making a Type I error—rejecting the null hypothesis when it was true. If you had already chosen the α level and you found that $p < \alpha$, you would reject the null hypothesis at the α level because your observation was more extreme than the critical value for α. But, in fact, you do not need to find the critical value for the significance of α. To reject H_0 and accept H_a, all you need to know is that $p < \alpha$. So the rule is easy:

$$\text{Reject } H_0 \text{ if } p < \alpha.$$

Small Samples

When the sample is small, the appropriate statistic for testing the mean is the *t* statistic. Example 8.10 illustrates the use of Minitab to find the *t* statistic.

EXAMPLE 8.10 ▪ USING MINITAB TO FIND THE *t* STATISTIC FOR A SMALL SAMPLE

To check compliance of GMC four-wheel-drive 1994 Sierras with federal standards, the EPA made fuel economy estimates. Tests of Sierras' 6.5-liter (395-cubic-inch) engines yielded the following observations: 21, 20, 19, 21, 20, and 18 mpg. Do these observations give evidence that this model is falling below the government standard mandating an average of at least 20 miles per gallon? Test at the .01 level.

Step 1. We state the null hypothesis:

$$H_0: \mu \geq 20 \text{ mpg}$$

Minitab *t*-statistic calculations

Step 2. We state the alternative hypothesis:

$$H_a: \mu < 20 \text{ mpg}$$

Step 3. We choose the test statistic. Because the sample is small ($n = 6$), we will use the t statistic. Whenever the sample is less than 30, the t statistic is the correct choice.

Step 4. We state a decision rule. The statistician is always a cynic. He or she asks the question "What would disprove the company's claim that Sierras meet government standards?" In this case, clear evidence that the miles-per-gallon estimate is *lower* would refute the company's claim. Thus, a t_{test} below the t_{critical} value would cause us to reject H_0.

Since we are using a computer, we need not look up the critical value to make that comparison. We will simply compare the p value to the chosen α. If the p value is less than .01, we will reject the null hypothesis.

Step 5. We calculate the test statistic. Once the data have been entered into the data sheet, we use the subcommand "SUBC" to control whether the test is one-tailed or two-tailed. For an upper-tail test, "Alternative +1" is the command. For a two-tailed test, "Alternative 0" is the command. For a lower-tail test, "Alternative −1" is the command, highlighted in color in the printout below.

```
MTB> name C1 'SIERRAS'
MTB> print C1
SIERRAS
    21      20      19      21      20      18
MTB> TTest 20.0 'SIERRAS';
SUBC> Alternative -1.

TEST OF MU = 20.000 VS MU L.T. 20.000
              N    MEAN    STDEV    SEMEAN      T     P VALUE
SIERRAS       6   19.833   1.169     0.477    -0.35    0.37
```

Step 6. We decide whether to reject or provisionally accept the null hypothesis. The p value of .37 is the probability of obtaining a mean as small as 19.833 when the true mean is 20.000. This value is well above our test level, $\alpha = .01$. As a result, we accept the claim. Our conclusion is that the data are consistent with the claim that the fleet, on the whole, averages at least 20 miles per gallon.

To find the t statistic for a single sample using Excel, you set up a worksheet by carrying out the following steps.

1. In cell A1, type the title of the worksheet: SMALL SAMPLE TEST - ONE TAIL.
2. Type Data in cell A3.
3. Fill in the data in column A.
4. In cells C3 to C12 of column C, type in n, Hsub0, mean, std dev, std err, alpha, df, Tcritical, and Ttest.
5. Go to cell D3. Using the function wizard f_x (a button on the tool bar), go to "Statistical Functions," and select "COUNT." On the first "values" line, enter A:A. This function will count all the data elements in column A. When it finishes, n (in this case, 6) will appear in cell D3.
6. In cell D4, opposite Hsub0, enter 20, the value of the null hypothesis.
7. Repeat the process in Step 5 for the mean and standard deviation.
8. Enter = D6/sqrt(D3) in cell D7.
9. Enter the alpha level 0.01 in D8.
10. Enter =D3-1 in cell D9.
11. Go to cell D10. Find t_{critical} using the TINV function. When asked for the alpha level, enter 0.02. (Excel requires that we double the true alpha for a one-tailed test.) When asked for the df, enter D9.
12. Enter =(D5-D4)/D7 in cell 11.

Here's the completed worksheet:

Excel t-statistic calculations

SMALL SAMPLE TEST - ONE TAIL		
Data	n	6
21	Hsub0	20
20	mean	19.833333
19	std dev	1.1690452
21	std err	0.4772607
20	alpha	0.01
18	df	5
	Tcritical	3.3649303
	Ttest	−0.3492151

The printout reveals that the mean mileage for the sample is 19.83 miles per gallon, with a standard deviation of 1.17 miles per gallon. Because no p value appears, we need to compare the t_{critical}, 3.365, with the test value, −.35. The test value is lower, so we should accept the null hypothesis. This result agrees with that of the previous analysis, made using p values from Minitab.

Another useful application for the computer is comparing means.

EXAMPLE 8.11 ▪ USING MINITAB TO COMPARE MEANS OF SMALL SAMPLES WITH THE t STATISTIC

Suppose we wish to compare the EPA's city-mileage estimates for 1994 Audis and Mercedes-Benzes. Assume that the gas mileages are approximately normally distributed. Is there a significant difference in the city mileages for these two brands of cars? Test at the .10 level. The data for the two models are shown in Table 8.5.

Table 8.5

ESTIMATED CITY MILEAGES FOR AUDIS AND MERCEDES-BENZES

Audi			Mercedes-Benz		
18	14	18	19	18	16
19	18	18	14	13	
18	20				

Step 1. We state the null hypothesis.

$$H_0: \mu_{\text{Audi}} = \mu_{\text{Mercedes}}$$

Step 2. We state the alternative hypothesis:

$$H_a: \mu_{\text{Audi}} \neq \mu_{\text{Mercedes}}$$

Step 3. We choose the test statistic. The appropriate statistic is the t statistic because we are comparing two small samples. (See Section 3.)

Step 4. We state a decision rule. As the concern is the difference between the two cars and no indication is given of the expected direction of the difference, a two-tailed test is appropriate. A t_{test} smaller than the lower t_{critical} or larger than the higher t_{critical} will cause us to reject the claim that city-mileage ratings for Audis and Mercedes-Benzes are equal. Figure 8.15 illustrates the situation for this test.

Figure 8.15
Two-tailed t test

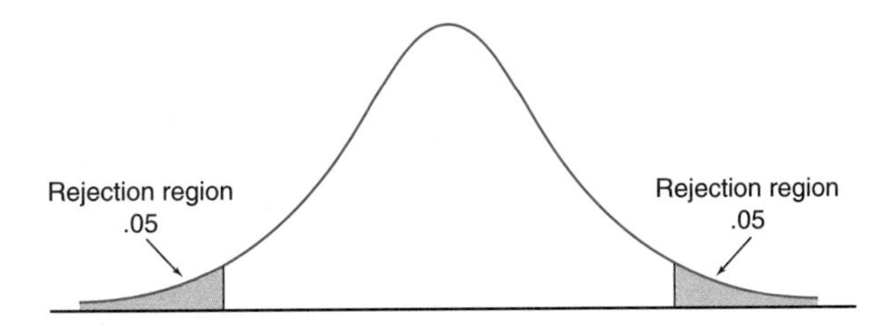

Step 5. We calculate the test statistic. The Minitab printout looks like this:

Minitab two-sample t-statistic calculations

```
MTB > TwoSample 90 'AUDI' 'MERCEDES';
SUBC> Alternative 0;
SUBC> Pooled.

TWO SAMPLE  T FOR AUDI VS MERCEDES
              N     MEAN     STDEV    SEMEAN
AUDI          8     17.88     1.73      0.61
MERCEDES      5     16.00     2.55      1.1

90 PCT CI FOR MU AUDI - MU MERCEDES: (-0.24, 4.0)
TTEST MU AUDI = MU MERCEDES (VS NE): T=1.59 P=0.14 DF=11
POOLED STDEV=2.06
```

(See Example 8.10, Step 5, for an explanation of the command "Alternative 0.")

Step 6. We decide whether to reject or provisionally accept the null hypothesis. At the 90% level of confidence, we conclude that there is no statistically significant difference between the Audi and the Mercedes-Benz with respect to city gasoline mileage.

Let's look more carefully at a few of the items in the computer printout in Example 8.11. The third line indicates that this is a pooled t test. In the second line, "Alternative 0" means that we are checking against the H_0 of $\mu_1 = \mu_2$, or $\mu_1 - \mu_2 = 0$, so this is a two-tailed t test. The 90% confidence interval for $\mu_1 - \mu_2$ includes 0, so we cannot reject H_0: $\mu_1 = \mu_2$ at the 10% level. The fact that the p value, which is .14, is greater than α, which is .10, also requires that we accept the null hypothesis.

It should be noted that Excel also incorporates the two-sample t test as part of its data analysis tool kit. Data are entered on the standard Excel worksheet. Once the data analysis tool kit has been chosen and the appropriate t test has been selected, the program asks the user to identify the variables and observations to be compared. For the data in Example 8.11, the following analysis would be produced:

Excel two-sample *t*-statistic calculations

t-Test: Two-Sample Assuming Equal Variances

	Audi	Mercedes
Mean	17.875	16
Variance	2.982143	6.5
Observations	8	5
Pooled Variance	4.261364	
Hypothesized Mean Difference	0	
df	11	
t Stat	1.593255	
P(T<=t) one-tail	0.069705	
t Critical one-tail	1.795884	
P(T<=t) two-tail	0.13941	
t Critical two-tail	2.200986	

The results are somewhat different from those produced by Minitab, but they yield the same conclusions. Note that variances are used instead of standard deviations. Results are presented in terms of one-tailed and two-tailed tests. The choice of the appropriate p value depends on the null and alternative hypotheses being tested.

EXERCISES 8.36–8.38

Applying the Concepts

If you use a computer for the following exercises, include a printout of the results, with notes showing what test you used, why you chose that particular test, and how you drew your conclusion.

8.36 Ford Motor Company claims that its 1993 cars and trucks contain over 85% domestic content. Assume that Ford makes equal numbers of each of the vehicles listed below. Test the company's claim at the .01 level.

Vehicle	% Domestic	Vehicle	% Domestic	Vehicle	% Domestic
Taurus	91	F-Series	96	Town Car	95
Tempo	94	Explorer	85	Continental	97
Festiva	22	Ranger	89	Mark VIII	95
Crown Victoria	72	Aerostar	91	Grand Marquis	72
Escort	85	Econoline	97	Sable	94
Mustang	92	Bronco	98	Capri	19
Probe	80	Villager	80	Cougar	94
Thunderbird	94	Tracer	81	Topaz	94

8.37 Randomly selected students at St. John's University were asked to indicate their political preferences and what they thought would be a comfortable income. Compare the responses of Republicans and Democrats. Is there a significant difference? Test at the .05 level. Data are given in thousands of dollars.

Republicans				Democrats			
45	60	80	150	65	75	70	50
52	65	75	95	85	70	25	60
60	100	80	40	50	75	40	80
33	80	70	50	40	35	60	75
35	75	65	75	25	60	70	
75	83	70	100				
40	100	100	100				
30	65	50					

8.38 The following data from the U.S. Bureau of Labor Statistics reflect labor force participation rates for 1992 and the first 11 months of 1993. Is the labor force participation rate significantly lower in New York City than in New York State as a whole? Assume that the participation rates are approximately normally distributed. Test at the .05 level.

New York State					New York City				
61.1	60.9	61.2	61.3	61.6	55.5	55.7	56.7	55.9	55.3
63.2	63.0	62.4	61.6	61.0	56.6	57.2	56.9	57.1	55.7
61.6	61.8	61.7	62.0	62.1	56.0	57.1	56.3	56.8	57.0
61.8	61.6	61.9	62.5	61.8	56.1	55.2	55.7	57.0	56.0
60.4	61.3	61.1			54.7	56.5	54.9		

FORMULAS

Test Statistic for a Single Mean

For a large sample, use the single-mean z statistic, given by

$$z_{\text{test}} = \frac{\bar{x} - \mu_x}{s_{\bar{x}}}$$

For a small sample, use the single-mean t statistic, given by

$$t_{\text{test}} = \frac{\bar{x} - \mu_x}{s_{\bar{x}}}$$

where degrees of freedom is given by $n - 1$ with $n =$ sample size.

Test Statistic for Two Means

For large samples, use the two-mean z statistic, given by

$$z_{\text{test}} = \frac{\bar{x}_1 - \bar{x}_2}{\sqrt{\frac{s_1^2}{n_1} + \frac{s_2^2}{n_2}}}$$

For small samples with equal variances, use the pooled t statistic, given by

$$t_{\text{test}} = \frac{\bar{x}_1 - \bar{x}_2}{\sqrt{s_p^2\left(\frac{1}{n_1} + \frac{1}{n_2}\right)}}$$

where s_p^2, the estimate of the population variance, is

$$s_p^2 = \frac{(n_1 - 1)s_1^2 + (n_2 - 1)s_2^2}{n_1 + n_2 - 2}$$

and degrees of freedom is given by $n_1 + n_2 - 2$.

For small samples with unequal variances, use the unequal variance t statistic, given by

$$t_{\text{test}} = \frac{\bar{x}_1 - \bar{x}_2}{\sqrt{\frac{s_1^2}{n_1} + \frac{s_2^2}{n_2}}}$$

where degrees of freedom is given by

$$\text{df} = \frac{(v_1 + v_2)^2}{\frac{v_1^2}{n_1 - 1} + \frac{v_2^2}{n_2 - 1}}$$

with $v_1 = s_1^2/n_1$ and $v_2 = s_2^2/n_2$ and the result rounded down to the nearest integer.

Quality Control Limits (based on three-sigma limit)

$$\text{Upper control limit (UCL)} = \mu + 3\sigma$$

$$\text{Lower control limit (LCL)} = \mu - 3\sigma$$

$$\text{Center control line (CCL)} = \mu$$

If μ is not available, use x or a target value for μ.

To estimate σ, use historical values to calculate $s/\sqrt{n}$, where n = sample size, or use current data to come up with a point estimate.

NEW STATISTICAL TERMS

- α (probability of a Type I error)
- alternative hypothesis
- β (probability of a Type II error)
- calculated value
- critical value
- decision rule
- degrees of freedom
- null hypothesis
- one-tailed test
- p value
- process mean
- quality control
- rejection region
- robust statistic
- significance level
- statistical process control

t statistic
test statistic
three-sigma limit
two-tailed test
Type I error
Type II error
z statistic

EXERCISES 8.39–8.42

Supplementary Problems

8.39 Each year Mercedes-Benz sponsors a marathon. Qualifiers are held at the Central Park Delacorte Oval. On August 4, 1990, the men finished in between 4 minutes 25 seconds and 10 minutes 31 seconds. The average time was 5 minutes 16 seconds. On August 18, 1990, another qualifier was held in which a total of 62 men competed. Their average time was 5 minutes 20 seconds; the standard deviation was 58 seconds (*Running News,* October/November 1990). Is the second group significantly different from the first? Test at the .10 level.

8.40 A group of accounting students were asked what they would consider to be a comfortable income. A group of finance majors were asked the same question. Data for each group (in thousands of dollars) are below. Is there a significant difference in perception? Test at the .05 level.

	Accounting Majors	Finance Majors
Average	66	74
Standard deviation	18	28
n	40	50

8.41 Do homes in East Hampton, New York cost less than homes in Great Neck, New York? Data on home sales in 1990 for both locales (in thousands of dollars) are found below. Test at the .01 level.

Great Neck				East Hampton			
500	500	667	931	190	115	450	700
352	435	415		90	150	620	197
				112	450		

8.42 A production process is set to fill toothpaste tubes with an average of 6.4 ounces of toothpaste. The process has a standard deviation of .1 ounce. Samples of four tubes are drawn from the production line.

Sample Number	Observed Value
1	6.3, 6.5, 6.6, 6.1
2	6.2 ,6.3, 6.2, 6.3
3	6.1, 6.2, 6.3, 6.4
4	6.5, 6.5, 6.6, 6.4
5	6.3, 6.3, 6.1, 6.2
6	6.1, 6.6, 6.5, 6.3
7	6.3, 6.2, 6.3, 6.5
8	6.4, 6.3, 6.2, 6.1
9	6.1, 6.1, 6.0, 6.2
10	6.4, 6.5, 6.5, 6.6

a. Chart the UCL, LCL, and CCL.

b. Given the data above, test to see if the process is in control.

9

Hypothesis Testing and Decision Making for Categorical Data

OUTLINE

LEARNING OBJECTIVES

1. Conduct hypothesis tests of single proportions.
2. Apply tests of proportions to quality control.
3. Compare multiple proportions.
4. Understand the nature of nonparametric tests.
5. Use the chi-square statistic to test for "goodness of fit."
6. Perform contingency table analysis.
7. Make use of the computer in contingency table analysis.

BUILD YOUR KNOWLEDGE OF *STATISTICS*

Statistics on drug use in America indicate that 10% of all employees nationwide use drugs in the workplace. Rates of illicit drug use are 18% for adults aged 18–20 years, 12% for adults aged 21–25 years, and 8% for adults aged 26–34 years (U.S. Department of Health and Human Services, *National Household Survey on Drug Abuse,* 1996).

With 26.2% of 12th graders admitting to illicit drug use in the last 30 days (National Institute of Drug Abuse, *"Monitoring the Future" Study,* 1997), it is no surprise that this habit carries over to the workplace. Department of Transportation regulations require both pre-employment and random drug testing of commercial drivers. A person who refuses to be tested will not be permitted to operate a commercial motor vehicle, and a worker who refuses to be tested can be fired for insubordination.

To comply with the requirement that it test its employees for drug use, a trucking company sets up a plan for randomly selecting employees. An employee who is selected to be tested complains that the employer has a vendetta against younger employees. In order to determine whether the employee's complaint has merit, the age distribution of the employees tested must be compared with the overall age distribution of the employees in the firm. For such a goodness-of-fit test, the chi-square statistic is the appropriate tool. If the distribution of tested employees' ages differs significantly from that of all employees' ages—for example, more are from the 18–25 group—it is possible that younger adults were targeted.

In this chapter, we explore tests of categorical data. We begin by looking at single proportions and quality control applications and then move to testing for two or more proportions. For example, are younger people more likely than middle-aged or older folks to leave their Christmas shopping to the last minute? Sometimes, we suspect a pattern where none should exist—in our drug-testing example, for instance. In Section 3, this issue is addressed using nonparametric methods. If, on the other hand, we want to pin down a connection that we believe exists among categorical data, contingency table analysis will help us deal with the relationship between categorical variables.

SECTION 1 ▪ TESTS OF SINGLE PROPORTIONS

In the opening to Chapter 8, we saw how a preponderance of overpriced items in Boston retail businesses signaled a problem that needed correction. Accountants, politicians, and businesspeople are all concerned with proportions or percentages of one sort or another. Accountants worry about clerical accuracy: What proportion of entries are correct? Politicians are concerned about public opinion: What proportion of the voters favor them? Manufacturers must think about matters of quality control: What proportion of the output will pass final inspection? In Example 9.1, we look at a procedure for testing claims about proportions that resembles the process for testing claims about means.

EXAMPLE 9.1 ▪ USING A *z* TEST TO CHECK FOR PROPORTIONAL DIFFERENCES IN ATTITUDES TOWARD INDUSTRIAL MIGRATION

Proportions: two-tailed test

Development agencies want to know what community features will attract new industry. In 1979, a survey was conducted for the Office of Economic Development of Suffolk County, New York to find out what factors influenced the relocation of businesses, or "industrial migration." It was found that a company's occupancy costs were the most important factor. Of the CEOs surveyed, 63% said that these costs were very important. Suppose another survey were conducted today, and only 55% said occupancy costs were the most important factor. Does this difference signal a shift in the population percentage (π), or might it be due merely to sampling error? Let's test this hypothesis with 90% confidence. The total number of CEOs polled was 150.

We use the hypothesis-testing steps outlined in Chapter 8.

Step 1. We state the null hypothesis:

$$H_0: \pi = .63$$

In words, the hypothesis claims that there was no change in π—the variation is due to sampling error.

Step 2. We state the alternative hypothesis:

$$H_a: \pi \neq .63$$

We are interested in whether the new proportion is significantly different from the old proportion. At the start of our survey, when we determine the critical value for rejection of H_0, we do not yet know the current proportion, which could be either higher or lower than 63%. Stating the hypothesis as we have allows for either possibility, and both possibilities must be envisioned when we set the critical value for z in Step 4. (If we wished to test for a decrease only, then the alternative hypothesis would be $\pi < .63$ and a one-tailed test would be used. We'll look at this type of situation in Example 9.2.)

Step 3. We choose the test statistic. We are testing a percentage that is binomial in nature. That is, a respondent can answer only yes or no to the question of whether occupancy costs are the most important factor in industrial migration. For large samples, we approximate the binomial distribution by a normal distribution. As a result, the z statistic may be used.

$$z_{\text{test}} = \frac{p - \pi}{\sigma_\pi}$$

We are interested in how probable it is that z will be beyond our test cutoff when the null hypothesis is true. Since π is known under the null hypothesis ($\pi = .63$), we use this value in calculating σ_π, which is given by

$$\sigma_\pi = \sqrt{\frac{\pi(1 - \pi)}{n}}$$

(This formula is from Section 2 of Chapter 6.)

Step 4. We state a decision rule. Looking at H_a, we see that this test should be two-tailed, with rejection regions as shown in Figure 9.1. Centering the 90% leaves 5% in the upper tail and 5% in the lower tail. In looking up the z statistic in Table IV in Appendix B, we seek that value which covers .4500 (45%) on either side of the mean. Since .4500 falls exactly halfway between .4495 and .4505, which correspond in the table to z values of 1.64 and 1.65, respectively, we'll split the difference. We make $z_{\text{critical}} = 1.645$. If our calculated value is either below -1.645 or above $+1.645$, we will reject the null hypothesis.

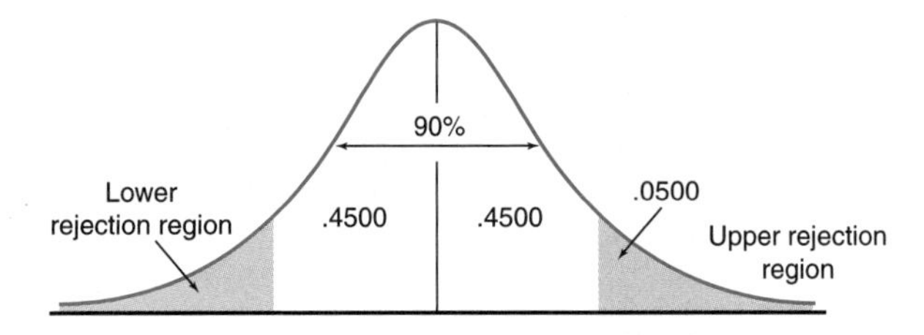

Figure 9.1
Testing for a proportion

Step 5. We calculate the test statistic. Let's review the data:

$$p = .55 \qquad \pi = .63 \qquad n = 150$$

$$z_{\text{test}} = \frac{.55 - .63}{\sqrt{\frac{(.63)(.37)}{150}}} = \frac{-.08}{.0394} = -2.03$$

Step 6. We decide whether to reject or provisionally accept the null hypothesis. Because -2.03 falls beyond -1.645, we reject the claim that, in the underlying population, 63% of manufacturers moving to Suffolk County, New York would still cite occupancy costs as the critical factor. We thus conclude that the percentage citing occupancy costs as a critical factor has indeed changed. The sample data suggest a decrease, but the test we set up in this example was designed simply to detect any change from the existing baseline of CEOs' opinions. The data could have departed in either direction, so a two-tailed test was in order.

Example 9.2 illustrates a situation in which an analyst has an idea about the direction of the outcome or is interested in one particular direction.

EXAMPLE 9.2 ■ USING THE z TEST FOR PROPORTIONS (ONE-SIDED) TO CUT ONLINE COSTS

The XYZ Company claims that use of its electronic quotes service will slash online costs for at least 90% of its clients. To test that claim, the Better Business Bureau asks a random sample of investors whether the XYZ Company reduced their online costs. A total of 84 out of 95 investors reported a reduction of online costs. Can the difference between the claim (90% would be 85.5 clients) and the evidence be attributed to random variation, or should it be attributed to a failure on the part of the XYZ Company to live up to its claim? Because the Better Business Bureau would be publicly confronting the company, it should insist on a high level of proof before rejecting the null hypothesis; thus, it decides to test at the .01 level.

Step 1. We state the null hypothesis:

$$H_0: \pi \geq .90$$

Here, π represents the population proportion of those who would save by switching to XYZ's service. The null hypothesis states that online costs will be reduced for at least 9 out of 10 clients.

Proportions: one-tailed test

Step 2. We state the alternative hypothesis:

$$H_a: \pi < .90$$

The alternative hypothesis says that online costs will be reduced for fewer than 9 out of 10 clients.

Step 3. We choose the test statistic. The normal curve may be used to approximate the binomial distribution when $np \geq 5$ and $n(1 - p) \geq 5$. In this instance, $np = 95(.9) = 85.5$, and $n(1 - p) = 95(.1) = 9.5$. Therefore, the normal curve may be used.

$$z_{test} = \frac{p - \pi}{\sqrt{\frac{\pi(1 - \pi)}{n}}}$$

Step 4. We state a decision rule. We are willing to give XYZ Company the benefit of the doubt unless we see strong evidence that does not live up to its claim. If the evidence goes beyond the minimum claim of savings for 90% of investors, it doesn't matter. Thus, the rejection region is the lower tail shown in Figure 9.2, with a probability of .01. Hence, the probability of falling between $z_{critical}$ and 0 must be .49. In Table IV, the corresponding z value, $z_{critical}$, is -2.33. It actually corresponds to an area of .4901, but this is close enough. If the calculated value falls below -2.33, we will reject the null hypothesis.

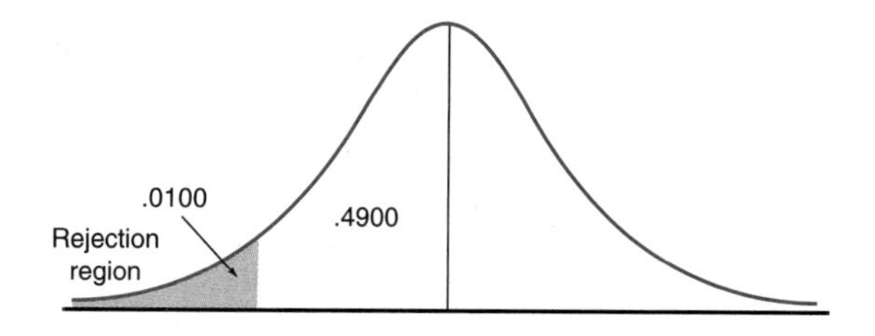

Figure 9.2
One-tailed test of a proportion

Step 5. We calculate the test statistic. Let's review the relevant information:

$$p = \frac{84}{95} = .88 \qquad \pi = .90 \qquad n = .95$$

So,

$$z_{\text{test}} = \frac{.88 - .90}{\sqrt{\frac{(.90)(.10)}{95}}} = \frac{-.02}{.0308} = -.65$$

Step 6. We decide whether to reject or provisionally accept the null hypothesis. As −.67 is not less than −2.33, we cannot reject the claim. It could, in fact, be true. If you set up a worksheet for this calculation in Excel, you will get a p value of .2514 for this test. This probability is calculated under the assumption that $\pi = .90$. It means that we have about a 1 in 4 chance of observing such an extreme value of the z test under H_0, and so our data do not provide strong evidence against H_0.

Earlier, the concept of quality control was introduced in the context of Xbar control charts. In instances where it is not possible to quantify the output from a process, it may still be possible to characterize the process as being satisfactory or unsatisfactory. Consider a dry cleaner that advertises one-day service. For each lot of dry cleaning, the business either provides delivery in one day or fails to do so. Computer chips manufactured by Intel either work or don't work; the key to economic success for Intel is to achieve a high proportion of working chips. In order to improve performance, *p charts* are used. Like the chart in Section 4 of Chapter 8, this chart has a center control line, an upper control limit, and possibly a lower control limit.

> A ***p* chart** is a control chart that tracks *p* over time for random samples taken from a production process.

EXAMPLE 9.3 ■ USING *p* CHARTS IN MICROCHIP PRODUCTION

Suppose Intel wishes to monitor its production process. Samples of 200 chips are drawn from the production line each day for a month. The results appear in Table 9.1 on page 306, where p is the proportion of defectives. We will construct a control chart for this process, given that the average percentage of defectives is 3%, and then determine whether this process is in control.

Table 9.1

RECORD OF DAILY QUALITY CONTROL CHECKS ON MICROCHIP PRODUCTION FOR A 30-DAY PERIOD

Constructing a p chart

Sample No.	No. of Defectives	p	Sample No.	No. of Defectives	p
1	6	.03	16	8	.04
2	6	.03	17	6	.03
3	10	.05	18	6	.03
4	4	.02	19	8	.04
5	4	.02	20	2	.01
6	8	.04	21	6	.03
7	6	.03	22	6	.03
8	8	.04	23	8	.04
9	4	.02	24	8	.04
10	6	.03	25	6	.03
11	8	.04	26	2	.01
12	6	.03	27	10	.05
13	10	.05	28	6	.03
14	4	.02	29	2	.01
15	8	.04	30	6	.03

Our control chart will be a p chart, plotting p against time. The average p is μ, which is equal to .03. The upper control limit (UCL) is at 3σ (that is, 3 standard deviations), and the lower control limit (LCL) is at -3σ. Each sample p is represented as an individual value in the chart in Figure 9.3. The standard deviation for the process is calculated as the standard deviation for a binomial proportion:

$$s_p = \sqrt{\frac{p(1-p)}{n}} = \sqrt{\frac{(.03)(.97)}{200}} = .01206$$

We can see that there are not any points outside the control limits, nor is there a trend in any direction. The chart shows at a glance that the process is in control.

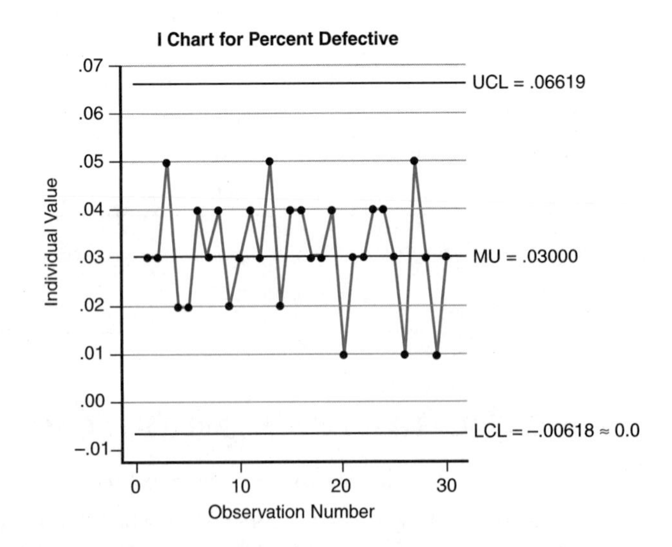

Figure 9.3
Control chart for the proportion defective

EXERCISES 9.1–9.12

Skill Builders

9.1 You are given H_0: $\pi = .35$, H_a: $\pi \neq .35$, $p = .32$, and $n = 40$. Indicate the type of test used and the critical value at the .05 level.

9.2 You are given H_0: $\pi \leq .50$, H_a: $\pi > .50$, $p = .52$, and $n = 100$. Test H_0 at the .10 level. Indicate the type of test used and the critical value.

9.3 You are given H_0: $\pi \geq .80$, H_a: $\pi < .80$, and $n = 50$. If H_0 is tested at the .01 level, what is the maximum value p can have before H_0 has to be rejected?

9.4 You are given H_0: $\pi = .60$ and H_a: $\pi \neq .60$. What sample size is necessary to reject H_0 if $p = .58$? The probability of a Type I error should be no larger than .10.

Applying the Concepts

9.5 It is claimed that a production process yields no more than 2% defectives. In a sample of 65 observations, 3 are defective. Test (at the .05 level) the hypothesis that the process is producing more defectives than it should. Describe the type of test you used and how you determined the critical value.

9.6 In 1990, the market share of a popular cola company was 37%. A new survey of 200 consumers indicates that 40% favor this cola. Does this finding give evidence of an increase in market share? Test whether an increase can be claimed at the .05 level.

9.7 Acceptance sampling is a procedure used by purchasing agents to ensure that shipments received conform to required specifications. A shipment of washers will be rejected if it has more than 5% defectives. Inspection of a lot of 80 washers reveals 2 defectives. Should the shipment be accepted? Test at the .10 level, and outline your reasoning.

9.8 A summer survey of tourists to Suffolk County, conducted in 1979, showed that 59% of the tourists came to visit the beaches. If a survey were conducted this summer and 160 of the 250 people surveyed responded similarly, could you conclude that there had been no change in the preferences of Suffolk County tourists? Test at the .01 level.

9.9 A random sample of 54 National Basketball Association draft players revealed that 7.4% had not completed college. Use the data to test the hypothesis that at least 10% of all NBA players failed to complete college. Test at the .05 level.

9.10 Prodigy Services Company regularly conducts opinion surveys of its membership concerning issues of the day. While these are not scientific surveys, they provide interesting insights into the Prodigy-using public's view of the world. A survey conducted by Prodigy in 1994 concerned potential Oscar hosts. Perform your calculations as though it were a scientific survey.

Suppose that, of 828 persons responding to the poll, 42% preferred Robin Williams. Could such data be claimed to be consistent with the hypothesis that over half of the Oscar-viewing population would prefer Robin Williams? Test at the .10 level.

9.11 The Long Island Railroad operates 700 commuter trains per day. The railroad claims an on-time performance record of over 90%. Suppose that, over the past 20 working days, the proportion of late trains was as noted below.

Day	p	Day	p	Day	p
1	.08	8	.06	15	.11
2	.02	9	.07	16	.05
3	.11	10	.08	17	.04
4	.15	11	.12	18	.05
5	.02	12	.15	19	.18
6	.14	13	.16	20	.07
7	.19	14	.13		

Using three-sigma limits, construct a control chart for the proportion of late trains. Is the process in a state of statistical control? As an analyst for the railroad, what would be your next step in investigating on-time performance?

9.12 The Fairfield Company produces automotive batteries. Each day, 100 batteries are randomly selected for inspection. The number of defects recorded each day for 30 days follows.

Day	No. of Defects	Day	No. of Defects	Day	No. of Defects
1	1	11	4	21	2
2	6	12	6	22	5
3	3	13	3	23	2
4	4	14	5	24	3
5	3	15	4	25	4
6	3	16	7	26	6
7	7	17	5	27	6
8	4	18	2	28	7
9	5	19	5	29	8
10	3	20	1	30	9

Using three-sigma limits, construct a control chart for the proportion of defective batteries. Is the production process in statistical control?

SECTION 2 ▪ GENERAL ANALYSIS OF PROPORTIONS, COUNTS, AND CHI-SQUARE

Hypothesis testing gives us a way to decide between the null hypothesis, H_0 (that what we observe arises from random effects rather than systematic differences), and the alternative hypothesis, H_a (that what we observe reflects real differences between the populations represented by our data). In Chapter 8, we looked at this decision problem using methods based on $\bar{x}$, the sample mean, and the distributions of z and t variables. In that chapter, we dealt with data measuring continuous quantitative variables. In this chapter, we have already seen how familiar tools such as z statistics can be used to test hypotheses about categorical data—specifically, questions about single proportions. Now we will consider comparisons of two proportions and how new tools such as the chi-square statistic can deal with more general questions about categorical data.

A salesperson claims that her company's widgets are more likely to meet specifications than are those of a rival company. But a purchasing agent claims that the company's edge over the competition in percentage of acceptable widgets is due just to the luck of the draw and that there is no real underlying difference. How do we decide between these two points of view? A politician claims that he is actually ahead with the voters at large, despite a poll giving the edge to his opponent. How do we decide whether to give him the benefit of the doubt or believe the poll supporting his opponent? Worse yet, how do we sort things out if there are several candidates competing for the allegiances of several constituencies? Is the picture given by the polls unclear, or can we say that politician A leads among Republicans while politician B leads among Democrats?

Comparing Two Proportions

In the simplest case, such as the difference of opinion between the salesperson and the purchasing agent, we are comparing two proportions. As usual, a comparison between proportions should be made using either a one-tailed test or a two-tailed test, depending on the goal. Although the data are categorical, the proportions can be treated as if they were continuous quantitative variables, using the normal approximation.

When two large samples are randomly and independently selected from large populations, the distribution of the differences between the sample proportions is approximately normal. As a result, the z statistic is the appropriate test statistic to use. The test of differences between two proportions is two-tailed and takes the following form:

$$H_0: \pi_a = \pi_b$$
$$H_a: \pi_a \neq \pi_b$$

The general form for a z test statistic is

$$z_{\text{test}} = \frac{\text{Observed value} - \text{Expected value}}{\text{Standard error of test statistic}}$$

Here, we have

$$z_{\text{test}} = \frac{(p_a - p_b) - (\pi_a - \pi_b)}{s_{p_a - p_b}}$$

where

$$s_{p_a - p_b} = \sqrt{p(1 - p)\left(\frac{1}{n_a} + \frac{1}{n_b}\right)}$$

and

$$p = \frac{n_a p_a + n_b p_b}{n_a + n_b}$$

With H_0 as above, we evaluate z_{test} with $\pi_a = \pi_b$, so our test statistic becomes

$$z_{\text{test}} = \frac{p_a + p_b}{s_{p_a - p_b}}$$

In a one-tailed test, we might be interested in the conjecture that $\pi_a \geq \pi_b + 5$. In that case, we would have

$$z_{\text{test}} = \frac{p_a + p_b - 5}{s_{p_a - p_b}}$$

We would reject the null hypothesis only if $z_{\text{test}} > 2.33$, which is the critical value for z for a one-tailed test at the $\alpha = .10$ level (assuming that is the level at which we wish to test the hypothesis).

EXAMPLE 9.4 ▪ USING THE z TEST TO DETERMINE WHETHER GENDER IS A FACTOR IN SAVING BEHAVIOR

Comparing two proportions

The Prodigy Services Company conducted a survey concerning saving behavior. The survey asked "Do you save as much as you think you should for retirement?" We will assume that the data came from a simple random sample, in which males and females were independently chosen. (The test described here would not be appropriate if married couples were selected. Because saving is a joint decision for a married couple, data from a husband and a wife do not represent independent pieces of information. For the z test to be valid, the data gathered from the subjects must be independent.) Of the 250 males, 53% responded positively; of the 150 females, 49% responded positively. Is there a statistically significant difference between the responses of males and females? In asking this type of question, we are looking for general guidance—perhaps for marketing purposes—so we should be willing to accept a 1 in 20 chance of a Type I error; therefore, we test at the .05 level.

Step 1. We state the null hypothesis:

$$H_0: \pi_m = \pi_f$$

In words, the hypothesis states that no significant difference exists.

Step 2. We state the alternative hypothesis:

$$H_a: \pi_m \neq \pi_f$$

In words, the hypothesis says that a significant difference exists.

Step 3. We choose the test statistic:

$$z_{\text{test}} = \frac{(p_m - p_f) - (\pi_m - \pi_f)}{s_p} = \frac{p_m - p_f}{s_p}$$

As before, z_{test} can be simplified because we have $\pi_m = \pi_f$ under H_0. Under H_0, the variance of p is estimated by $p(1 - p)/n$, so for $s_{p_m - p_f}$, we use

$$s_p = \sqrt{\frac{p(1-p)}{n_f} + \frac{p(1-p)}{n_m}} = .0516$$

Step 4. We state a decision rule. If z_{test} falls outside the z_{critical} interval—that is, if the absolute magnitude of z_{test} exceeds z_{critical}—then we reject H_0. This is a two-tailed test at the .05 level, so the critical value of z is 1.96.

Step 5. We calculate the test statistic:

$$z_{\text{test}} = \frac{.53 - .49}{.0516} = .78$$

and

$$p = \frac{250(.53) + 150(.49)}{250 + 150} = .515$$

Step 6. We decide whether to reject or provisionally accept the null hypothesis. Because .78 is more than −1.96 and less than +1.96, we conclude that there is no statistically significant difference between the attitudes of men and women toward putting away money for retirement.

The procedure applied above in the two-tailed test may also be applied in one-tailed tests. Suppose we wished to test the belief that a greater percentage of males than females responded positively to the survey question. The only changes would be in the statement of the null and alternative hypotheses and the location of the rejection region.

As long as only two proportions are being compared, the z statistic may be used. Frequently, however, more than two proportions must be compared. Then, the z statistic can no longer be used.

Testing for Several Proportions

When more than two proportions are being compared, we are dealing with one-dimensional count data. One way to evaluate one-dimensional count data is with the **chi-square statistic, χ^2.** When we are comparing more than two proportions, the null hypothesis is, as usual, that there are no real differences between the underlying proportions in the groups. The alternative hypothesis is that the observed results reflect underlying differences between the proportions in the different groups. To test the null hypothesis, we pool the numbers from all the groups to generate **expected values** for the underlying population, under the assumption that there are no differences between the groups. These expected values are then compared with observed values. If there is no statistically significant difference between these two sets of values, the null hypothesis is accepted, and the observed variations are attributed to sampling fluctuations.

EXAMPLE 9.5 ▪ USING THE CHI-SQUARE TEST TO COMPARE AGE GROUPS AND CHRISTMAS SHOPPING BEHAVIOR

Comparing more than two proportions

Let's use the chi-square statistic to look at some data on Christmas shopping. A Prodigy survey asked people to respond to the statement "I will begin my shopping at the last minute." Suppose the counts were as follows:

ADMITTED LAST-MINUTE SHOPPERS BY AGE GROUP

Last-Minute Shoppers \ Age	Under 25	25–44	Over 44
Yes	440	397	371
No	562	641	623

Is there a statistically significant difference between age groups? We will test at the .05 level.

Step 1. We state the null hypothesis:

$$H_0: \pi_a = \pi_b = \pi_c$$

In words, all proportions are equal; our data can be attributed to chance variation.

Step 2. We state the alternative hypothesis:

$$H_a: \text{Not all } \pi_i \text{ are equal.}$$

In words, there is systematic variation here.

Step 3. We choose the test statistic:

$$\chi^2 = \sum\left[\frac{(f_o - f_e)^2}{f_e}\right]$$

where f_o are observed values and f_e are expected values. While this is a departure from the standard form for a test statistic, it has similar elements: f_o (observed) $- f_e$ (expected). But in this case, the numerator is squared and divided by f_e, the expected value. In order to use the chi-square test, we should have $f_e > 5$ for each category.

Step 4. We state a decision rule. The chi-square distribution is a one-tailed distribution, as shown in Figure 9.4. There are tables of the critical values for the chi-square distribution, above which the null hypothesis should be rejected. A portion of the table of critical values for chi-square is reproduced in Table 9.2.

Figure 9.4
Chi-square distribution

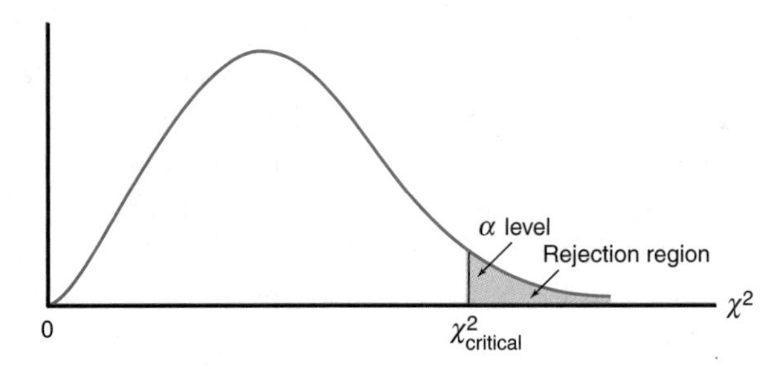

The critical value for the chi-square statistic in this type of problem is a function of the number of rows and columns in the table and the alpha level for the hypothesis test. The formula for degrees of freedom (df) is (rows $-$ 1) $\times$ (columns $-$ 1). For this test, the table has 2 rows and 3 columns, so the chi-square statistic has df $= 1 \times 2 = 2$. The test is at the .05 level, so the corresponding critical value is 5.991.

Step 5. We calculate the test statistic. When the chi-square statistic is calculated, the observed frequency, f_o, is compared with the expected frequency, f_e, for each cell in the table containing data. The null hypothesis implies that age can be ignored in considering the likelihood that someone responds positively. Therefore, the overall percentage (39.8%) of last-minute shoppers who responded positively should be used to compute the expected frequency for those responding positively within each age

Table 9.2

PORTION OF THE CHI-SQUARE DISTRIBUTION TABLE

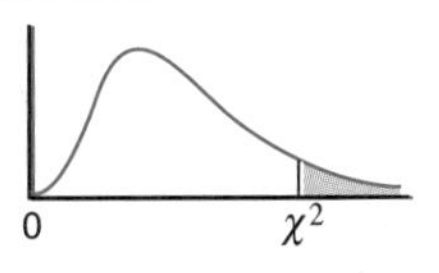

RIGHT-TAIL AREAS

Degrees of Freedom (df)	.995	.990	.975	.950	.900	.100	.050	.025	.010	.005
1	0.000	0.000	0.001	0.004	0.016	2.706	3.841	5.024	6.635	7.879
2	0.010	0.020	0.051	0.103	0.211	4.605	5.991	7.378	9.210	10.597
3	0.072	0.115	0.216	0.352	0.584	6.251	7.815	9.348	11.345	12.838
4	0.207	0.297	0.484	0.711	1.064	7.779	9.488	11.143	13.277	14.860
5	0.412	0.554	0.831	1.145	1.610	9.236	11.070	12.833	15.086	16.750
6	0.676	0.872	1.237	1.635	2.204	10.645	12.592	14.449	16.812	18.548
7	0.989	1.239	1.690	2.167	2.833	12.017	14.067	16.013	18.475	20.278

group. The overall percentage is arrived at by obtaining the row total for all positive responses and dividing that total by the overall number of observations.

ADMITTED LAST-MINUTE SHOPPERS BY AGE GROUP

Age / Last-Minute Shoppers	Under 25	25–44	Over 44	Row Total	Percentage Distribution
Yes	440 (cell 11)	397 (cell 12)	371 (cell 13)	1208	39.8
No	562 (cell 21)	641 (cell 22)	623 (cell 23)	1826	60.2
Column Total	1002	1038	994	3034	100.0

The expected value for each cell may now be calculated. The expected value for cell 11 is .398 × 1002; for cell 12, the expected value is .398 × 1038; for cell 22, the expected value is .602 × 1038; and so on. Continuing in this fashion, we fill in the worksheet shown in Table 9.3.

Table 9.3

WORKSHEET FOR CALCULATING CHI-SQUARE VALUES

Cell Number	Observed	Expected	$\lvert f_o - f_e \rvert$	$(f_o - f_e)^2$	$\frac{(f_o - f_e)^2}{f_e}$
11	440	399	41	1681	4.213
12	397	413	16	256	0.620
13	371	396	25	625	1.578
21	562	603	41	1681	2.788
22	641	625	16	256	0.410
23	623	598	25	625	1.045
				Calculated Value of Chi-Square:	10.654

Note that the expected count for each cell is greater than or equal to 5, which is a requirement for using the chi-square statistic here. If you are carrying out a chi-square test using Excel or other software, the software will supply a warning if the requirements of the test are violated.

Step 6. We decide whether to reject or provisionally accept the null hypothesis. Because 10.654 exceeds 5.991 (the critical value), we reject the null hypothesis. We conclude that not all the proportions are equal. The data suggest that there is a difference in the behavior of different age groups with respect to last-minute shopping. In fact, it appears that we learn with age to get our shopping done before the last minute (or, perhaps, older people belong to a generation that has always shopped ahead of time).

Let's look at Step 6 of Example 9.5 again, this time using the p value provided by Excel. The p-value, chi-square value, and degrees of freedom have been highlighted in color in the printout below. (Details of the procedure will be found in Section 6 of this chapter.) Note that the p value is less than .05, so we would reject the null hypothesis at the .05 level using the computer approach as well.

Actual

	under 25	25–44	over 44
Yes	440	397	371
No	562	641	623

Expected

	under 25	25–44	over 44
Yes	399	413	396
No	603	625	598

p value	0.004859
chitest (b3:d4,b8:d9) equals	10.65
with 2 degrees of freedom	

EXERCISES 9.13–9.22

Skill Builders

9.13 You are given H_0: $\pi_a = \pi_b$, H_a: $\pi_a \neq \pi_b$, $p_a = .37$, $p_b = .40$, $n_a = 100$, and $n_b = 50$. Test H_0 at the .05 level. Describe your test and how you found your critical value.

9.14 You are given H_0: $\pi_a = \pi_b = \pi_c$ and H_a: Not all π_i are equal. Test H_0 with the data below at the .05 level, giving the test used and your critical value.

	A	B	C
Pro	17	57	35
Con	18	68	36

9.15 If $n_a = 70$, $n_b = 65$, $p_a = .57$, and $p_b = .53$, can we conclude that $\pi_a > \pi_b$? Test at the .10 level. Indicate the type of test you used and your critical value.

9.16 In four groups, each having 50 individuals, 20, 25, 24, and 29 respondents replied positively to a question. Test the hypothesis that there exists a statistically significant difference between the groups. Test at the .10 level. What test did you use, and what was your critical value?

9.17 H_0 is that $\pi_a \geq \pi_b$, and H_a is that $\pi_a < \pi_b$. Given $p_a = .47$, $p_b = .54$, $n_a = 105$, and $n_b = 50$, test H_0 at the .01 level, indicating the critical value you used.

Applying the Concepts

9.18 It is claimed that the majority of Republicans favor new welfare legislation and that significantly fewer Democrats feel the same way. Suppose a random sample of 100 Republicans was drawn and 65 responded positively. Another random sample of 85 Democrats produced 31 in favor. Can we conclude that the proportion of Republicans favoring the legislation is greater than the proportion of Democrats? Test the claim at .01 level. State the H_0 and the H_a you used.

9.19 Management of an electronics firm wanted to compare three production processes. One of the criteria for comparison was the percentage of defectives produced by each process. For Process A, a random sample of 50 units was drawn and 4 defectives were found. For Process B, a random sample of 35 units had 3 defectives. Process C produced 2 defectives in a random sample of 25. Is there a significant difference between processes? Test at the .05 level, indicating what test you used and your critical value.

9.20 A prescription drug for hypertensives was recently put on the market by a major pharmaceutical firm. When any drug is marketed, adverse reactions are noted. Two of the adverse reactions were dizziness and vertigo. Data are presented for the group given the drug and a control group.

Group	Drug	Placebo
Adverse Reactions	**339**	**336**
Dizziness	64	30
Vertigo	7	3

Is there a statistically significant difference in the proportion experiencing dizziness between the drug and the placebo group? Is there a statistically significant difference between the two groups with respect to vertigo? Test both at the .05 level.

9.21 A Prodigy poll asked about favorite snacks at the movie theater. Suppose there were 10,000 responses and they represented a random sample of the public (which they do not). Do the data suggest a significant difference between males and females in their favorite movie snacks?

Sex	Males	Females
Preferred Snack	**5322**	**4678**
"Unhealthy Popcorn"	1916	1544
"A Smuggled Snack"	1809	1684

Is there a significant difference between males and females with respect to unhealthy popcorn? (*Hint:* Ignore the data concerning a smuggled snack.) Is there a significant difference between males and females with respect to a smuggled snack? Test both at the .10 level.

9.22 Prodigy also divided the data on movie snacks by region. Using the general assumptions of Exercise 9.21, consider the following data:

Region	Northeast	Mid-Atlantic	South	Midwest	Mountain	West Coast
Preferred Snack	**3987**	**2124**	**1555**	**1003**	**378**	**953**
"Unhealthy Popcorn"	1196	743	591	391	132	324
"A Smuggled Snack"	1395	701	513	331	159	372

Is there a significant difference among regions in the percentage preferring unhealthy popcorn? Is there a significant difference among regions in the percentage preferring a smuggled snack? Test at the .05 level.

SECTION 3 ▪ GOODNESS-OF-FIT TESTS

The phrase *goodness of fit* refers to how closely an observed distribution fits some specified distribution. One of the main uses of goodness-of-fit tests is to see whether some randomly selected group is likely to represent a larger population—for example, whether CEOs represent the broader population in general in various characteristics such as religion, race, or sex or whether they represent certain favored groups.

So far, we have discussed using the null hypothesis to assign a p value to the observed statistic. If the p value is large, we let the null hypothesis stand, but if the p value is less than α, the probability of a Type I error, we are willing to accept its alternative. Thus far, all of these tests have been based on assumptions about population parameters, such as μ, π, and σ. For this reason, they are called *parametric tests.* In general, the study of statistical methods involving distribution specified by a small number of parameters such as μ, π, σ, and the like is called **parametric statistics** and its methods *parametric methods.* The goodness-of-fit method described in this section allows us to assign a p value to the sample we observe under a null hypothesis, thereby permitting specification of the underlying population without the use of parameters. This type of test is a *nonparametric test*. Once the p value has been calculated, we can proceed with the hypothesis testing as before, accepting or rejecting the null hypothesis according to whether p is above or below the critical value for the level of the test chosen. The goodness-of-fit method uses the chi-square distribution, and it applies generally to categorical data, even when there is no underlying parametric model for the population distribution. It is part of the broad area of **nonparametric statistics.**

The goodness-of-fit test is used by personnel management in evaluating fairness of procedures used in hiring, promotion, and dismissal, as well as in selecting personnel for drug screening. Questions of bias or fairness arise frequently in today's litigious atmosphere. Was an employee or group singled out for certain treatment based on some sinister motive, or was selection random and fair? Example 9.6 utilizes a goodness-of-fit test to address the issue of random selection for drug testing.

EXAMPLE 9.6 ▪ USING A GOODNESS-OF-FIT TEST TO DETERMINE FAIRNESS IN DRUG TESTING

In 1991, legislation was passed mandating drug and alcohol testing for safety-related jobs. This law applies to mass transit, trucking, airlines, and railroads. Suppose an employee of a trucking company, selected for a drug test, claimed the test was invalid because the selection procedure was not fair. To test the employee's claim that the procedure was not random (that is, each employee did not have an equal chance of being chosen), the chi-square distribution could be used. Categorical variables such as age, sex, and race, which are not random, might be influencing the choice. It would be expected that if random selection were used, the distribution of employees selected would not deviate greatly from the overall distribution of employees in the company. To illustrate this point, let's consider the age distribution of employees in the company.

Goodness of fit

If the distribution of selected employees doesn't match that of employees overall, it is possible that the drug-testing selection procedure was biased. Perhaps management is out to harass younger employees. Let's see how this works with data for Company X.

Company X employs 5452 employees in safety-related activities. A total of 1222 employees had been selected for drug and alcohol tests prior to Tracy's being selected. It is Tracy's contention that the selection process was unfair and represents age discrimination. As a legal remedy, Tracy asks that the results obtained be thrown out. The data are presented in Table 9.4 and Figure 9.5.

Table 9.4

AGE DISTRIBUTIONS FOR COMPANY X

Age Group	All Employees	Employees Selected for Testing
18–24	2715	631
25–29	851	237
30–39	794	217
40–49	663	72
50 & over	429	65
Totals	5452	1222

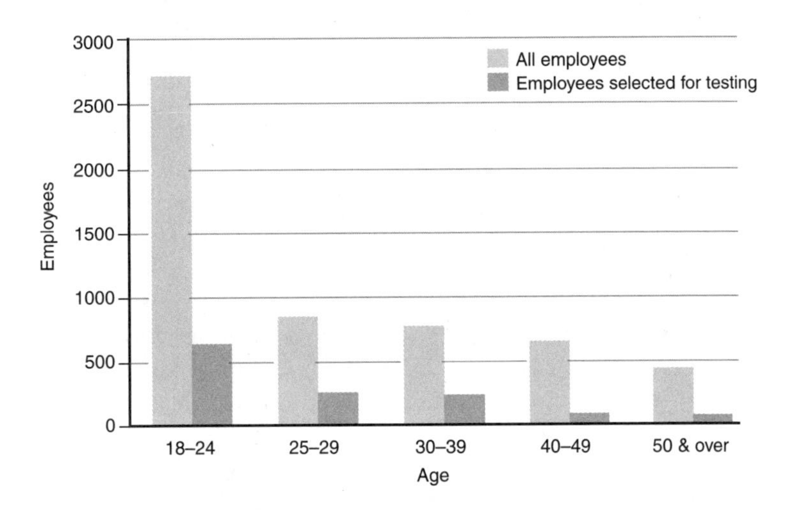

Figure 9.5 *Age distributions of employees at Company X*

Step 1. We state the null hypothesis:

H_0: The selected pattern matches the overall demographics.

(That is, it is a random selection.)

Step 2. We state the alternative hypothesis:

H_a: The selected pattern does not match the overall demographics.

(This suggests that bias may be a possibility.)

Step 3. We choose the test statistic:

$$\chi^2 = \sum\left[\frac{(f_o - f_e)^2}{f_e}\right]$$

Step 4. We state a decision rule. The critical value for the chi-square distribution has degrees of freedom equal to the number of categories less 1 (df = 4). When no alpha level is mentioned, we operate at the .05 level. This gives us a critical value for chi-square of 9.488. Note that this is a one-tailed test designed to see whether any category has been singled out for over- or under-selection.

Step 5. We calculate the test statistic. Once again, observed frequencies are compared with expected frequencies. Observed values for the selected individuals are presented in Table 9.5. Expected frequencies must be calculated from the percentage distribution of all employees. Notice that all observed and expected values are well above 5, so use of the chi-square test is justified.

Table 9.5

WORKSHEET FOR CALCULATING CHI-SQUARE IN A GOODNESS-OF-FIT TEST

Age Categories	% Overall Distribution	Observed Values	Expected Values	$\lvert f_o - f_e \rvert$	$(f_o - f_e)^2$	$\frac{(f_o - f_e)^2}{f_e}$
18–24	.50	631	611	20	400	.65
25–29	.16	237	196	41	1681	8.58
30–39	.15	217	183	34	1156	6.32
40–49	.12	72	147	75	5625	38.27
50 & over	.08	65	98	33	1089	11.11
Totals	1.01*	1222	1235	—	—	64.93

*The total percent distribution does not add to exactly 1.00 because of rounding. This yields a result in the "Expected Values" column that is slightly different from the 1222 observed values. The last column total (64.93) is the calculated value of chi-square.

Step 6. We decide whether to reject or provisionally accept the null hypothesis. Because 64.93 exceeds the critical value of 9.488, we reject the null hypothesis at the .05 level. The fact that we reject the hypothesis that the observed values arose from unbiased random sampling, however, does not legally prove that there was bias in the selection of employees for drug testing, nor does it establish how bias may have entered the selection process. The p value generated by Excel is 2.66765E-13. As this p value is less than $\alpha = .05$, the null hypothesis once again is rejected at this level.

Figure 9.6 shows chi-square probability distribution functions for 6, 10, 20, and 40 degrees of freedom. Notice how the node moves to the right, tracking the degrees of freedom as the number of degrees of freedom increases. Also, the curve becomes increasingly normal in appearance. This means that when the number of degrees of freedom is large, with chi-square values close to zero, values in the left-hand tail of the distribution are improbable. Thus, very small values for the chi-square statistic can be as much a signal of nonrandom effects as very large ones. If $\chi^2 = 0$, the fit to the underlying population is exact—something that is extremely unlikely to happen by chance alone. It suggests that someone has altered the results.

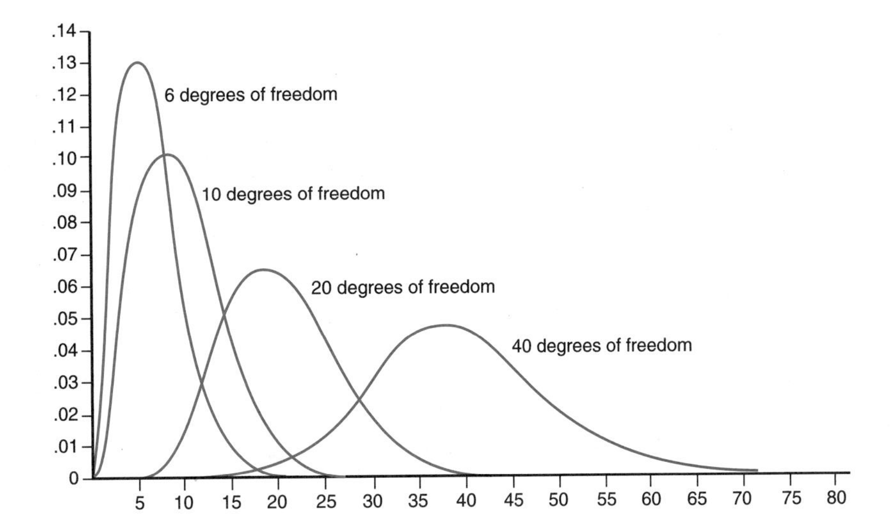

Figure 9.6
Chi-square probability density functions for various numbers of degrees of freedom

Too-good-to-be-true results are not a common worry, but they do arise and are especially troublesome when a process is supposed to be random (for example, random selection for the draft or for drug testing). A classic study by R. A. Fischer showed that the experimental data reported by Gregor Mendel in his theory of the laws of genetics fit his theory too closely to be consistent with the random nature of inheritance. The probability of such close agreement by chance alone was about 1 in 25,000, according to Fischer's calculations. So we should be worried about small values of χ^2, as well as large ones, and we should think of the χ^2 test as potentially two-tailed, especially when the number of degrees of freedom is large.

SECTION 4 ▪ CONTINGENCY TABLE ANALYSIS USING CHI-SQUARE

An exciting part of statistics is the evaluation of relationships between variables. Is there an association between smoking and cancer? Does political affiliation color one's views on legal gambling? Is there a bias against unattractive applicants in hiring? Do public school children worry more about crime in their neighborhood than do those attending private or parochial school? In each of these cases, the variables being studied are categorical. They are not measurements, but labels. Statistical techniques based on numerical data cannot answer such questions. We need to return to the idea of contingency tables, seen first in Example 4.7. These tables show how the observations are broken down into various categories. **Contingency table analysis** using the chi-square statistic permits us to answer the types of questions asked at the beginning of this paragraph. It allows us to analyze associations between categorical variables. In this section, you will learn to apply such tests.

Example 9.7 presents the result of a study of potential discrimination in the workplace.

EXAMPLE 9.7 ■ DETERMINING WHETHER UNATTRACTIVE PEOPLE ARE DISCRIMINATED AGAINST IN THE WORKPLACE

A recent study by Jeff Biddle and Daniel Hammermesh, published by the National Bureau of Economic Research (Working Paper 5366), suggests that more attractive people progress further in the workplace than others. Biddle and Hammermesh tested their contention using a sample of 2000 law school graduates. Based on a collection of head-and-shoulder photos, individuals were rated on a scale from 1 to 5 for attractiveness. Fifteen years after graduation, the average salaries of those in the top third in attractiveness were compared with those of the rest of the sample.

Suppose that the results obtained by Biddle and Hammermesh look like this:

Contingency table analysis

Appearance / Salary	Attractive	Unattractive
Higher	403	112
Lower	297	1188

Can we conclude that Biddle and Hammermesh's hypothesis—that better-looking attorneys are more successful—is valid?

Step 1. We state the null hypothesis:

H_0: There is no relationship between appearance and salary.

Step 2. We state the alternative hypothesis:

H_a: There is a relationship between appearance and salary.

Step 3. We choose the test statistic:

$$\chi^2 = \sum\left[\frac{(f_o - f_e)^2}{f_e}\right]$$

Step 4. We state a decision rule. As with all test statistics, there is a critical value. Calculated values of the test statistic above the critical value will cause us to reject the null hypothesis. Values below the critical value suggest that the null hypothesis is feasible. The table of results above is a 2-by-2 table (that is, two rows by two columns). Degrees of freedom are calculated by multiplying the number of rows less 1 by the number of columns less 1:

$$\text{Degrees of freedom} = (\text{Rows} - 1)(\text{Columns} - 1)$$

Here df = $(2 - 1)(2 - 1) = 1$; a 2-by-2 table has 1 degree of freedom. At the .05 level for a Type I error, with 1 degree of freedom, χ^2_{critical} equals 3.841.

Step 5. We calculate the test statistic. To compare observed frequencies with expected frequencies, we must calculate row and column totals. The percentage distribution of the row totals is calculated and applied to the column totals. The result is shown below, where the expected values are in parentheses.

Appearance / Salary	Attractive	Unattractive	Row Total	% Distribution
Higher	403 (180)	112 (335)	515	25.75%
Lower	297 (520)	1188 (965)	1485	74.25%
Column Total	700	1300	2000	100.00%

To obtain 180, we multiply 700 by 25.75%. To obtain 335, we multiply 1300 by 25.75%. To obtain 520, we multiply 700 by 74.25%. To obtain 965, we multiply 1300 by 74.25%. All of the f_e's are well above 5, so use of the chi-square test is appropriate. Table 9.6 contains a worksheet for calculating the chi-square value for a contingency table analysis. The calculated value of chi-square is 571.87.

Table 9.6

WORKSHEET FOR CALCULATING CHI-SQUARE IN A CONTINGENCY TABLE

Cell No.	Observed Value	Expected Value	$\lvert f_o - f_e \rvert$	$(f_o - f_e)^2$	$\frac{(f_o - f_e)^2}{f_e}$
11	403	180	223	49,729	276.27
12	112	335	223	49,729	148.44
21	297	520	223	49,729	95.63
22	1188	965	223	49,729	51.53
Totals	2000	2000	—	—	571.87

Step 6. We decide whether to reject or provisionally accept the null hypothesis. Since 571.87 exceeds 3.841, we reject the null hypothesis. It is highly unlikely that the observations are the result of chance in the absence of systematic differences in salary between the attractive and unattractive groups. The data suggest that there is a relationship between appearance and salary, or that unattractive people are discriminated against. The p value for this result is 2.1871E-126, which is less than $\alpha = .05$.

EXERCISES 9.23–9.31

Skill Building

9.23 Past history has indicated that a data set has the distribution indicated below:

Category	No. of Observations
A	100
B	50
C	25
D	25

A recent survey revealed the following distribution:

Category	No. of Observations
A	90
B	57
C	33
D	20

Has there been a statistically significant change in the distribution of values? Test at the .05 level. Indicate the type of test you used and your critical value.

9.24 A contingency table contains data on variables X and Y.

Variable *X* / Variable *Y*	Category A	Category B
Category 1	43	76
Category 2	12	100

Is there an association between variables X and Y? Test at the .05 level. What test and critical value did you use?

9.25 It is claimed that in Shangri-la the distribution of rainy days over the seasons is as follows: 40% in winter, 40% in spring, 15% in fall, and 5% in summer. One thousand events are observed with the following distribution: winter, 351; spring, 357; fall, 197; and summer, 95. Do the data conform to the claim? Test at the .10 level.

9.26 Variables A and B can take on high, moderate, or low values. Is there an association between them? Test at the .01 level.

Variable *A* / Variable *B*	High	Medium	Low
High	123	45	67
Medium	89	98	76
Low	54	32	10

Specify your test and critical value.

Applying the Concepts

9.27 The 1990 distribution of tax returns in the United States is shown below. Suppose a random sample taken sometime later revealed the results in the last column of the table. Can we conclude that there has been a change in the distribution of tax returns? Test at the .05 level. Data for 1990 are in thousands.

Type of Tax Return	1990	Sample Data
Individual Tax Returns	112,492	228
Estimated Tax Forms	36,188	82
Fiduciary Forms	3,353	8
Partner & Corporate Tax Forms	6,052	9
Other Tax Forms	40,584	84
Total Tax Forms	201,715	411

9.28 Suppose a study of U.S. senators in the 104th Congress revealed the distribution below with respect to American Conservative Union (ACU) ratings. Could we conclude there exists a relationship between party affiliation and voting behavior? Test at the .01 level.

ACU Rating / Party Affiliation	Low	High
Democrat	46	5
Republican	7	42

9.29 Suppose a study of U.S. senators in the 104th Congress revealed the distribution below with respect to Americans for Democratic Action (ADA) ratings.

Could we conclude that there exists a relationship between party affiliation and ADA rating? Test at the .01 level.

ADA Rating / Party Affiliation	Low	High
Democrat	5	45
Republican	36	14

9.30 Is there an association between the type of heart surgery performed and the sex of the patient? Data gathered by the American Heart Association for 1992 appear below. Test the association at the .10 level. (Data are in thousands of treatments.)

Sex / Type of Surgery	Male	Female
Cardiac Catherization	675	409
Bypass	347	122
Angioplasty	262	136
Pacemaker	61	52
Other Procedures	473	213

9.31 Is there an association between "gift action" and income level? In a survey, 1000 people were asked what they would do with an unwanted gift. The choices were to keep it or to rewrap it and give it to someone else. Is there a relationship between people's income level and their "gift action"? Test at the .05 level.

Action / Income	Rewraps	Keeps
High	155	345
Low	115	385

Source of Data: *U.S. News & World Report,* November 26, 1990.

SECTION 5 ■ USING THE COMPUTER TO SOLVE CHI-SQUARE PROBLEMS

Chi-square problems can involve many calculations, particularly with larger data sets. In such cases, a computer should be used for speed, convenience, and accuracy.

Using Excel

In 1977, a study was conducted concerning the relationship between per capita pupil expenditure by school districts in Suffolk County, New York and district performance. The study focused on schools because they account for the largest portion of the real property tax bills in Suffolk County.

Suppose the data counts in each cell looked like this:

SCHOOL PERFORMANCE VERSUS PER PUPIL COSTS FOR 94 SCHOOL DISTRICTS

Performance / Costs per Student	High	Medium	Low	Row Total
High	6	14	10	30
Medium	10	19	6	35
Low	7	14	8	29
Column Total	23	47	24	94

To test using Excel, you start out in the same manner as you would by hand—by stating the null hypothesis and alternative hypothesis.

H_0: There is no systematic relationship between school districts' performance and their expenditure per student; the data observed can be attributed to sampling variation.

H_a: There is a systematic relationship between school districts' performance and their expenditure per student.

Then, to calculate the p value for a chi-square test, you take the following steps:

1. Double-click on "Sheet 1" on the bottom of the screen, and type "CHI-SQUARE TEST."
2. Following the format illustrated below, enter the data in columns A through D.
3. Complete the sums for the "OBSERVED" tables by going to the "COL. TOTALS" row and pressing the "Σ" key.
4. You are now ready to fill in the "EXPECTED" table by prorating each column total by the percent distribution in the total column.
   Row 1. Use the multiplier E3/E6 against 23, 47, and 24.
   Row 2. Use the multiplier E4/E6 against 23, 47, and 24.
   Row 3. Use the multiplier E5/E6 against 23, 47, and 24.
5. Go to the function wizard, f_x, and follow the instructions for "CHITEST." Enter the range for actual values: B3:D5. Then enter the range for expected values: B9:D11.
6. Go to "Finish" and record the p value.

The resulting printout is shown below.

Using Excel for χ^2

		OBSERVED		
COST	HIGH	MEDIUM	LOW	TOTAL
HIGH	6	14	10	30
MEDIUM	10	19	6	35
LOW	7	14	8	29
COL.TOTALS	23	47	24	94
		EXPECTED		
COST	HIGH	MEDIUM	LOW	
HIGH	7.340426	15	7.659574	
MEDIUM	8.56383	17.5	8.93617	
LOW	7.095745	14.5	7.404255	
P-VALUE	0.657718			

Using Minitab

To use Minitab, enter the school district data into a worksheet in columns 2 through 4, cross-tabulating performance data by cost level. Then use the chi-square command to find the chi-square test statistic. Only now will the computer take over. If you type (or select from the statistical procedures) "Chi-Square over C2 - C4," the Minitab printout will look like this:

Using Minitab for χ^2

```
MTB > ChiSquare 'High'-'Low'.
Chi-Square Test
        High    Medium    Low    Total
   1    6       14        10     30
        7.34    15.00     7.66
```

```
2       10        19        6       35
        8.56      17.50     8.94
3       7         14        8       29
        7.10      14.50     7.40
-------------------------------------
Total   23        47        24      94

ChiSq = 0.245 + 0.067 + 0.715 +
        0.241 + 0.129 + 0.965 +
        0.001 + 0.017 + 0.048 = 2.427
df = 4, p = 0.658
```

The expected counts are printed below the observed counts. The value of the chi-square statistic in this instance is 2.427. The critical value for a chi-square statistic at the .05 level with 4 df is 9.488. Because 2.427 is less than 9.488, we accept the null hypothesis. The data can be explained on the basis of random variation; there is no compelling evidence of a relationship between teaching expenditures and student performance.

The *p* value of .658 is much greater than α, the confidence level of the test. Again, this suggests that we should not reject the null hypothesis. The conclusions are the same as those reached in Excel. Note the slight discrepancy between the two *p* values, though it is so small that it would never affect the conclusions.

EXERCISES 9.32–9.36

Applying the Concepts

9.32 A study of 150 case outcomes in the juvenile justice system produced the following data on types of legal representation versus case outcome.

Representation / Case Outcome	Guardian/Attorney	No Counsel	Row Total
Probation	90	36	126
Institutionalization	18	6	24
Column Total	108	42	150

Is there a relationship between the type of representation and the case outcome? Test at the .05 level. Find the *p* value for your result. Interpret your finding.

9.33 A study of the quality of "ride" in automobiles gave the following results for various types of vehicles.

Ride Quality / Type of Vehicle	Better than Expected	Just as Expected	Worse than Expected	Row Total
Sport Vehicle	38	42	23	103
Sedan	29	31	5	65
Van	34	36	7	77
Column Total	101	109	35	245

The manufacturer contends that sedans exhibit better ride quality than other types of vehicles. Do the data support that contention? Test at the .01 level. What is the *p* value for this result? Clearly state your conclusions.

9.34 A study of college students' snacking and drinking preferences was performed at St. John's University.

Snack Choice / Beverage Choice	M&Ms	Pretzels	Chips	Row Total
Milk	18	15	6	39
Beer	8	5	8	21
Soda	17	6	22	45
Column Total	43	26	36	105

Is there an association between the type of beverage chosen and the preferred snack? Test at the .10 level. What is the p value for your analysis? Interpret this p value.

9.35 Data concerning gender and placement for juvenile offenders suggest that males are more likely to be placed with the State Division for Youth (DFY) than females are. Do the following data bear out this theory?

Gender / Placement	Male	Female	Row Total
Private/Residential	95	25	130
DFY	23	11	34
Column Total	118	36	164

Test the hypothesis that there exists no relationship between placement and gender at the .05 level. Compare the p value with the alpha level. What does this comparison show?

9.36 The following data were compiled relating number of prior court referrals for youth and detention decision outcomes:

Priors / Decision	0–1	2–4	5	More than 5	Row Total
Not Detained	1794	620	723	360	3497
Detained	204	172	368	299	1042
Column Total	1998	792	1091	659	4539

Is there a relationship between detention decision outcomes and prior court referrals? Test at the .05 level. Find the p value for your result, and interpret it.

FORMULAS

Test Statistic for a Single Proportion

$$z_{\text{test}} = \frac{p - \pi}{\sigma_\pi}$$

where the standard error for the proportion is

$$\sigma_\pi = \sqrt{\frac{\pi(1 - \pi)}{n}}$$

and the standard deviation for a binomial proportion is

$$s_p = \sqrt{\frac{p(1-p)}{n}}$$

Test Statistic for Comparison of Two Proportions with Independent Samples

$$z_{\text{test}} = \frac{(p_a - p_b) - (\pi_a - \pi_b)}{s_{p_a - p_b}}$$

When you are testing the null hypothesis that $\pi_a - \pi_b = 0$, the standard deviation for the comparison of two proportions is

$$s_{p_a - p_b} = \sqrt{p(1-p)\left(\frac{1}{n_a} + \frac{1}{n_b}\right)}$$

where

$$p = \frac{n_a p_a + n_b p_b}{n_a + n_b}$$

Chi-Square Test Statistic

$$\chi^2 = \sum\left[\frac{(f_o - f_e)^2}{f_e}\right]$$

where f_o are observed values and f_e are expected values. (Note that you should have $f_e > 5$ for all cells.)

NEW STATISTICAL TERMS

chi-square statistic
contingency table analysis
expected values
goodness of fit
nonparametric statistics
p chart
parametric statistics

EXERCISES 9.37–9.46

Supplementary Problems

9.37 A total of 300 consumers were asked if they had had difficulty obtaining replacement parts for household appliances. Of these, 63 replied that they had had difficulty. Test the claim that more than 20% of consumers experience difficulty in finding replacement parts. Test at the .10 level.

9.38 A fast food restaurant conducts a survey of its customers. In the past, 30 percent of its customers believed that the service was excellent. If 200 customers are surveyed today and 50 believe that the service is excellent, can we conclude at the .01 level that there is no change in people's perception?

9.39 A survey by the Gallup Poll appearing in *American Medical News* (December 27, 1993) asked smokers' opinions of their likelihood of dying from a smoking-related disease if they didn't quit. A total of 62% responded that it was somewhat or very likely that they would die from a smoking-related disease. Assume that the survey had 900 respondents. Test the hypothesis that over two-thirds of the smoking public believe that they will die from a smoking-related disease. Test at the .10 level.

9.40 A *Psychology Today* (April 1986) survey of males and females found that 17% of the females surveyed were dissatisfied with their height and 20% of the males surveyed were dissatisfied with their height. If the survey consisted of 100 females and 150 males, do the data show a statistically significant difference at the .05 level?

9.41 In order to assess the influence of cable television in various markets, a survey was conducted in a number of cities. Suppose the following results were obtained.

City / Households	Dallas	Philadelphia	Fargo	Row Total
Served	363	165	756	1284
Not Served	637	335	1244	2216
Column Total	1000	500	2000	3500

Is there a significant difference in the percentage of households served between markets? Test at the .01 level. What would this study suggest for a cable company that served all three markets?

9.42 Business students at St. John's University were surveyed concerning the imposition of a flat tax to replace the current income tax system. Democrats surveyed opposed the flat tax: 54 against to 27 in favor. Republicans favored it: 18 against and 63 in favor. Can we conclude that there exists a relationship between party affiliation and voting behavior at the .01 level?

9.43 The program ratio is a statistic used to evaluate charities. It represents the percentage of total expenses devoted to programs. The higher the ratio, the less spent proportionately on administration, fund raising, and so on. A high program ratio has been defined as 85% or more going to programs. Data have been collected for a variety of charities (as classified in *U.S. News & World Report,* December 4, 1995).

Type of Charity / Program Ratio	Human Services	Health	Global Aid	Religious	Youth	Row Total
High	6	4	4	5	1	20
Low	2	6	6	5	9	28
Column Total	8	10	10	10	10	48

Is there an association between the type of charity and the program ratio? Test at the .05 level.

9.44 A producer of drug containers wishes to monitor the production process. Samples of 1000 containers were drawn from the production line each day over a two-month period (a total of 43 working days). The results appear below.

Day	No. of Defectives	% Defective	Day	No. of Defectives	% Defective
1/2	0	.000	2/1	24	.024
1/3	17	.017	2/2	6	.006
1/4	15	.015	2/3	29	.029
1/5	26	.026	2/6	23	.023
1/6	5	.005	2/7	24	.024
1/9	14	.014	2/8	12	.012
1/10	17	.017	2/9	20	.020
1/11	21	.021	2/10	14	.014
1/12	19	.019	2/13	22	.022
1/13	23	.023	2/14	29	.029
1/16	31	.031	2/15	29	.029
1/17	15	.015	2/16	28	.028
1/18	17	.017	2/17	31	.031
1/19	23	.023	2/20	31	.031
1/20	25	.025	2/21	33	.033
1/23	6	.006	2/22	34	.034
1/24	29	.029	2/23	16	.016
1/25	0	.000	2/24	15	.015
1/26	36	.036	2/27	17	.017
1/27	14	.014	2/28	25	.025
1/30	11	.011	2/29	5	.005
1/31	7	.007			

Construct a control chart for this process, using the historical average percentage of defectives, which is 2%. Is this process in control?

9.45 The Wenco Corporation, a Wendy's franchisee, operates numerous Wendy's on Long Island. Customer response cards can be found at each establishment. While respondents are not a random sample, let's treat the data as though they were. Suppose that, in a survey of 100 respondents, a total of 68% felt that service was either good or excellent. Could it be concluded that over two-thirds of all Wenco patrons feel that way? Test at the .01 level.

9.46 The Lincoln Mercury Division of the Ford Motor Company gives its leaseholders a detailed questionnaire asking about all aspects of the transaction. When asked "Was your vehicle ready for pick up at the agreed-upon time?" suppose that, of 250 leaseholders responding, 235 answered yes. Can we conclude that at least 95% of leaseholders found their vehicles ready on time? Test at the .05 level.

Using the database in Appendix A and on the CD in the back of the book, test this hypothesis against the actual data. Interpret your results carefully.

10

Regression and Correlation

OUTLINE

Section 1 **Scatterplots**
Association and Causation | Errors in Specification, Errors in Measurement, and Sampling Error

Section 2 **Simple Linear Regression**

Section 3 **The Standard Error of Estimate**

Section 4 **Coefficients of Correlation and Determination**

Section 5 **Forecasting with Simple Regression Models**
The Uses and Limitations of Predictions Based on Models | Prediction Intervals | Confidence Intervals

Section 6 **Assessing the Significance of Slope Estimates**
Two-Tailed Tests | One-Tailed Tests | Confidence Intervals for Slope Estimates

Section 7 **Using the Computer for Regression Analysis**
Using Excel for Regression Analysis | Using Minitab for Simple Linear Regression

LEARNING OBJECTIVES

1. Draw a scatterplot for bivariate data.
2. Compute the coefficients for a linear regression line.
3. Measure variation about the regression line.
4. Calculate the coefficients of correlation and determination.
5. Make a forecast using regression analysis.
6. Test the significance of regression coefficients.
7. Use the computer for regression analysis.

BUILD YOUR KNOWLEDGE OF *STATISTICS*

On November 9, 1965, at 5:16 P.M., the East Coast of the United States was thrown into darkness. Within a few minutes, 30 million people were without electricity, some for as long as 13 hours ("The Great Northeast Blackout of 1965," Central Maine Power Company 5/19/97 Web page, www.cmpco.com/aboutCMP/powersystem/blackout.html). Better demand forecasting might have helped avoid the blackout.

Forecasting demand for energy is important to utilities and everyone who depends on them. Forecasting is important to business planning as well. If a business can predict sales accurately, it can determine future personnel requirements, raw material needs, and cash flow. Good forecasts strengthen planning, and sound business practices are based on good planning.

This chapter examines regression and correlation. Regression analysis is a forecasting tool that relates one variable to at least one other variable. The demand for electricity, for example, could be related to population, weather conditions, time of day, and any number of additional variables, as a change in the value of any of these variables could have an impact on demand. Simple regression analysis addresses the linear relation between just two variables. We will look at how to construct a model relating two variables and how to use such a model to predict one variable, given the value of the other. We will also examine several measures of the effectiveness of a model, including the coefficient of correlation. And we will look at how to do all of these tasks on the computer.

Chapter 11 will expand on the foundation established in this chapter, with a look at multiple regression analysis and time series analysis.

SECTION 1 ■ SCATTERPLOTS

If there are two variables associated with each item in a sample, a scatterplot will show how these variables are related.

> A **scatterplot** is a plot of the data points (*X*, *Y*) for each case studied, where *X* is the independent variable, measured along the horizontal axis, and *Y* is the dependent variable, measured along the vertical axis.

Figure 10.1 shows a scatterplot of data provided by the New York City Board of Education on reading and mathematics achievement for city school districts. Each point represents a single school district. The horizontal coordinate, *X*, is the percentage of students scoring at or above grade level in reading, and the vertical coordinate, *Y*, is the percentage scoring at or above grade level in mathematics. (Software packages such as Excel and Minitab will produce scatterplots for you once the data are entered.) The scatterplot in Figure 10.1 suggests a trend: Higher percentages in reading are associated with higher percentages in math.

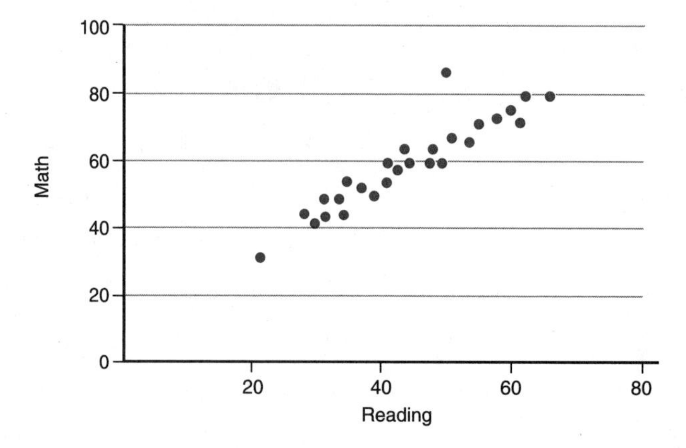

Figure 10.1
Scatterplot showing percentage of students at or above grade level in reading and math

In regression analysis, one variable, called the predictor variable, is used to predict the level of another variable, called the response variable. By convention, the horizontal variable is the predictor variable, and the vertical variable is the response variable.

> An **independent,** or **predictor, variable** is used to estimate or predict the values of other variables. It is usually denoted by *X* and shown on the horizontal axis.
>
> A **dependent,** or **response, variable,** is estimated or predicted as a function of other variables. It is usually denoted by *Y* and shown on the vertical axis.

How are scatterplots used to make predictions? Suppose a friend told you that he had just checked on the reading skills for a school district in New York City and found that 60% of the students were at or above grade level in reading. Suppose he went on to say "I'll bet they don't do that well in mathematics." Should you take that bet? The trend exhibited in Figure 10.1 suggests that you should. The trend shows that the percentage of students performing at or above grade level in mathematics runs a bit more than 10 points higher than the percentage performing at or above grade level in reading. Given the scatterplot, you would expect a school in which 60% of the students were at or above grade level in reading to have 70% to 75% of its students at or above grade level in mathematics.

These observations can be summed up informally by sketching a trend line on the scatterplot, as shown in Figure 10.2. Later in this chapter, we will explore how the regression line captures in a more precise way the variation of Y with X that we have informally indicated with our trend line. The **trend line** is simply a graphical attempt to express the pattern of variation of Y with X seen in a scatterplot.

Figure 10.2
Scatterplot with trend line

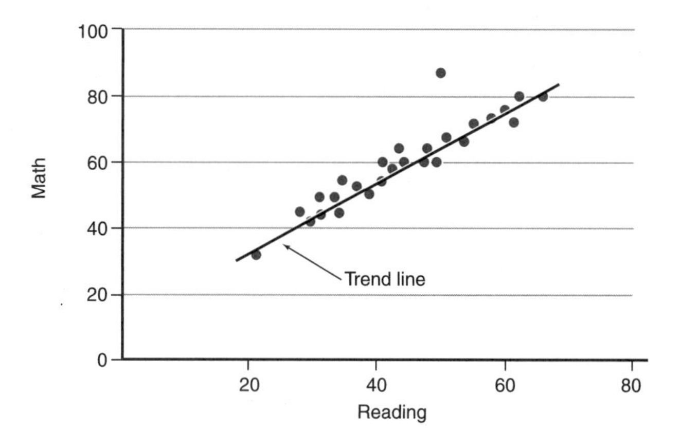

Figure 10.2 shows some scatter of the data about the trend line, even though the linear trend is quite strong. Scatterplots reflect both systematic relationships between the variables and random variations due to uncontrolled differences among the observations, sampling errors, or other factors. We will discuss some of those factors later in this section.

It is essential to be aware of trend and scatter in business situations. By taking advantage of a trend, a businessperson may increase sales or decrease costs. A businessperson may take advantage of scatter by soliciting competitive bids and choosing the lowest cost supplier from among them. The variation seen in the scatterplot in Figure 10.2 is typical of data from practical studies. This scatterplot shows a stochastic relationship—one in which chance deviations combine with systematic effects.

> A **stochastic relationship** exists between two variables when at least one of the variables has a random component leading to scatter in the data.

Association and Causation

From Figure 10.2, it is clear that there is an association between reading and mathematics achievement and that a simple trend line allows us to predict mathematics performance, given reading performance. But this does not mean that better schoolwide reading achievement *causes* better schoolwide mathematics achievement. It is more likely that schools scoring high on both measures are those with better students, better parental support, and, perhaps, better staffing and funding.

Be forewarned:

Association does not imply causation.

This point is important in planning: You need to look beyond scatterplots and trend lines in order to decide how to intervene to improve a situation. In the case at hand, pouring more effort into teaching reading may or may not raise math scores. We now turn to two cases in which the predictor variable clearly does play a large role in determining the response variable.

The regression model of this chapter is a formal means of expressing two essential ingredients in a stochastic relationship: (1) a tendency of the response variable, Y, to vary with the predictor variable, X, and (2) a scattering of points around the observed trend, the regression line. Let's take a closer look at such relationships by considering the association between heating costs and outside temperature. The scatterplot in Figure 10.3, where X is the average temperature for a month and Y is the heating cost for the same month for the same dwelling or group of dwellings, illustrates the relationship between monthly heating bills and outside temperature. The trend line is an informal description of how Y changes with X. (Later, we will replace this rough trend line with the regression line for Y on X.) The relationship is a **negative,** or **inverse, relationship;** that is, as outside temperature rises, heating costs tend to fall.

Figure 10.3
Monthly heating costs versus outside temperature

An imperfect inverse relationship

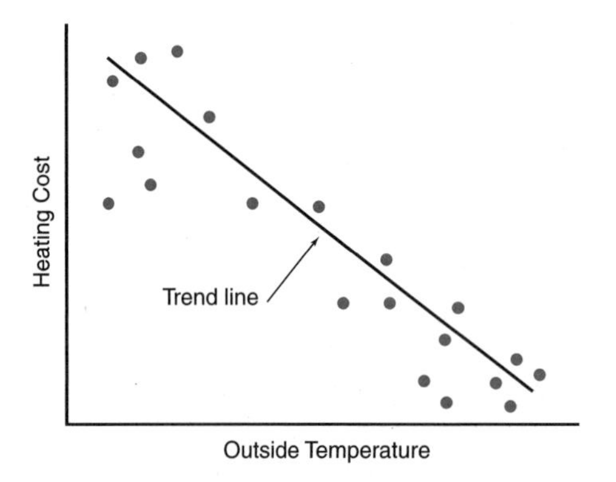

In the scatterplot in Figure 10.4, each dot corresponds to a data point (X, Y), for a hypothetical taxpayer who has received income X and has paid income tax Y. The trend line indicates the general linear trend of the variation in tax paid, Y, as income, X, varies. As income earned increases, tax paid tends to increase. When the two variables move in the same direction, the relationship is considered a **positive,** or **direct, relationship.**

Figure 10.4
Income tax paid versus income earned

An imperfect direct relationship

STATISTICS TOOL KIT

Drawing Scatterplots

The scatterplot in Figure 10.1 can also be drawn using Excel. Enter the data, and the software will do the rest. The internal steps carried out by the software include determining the ranges of the X and Y variables, setting up the horizontal and vertical scales, and plotting the data points. Although you could do these steps by hand, the computer is particularly useful for large data sets, easily carrying out the more involved calculations of regression and correlation.

You should first practice with small data sets, drawing scatterplots by hand and working through calculations using worksheets, so that you understand the ideas. However, practical work will eventually require that you become familiar with statistical calculators or software such as Excel or Minitab. Section 7 of this chapter discusses performing regression analysis on the computer. You may also consult your calculator or software manual for specific procedures to use with the tools you have at your disposal.

In Figure 10.3, temperature is the predictor variable, and the amount of the heating bill is the response variable. In Figure 10.4, income earned is the predictor variable, while tax paid is the response variable.

Here, we can see how decreases in the outside temperature would cause heating bills to rise. Similarly, when we observe that changes in tax paid are associated with changes in income earned, we know this is a cause-and-effect relationship, even though the relationship is not exactly linear. (Changes in filing status and other factors affect the tax computation as well, causing the scatter we observe in the data.) But when statistics showed an association between lung cancer and cigarette smoking, suggesting a possible causal relationship, further experiments were needed to prove the link.

Errors in Specification, Errors in Measurement, and Sampling Error

Looking at the scatterplot of taxes paid versus income earned in Figure 10.4, we expect the tax to increase, on average, with the income and to do so at approximately the federal tax rate. But the taxpayer has many choices to make that affect just how much of his or her total income from line 22 of Form 1040 makes its way to line 37, where we find the taxable income that determines what this taxpayer will have to pay. The scatter is related, in part, to not specifying all the variables in addition to total income that legitimately affect the taxpayer's bottom line. Such variability is attributed to **errors in specification.** If all these factors were taken into account, much of the variation observed would be eliminated.

The inverse relationship between monthly heating costs and average monthly temperatures shows scatter not only because there are great differences in the efficiency with which buildings retain heat (and great differences in where people set their thermostats), but also because a regional monthly average temperature does not accurately reflect the micro-climates around the region. Thus, the average temperature is only a rough guide, and the scatter can be attributed, in part, to **errors in measurement.**

Another source of scatter is sampling error, arising from the natural variation of the population and the sample studied and—unlike errors in specification and measurement—not attributable to specific known neglected variables.

> **Sampling error** is that variation of measurements from the population parameter found from sample to sample in any statistical study and arising from the variations in the population and the nature of random sampling. Sampling error is normal variation and characteristic of all data sets.

EXERCISES 10.1–10.15

Skill Builders

10.1 Use the numbers below:

X	56	90	2	8	41	24	97	95	43	62
Y	46	83	12	17	32	34	87	84	43	53

a. Putting X on the horizontal axis and Y on the vertical axis, construct a scatterplot.
b. Is there a linear relationship between X and Y?
c. If so, is the relationship positive or negative?

10.2 Consider this data set:

X	99	13	54	73	8	44	10	7	59	0
Y	0	99	7	8	73	54	88	93	10	84

a. Plot the data on a scatterplot.
b. Is there a linear relationship between X and Y?
c. If so, is the relationship inverse or direct?

10.3 Use the numbers below:

X	−.04	.29	−.04	.28	.79	−.87	.16	−.91	.80
Y	−.56	.32	.72	.52	−.20	.72	.02	.31	−.28

a. Putting X on the horizontal axis and Y on the vertical axis, construct a scatterplot.
b. Is there a linear relationship?
c. If there is a linear relationship, is it positive or negative?

Applying the Concepts

10.4 The following scatterplot reflects the Olympic male high jump records from 1956 to 1992 (*The Official Olympic Companion,* 1996). The scale on the vertical axis is expressed in meters (1 meter = 39.37 inches).

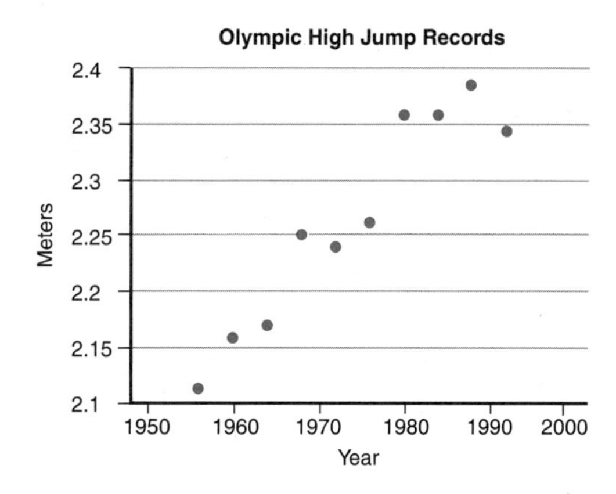

a. Describe this scatterplot in words.
b. Sketch the trend line.
c. In 1968, Dick Fosbury introduced his flop. How did that affect the high jump?

10.5 The data below reflect grades on two exams in an elementary statistics class:

Student	A	B	C	D	E	F	G	H	I	J
Exam 1 Grade	69	76	86	99	65	64	55	88	74	70
Exam 2 Grade	81	65	97	100	85	53	92	100	69	99

a. Putting "Exam 1" on the *X* axis and "Exam 2" on the *Y* axis, plot the data on a scatterplot.
b. Is there a linear relationship between Exam 1 and Exam 2 grades?
c. If so, is the relationship positive or negative?

10.6 The annual operating profits and net revenues of Woolworth Corporation's various subdivisions for 1989 (in millions of dollars) were as follows.

Subdivision	Revenues	Profits
Kinney U.S.	2357	261
Kinney Canada & Australia	685	67
Richman Brothers	293	10
Other Specialty Operations	442	12
Woolworth U.S.	2169	123
Woolworth Canada	1867	106
Woolworth Germany	1123	78

a. Which variable is the predictor variable?
b. Construct a scatterplot of the data.
c. Is there a linear relationship between revenues and profits?
d. If so, is the relationship direct or inverse?
e. What does this scatterplot suggest about the performance of the various subdivisions? What conclusions would you draw as a manager?

10.7 Attendance at New York Yankees games (in millions) for the years 1976 through 1990 appears below, along with the number of wins for the previous year.

Year	Attendance	Previous Year's Wins
1976	2.0	83
1977	2.1	97
1978	2.3	100
1979	2.5	100
1980	2.6	89
1981	2.4	103
1982	2.0	89*
1983	2.3	79
1984	1.8	91
1985	2.2	87
1986	2.3	97
1987	2.4	90
1988	2.6	89
1989	2.2	85
1990	2.0	74

*Adjusted figure. There was a strike that season. This event may account for the drop in attendance in 1983.

a. Identify the independent variable.
b. Construct a scatterplot of the data.
c. Is there a linear relationship?
d. Is the relationship, if any, direct or inverse?
e. What conclusions, if any, can you draw for the team's management?

10.8 A cruise vacation is generally planned from six months to a year in advance. An analyst of the cruise ship industry compiled the following data. (The Consumer Confidence Index is for the same quarter, a year earlier.)

Year: Quarter	No. of Cruise Passengers	Consumer Confidence Index
1988: I	761,597	90.97
II	820,879	100.86
III	907,621	110.68
IV	711,177	107.94
1989: I	766,929	112.45
II	859,461	117.23
III	903,083	114.68
IV	751,003	116.40
1990: I	799,853	117.88
II	968,820	116.83
III	1,046,024	117.38
IV	842,830	115.03

a. Identify the response variable.
b. Draw a scatterplot of the data.
c. Does a linear relationship exist?
d. If a linear relationship exists, is it positive or negative?
e. What implications does the pattern have for managers of a cruise ship line? When should such lines advertise most heavily?

10.9 In order to estimate how profits and sales are related, the accountant for a chain of beauty salons compiled sales and payroll figures for the 110 stores in the chain. The accountant knows that rent and other costs are essentially constant. A small portion of the relevant data set appears below. (All figures are expressed in thousands of dollars.)

Sales	Payroll Expenses
16	9
453	242
3521	2205
342	200
283	164
3179	1756
226	115
139	64
290	149

a. Which is the predictor variable; that is, which variable is being used to drive the accountant's model?
b. Construct a scatterplot.
c. Is there a linear relationship?
d. If so, is the relationship inverse or direct?
e. What conclusions, if any, can the accountant draw from this model?

10.10 Between 1980 and 1989, Anheuser-Busch spent over $13 billion on advertising. Sales climbed from $3.8 billion to $10.3 billion per annum. The data (in millions of dollars) are as follows:

Year	Sales	Advertising
1980	3,822.4	428.6
1981	4,435.9	518.6
1982	5,251.2	758.8
1983	6,714.7	1,226.4
1984	7,218.8	1,338.5
1985	NA	NA
1986	7,756.7	1,498.2
1987	8,478.8	1,709.8
1988	9,705.1	1,834.5
1989	10,283.6	1,876.8

a. Which is the predictor variable?
b. Plot the data on a scatterplot.
c. Is the relationship linear?
d. If so, is the relationship direct or inverse?
e. What conclusions should the managers at Anheuser-Busch draw from the data?

10.11 *The World Almanac and Book of Facts 1993* gives historical statistics on the economy under presidential administrations from 1945 to 1992. Among the data presented are the GDP and the misery index. The GDP is the gross domestic product; the misery index is the sum of unemployment and inflation rates.

President	Change in GDP	Misery Index
Truman (1945–1952)	25.0%	3.8
Eisenhower (1953–1956)	9.4	6.7
Eisenhower (1957–1960)	7.3	6.9
Kennedy/Johnson (1961–1964)	20.4	6.1
Johnson (1965–1968)	19.4	8.1
Nixon (1969–1972)	12.4	8.9
Nixon/Ford (1973–1976)	7.6	12.5
Carter (1977–1980)	11.5	18.2
Reagan (1981–1984)	10.1	11.4
Reagan (1985–1988)	14.0	9.7
Bush (1989–1992)	2.5	10.5

a. Which variable would you make the response variable?
b. Plot the data on a scatterplot.
c. Is there a linear relationship?
d. If so, is the relationship positive or negative?
e. Do you see any variations in the treatments of different presidential terms or in the historical context that may account for the variations in these statistics?

10.12 The ratio of a stock's price to the company's earnings per share is called the price/earnings multiple. The stock's yield is the annual dividend as a percentage of the price per share. Both measures are used in evaluating stocks for purchase. According to *Business Week* (December 28, 1992), the average P/E ratio for the aerospace industry in 1992 was 19 (that is, 19:1); the average yield was 3.06%. Data for the major firms follow:

Firm	P/E Ratio	Yield
Boeing	7	2.85
GenCorp	10	5.65
General Dynamics	14	1.65
Grumman	8	4.42
Lockheed	9	4.25
Martin Marietta	9	2.67
McDonnell Douglas	6	3.03
Northrop	6	4.19
Rohr	125	0.00
Sequa	19	1.65
Sundstrand	13	3.15
Thiokol	6	2.27
United Technologies	11	4.01

a. Plot the data on a scatterplot. Let "P/E Ratio" be the independent variable.
b. Is there a linear relationship between the P/E ratio and the yield?
c. If there is a linear relationship between the variables, is it positive or negative?

10.13 A 30-year fixed-rate mortgage is one of the more popular ways to finance the purchase of a home. Some lending institutions charge points—a sum that must be paid when the money is loaned, in addition to the long-term interest rate. The data for lending institutions on September 24, 1992 appear below (*Money,* November 1992).

Interest Rate	Points	Interest Rate	Points
7.88	1.00	7.88	2.50
7.75	2.38	7.63	3.00
7.63	2.50	7.75	2.25
7.88	2.38	7.63	2.50
7.63	1.75	7.75	2.00
7.88	3.00	7.75	3.00
7.63	2.75	7.63	3.00
7.88	3.00	7.75	3.00
7.75	1.50	7.75	1.50
7.88	2.00	7.75	2.50
7.63	2.38	7.63	2.38
7.75	2.00	7.75	2.50

a. Plot the data on a scatterplot. Let "Interest Rate" be the predictor variable.
b. Is there a linear relationship between the interest rate and the points charged?
c. Does a higher interest rate imply lower points?
d. What conclusions can you draw from the data about shopping for loans?

10.14 Various economists were asked to predict unemployment and inflation (as measured by the Consumer Price Index) for 1993. Their predictions appear below (*Business Week,* December 28, 1992). (These data can also be found on the CD in the back of the book.)

Predicted Increase in CPI	Predicted Unemployment Rate	Predicted Increase in CPI	Predicted Unemployment Rate
3.0	6.6	2.8	6.3
3.7	6.8	3.4	6.5
2.3	6.4	2.8	6.8
2.7	6.7	3.7	6.7
3.4	6.8	3.9	6.4
3.2	6.8	3.4	7.0
2.0	6.2	3.0	6.8
2.8	6.8	2.2	7.6
3.4	6.8	2.6	6.9
3.2	6.8	2.6	7.0
2.6	6.7	3.3	6.5
2.7	7.2	3.4	7.1
4.0	7.1	3.0	7.0
3.9	6.8	3.2	6.9
3.0	6.8	2.8	6.4
2.7	6.9	2.9	6.7
3.5	6.7	2.8	6.7
3.0	6.8	2.9	7.0
2.6	6.9	3.3	7.2
3.0	7.1	3.5	6.7
2.8	7.0	2.3	7.4
2.9	7.4	2.0	7.0
3.7	7.5	3.7	6.8
2.8	7.2		

a. Plot "Predicted Increase in CPI" on the Y axis and "Predicted Unemployment Rate" on the X axis.

b. Is there a linear relationship between these variables?
c. Does a higher unemployment rate imply lower inflation for these economists?

10.15 *The Chronicle of Higher Education* (May 6, 1992) provided information on staff and expenditures for major libraries. Data for non-university libraries appear below:

Library	Total Staff	Total Expenditures
Boston Public Library	588	$ 28,421,160
Canada Institute for Scientific & Technical Information, Ottawa	225	22,198,937
Center for Research Libraries, Chicago	72	2,900,444
Library of Congress	5045	307,102,000
Linda Hall Library, Kansas City	61	3,548,000
National Agricultural Library, Beltsville, Md.	235	16,798,000
National Library of Canada, Ottawa	498	31,585,432
National Library of Medicine, Bethesda, Md.	289	23,670,336
New York Public Library	801	38,149,640
New York State Library, Albany	208	9,960,650
Newsberry Library, Chicago	107	5,956,168
Smithsonian Institute	125	5,606,502

a. Plot "Total Staff" on the horizontal axis and "Total Expenditures" on the vertical axis. (*Hint:* Consider making breaks on both axes, rather than starting with zero.)
b. Is there a linear relationship?
c. If there is a linear relationship, is it direct or inverse?

SECTION 2 ▪ SIMPLE LINEAR REGRESSION

We have seen how a line drawn on a scatterplot can be used to predict the response variable, Y, given the value of the predictor variable, X. Let's now look at how we can find the best line for such predictions, called the **simple linear regression line,** from the data in the scatterplot. Figure 10.5 shows the simple linear regression line and the data points for the test scores versus study time data.

The slope, b_1, and the intercept, b_0, in the regression line equation are calculated from the sample data. As statistics, they are represented by Roman letters, following the notational conventions established in Chapter 1. Underlying these statistics are population parameters, β_0 and β_1, denoted by Greek letters.

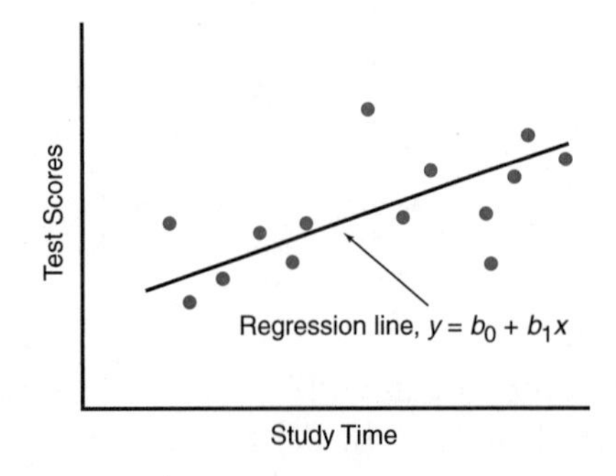

Figure 10.5
Scatterplot of test scores and study time

The probability model that underlies regression assumes that the values of x_i are normally distributed and that the pairs of values (x_i, y_i) that we observe can be thought of as being determined by the formula

$$y_i = \beta_0 + \beta_1 x_i + u_i$$

where u_i is random disturbance. The disturbances are independently drawn from a second normal population and thus have a common variance. If we observe the point (x_i, y_i), then the regression line predicts the value $b_0 + b_1 x_i$ for the response variable, Y. The random term u_i, called the *i*th *residual,* is the vertical distance between (x_i, y_i) and the regression line:

$$\text{Observed} - \text{Predicted} = y_i - y_c = y_i - (b_0 + b_1 x_i) = u_i$$

It makes up the difference, as shown in Figure 10.6.

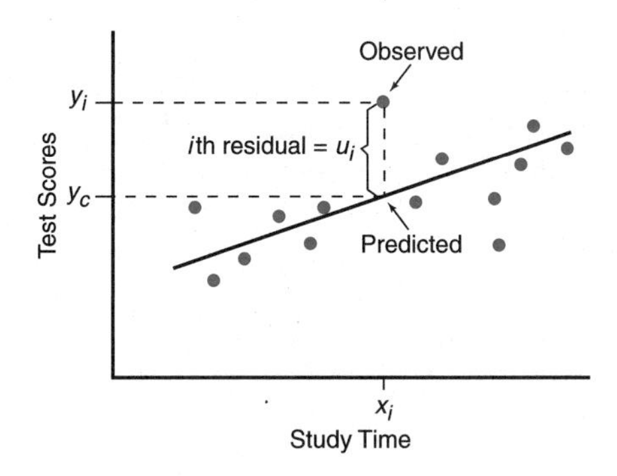

Figure 10.6
Regression line and deviations for test scores and study time

Only one line minimizes the sum of the squares of the vertical distances: the **least squares regression line.** The process of minimizing the sum of the squares consists of finding the minimum for this formula:

$$\text{Sum of the squares of the residuals, } s = \Sigma(y_i - y_c)^2$$

where $y_c = b_0 + b_1 x_i$ = Predicted y_i and b_0 and b_1 are the estimates of β_0 and β_1.

Minimization of s yields two equations called the *normal equations,* whose simultaneous solution results in a simple least squares regression line:

The normal equations

$$\sum y_i = nb_0 + b_1 \sum x_i \tag{1}$$

$$\sum x_i y_i = b_0 \sum x_i + b_1 \sum x_i^2 \tag{2}$$

Filling in the values for x_i and y_i from the data, we get two equations in two unknowns, b_0 and b_1. These equations can be solved for b_0 and b_1 to find the equation for the regression line of y on x: $y = b_0 + b_1 x$.

As a shortcut, we can use the following two formulas, which give the general solution to normal equations (1) and (2):

Calculating b_0 and b_1

$$b_1 = \frac{SS_{xy}}{SS_{xx}} \tag{3}$$

$$b_0 = \bar{y} - b_1 \bar{x} \tag{4}$$

where SS_{xx} and SS_{xy} can be computed from the formulas

$$SS_{xy} = \sum(x - \bar{x})(y - \bar{y}) = \sum xy - \frac{\left(\sum x\right)\left(\sum y\right)}{n} \tag{5}$$

and

$$SS_{xx} = \sum(x - \bar{x})^2 = \sum x^2 - \frac{\left(\sum x\right)^2}{n} \tag{6}$$

This approach is known as the sum of squares method.

As the formulas show, SS_{xx} is $\Sigma(x - \bar{x})^2$. It is the sum of the squared deviations of x_i from $\bar{x}$, or the sum of squares for x_i deviations from $\bar{x}$. SS_{xy} is $\Sigma(x - \bar{x})(y - \bar{y})$, the sum of the products of the deviations of x_i from $\bar{x}$ and the deviations of y_i from $\bar{y}$. (The formulas using x_i and y_i have been included here in order to be clear about how the formulas relate to x_i and y_i of the data. Henceforth, for convenience of notation, we will use ΣX for Σx_i and ΣXY for $\Sigma x_i y_i$; that is, we will use the symbols for the random variables X and Y as substitutes for the specific data values.)

In Example 10.1, we will use the second set of formulas to derive coefficients for a regression line to fit data on bingo games.

EXAMPLE 10.1 ■ FITTING A REGRESSION LINE TO BINGO DATA

Gambling is a business. Every local church or senior center that depends on a weekly bingo game to fund programs understands that revenues depend on the number of players. This information can serve as a guide in deciding how much advertising to do to draw new players or whether to offer special payoffs for the same purpose. The following data set, given in the form (number of players, net revenues), comes from past bingo sessions at the Suffolk Community Center.

(98, $585)	(98, $758)	(153, $2065)	(112, $1091)	(100, $785)
(110, $1080)	(122, $1384)	(150, $2136)	(105, $967)	(107, $966)
(128, $1479)	(110, $1043)	(118, $1143)	(90, $502)	(109, $1037)
(140, $1820)	(119, $1345)	(139, $1667)	(113, $1159)	(125, $1455)
(160, $2186)	(128, $1359)			

Figure 10.7 shows a scatterplot of the data.

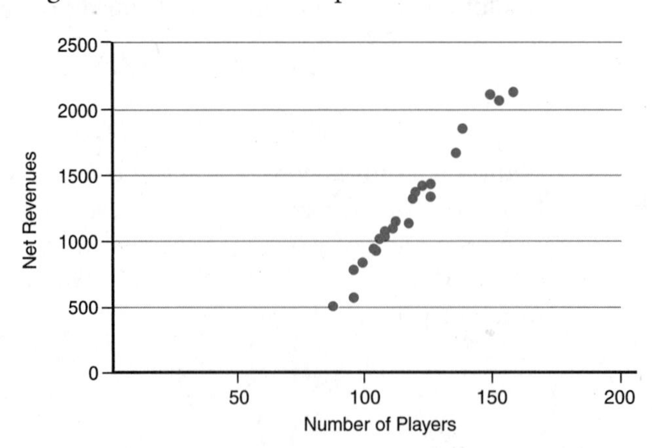

Figure 10.7
Scatterplot of the bingo data

We need five values to solve for the least squares line: $\Sigma X, \Sigma Y, \Sigma XY, \Sigma X^2$, and n. The values can be calculated using a worksheet, such as the one in Table 10.1.

Table 10.1

WORKSHEET FOR FINDING THE REGRESSION COEFFICIENTS FOR THE BINGO DATA

Players, X	Revenues, Y	XY	X^2
98	585	98×585	98×98
98	758	98×758	98×98
.	.	.	.
.	.	.	.
.	.	.	.
113	1159	113×1159	113×113
$\Sigma X = 2634$	$\Sigma Y = 28{,}035$	$\Sigma XY = 3{,}544{,}391$	$\Sigma X^2 = 322{,}968$

Substituting $\Sigma XY = 3{,}544{,}391$, $\Sigma X = 2634$, $\Sigma Y = 28{,}035$, $\Sigma X^2 = 322{,}968$, and $n = 22$ into equations (5) and (6), we have

$$SS_{xy} = 3{,}544{,}391 - \frac{(2634)(28{,}035)}{22} = 187{,}836.91$$

and

$$SS_{xx} = 322{,}968 - \frac{(2634)^2}{22} = 7606.36$$

Then, equations (3) and (4) give

$$b_1 = \frac{SS_{xy}}{SS_{xx}} = \frac{187{,}836.91}{7606.36} = 24.69$$

$$b_0 = \bar{y} - b_1\bar{x} = 1274.32 - 24.69(119.73) = -1681.82$$

Thus, the equation of the regression line is $Y = -1681.82 + 24.69X$.

The slope of the regression line, $b_1 = 24.69$, will help us decide how much we should spend to attract new players. It reflects the increase in dollar income per player added. Because this amount is just shy of $25 per player, profits will increase if additional players can be drawn at a cost of $10 or even $15 per player.

An initial assessment of the "goodness of fit" of the regression line may be made by drawing the line on the scatterplot. If the line falls near the points and the points are scattered in an apparently random way around the line, the line exhibits a good fit. Example 10.2 contains the plot of the regression line for the bingo sessions; it appears to fit the data quite well.

EXAMPLE 10.2 ▪ PLOTTING THE REGRESSION LINE FOR THE BINGO DATA ON A SCATTERPLOT

A regression line can be plotted by finding two points on it. We take two X values within the range of the data and find the values of Y that the regression predicts.

If we let $X = 90$ in the equation of the regression line for the bingo data, we have

$$Y_c = -1681.82 + 24.69(90) = 540.28$$

Rounding to 540, we call ($X = 90$, $Y = 540$) point A.

Then if we let $X = 150$, we have

$$Y_c = -1681.82 + 24.69(150) = 2021.68$$

Rounding to 2022, we call ($X = 150$, $Y = 2022$) point B. We plot point A and point B in Figure 10.8 and connect them with a ruler.

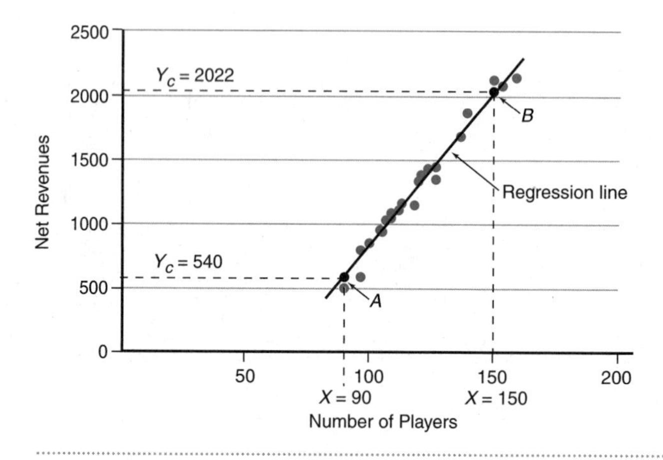

Figure 10.8
Regression line for the bingo data

The regression line that we have defined here is the mathematical embodiment of the trend line from Section 1 of this chapter. In this particular case, it allows us to predict the average revenue for various levels of attendance. The scatter of the data about the average predicted by the regression line will be discussed in Section 3, where you will learn about the part it plays in determining the standard error of the estimate in the regression.

If you use a computer to find the regression line (as described in Section 7 of this chapter), you will not see the calculations shown in Example 10.2. What you will get is a printout of the results, as shown below, where the values of b_0 and b_1 are highlighted. A picture of the regression line is superimposed on the scatterplot, as shown in Figure 10.9 on page 347.

SUMMARY OUTPUT

Regression Statistics	
Multiple R	0.989441
R Square	0.978993
Adjusted R	0.977943
Standard E	70.56081
Observation	22

ANOVA

	df	SS	MS	F	Significance F
Regression	1	4640629	4640629	932.0727	298E-18
Residual	20	99576.55	4978.828		
Total	21	4740206			

	Coefficients	Standard Error	t Stat	*p* Value	Lower 95%	Upper 95%
Intercept	−1685.375	98.02648	−17.19305	1.9E-13	−1889.854	−1480.895
Players	24.70017	0.809049	30.52987	2.98E-18	23.01252	26.387814

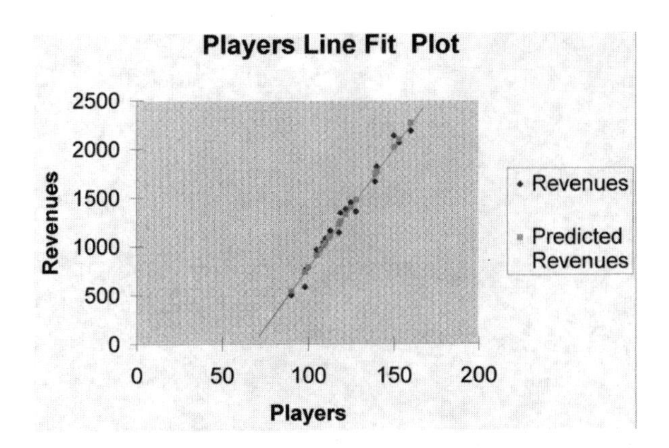

Figure 10.9
Excel plot of the bingo regression results

STATISTICS TOOL KIT

Regression Assumptions, Regression Lines, and Residuals

Regression is a model for predicting how the means of the Y vary with those of X. In simple linear regression, this relation is assumed to be of the form $\mu_y = \beta_0 + \beta_1 x$. The line $y = \beta_0 + \beta_1 x$ is the population regression line, and β_0 and β_1 are the population parameters, corresponding to the b_0 and b_1 that we found when we fitted the data using a least squares regression line. In our model, the sample data are obtained from $y = \beta_0 + \beta_1 x_i + u_i$, where the values of u_i are independently drawn random samples from a normal population with a mean of zero and variance σ^2.

We cannot look directly at the population parameters β_0 and β_1, nor at the values of u_i. We can only look at our values of x_i and y_i, the least squares line, and its coefficients, b_0 and b_1. From these, we defined the **residuals,** the differences between the observed and expected values for the data we have:

$$e_i = y_i - (b_0 + b_1 x_i)$$

If our regression assumptions are correct, the values of e_i will be normally distributed, though they are not necessarily independent nor do they necessarily all have exactly the same variance. However, if the underlying regression model is correct, a plot of the residuals against either the values of x_i or those of y_i will show a good random scatter above and below zero. Any systematic pattern in the signs of the residuals is reason to suspect nonlinear effects or other violations of the assumptions of our model. We will return to this issue in Chapter 11.

For the moment, we will simply consider three types of residual scatterplots, each containing 100 residuals. The first is normal, in accordance with the model; the other two show violations of the model. Such scatterplots can help you check the validity of the regression assumptions.

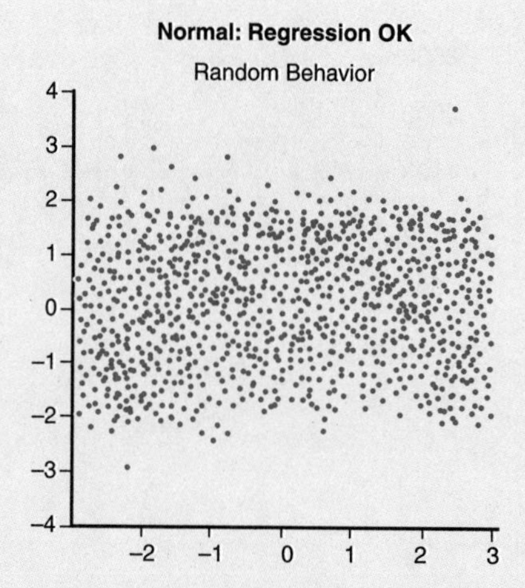

Abnormal: Regression Doubtful

Quadratic Behavior

Discontinuous Behavior

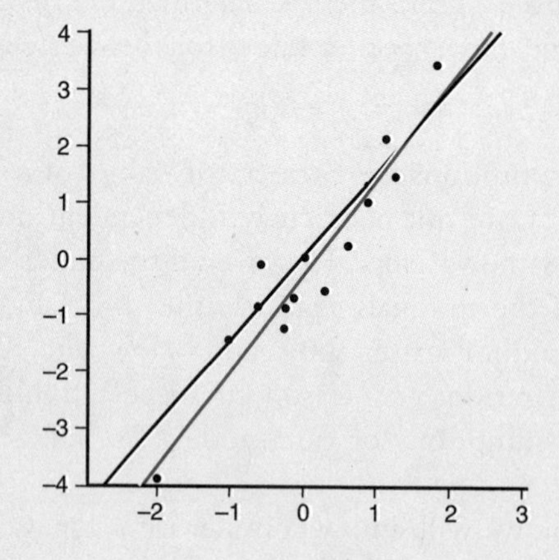

The *sample* regression line $y = b_0 + b_1x$ may depart from the *population* regression line $y = \beta_0 + \beta_1x$. However, b_0 and b_1, as we have defined them, are unbiased and consistent estimators for β_0 and β_1, so these departures will average out to zero for a fixed value of x. The figure to the left shows a sample of size $n = 15$, drawn from a population satisfying our hypotheses, with $\mu = 0 + 1.5x$. The population regression line is shown in color, and the sample points and sample regression line are shown in black. The match is fairly close in the range from $x = -3$ to $x = 3$, as you can see, but the lines eventually diverge. The sample regression line is $y = b_0 + b_1x = -.284 + 1.702x$, so b_0 is a bit less than β_0 while b_1 is a bit larger than β_1.

EXERCISES 10.16–10.29

Skill Builders

10.16 Use the numbers below:

X	5	6	2	4	9	0	8	9	1	7
Y	2	4	4	5	9	6	5	2	3	9

a. Calculate the regression line of *Y* as a function of *X*.
b. Plot the data on a scatterplot.
c. Plot the regression line on the scatterplot.
d. Does the line fit the data well?

10.17 Consider the following data set for the variables *X* and *Y*.

X	56	90	2	8	41	24	97	95	43	62
Y	46	83	12	17	32	34	87	84	43	53

a. Calculate the regression line of *Y* on *X*.
b. On the scatterplot produced earlier, plot the regression line.
c. Is the regression line a good fit for the data?

10.18 Use the exam data, repeated below:

Student	A	B	C	D	E	F	G	H	I	J
Exam 1 Grade	69	76	86	99	65	64	55	88	74	70
Exam 2 Grade	81	65	97	100	85	53	92	100	69	99

a. Compute the linear regression of Exam 2 grades as a function of Exam 1 grades.
b. On the scatterplot, plot the regression line.
c. How would you describe the fit of the data?
d. How might you use this line in a practical way?

10.19 Consider the numbers below:

X	−.95	−.43	.62	−.99	.08	−.59
Y	.18	.84	.64	.82	−.27	.22

a. Plot the data on a scatterplot.
b. Calculate the regression line of *Y* on *X*.
c. Draw the regression line on the scatterplot.
d. Describe the fit.
e. Would you use this line to forecast values of *Y*?

10.20 Use the data below:

X	435	821	81	294	922	982	410	642
Y	279	83	484	434	786	407	80	111

a. Draw a scatterplot of the data.
b. Calculate the regression line of *Y* on *X*.
c. Plot the line on the scatterplot.
d. Comment on the fit.

Applying the Concepts

10.21 Look once again at the data on sales and advertising for Anheuser-Busch:

Year	Sales	Advertising
1980	3,822.4	428.0
1981	4,435.9	518.6
1982	5,251.2	758.8
1983	6,714.7	1226.4
1984	7,218.8	1338.5
1985	NA	NA
1986	7,756.7	1498.2
1987	8,478.8	1709.8
1988	9,705.1	1834.5
1989	10,283.6	1876.8

a. Use the least squares method to calculate the regression line.
b. Plot the line on a scatterplot.
c. Do you think that the line is a good fit for the data? Why?
d. What conclusions, if any, do the data suggest about changes to the company's advertising budget?

10.22 Data for the subdivisions of Woolworth's appear below. (Revenues and profits are in millions of dollars.)

Subdivision	Revenues	Profits
Kinney U.S.	2357	261
Kinney Canada & Australia	685	67
Richman Brothers	293	10
Other Specialty Operations	442	12
Woolworth U.S.	2169	123
Woolworth Canada	1867	106
Woolworth Germany	1123	78

a. Calculate the regression line with "Revenues" as your predictor variable.
b. Plot the line on the scatterplot.
c. Is this line a good fit? Why?
d. What might be a good predictor of revenues?
e. On average, what increase in profits can be expected with a $1 million increase in revenue, based on this regression line?

10.23 Data on attendance (in millions) at New York Yankee games from 1976 to 1990 appear below, along with the number of wins for the previous year.

Year	Attendance	Previous Year's Wins
1976	2.0	83
1977	2.1	97
1978	2.3	100
1979	2.5	100
1980	2.6	89
1981	2.4	103
1982	2.0	89*
1983	2.3	79
1984	1.8	91

(cont.)

1985	2.2	87
1986	2.3	97
1987	2.4	90
1988	2.6	89
1989	2.2	85
1990	2.0	74

*Adjusted figure. There was a strike that season. This event may account for the drop in attendance in 1983.

a. Use the least squares method to calculate "Attendance" as a function of "Previous Year's Wins."

b. Plot the regression line on your scatterplot of the data.

c. Assess the fit of the line.

10.24 Between 1981 and 1989, net sales at Wal-Mart grew from $2,445 million to $25,811 million. U.S. disposable income, as measured by the Department of Commerce, grew from $2128 billion to $3726 billion. Data for both series appear below, with net sales in millions of dollars and disposable income in billions of dollars.

Year	Wal-Mart's Net Sales	U.S. Disposable Income
1981	2,445	2128
1982	3,376	2261
1983	4,667	2428
1984	6,401	2669
1985	8,401	2839
1986	11,909	3013
1987	15,959	3195
1988	20,650	3479
1989	25,811	3726

a. Calculate the least squares regression line of "Wal-Mart's Net Sales" versus "U.S. Disposable Income."

b. Draw a scatterplot of the data.

c. Plot the regression line on the scatterplot.

d. Assess the fit.

e. Considering disposable income as a "pie," do you think Wal-Mart's growth should be attributed to growth of the pie or to growth of its slice of the pie? Support your conclusions.

10.25 Consider these statistics on change in gross domestic product (GDP) and the misery index.

Change in GDP	25.0	9.4	7.3	20.4	19.4	12.4	7.6	11.5	10.1	14.0	2.5
Misery Index	3.8	6.7	6.9	6.1	8.1	8.9	12.5	18.2	11.4	9.7	10.5

a. Letting change in GDP be the X variable and the misery index be the Y variable, calculate the regression line.

b. Place the regression line on the scatterplot.

c. Does there appear to be a relationship between the change in GDP and the misery index?

10.26 *Consumer Reports* (September 1990) published an analysis of brands of videotapes. In the article, each brand was rated and the price paid was noted. The data are as follows:

Brand	Score	Price Paid
Fuji SXG-Pro	93	$8.81
Maxell RX-Pro	93	9.87
TDK X-Pro	92	6.73
Scotch EG	89	4.57
Scotch EXG	89	8.96
Fuji HQ	88	4.17
Memorex HS	87	4.77
TDK HS	87	4.33
Kodak HS	86	3.98
Sony ES	86	4.59
BASF EQ	85	4.41
JVC ER	85	3.99
Maxell EX	84	4.83
Polaroid	83	5.05

a. Plot the data on a scatterplot. Let "Score" be the predictor variable.
b. Is there a linear relationship?
c. If so, is the relationship positive or negative?
d. Calculate the regression line of price paid as a function of score.
e. Plot the line on the scatterplot.
f. Complete this paragraph: In an analysis of videotape brands, the relationship between price paid and rating was studied. A plot of the data indicated (a linear/no) relationship. A higher score would appear to be (positively/negatively) related to price paid. It would appear that you (get/do not get) what you pay for.

10.27 *The World Almanac and Book of Facts 1993* presented data on sales and income of manufacturing corporations by industry. Income was defined as sales less cost of goods. The data for nondurable manufacturing corporations for the first quarter of 1992 (in millions of dollars) are presented below.

Industry	Sales	Income
Food & Kindred Products	96,913	5085
Textile Mill Products	14,556	294
Paper & Related Products	30,785	796
Printing & Publishing	35,706	878
Chemicals, etc.	75,682	6268
Petroleum and Coal Products	62,473	2252
Rubber & Plastics	21,299	544
Other Nondurables	15,434	453

a. Plot the data on a scatterplot. Make "Sales" the predictor variable. Make "Income" the response variable.
b. Is there a linear relationship?
c. Calculate the regression line.
d. Draw the line on the scatterplot.
e. Does the line fit well or poorly?

f. Compare current data from a recent *World Almanac* with those listed above. How has each industry fared since the first quarter of 1992? Repeat parts a through e for the new data.

10.28 A comparison of school districts in Nassau County (*Newsday's Long Island Help Book,* 1992) provided data on each district. It has often been argued that smaller class size is preferable to larger class size. To address that argument, use the following data on average class size and percentage of students going on to four-year colleges.

School District	Average Class Size	Percentage of Students Going to 4-year Colleges
Baldwin	22.7	66
Bethpage	19.9	40
Carle Place	23.7	47
East Meadow	22.4	43
East Rockaway	22.1	34
East Williston	20.7	81
Farmingdale	22.4	42
Freeport	24.8	49
Garden City	22.7	79
Glen Cove	19.8	53
Great Neck	21.0	87
Hempstead	22.4	37
Herricks	23.0	79
Hewlett/Woodmere	23.2	79
Hickville	21.3	52
Island Trees	23.2	50
Jericho	19.9	85
Lawrence	20.3	64
Levittown	21.8	28
Locust Valley	20.2	63
Long Beach	24.5	58
Malverne	23.2	42
Manhasset	19.6	83
Massapequa	21.6	54
Mineola	21.4	45
North Shore	16.6	68
Oceanside	23.7	57
Oyster Bay/East Norwich	24.2	61
Plainedge	21.0	45
Plainview/Old Bethpage	22.5	61
Rockville Centre	20.7	70
Roosevelt	25.4	40
Roslyn	21.5	92
Seaford	23.0	43
Syosset	20.4	78
Uniondale	23.2	44
Wantagh	19.6	59
West Hempstead	22.1	40
Westbury	22.7	47

a. Plot the data on a scatterplot. Make "Average Class Size" the predictor variable.

b. Is there a linear relationship?

c. Based on the data, is there sufficient justification to believe that average class size is related to success?

d. Calculate the regression line of percentage of students on average class size.

e. Does the line fit the data well?

f. Complete this paragraph: In an analysis of data on Nassau County's school districts, it was found that there exists a (negative/positive) relationship between average class size and success. The relationship is (linear/nonlinear). The data suggest that class size (should be reduced for/has no effect on) educational success.

10.29 *Forbes* (November 11, 1991) compiled data on the number of years an executive had served as CEO (chief executive officer) in a small company and his or her salary plus bonus. Data for a random sample of these CEOs are below, in the form (years as CEO, compensation), with compensation in thousands of dollars.

(6, $391.8)	(6, $261.8)	(15,$252.0)	(8, $204.6)	(4, $205.3)
(10, $216.0)	(3, $212.0)	(9, $369.7)	(9, $475.0)	(7, $314.0)
(13, $421.8)	(4, $88.8)	(8, $615.2)	(2, $141.3)	(1, $141.3)
(11, $176.2)	(26, $245.9)	(13, $400.6)	(7, $198.6)	(7, $193.6)
(12, $323.7)	(4, $158.0)	(2, $184.6)	(22, $294.2)	(7, $268.5)
(19, $241.2)	(20, $299.5)	(2, $103.4)	(3, $310.0)	(20, $163.4)

a. Produce a scatterplot of the data. Make "Years as CEO" the independent variable.

b. Is there a linear relationship between longevity in the position and compensation?

c. Calculate the regression line of compensation as a function of years in the job.

d. Does the line fit the data well?

e. Complete this sentence: Years as a chief executive officer (is/is not) related to compensation; there is (a positive/a negative/no) relationship.

SECTION 3 ▪ THE STANDARD ERROR OF ESTIMATE

We can fit a regression line to any set of numbers—for example, the daily volume of New York City garbage collections versus closing levels for the stock market. However, if a regression line is to help us make predictions for business planning, it must match the observed relationship between the variables reasonably well. Its usefulness is measured by how close a fit it gives between observed and predicted values for Y. In this section, you will learn how to measure variation about the regression line.

The empirical regression line we have defined, $y = b_0 + b_1x$, is the line for which the sum of the squared errors (observed − predicted) is minimized.

$$\text{Sum of the squared errors (observed − predicted), SSE} = \Sigma(Y - Y_c)^2$$

The standard error of the estimate for the regression and the data is a measure derived from SSE, much like the standard deviation for a sample. You will recall that the standard deviation for a sample is given by the formula

$$\text{Standard deviation} = \sqrt{\frac{\sum(x - \bar{x})^2}{n - 1}}$$

Similarly, the formula for the standard error of the estimate of the variable Y as regressed on X is

Computing the standard error of the estimate ($s_{y|x}$)

$$\text{Standard error of the estimate, } s_{y|x} = \sqrt{\frac{\sum(Y - Y_c)^2}{n - 2}}$$

> The **standard error of the estimate** measures the scatter of the data points about the regression line. It is the standard deviation of prediction errors. In regression, $s_{y|x}$ plays the role that the sample standard deviation, s, does in descriptive statistics.

Just as s is an unbiased estimator of σ, $s_{y|x}$ is an unbiased estimator of the standard deviation of u_i in the basic regression model. The numerator is simply the squared error. The denominator is $n - 2$ because two parameters are fitted to the regression model that gives the slope and the intercept. If $n \leq 2$, the regression line either is undefined or runs through each data point and $s_{y|x}$ is not meaningful. The statistic $s_{y|x}$ comes into play when $n \geq 3$, as illustrated in Example 10.3, where this formula is applied to the bingo data set.

EXAMPLE 10.3 ▪ MEASURING VARIATION ABOUT THE REGRESSION LINE FOR THE BINGO DATA

Using the worksheet in Table 10.2, we can calculate the standard error of the estimate for the regression line.

Table 10.2

WORKSHEET FOR CALCULATING THE STANDARD ERROR OF THE ESTIMATE FOR BINGO NET VERSUS NUMBER OF PLAYERS

Number of Players, X	Observed Net, Y	Predicted Net, Y_c	Error, $(Y - Y_c)$	Error Squared
98	585	737.8	−152.8	23,237.84
98	758	737.8	20.2	408.04
.	.	.	.	.
.	.	.	.	.
.	.	.	.	.
113	1159	1107.8	51.2	2,621.44
				94,302.72

Then, from the formula,

$$s_{y|x} = \sqrt{\frac{94{,}302.72}{20}} = 68.67$$

The standard error of the estimate for net revenue as regressed on the number of players is $68.67. This is an unbiased estimator of the standard deviation of the normal disturbance terms, u_i, in the regression model.

Selected statistical software packages can produce standardized residuals for a plot of residuals, dividing the residuals, e_i, by the standard error of the estimate and then making what are usually very small adjustments, depending on the distribution of the values of x_i. For most purposes, you can think of the standard error of the estimate as the standard deviation of the residuals. Thus, roughly 68% of the data points should lie within a band about the regression line, of vertical height twice the standard error of the estimate, as shown in Figure 10.10. If data points fall outside this band in an unusual pattern, this is a signal that the regression assumptions may be wrong. A plot of the standardized residuals provides a quick and informal check of the correctness of your assumptions. It should always be considered if your software offers that option.

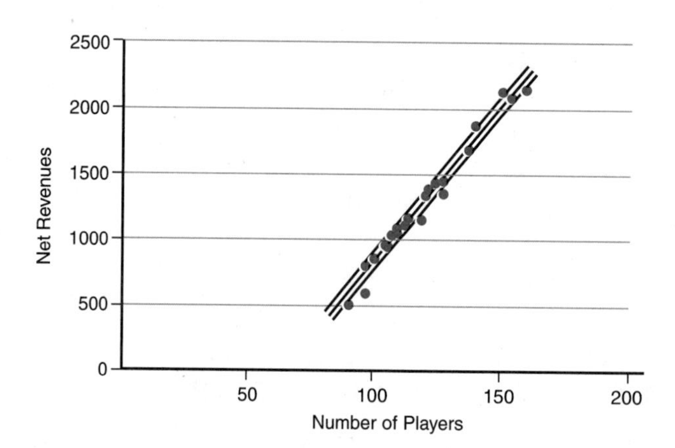

Figure 10.10
One-sigma bands for the bingo regression

Note that the bands in Figure 10.10 are valid for the data in this sample. As noted in the Statistics Tool Kit on regression assumptions, regression lines, and residuals, the regression line calculated from the data may diverge from the underlying population regression line; therefore, additional data may fall further from the regression line than these bands indicate. We will look at prediction bands for regressions in the next section.

The method just described for calculating standard errors is fine as long as there aren't too many observations. Imagine, however, a situation in which there were 100 observations. It would be necessary to calculate 100 predicted response values, 100 differences, and 100 squares and then sum the squares and extract the square root. Each step holds the potential for error. The following formulas, however, can be used to reduce the number of steps necessary and, consequently, the likelihood of making an error.

Shortcut formulas for the standard error of the estimate

1. Calculate $SS_{yy} = \sum Y^2 - \frac{\left(\sum Y\right)^2}{n}$.
2. Calculate $SSE = SS_{yy} - b_1 SS_{xy}$.
3. Calculate $s_{y|x} = \sqrt{\frac{SSE}{n-2}}$.

As noted earlier, when large data sets are involved, software is generally used to carry out the calculations.

EXAMPLE 10.4 ■ APPLYING THE SHORTCUT FORMULA TO THE BINGO DATA

To apply the shortcut formula to the bingo data set, we need only one additional figure:

$$\sum Y^2 = y_1^2 + y_2^2 + \cdots + y_{22}^2$$
$$= 585^2 + 758^2 + \cdots + 1159^2 = 40{,}458{,}369$$

Using the three formulas just given, as well as $\Sigma Y = 28{,}035$, $SS_{xy} = 187{,}836.91$, and $b_1 = 24.69$ from Table 10.1, we can quickly find the standard error of the estimate.

Step 1. We calculate SS_{yy}:

$$SS_{yy} = 40{,}458{,}369 - \frac{(28{,}035)^2}{22} = 4{,}732{,}858.77$$

Step 2. We calculate SSE:

$$SSE = 4{,}732{,}858.77 - 24.69(187{,}836.91) = 95{,}165.46$$

Step 3. We calculate $s_{y|x}$:

$$s_{y|x} = \sqrt{\frac{95{,}165.46}{20}} = 68.98$$

What $s_{y|x}$ provides is an estimate of the standard deviation of Y about its true mean for any fixed value of x. This is our measure of the scatter about the regression line of X. For the bingo data, $b_1 = \$24.69$ per player added. So if we added a player, we would expect to see, on average, about \$24.69 more in revenue. However, $s_{y|x} = \$68.98$, by our estimate. This is what we would expect for the standard deviation of the values of Y for a fixed value of x. Hence, the effects of recruiting a single extra player could easily be overcome by random variations. But recruiting 40 extra players would result in an expected gain of $40 \times \$24.69 = \$987.60$. So even with a scatter of \$68.98, we would expect to see a substantial increase in revenues if 40 more players were recruited. We will look at this issue further in Section 5 on forecasting, where we will take the uncertainties in b_1 into account as well.

SECTION 4 ■ COEFFICIENTS OF CORRELATION AND DETERMINATION

The *coefficient of correlation, r,* and the *coefficient of determination, r^2,* are two additional measures that indicate how well a regression model describes a data set. The coefficient of correlation (sometimes called the Pearson product moment correlation coefficient) tells the analyst whether a relationship is direct ($r > 0$) or inverse ($r < 0$) and how strong the relationship is. The coefficient of determination, r^2, measures explained variance, which is that fraction of the variance of Y that arises in response to the variation of X in accordance with the linear regression relationship.

The coefficient of correlation, r, varies between -1 and $+1$. A value of -1 indicates a perfect inverse relationship. An example of an inverse relationship is the ideal relationship between the quantity demanded of a product and the price of that product—the kind of relationship you might find in a demand curve that describes the quantity of a product purchased as a function of the price per unit. A value of $+1$ indicates a perfect direct relationship. An example of a direct relationship is the ideal relationship between quantity supplied and price—the kind of relationship you might find in a supply curve. See Figure 10.11.

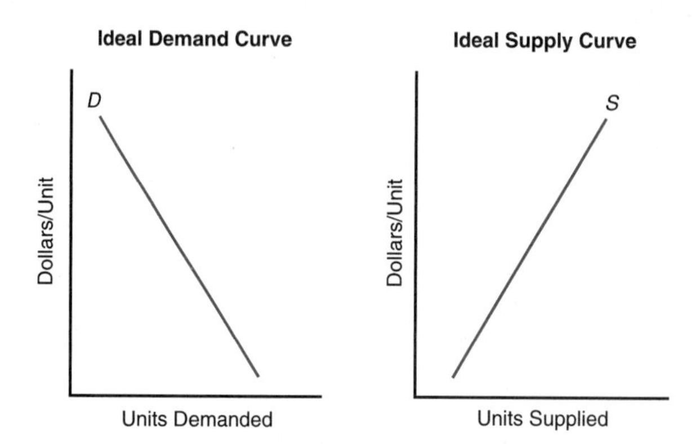

Figure 10.11
Inverse and direct relationships

If the linear relationship between the variables is nonexistent or weak, the coefficient of correlation will be at or near zero. A strong correlation will be indicated by r near either -1 or $+1$. Figure 10.12 shows examples of data sets with different correlation coefficients. It is important to note that, because r is blind to nonlinear relationships, a perfect relationship may exist between two variables and r may equal zero. The correlation coefficient can only "see" linear relationships.

The coefficient of determination plays a very important role in multiple regression analysis, which will be discussed in Chapter 11. It is used as a model-building tool.

The scatterplots in Figure 10.12 each have 500 points (x_i, y_i). For each plot, the values of x_i were independently chosen from a standard normal distribution. The values of y_i were generated using the formula $y_i = \beta_0 + \beta_1 x_i + u_i$, where the values of u_i were independently selected from a normal population with mean 0 and a standard deviation chosen so that the population standard deviation, σ_y, would be 1. For these particular samples, β_0 was chosen to be 0; hence, $\mu_y = 0$.

The procedure just described is a general model for producing bivariate normal data, except that the values of x_i could belong to any fixed normal population, β_0 could be arbitrary, and there is no necessity to ensure that $\sigma_y = 1$. These particular choices were made to ensure that each cloud of data points would fit neatly into a square six sigmas wide, centered on (0, 0). Figure 10.12 gives a general sense of what bivariate normal data look like. The only difference between these plots and a completely general bivariate normal data cloud is in the degrees of horizontal and vertical stretching and compressing.

Keep in mind that an r close to 1 does not necessarily imply a cause-and-effect relationship. Although it is known that, among children, shoe size and spelling abil-

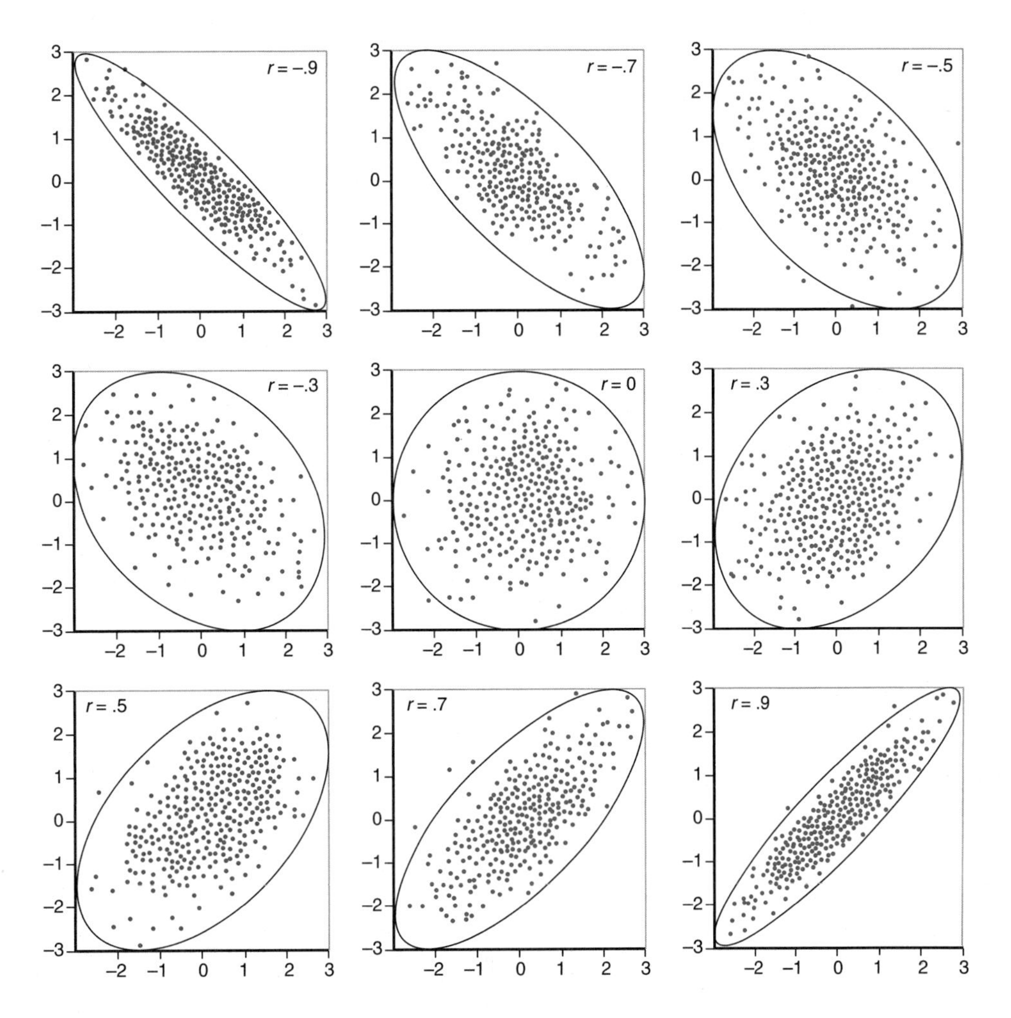

Figure 10.12 *Scatterplots of bivariate normal data having various values of* r

ity are highly correlated, no one believes that either in any way causes the other. The fact is that older children tend to have larger feet and to spell better. A value of r^2 close to 1 may suggest a causal relationship, but it does not prove one.

The correlation coefficient is found by taking the variance of Y from $y_c = b_0 + b_1x$ and dividing it by the square root of the product of the variances of X and Y:

$$r = \frac{SS_{xy}}{\sqrt{SS_{xx}SS_{yy}}}$$

Note that we have already calculated all the values needed to determine the coefficient of correlation.

> The **coefficient of correlation,** *r*, measures the strength of the relationship between two variables. If $r = \pm 1$, the relationship is perfectly linear.

Example 10.5 illustrates calculation of the coefficient of correlation for the bingo data set.

EXAMPLE 10.5 ■ CALCULATING THE COEFFICIENT OF CORRELATION FOR THE BINGO DATA SET

Calculating the correlation coefficient, r

For the bingo data set, we already know that SS_{xy}, the variance of y from y_c, is 187,836.91; SS_{xx}, the variance of x, is 7606.36; and SS_{yy}, the variance of y, is 4,732,858.77. Thus,

$$r = \frac{187{,}836.91}{\sqrt{(7{,}606.36)(4{,}732{,}858.77)}} = .99$$

The relationship is very strong: Revenues are closely tied to number of players. It is also direct: As the number of players increases, revenues increase. Note, however, that such a strong relationship is the exception and not the rule. When results near 1.00 appear, it is a good idea to first check your calculations and then check the definitions of the variables involved. It is possible that a perfect or nearly perfect relationship has appeared because you are regressing a derivative of a variable on itself. For example, gross domestic product regressed on national income would yield this type of result, because national income is derived from GDP. In the case of the bingo players, neither of the two variables is calculated from the other.

To answer the question "How much of the variation in revenues appears to be associated with the number of players?", the coefficient of determination should be calculated. The coefficient of determination, r^2, is simply the square of the coefficient of correlation.

> The **coefficient of determination,** r^2, is the fraction of the variance of Y that is accounted for by the variations of the predictor variable, X, and the regression $y = b_0 + b_1x$.

Example 10.6 shows calculation of the coefficient of determination for the bingo data set.

EXAMPLE 10.6 ■ CALCULATING THE COEFFICIENT OF DETERMINATION FOR THE BINGO DATA SET

Calculating the coefficient of determination, r^2

The required input for the bingo data set, $r = .99$, has already been calculated. We find the coefficient of determination as follows:

$$r^2 = .99 \times .99 = .98$$

The coefficient of determination indicates that the linear relationship between the number of players and revenues accounts for 98% of the variance in the revenues observed. It should be noted that the term *accounts for,* in the statistical sense, *does not imply causality,* though in this case it is reasonable to think of changes in the number of players as causing a change in income. (When we use several predictor variables for multiple regression in Chapter 11, we will use a different method for analysis of variance in order to determine the percentage of the variance accounted for by each of the predictor variables.)

EXERCISES 10.30–10.44

Skill Builders

(*Note:* These exercises utilize the same data as Exercises 10.16–10.29.)

10.30 You are given these partial regression results:

$$\Sigma X = 51 \qquad \Sigma Y = 49 \qquad \Sigma XY = 277$$

$$\Sigma X^2 = 357 \qquad \Sigma Y^2 = 297 \qquad n = 10$$

a. Find the standard error of the estimate for the regression line.
b. Calculate the value of the coefficient of correlation.
c. Is this a strong relationship? Explain.

10.31 Consider the intermediate regression results below:

$$\Sigma X = 518 \qquad \Sigma Y = 501 \qquad \Sigma XY = 33{,}888$$

$$\Sigma X^2 = 37{,}688 \qquad \Sigma Y^2 = 32{,}061 \qquad n = 10$$

a. Find the standard error of the estimate for the regression line.
b. Compute the coefficient of determination.
c. How much of the variation in Y is accounted for by the linear relationship between X and Y?

10.32 From the examination data in Exercise 10.5, the following results were obtained:

$$SS_{xy} = 749.4 \qquad SS_{yy} = 2634.9$$

$$SS_{xx} = 1548.4 \qquad n = 10$$

a. Compute the standard error of the estimate.
b. Find r.
c. Is this a strong relationship?

10.33 Consider these results:

$$\overline{X} = -.2333 \qquad \overline{Y} = .4050 \qquad n = 6$$

$$\Sigma X^2 = 2.8064 \qquad \Sigma Y^2 = 1.9413 \qquad \Sigma XY = -.3762$$

a. Calculate the standard error of the estimate.
b. Compute r and r^2.
c. What do r and r^2 tell you?

10.34 Use these partial results:

$$b_1 = .0794 \qquad b_0 = 287.451$$

$$SS_{xx} = 720{,}863.875 \qquad SS_{yy} = 422{,}396 \qquad n = 8$$

a. Find the standard error of the estimate.
b. Find r.
c. Explain the significance of r.

Applying the Concepts

10.35 The inputs needed for regression analysis of Anheuser-Busch's sales versus advertising costs appear below. Recall that X represents advertising costs and Y represents sales volume. (These figures are based on data in Exercise 10.10.)

$$\Sigma X = 66{,}519 \qquad \Sigma Y = 11{,}518.20 \qquad \Sigma XY = 94{,}062{,}341.32$$
$$\Sigma X^2 = 516{,}134{,}816.34 \qquad \Sigma Y^2 = 17{,}471{,}929.38 \qquad n = 9$$

Note: A good statistical calculator will yield all these results in just one pass through the data set!

a. Find the standard error of the estimate.
b. Compute r and r^2.
c. Is advertising closely associated with sales?
d. Complete this paragraph: About 68% of the data fall within a range of $ _____ of the regression line. Advertising (is/is not) closely associated with sales. Evaluating r gives a value of _____ for the strength of the relationship. Advertising explains _____ % of the sales.
e. Based on this analysis, what would be your recommendations to management about advertising budgets?

10.36 Use the information below concerning attendance at New York Yankee games. (The figures are derived from data in Exercise 10.7.)

$$SS_{xy} = 8.96 \qquad SS_{yy} = 930.40$$
$$SS_{xx} = .78 \qquad n = 15$$

a. Find the standard error of the estimate.
b. Compute r and r^2.
c. Is attendance closely associated with previous year's wins?
d. Would you use this regression, in light of your calculations? Why or why not?
e. Why are SS_{xx} and SS_{yy} never negative?

10.37 In a study of retail sales of new cars, both domestic and imported, the following data were gathered. (Retail sales are in millions of units, seasonally adjusted. Per capita personal income is in constant 1982 dollars, and interest rates are for U.S. Treasury notes with three-month maturities; both are for the previous quarter.)

Year: Quarter	Retail Sales	Per Capita Personal Income	Interest Rate
1987: I	9.68	10,909	5.34
II	10.20	10,982	5.53
III	11.20	10,779	5.73
IV	9.87	10,927	6.03
1988: I	11.00	11,097	6.01
II	10.63	11,268	5.76
III	10.37	11,320	6.23
IV	10.57	11,424	6.99
1989: I	9.90	11,458	7.70
II	10.27	11,553	8.54
III	10.67	11,492	8.44
IV	8.80	11,538	7.85

(cont.)

1990: I	9.81	11,541	7.63
II	9.53	11,586	7.76
III	9.73	11,564	7.77

a. Draw a scatterplot of the retail sales and per capita personal income data.
b. Is there a linear relationship between retail sales and per capita personal income?
c. Calculate the least squares regression line.
d. Calculate the standard error of the estimate.
e. Compute r and r^2.
f. Using the interest rate as the predictor variable and retail sales as the response variable, repeat parts a through e.
g. Compare the standard errors for the regressions. For which regression is the standard error of the estimate smaller?
h. Compare r and r^2 for the regressions. For which regression is r^2 higher?
i. Complete this paragraph: A study of retail sales of automobiles revealed that the relationship between per capita personal income and retail sales between 1987 and 1990 is a (weak/strong) one. While per capita personal income rose, retail sales (rose slightly/remained about level/fell slightly). The linear relationship between per capita personal income and retail sales explains _____% of the variation in retail sales. Interest rates exhibited a (weaker/stronger) relationship to retail sales than did per capita personal income. The relationship was (direct/inverse). A 1% increase in the interest rate is associated with a _____% (increase/decrease) in retail automobile sales.
j. Discuss the possible implications of this information for a manager of an automobile agency.

10.38 The money market is an important source of short-term funds, as well as a place in which temporary surpluses can be put to work. Money managers in corporations, financial institutions, and governments would all like to be able to predict the course of money market rates. Use regression to look at the data below, relating the interest rates for Treasury bills (T-bills) and federal funds from 1971 to 1984 to the money supply (in billions of dollars). (See D. C. Colander's *Economics* [Irwin, 1993] or C. R. McConnell's *Economics,* 10th ed. [McGraw Hill, 1987] for more on this subject.)

Year	T-Bill Rate	Federal Funds Rate	Money Supply
1971	4.511	4.66	903.1
1972	4.466	4.43	1023.1
1973	7.178	8.73	1142.6
1974	7.926	10.50	1250.3
1975	6.122	5.82	1367.0
1976	5.266	5.04	1516.7
1977	5.510	5.54	1705.5
1978	7.572	7.93	1911.2
1979	10.017	11.19	2119.6
1980	11.374	13.36	2327.8
1981	13.776	16.38	2599.4
1982	11.084	12.26	2853.5
1983	8.700	9.09	3155.5
1984	9.800	10.23	3523.4

a. Draw a scatterplot of T-bill rates versus money supply.
b. Is there a linear relationship?
c. Calculate the least squares regression line.
d. Calculate the standard error of the estimate.
e. Compute r and r^2.
f. Repeat parts a through e, using the federal funds rate instead of the T-bill rate.
g. Complete this paragraph: Money supply appears to have (no/some/a strong) relationship to interest rates as reflected in T-bills. It appears that there (exists/does not exist) a linear relationship between the two variables. An increase in money supply is (directly/inversely) related to the T-bill rate. The linear relationship between money supply and T-bill rates explains _____% of the variation in T-bill rates. (In contrast to/Supporting this) finding is the relationship between money supply and the federal funds rate. It appears that there (exists/does not exist) a linear relationship between money supply and the federal funds rate. The linear relationship between money supply and the federal funds rate explains _____% of the variation in the latter.

10.39 Industry revenues are useful predictors of the revenues of particular companies. Cray Research was a manufacturer of supercomputers. Data (in millions of dollars) on its revenues and its engineering and development expenditures for 1982 through 1988, as well as on industry revenues, are found below.

Year	Cray Revenues	Eng. & Dev. Expenditures	Industry Revenues
1982	141.15	29.15	11,366
1983	169.69	25.54	12,989
1984	228.75	37.54	17,227
1985	380.16	49.17	15,607
1986	596.69	87.63	15,800
1987	687.34	108.83	19,100
1988	756.31	117.76	24,000

a. Draw a scatterplot relating Cray's revenues to its engineering and development expenditures.
b. Is there a linear relationship?
c. Calculate the least squares regression line.
d. Calculate the standard error of the estimate.
e. Compute the coefficient of correlation, r, and the coefficient of determination, r^2.
f. Repeat parts a through e, relating Cray's revenues to industry revenues.
g. Considering r, r^2, and the standard errors, decide which regression provides a better forecast of Cray's revenues.
h. Complete this paragraph: Examining Cray's revenues relative to its engineering and development expenditures reveals (a linear/a nonlinear/no) relationship. There is a (positive/negative) relationship. For each dollar spent on engineering and development, revenues will (increase/decrease) by $_____. Industry revenues exhibit a (positive/negative) relationship to Cray's revenues. Industry revenues explain _____% of the variation in

Cray's revenues. Each dollar spent in the industry translates to $_____ in revenues for Cray. If I were asked to forecast Cray's revenues, I would use (engineering and development expenditures/industry revenues) as the predictor variable.

10.40 Reexamining the cruise ship data, consider the following partial results. (The original data are presented in Exercise 10.8.)

$$\Sigma X = 1338.33 \qquad \Sigma Y = 10{,}139{,}277 \qquad \Sigma X^2 = 149{,}992.63$$
$$\Sigma Y^2 = 8.67274427\text{E}12 \qquad \Sigma XY = 1{,}134{,}481{,}303.67 \qquad n = 12$$

a. Calculate the least squares regression line.
b. Draw the line on the scatterplot.
c. Find the standard error of the estimate.
d. Is the line a good fit for the data?
e. Find the coefficient of correlation and the coefficient of determination.
f. Is the relationship weak or strong?
g. Is the Consumer Confidence Index linearly related to the number of cruise ship passengers?

10.41 Partial results for the regression of the misery index as a function of a change in GDP appear below. (The original data were presented in Exercise 10.11.)

$$SS_{yy} = 538.09 \qquad SS_{xx} = 150.05$$
$$SS_{xy} = -125.62 \qquad n = 11$$

a. Find the standard error of the estimate.
b. Draw the 2σ bands around the line.
c. Did the misery index fall outside the bands during any administration?
d. Calculate r and r^2.
e. Is change in GDP linearly related to the misery index?

10.42 Sales and profit data (in millions of dollars) for the first quarter of 1992 for corporations in durable manufacturing appear below:

Industry	Sales	Income After Tax
Stone, Clay & Glass	10,978	−575
Fabricated Metals	32,448	1060
Machinery (excl. electrical)	62,245	1632
Transportation Equipment	87,289	722
Instruments	25,492	2138
Electrical & Electronic Equipment	52,740	1847
Primary Metals	28,193	260
Other Durables	25,415	678

a. Construct a scatterplot of the data. Let "Sales" be the independent variable.
b. Calculate the regression line.
c. Draw the regression line on your scatterplot.
d. Calculate the standard error of the estimate.
e. Find r and r^2.
f. Are sales a good predictor of income after taxes? Why?

10.43 In January 1993, *Consumer Reports* published an article on frozen light entrees. Data on various light lasagna entrees follow:

Entree	Calories	Fat (g)	Cholesterol (mg)
Budget Gourmet Three Cheese Lasagna	390	17	70
Kraft Lasagna	347	14	45
Swanson	400	12	77
Weight Watchers Italian Cheese Lasagna	290	7	20
Budget Gourmet Light	290	11	40
Weight Watchers Meat Sauce Lasagna	270	6	5
Weight Watchers Garden Lasagna	260	7	15
Stouffers Single Serving	360	13	46
Healthy Choice Zucchini	240	6	15
Lean Cuisine Lasagna	260	5	25

a. Draw a scatterplot relating calories to fat. Let "Fat" be the independent variable.

b. Calculate the regression line for the data.

c. Find the standard error of the estimate for the regression.

d. Calculate r and r^2.

e. Repeat the process, relating calories to cholesterol. Let "Calories" be the independent variable.

f. Complete this paragraph: Examination of the relationship between calories and fat in frozen light lasagna entrees reveals a (positive/negative) relationship. Cholesterol content is (positively/negatively) associated with calories. It appears that (fat content/cholesterol content) is more closely associated with calories.

10.44 Partial results of the regression relating Wal-Mart's net sales, Y, to U.S. disposable income, X, appear below. (The original data are found in Exercise 10.24.)

$$n = 9 \qquad \Sigma X^2 = 75{,}991{,}882 \qquad \Sigma Y^2 = 1{,}639{,}850{,}075$$
$$\Sigma XY = 319{,}986{,}238 \qquad \Sigma X = 25{,}738 \qquad \Sigma Y = 99{,}619$$

a. Calculate the standard error of the estimate.

b. Assess the fit of the regression line to the data.

c. Are sales closely correlated with disposable income?

d. Complete this paragraph: Disposable income proves to have (little/moderate/much) influence on Wal-Mart's sales. There is a (positive/negative) relationship between disposable income and sales. The (strong/weak) association between these variables is borne out by the fact that ________% of the variation in sales is explained by its linear relationship with disposable income. Disposable income appears to be a (good/poor) predictor of sales for Wal-Mart.

SECTION 5 ■ FORECASTING WITH SIMPLE REGRESSION MODELS

If regression yields a value close to 1 for r^2, most of the variation in the mean of y can be attributed to changes in x. The next logical step is to use the regression line to make predictions for a given x. This section explores procedures for predicting individual and average values for the response variable.

The Uses and Limitations of Predictions Based on Models

One of the purposes of regression analysis is to predict values for the response variable, based on the values of the predictor variable. In the case of the bingo data, we were interested in knowing what revenues could be expected, given a certain number of players. If revenues fell short of expectations, organizers might want to look into whether someone was dipping into the "take"; if revenues exceeded expectations, they might want to know what particularly good coincidence of factors was contributing to the high net revenues. A further look would be in order in either case, to determine what went wrong or to lock in what went right.

Using linear regression for forecasting is one application of the more general idea of mathematical modeling, which is discussed in more detail in Chapter 11.

> A **mathematical model** is a simplified representation of the relationships between variables, expressed through the use of equations.

Projections, generally speaking, should not be made too far outside the range of the data. If the predictor variable for the bingo data has a range of, say, 90 to 150 players, projections should not be made for less than 55 players or more than 195 players. Going no more than 50% beyond the range in either direction represents a prudent limit on projections or forecasts. Beyond that range, it is likely that the nature of the relationship between variables will change. Indeed, with respect to the bingo players, any number of players below 80 would yield negative revenues, while the bingo hall might have an occupancy limit of 200 players. In many economic relationships, economies or diseconomies of scale come into play as we move significantly outside the range of existing data. These economies or diseconomies of scale change the nature of the relationship.

Prediction Intervals

Because all data used for forecasting may be thought of as sample data, even projections within the range of the model's validity are subject to sampling error. Recall that sampling error is the variation in the values of observed statistics due to the random nature of the sample (see Chapter 6). Point estimates will, most likely, be incorrect. However, point estimates do represent the center of a range. The range for an estimate of an individual value is called a **prediction interval.**

Prediction intervals are defined for a fixed value x_0 of the variable X and a given probability, $1 - \alpha$. They are centered on the value of Y predicted by the regression equation, $y_c = b_0 + b_1x_0$, when $X = x_0$. They are made wide enough so that y will fall in the given interval with probability $1 - \alpha$ when $X = x_0$. Remember that our model is $y_i = \beta_0 + \beta_1x_i + u_i$, where β_0 and β_1 are population parameters, the values of x_i are normally distributed, and the values of u_i are normally distributed random fluctuations that move y_i away from the predicted values of $\beta_0 + \beta_1x_i$. In typical regression situations, the values of x_i might be evenly spaced within a range, or the data might come from a bivariate normal distribution, as in Figure 10.12. In either case, there is good coverage of the range of the predictor variable.

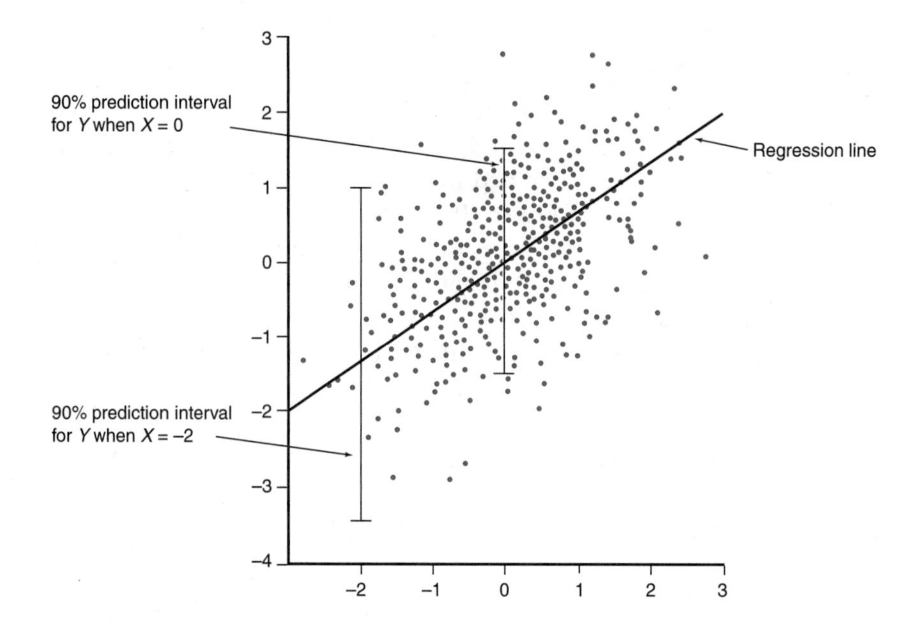

Figure 10.13
90% prediction intervals about a regression line

In Figure 10.13, the prediction interval is narrower near the center of the cloud of data points. Here, $\alpha = 10\%$ and $1 - \alpha = 90\%$ (see also Figure 10.14, later in this section).

If prediction intervals are to be wide enough to have probability $1 - \alpha$ of including points, three independent sources of variance must be taken into account in determining their width: the variance of the values of u_i, the variance arising from the approximation of β_0 by b_0, and the variance arising from the approximation of β_1 by b_1. The result may look complex, but the idea is simple. In practice, such calculations are usually done by computer.

The prediction interval for the predicted variable, Y, given that the predictor variable, X, is equal to x_0, is an interval about $y_c = b_0 + b_1 x_0$ that has probability $1 - \alpha$ of containing the observed values of Y for this X. It is given by

$$y_c \pm t_{\alpha/2,n-2}\, s_{y|x}\sqrt{1 + \frac{1}{n} + \frac{(x_0 - \bar{x})^2}{SS_{xx}}} \tag{7}$$

where n is the size of the sample used to determine b_0 and b_1, $t_{\alpha/2,n-2}$ is the critical value for t at a level $\alpha/2$ and with $n - 2$ degrees of freedom, $s_{y|x}$ is the standard error of the estimate for y regressed on x, and SS_{xx} is the sum of squares for x.

This formula gives an interval centered on the predicted value, y_c. The width of this interval depends on the critical value $t_{\alpha/2,n-2}$ for the t distribution ($n - 2$ is the proper number of degrees of freedom for a regression based on n data points and fitting two parameters, β_0 and β_1, with b_0 and b_1). Such a t-distributed variable has probability $1 - 2(\alpha/2) = 1 - \alpha$ of falling between $-t_{\alpha/2,n-2}$ and $+t_{\alpha/2,n-2}$, and this is what we want. The easiest way to understand the product by which the critical value $t_{\alpha/2,n-2}$ is multiplied is to distribute $s_{y|x}$ under the square root sign, as $s^2_{y|x}$. You then have the sum of three terms. The first, $s^2_{y|x}$, is the variance of the values of u_i. The second, $s^2_{y|x}/n$, is the variance arising from sampling distribution errors when the sample estimate, b_0, is used in place of the population parameter, β_0. The third, $\frac{(x_0 - \bar{x})^2 s_{y|x}}{SS_{xx}}$, is the variance arising from using b_1 in place of the parameter β_1. This third term grows larger as the difference between x_0 and x increases, as you would expect, since it represents deviations arising from having the wrong slope for the regression line. The other variance terms remain constant as x varies. As a result, these intervals become larger the farther x_0 is from x, as we will see when we graph these intervals later.

In Example 10.7, we work out a prediction interval, drawing on quantities calculated in earlier examples.

EXAMPLE 10.7 ▪ FINDING A PREDICTION INTERVAL FOR NET REVENUES FROM BINGO

The determination of a prediction interval for bingo revenues is an extension of our work in Example 10.1, where we had $b_0 = -1681.82$ and $b_1 = 24.69$. We also found in Example 10.1 that $SS_{xx} = 7606.36$, $\Sigma x = 2634$, and $n = 22$, so $\bar{x} = 119.73$. From Example 10.3, we have $s_{y|x} = 68.67$ for the standard error of the estimate. This is all the information we need for point and interval estimates of y. We will work them out for $x_0 = 100$ players. The regression $y_c = b_0 + b_1 x_0$ gives

$$y_c = -1681.82 + 24.69x_0 = 787.18$$

Thus, a point estimate of revenues would be \$787. To find the margin of error, allowing for sampling variation (sampling error), we turn to formula (7) above. We need to find a t value, which is a function of the sample size and the confidence level we wish to assign to our prediction. At 95% confidence, $1 - \alpha = .95$, so $\alpha = .05$. For a two-variable regression with 22 observations, the critical value for t is $t_{.025,20}$, where the first subscript is $\alpha/2$ and the second is $n - 2$. Substituting into expression (7), we have

Calculating prediction intervals

$$787.18 \pm 2.086(68.67)\sqrt{1 + \frac{1}{22} + \frac{(100 - 119.73)^2}{7606.36}} = 787.18 \pm 150.01$$

Thus,

$$P(637 \leq Y_{x=100} \leq 937) = .95$$

It may be expected, with 95% confidence, that for 100 players net revenues in any given session will fall between \$636 and \$938. This is the prediction interval for the estimate.

Confidence Intervals

Usually, we are not concerned with a single instance, but with average tendencies. That is, we would be more interested in the average revenue brought in by bingo sessions involving 100 players than in the highs and lows for such sessions. Calculation of the confidence interval for the mean is almost identical to that of the prediction interval. The only difference is that one source of variance—that arising from u_i—is absent. The reason the term does not show up in the confidence interval for the mean of Y is that the values of u_i, being normally distributed with a mean of zero, contribute zero to the mean of Y. Therefore, the first term under the radical in formula (7), the 1, is eliminated.

The following formula is for a $1 - \alpha$ confidence level for the response variable, Y, given that the predictor variable, X, is equal to x_0 in the regression.

$$y_c \pm t_{\alpha/2,n-2} s_{y|x} \sqrt{\frac{1}{n} + \frac{(x_0 - \bar{x})^2}{SS_{xx}}} \tag{8}$$

> The **confidence interval** is the expected range for the population average value of the response variable.

In Example 10.8, we find a confidence interval for the same 100-player bingo session studied in Example 10.7.

EXAMPLE 10.8 ▪ FINDING THE CONFIDENCE INTERVAL FOR THE BINGO REVENUES

From Example 10.7, we have the regression equation

$$y_c = -1681.82 + 24.69x_0 = 787.18$$

To find the confidence interval for bingo revenues for $x_0 = 100$ players, we substitute values from Example 10.7 into formula (8):

Calculating confidence intervals

$$787.18 \pm 2.086(68.67)\sqrt{\frac{1}{22} + \frac{(100 - 119.73)^2}{7606.36}} = 787.18 \pm 44.53$$

Thus,

$$P(743 \leq \mu_y \leq 832) = .95$$

On average, we may expect with 95% confidence that 100-player sessions will bring in revenues of between \$743 and \$832. The implications of this confidence interval are very useful. Specifically, if bingo games are conducted for, say, 40 weeks per year and we can count on 100 players per session, then, based on average revenues, we may expect the games to gross between \$29,706 and \$33,280.

As noted earlier and shown in Figure 10.14, the farther we move from x with our predictions, the wider the confidence and prediction bands are about the regression line. In contrast, the larger SS_{xx} is, the narrower the confidence and prediction bands are. The reason is that the more spread out the x values are from $\bar{x}$, the more spread out the data points are horizontally and the better our estimate will be for β_1, the slope of the regression line. These observations are consistent with common sense. In our model, the data are concentrated about $(\bar{x}, \bar{y})$. As we move farther from the area of greatest certainty, where the data are concentrated, our estimates are likely to become less precise.

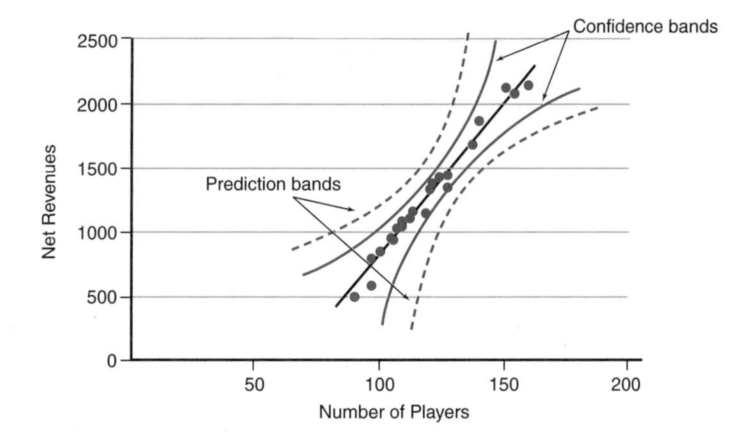

Figure 10.14
Prediction and confidence bands for the bingo revenue projection

EXERCISES 10.45–10.54

Skill Builders

In these exercises, the coefficients given are b_0 and b_1.

10.45 Consider the following results from a regression analysis:

$$Y = 3.47 + .28X$$

$$s_{y|x} = 2.48 \qquad n = 10 \qquad SS_{xx} = 96.9 \qquad \bar{x} = 5$$

a. If $x_0 = 6$, find the point estimate for Y.
b. Calculate the 95% prediction interval for the value of Y, given $x_0 = 6$.
c. Calculate the 95% confidence interval for Y, given $x_0 = 6$.
d. What is the difference in the interpretation of parts a, b, and c?

10.46 You are given this equation and the accompanying results:

$$Y = 12.29 + .73X$$

$$SSE = 1167.47 \qquad n = 10 \qquad SS_{xy} = 7936.2$$

a. For $x_0 = 41$, calculate the point estimate for Y.
b. Find the 90% prediction interval for Y when $x_0 = 41$.
c. Find the 90% confidence interval for Y when $x_0 = 41$.
d. Repeat parts a through c for $x_0 = 8$ and $x_0 = 95$.
e. Draw the forecasting bands around a plot of the regression line.
f. Let $x_0 = 150$, and repeat parts b and c.
g. Extend the bands in part e to cover your results in part f. What does this imply?

10.47 Use the results below:

$$r = .1238 \qquad SS_{xy} = .1907 \qquad SS_{yy} = .9572 \qquad n = 10$$

a. At $x_0 = -.43$, find a point estimate for y when $X = x_0$.
b. Calculate the 95% prediction interval for y when $x_0 = -.43$.
c. Calculate the 95% confidence interval for y when $X = x_0$.

10.48 You are given these partial results:

$$b_1 = .0794 \qquad SS_{xx} = 720{,}863.875$$
$$SS_{yy} = 422{,}396 \qquad n = 8 \qquad b_0 = 287.451$$

a. Assume $x_0 = 100$. Find the point estimate for y.
b. Find the 90% prediction interval at x_0.
c. Calculate the 90% confidence interval at x_0.

Applying the Concepts

10.49 The results from the statistics examination data are repeated below. (The original data can be found in Exercise 10.5.)

$$Y = 49.0 + .484X$$
$$\bar{x} = 74.6 \qquad n = 10$$
$$SS_{xy} = 749.4 \qquad SS_{yy} = 2634.9 \qquad SS_{xx} = 1548.4$$

a. If John Mills received a grade of 86 on Exam 1, estimate his grade on Exam 2.
b. Calculate the 99% prediction interval for his grade on Exam 2.
c. Find the 99% confidence interval for the average grade on Exam 2 for all those receiving an 86 on Exam 1.

10.50 In relating Woolworth's revenue and profit data for subdivisions (see Exercise 10.6), we found

$$\text{Profits} = 18.6 + .0880 \times \text{Revenues}$$
$$s_{y|x} = 44.58 \qquad SS_{xx} = 4{,}349{,}823.71 \qquad n = 7$$

a. What is the point estimate of profits when revenues are $1.5 billion?
b. Calculate the 95% prediction interval for an individual observation when revenues are $1.5 billion.
c. Calculate the 95% confidence interval for the mean when revenues are $1.5 billion.
d. Repeat parts a through c for revenues of $3 billion.
e. Complete this paragraph: When revenues are $1.5 billion, profits are estimated to be $_____. The range for an individual year is from as low as $_____ to as high as $_____. On average, profits can be expected to fall in the $_____ to $_____ range.
f. What events in 1997 might have affected these predictions? (*Hint:* The problem with extrapolation of regression is not solely moving out of range, but also supposing that old relations hold in new times.)

10.51 Our beauty salon accountant wishes to predict sales for a shop with a payroll of $150,000. (Exercise 10.9 contains the full data set.) He arrived at the following equation, in which payroll is in thousands:

$$\text{Sales} = 22.7 + 1.70 \times \text{Payroll}$$
$$s_{y|x} = 114.5 \qquad \Sigma X = 7536 \qquad \Sigma X^2 = 15{,}038{,}048 \qquad n = 10$$

a. What is the point estimate for sales?
b. Calculate the 90% confidence interval for average sales.

c. Calculate the 90% prediction interval for sales.

d. Complete this paragraph: With a payroll of $150,000, it is expected that sales will be about $____. We can say with 90% confidence that sales could be as high as $____ or as low as $____. On average, however, we can say with 90% confidence that sales will fall between $____ and $____.

10.52 Consider the Anheuser-Busch data once again. (The full data set is found in Exercise 10.10.) The necessary equation and other useful information are given below:

$$\text{Sales} = 2094 + 4.01 \times \text{Advertising}$$

$$s_{y|x} = 392.3 \qquad SS_{xx} = 2{,}467{,}795 \qquad n = 9$$

a. Suppose advertising expenditures are increased from $1876.8 million to $2000 million. How much may you expect sales to increase?

b. Find the 99% prediction interval for the new level of sales.

c. Find the 99% confidence interval for the average new level of sales.

d. Compare results at the old level of advertising and the new level. Is it conceivable that sales will not increase at all? Comment on your assumptions.

10.53 An investor wanted to relate dividends paid to net earnings for the General Electric Company for 1980 through 1984. The data she collected appear below. (All dollar amounts are in millions.)

Year	Dividends, *Y*	Net Earnings, *X*
1980	$670	$1514
1981	715	1652
1982	760	1817
1983	852	2024
1984	930	2280

a. Draw a scatterplot of the data.

b. Calculate the least squares regression line.

c. Calculate the standard error of the estimate.

d. Calculate r and r^2.

e. If net earnings are $1.9 billion, calculate average dividends with 95% confidence.

f. Complete this paragraph: A graphic analysis of the data indicates a (positive/negative) relationship between dividends and net earnings. An increase of $1 million in net earnings can be expected to yield an increase in dividends of $____. About 68% of the data can be expected to fall within $____ of the regression line. The correlation between net earnings and dividends can be said to be (weak/moderate/strong). The linear relationship between net earnings and dividends can be said to account for ____% of the variation in dividends.

10.54 Information on the New York Yankee attendance problem appears below. (See Exercise 10.7 for the full data set.)

$$SS_{xx} = .78 \qquad SS_{yy} = 930.4 \qquad SS_{xy} = 8.96 \qquad n = 15$$

a. Calculate $s_{y|x}$.

b. Find r^2.

c. Predict the approximate attendance with 95 wins in the previous season.
d. Calculate the 90% prediction interval for 95 wins.
e. Complete this paragraph: A study of the relationship between previous wins and current attendance suggests a (strong/moderate/weak) relationship. The linear relationship between previous wins and attendance explains ____% of the variation in attendance. Based on this analysis, it is expected that, on average, 95 wins in the previous season will generate an attendance of between ____ and ____.

SECTION 6 ▪ ASSESSING THE SIGNIFICANCE OF SLOPE ESTIMATES

When we conducted tests of hypotheses in Chapters 8 and 9, we examined a variety of claims concerning estimates of parameters. The procedures we used to make decisions about those claims can be extended to questions associated with regression analysis. Of frequent interest in regression analysis is the value of the slope parameter. Does our estimate of β_1 indicate a slope significantly different from zero? That is, do changes in the predictor variable, X, have linear effects on Y? If β_1 is nonzero, what is the likely range for its value? In this section, you will learn how to test the slope estimate (b_1) for significance. The same questions may be applied to the intercept, β_0. However, because β_0 does not have the significance of β_1, which tells us how (or if) Y varies with X, we will focus on studying β_1.

Consider the case of the relationship between revenues and sales for Woolworth's subdivisions. If the slope coefficient, β_1, differs significantly from zero, revenues can be used to predict profits for the corporation. If β_1 does not differ significantly from zero, revenues are not a good predictor of profits.

Two-Tailed Tests

The hypothesis that the true slope of the regression line is zero is to be tested against the hypothesis that it is not. A two-tailed test assumes no previous knowledge about the nature of the relationship, if any.

Null hypothesis, H_0: $\beta_1 = 0$
Alternative hypothesis, H_a: $\beta_1 \neq 0$
Test statistic: $t_{test} = \dfrac{b_1}{s_{b_1}}$
Decision rule: If $t_{test} > t_{critical}$, reject H_0.

The test statistic here takes the usual form of observed value minus expected value, divided by standard error, with observed value $= b_1$ and expected value $= 0$ from H_0. The only new element in this equation is the standard error of b_1, s_{b_1}. The standard error of the slope of the regression line is computed using this formula:

$$s_{b_1} = \frac{s_{y|x}}{\sqrt{SS_{xx}}}$$

The variability of β_1 is a direct function of the standard error of the regression line. The greater the variance of y about y_c, the less precise the estimate of β_1 is. The greater the spread of x values from $\bar{x}$ (here part of the denominator), the more precise the estimate of β_1 is.

EXAMPLE 10.9 ■ **TESTING THE BINGO DATA FOR SIGNIFICANCE OF THE REGRESSION SLOPE $\beta_1 \neq 0$**

Let's apply the hypothesis-testing procedure to the bingo data.

Testing the slope coefficient for significance (two-tailed test)

H_0: $\beta_1 = 0$
H_a: $\beta_1 \neq 0$

Test statistic: $t_{\text{test}} = \dfrac{b_1}{s_{b_1}}$

As always when H_0 has "= 0," this is a two-tailed test.

Rejection region: $\alpha = .05$
Decision rule: If $t_{\text{test}} > t_{\text{critical}}$, then reject H_0.

We find the critical value for the t statistic by looking in Table V in Appendix B.

$$t_{\text{critical}} = t_{\alpha/2,n-2} = t_{.025,20} = 2.086$$

Since $s_{y|x} = 68.67$ and $\text{SS}_{xx} = 7606.36$,

$$s_{b_1} = \frac{68.67}{\sqrt{7606.36}} = .79$$

Then

$$t_{\text{test}} = \frac{24.69}{.79} = 31.25$$

Since t_{test} exceeds t_{critical}, we reject H_0. We may conclude that there exists a relationship between the predictor and response variables, because the slope differs from zero.

One-Tailed Tests

Often, an investigator believes not only that there exists a relationship between a predictor variable and a response variable, but also that the relationship is direct or that the relationship is inverse. In the bingo example, we might have supposed a direct relationship. Similarly, any relationship between study time and examination grades might be supposed to be direct. In these two cases, the form of the hypothesis test would be

$$H_0: \beta_1 \leq 0$$
$$H_a: \beta_1 > 0$$

On the other hand, the relationship between the price of a commodity and the quantity purchased might be supposed to be inverse. It would be expected that as price rises, quantity purchased falls. The form of the hypothesis test would be

$$H_0: \beta_1 \geq 0$$
$$H_a: \beta_1 < 0$$

We use a one-tailed hypothesis test in Example 10.10.

EXAMPLE 10.10 ▪ TESTING THE SIGNIFICANCE OF A REGRESSION SLOPE: QUALITY VERSUS PRICE

We will use a one-tailed hypothesis test to determine the relationship between prices of cordless telephones and their overall ratings, as reported in *Consumer Reports 1993 Buying Guide.* First we establish the hypotheses:

$$H_0: \beta_1 \leq 0$$
$$H_a: \beta_1 > 0$$

The null hypothesis holds that quality does not increase with price. The alternative hypothesis holds that there is a positive relationship between the two: the higher the quality, the higher the price. A person manufacturing phones would want to look at Price $= \beta_0 + \beta_1 \times$ Quality to assure herself that if she built a better phone, she could charge more for it. A consumer would be interested in Quality $= \beta_0 + \beta_1 \times$ Price to assure himself that paying more for a cordless phone would yield greater quality.

Testing the slope coefficient for significance (one-tailed test)

Test statistic: $t_{\text{test}} = \dfrac{b_1}{s_{b_1}}$, where $s_{b_1} = \dfrac{s_{y|x}}{\sqrt{SS_{xx}}}$

Rejection region: $\alpha = .05$

Decision rule: If $t_{\text{test}} > t_{\text{critical}}$, reject H_0.

Looking up the critical value in the t-statistic table for 16 degrees of freedom and an α value of .05 gives

$$t_{\text{critical}} = t_{.05,16} = 1.746$$

Given that $b_1 = .090$, $s_{y|x} = 5.39$, and $SS_{xx} = 22{,}216.44$, we have

$$s_{b_1} = \frac{5.39}{\sqrt{22{,}216.44}} = .036$$

Then

$$t_{\text{test}} = \frac{.090}{.036} = 2.50$$

As $t_{\text{test}} > t_{\text{critical}}$, we reject H_0. A positive relationship exists between price and the quality of cordless phones.

Confidence Intervals for Slope Estimates

To find a confidence interval for a slope estimate, we follow standard procedures. Given a point estimate for a parameter, we may add or subtract a margin of error to arrive at an interval estimate with a certain level of confidence. For β_1, the formula looks like this:

$$P(b_1 \pm t_{\alpha/2,n-2}s_{b_1}) = 1 - \alpha$$

The critical value for the t statistic can be found by looking in Table V for the value indicated by the subscript $\alpha/2,n - 2$, where α is the level of Type I error and the number of degrees of freedom is $n - 2$, the sample size less 2. This is the number of degrees of freedom for a regression that fits two parameters.

EXAMPLE 10.11 ▪ FINDING CONFIDENCE INTERVALS FOR β_1: QUALITY VERSUS PRICE

For the relationship of the quality of cordless phones to their price, we have $b_1 = .090$. For a 95% confidence interval, we need the critical value for the t statistic at $.05/2 = .025$ for 16 degrees of freedom:

$$t_{\alpha/2,n-2} = t_{.025,16} = 2.120$$

Finding the confidence interval for the slope

We know that $s_{b_1} = .036$, so the 95% confidence interval is given by $.090 \pm 2.120(.036)$, or (.014, .163). Thus,

$$P(.014 \leq \beta_1 \leq .163) = .95$$

Simply put, it is estimated with 95% confidence that the true slope lies between .014 and .163. Practically speaking, this means that a $1 increase in price contributes to a point increase of between .014 and .163 in the quality rating score (on a 100-point scale). This observation suggests that although a positive relationship exists, it is not a strong one. Thus, other factors should be sought that might be better predictors of quality rating scores than price is.

EXERCISES 10.55–10.64

Skill Builders

10.55 You are given these results:

$$Y = 3.47 + .28X$$

$$s_{y|x} = 2.48 \qquad n = 10 \qquad SS_{xx} = 96.9$$

a. Test whether β_1 differs significantly from zero. Let $\alpha = .05$.
b. Construct a 95% confidence interval for the true slope.
c. Explain how your result in part b confirms your findings in part a.

10.56 Consider the regression equation and accompanying results below:

$$Y = 12.29 + .73X$$

$$SSE = 1167.47 \qquad n = 10 \qquad SS_{xy} = 7936.2$$

a. Does the slope differ significantly from zero? Test at the .01 level.
b. Construct a 99% confidence interval from the slope.
c. If X increases by 1 unit, what range of changes may you predict for Y with 99% confidence?

10.57 Use the partial regression results below:

$$\overline{X} = 74.6 \qquad SS_{xx} = 1548.1 \qquad SS_{yy} = 2634.1$$

$$\overline{Y} = 85.1 \qquad n = 10 \qquad SS_{xy} = 749.4$$

a. Find the slope coefficient.
b. Test at the .10 level to see whether $\beta_1 \neq 0$.
c. Calculate the 90% confidence interval for β_1.

10.58 The partial results of a regression calculation are given below:

$$SS_{xx} = 1.9551 \qquad n = 6 \qquad SS_{xy} = .1907 \qquad SS_{yy} = .9572$$

a. Calculate b_1, the estimate of β_1.
b. Is β_1 significantly greater than zero? Use $\alpha = .05$.
c. Construct a 95% confidence interval for β_1.
d. If X increases by 1 unit, what range of changes may we predict with 95% confidence?

10.59 Determine whether the partial regression results below are true.

$$Y = 287.451 + .079X$$

$$n = 25 \qquad SSE = 417{,}897.089 \qquad SS_{xx} = 720{,}863.875$$

a. Test $H_0: \beta_1 = 0$ against $H_a: \beta_1 \neq 0$ at the .01 level.
b. Find the 99% confidence level for β_1.
c. How does the result in part b confirm your result in part a?

Applying the Concepts

10.60 In relating Woolworth's revenue and profit data for subdivisions, the following equation was calculated. (All data are in millions of dollars.)

$$\text{Profits} = -18.6 + .0880 \times \text{Revenues}$$

$$s_{y|x} = 44.58 \qquad SS_{xx} = 4{,}349{,}823.71 \qquad n = 7$$

a. To find whether β_1 differs significantly from zero, test $H_0: \beta_1 = 0$ at the .05 level. What is H_a?
b. Construct a 95% confidence interval for β_1.
c. Complete this statement: It can be expected with 95% confidence that a $1 million change in revenues will have an impact on profits of between $____ and $____.

10.61 In the study of the General Electric Company data, the regression equation relating dividends paid to net earnings was found to be

$$\text{Dividends} = 142.93 + .35 \times \text{Net earnings}$$

$$s_{y|x} = 8.64 \qquad SS_{xx} = 368{,}091.2 \qquad n = 5$$

a. The coefficients in the equation are b_0 and b_1. Test the hypothesis that $\beta_1 = 0$ at the .10 level, clearly stating H_0 and H_a.
b. Construct a 90% confidence interval for β_1.
c. Complete this paragraph: On average, a ____ unit change in net earnings will cause a .35-unit change in ____. Ninety percent of the time, we may expect that a 1-unit change in net earnings will effect a change of between $____ and $____ in net earnings for General Electric.

10.62 Analysis of the cruise travel data set yielded the following regression equation. (The original data are given in Exercise 10.8.)

$$\text{Passengers} = 285{,}334 + 5108 \times \text{Confidence}$$

$$s_{b_1} = 3452 \qquad n = 12$$

a. The coefficients in the equation are b_0 and b_1. Test the hypothesis that $\beta_1 = 0$ at the .01 level. What is H_a?
b. Construct a 99% confidence interval for β_1.
c. Interpret your results.

10.63 In order to predict fuel consumption by passenger airlines, the following data were collected on fuel consumption and miles traveled by revenue passengers for 11 airlines during 1984 (*Aviation Daily,* March 12 and 19, 1984). (Fuel consumption is in millions of gallons, and revenue passenger miles are in millions.)

Airline	Fuel Consumed	Revenue Passenger Miles
American	1222	33,680
Delta	1175	25,279
Eastern	1153	26,605
Northwest	1021	9,823
Ozark	505	2,779
Pan American	1126	6,009
Republic	460	8,594
TWA	1346	14,756
United	1491	45,308
USAir	349	8,432
Western	376	8,764

a. Plot the data on a scatterplot. Let "Revenue Passenger Miles" be the predictor variable.
b. Calculate the least squares regression line of fuel on passengers.
c. Is this relationship linear or nonlinear?
d. Plot the regression line on the scatterplot.
e. Compute the standard error.
f. Calculate r and r^2.
g. An airline expects to fly 10 billion passenger miles next year. Forecast its average consumption of fuel with 95% confidence.
h. Test the hypothesis H_0: $\beta_1 = 0$ at the $\alpha = .05$ level.
i. Construct a 90% confidence interval for β_1.
j. Complete this paragraph: The relationship between fuel consumed and revenue passenger miles (appears/does not appear) to be linear. Revenue passenger miles explains _____% of the variation in fuel consumption. For an airline flying 10 billion passenger miles next year, it is expected that consumption will average between _____ and _____ million gallons. A 1-million-unit increase in revenue passenger miles can be expected to increase the demand for fuel between _____ and _____ million gallons.

10.64 Colt Industries is a conglomerate supplying a large variety of industrial and consumer goods. Given the wide range of products and services, an analyst for a brokerage house thought it might be best to predict sales based on GNP. The data for the 1973–1981 period appear below. (Sales are in millions of dollars; GNP is in billions of dollars.)

Year	Sales	GNP
1973	914	1326
1974	1219	1434
1975	1104	1549
1976	1346	1718
1977	1525	1918
1978	1808	2156
1979	2141	2414
1980	2166	2626
1981	2243	2922

a. Make a scatterplot for these data.
b. Calculate the least squares regression line of sales on GNP.
c. Is this relationship linear?
d. Plot the regression line on the scatterplot.
e. Compute the standard error of the estimate.
f. Calculate r and r^2.
g. Estimate what sales will be when GNP is $3000 billion, using a 99% confidence interval.
h. Why shouldn't this equation be used to estimate sales when GNP is $5000 billion?
i. Construct a 99% confidence interval for β_1.
j. Interpret *all* your results. Write at least one sentence about each finding in parts a through i. Assume that your audience consists of the Board of Directors and CEO of Colt Industries.

SECTION 7 USING THE COMPUTER FOR REGRESSION ANALYSIS

Regression analysis involves many calculations, so the computer is particularly helpful. To examine how computers are used in regression analysis, we will use data on the relationship between the salaries of statistics professors and the number of years on staff.

Using Excel for Regression Analysis

The first step in any regression analysis is to examine whether the relationship is linear. To generate a scatterplot in Excel, enter the data and then go to the chart wizard key on the standard toolbar. Choosing "XY (Scatter)" will allow you to view the plot. Click through Step 2 of the chart wizard, going on to Step 3. In Step 3, add a title and labels and decide whether to include major X and Y gridlines. If you eliminate the legend, follow through to Step 4, and click "Finish," your scatterplot will look like the one in Figure 10.15.

To produce the regression using Excel, go to "Data Analysis Add In," located on the menu under "Tools." After selecting "Regression," fill in "Input Y Range" and "Input X Range." These are cell ranges, not data ranges. If data and labels occupy cells A1 to B10, the input Y range and input X range will look like this: $A$1:$A$10 and $B$1:$B$10. The resulting printout appears on page 381.

Figure 10.15
Excel scatterplot of professors' salaries

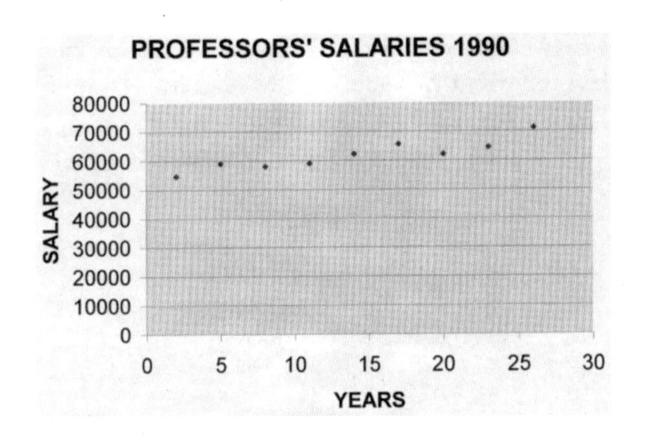

SUMMARY OUTPUT

Regression Statistics	
Multiple R	0.908409497
R Square	0.825207815
Adjusted R Square	0.800237503
Standard Error	2169.360962
Observations	9

ANOVA

	df	SS	MS	F	Significance F
Regression	1	155526000	1.56E+08	33.04756	0.000699348
Residual	7	32942888.89	4706127		
Total	8	188468888.9			

	Coefficients	Standard Error	t Stat	p Value	Lower 95%	Upper 95%
Intercept	54275.55556	1493.670912	36.33702	3.11E-09	50743.58762	57807.52349
Years	536.6666667	93.35443198	5.7487	0.000699	315.9186707	757.4146626

The Excel printout begins with a presentation of some overall summary statistics. Under "Regression Statistics" are found Multiple R, R Square, Adjusted R Square, Standard Error, and Observations. (The simple r is found next to the Multiple R listing.) The coefficient of correlation is approximately .91. The coefficient of determination, R Square, is .83, telling us that number of years on staff explains 83% of the variation in salary. The standard error is approximately $2169. We will address Adjusted R Square in Chapter 11 and explore the analysis of variance (ANOVA) there as well.

The equation, Salary $= 54275.56 + 536.67 \times$ Years, appears in the column labeled "Coefficients." Next to each of the coefficients is the standard error for the coefficient. Dividing each coefficient by its respective standard error yields its t statistic. The p value for the statistic appears to the right of the statistic. Remember that a p value less than the chosen α indicates that the null hypothesis should be rejected. Here, the p values are very small—3.11E-09 for the intercept and .000699 for the slope. Finally, 95% confidence intervals are given for the intercept and slope coefficients.

Using Minitab for Simple Linear Regression

Minitab generates a scatterplot with the "PLOT" command. Always check—or at least spot-check—the data to guard against the possibility of using wrong or corrupted data in later analysis. The first part of a Minitab run is shown below. The

Minitab data file in which the data were stored has been reproduced so that you can review the data that went into the scatterplot and the regression.

A sample Minitab run

```
MTB > # SALARY SURVEY OF STATISTICIANS
MTB > # SOURCE: AMSTAT NEWS DECEMBER 1990
MTB > # FULL PROFESSORS:YEARS ON STAFF VS. SALARY
MTB > PRINT C1 C2
ROW     YEARS     SALARY
  1        2      54700
  2        5      59000
  3        8      58000
  4       11      59000
  5       14      62200
  6       17      65600
  7       20      62100
  8       23      64400
  9       26      71100
MTB > Plot 'SALARY'*'YEARS';
SUBC> Symbol.
```

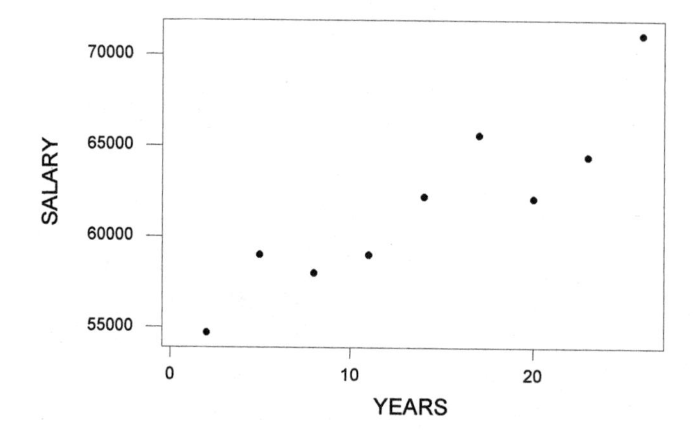

Figure 10.16
Minitab scatterplot of professors' salaries

Figure 10.16 illustrates the regression of salary on years. In the printout on page 383, salary is predicted for 10, 15, and 30 years on staff. Minitab produces a user-friendly printout, starting with the regression equation in its usual form. This equation is followed by a detailed analysis of the regression coefficients and of the parameters they estimate. Concentrating on the slope, it gives a standard error, s_{β_1}, of 93.35 (in the "Stdev" column). Dividing the coefficient by its standard error yields 5.75 in the next column for the t ratio, which is our test statistic for β_1. This t ratio is equivalent to the t_{test} used in hypothesis testing. Finally, the row contains a p value of .000, or zero to three decimal points, telling us that in this instance the slope coefficient *does* differ significantly from zero. All this information is contained in the row labeled "YEARS."

The next line of the printout shows "s = 2169." The variable s is the standard error of the estimate for this equation. Next to s is the coefficient of determination: "R-sq = 82.5%." Lastly, the computer prints out predicted values for 10, 15, and 30 years on staff, with both confidence intervals for the averages and prediction intervals for individual observations. These were requested by the command lines of the form "SUBC> PREDICT 10."

```
MTB > REGRESS C2 ON 1 C1;
SUBC> PREDICT 10;      } These subcommands get predictions of salaries
SUBC> PREDICT 15;      } after 10, 15, and 30 years of service.
SUBC> PREDICT 30.

The regression equation is
SALARY = 54276 + 537 YEARS
Predictor      Coef      Stdev     t-ratio      p
Constant      54276       1494       36.34    0.000
YEARS        536.67      93.35        5.75    0.000  <-- The p value is zero to three decimal
                                                         places, so the coefficient 537 is
                                                         significantly different from 0.
                 ┌─ percentage of variation in salaries accounted for
                 ▼
s = 2169      R-sq = 82.5%     R-sq(adj) = 80.0%
 ▲
 └─ standard error of the regression estimate

Analysis of Variance
SOURCE        DF          SS            MS          F         p
Regression     1   155526000     155526000      33.05     0.000
Error          7    32942888       4706127
Total          8   188468896

                           ┌─ confidence interval
                           ▼
Fit       Stdev.Fit      95% C.I.              95% P.I.  <-- prediction interval
59642        814      (57717, 61567)      (54162, 65123)
62326        729      (60601, 64050)      (56912, 67739)
70376       1660      (66450, 74301)      (63915, 76836)
 ▲
 └─ salaries point estimate
```

EXERCISES 10.65–10.74

Applying the Concepts

Solve all of the following problems using computer software. All of the data sets are available on the CD in the back of the book and in Appendix A.

10.65 Does evidence suggest that research and development expenditures increase company revenues? Consider the following data for Cray Research, a supercomputer company. (Revenues and R&D costs are in millions of dollars.)

Year	Revenues	R&D Costs
1976	.509	.155
1977	11.394	.476
1978	17.177	1.370
1979	42.715	2.960
1980	60.748	5.385
1981	101.648	16.250
1982	141.149	23.375
1983	169.690	25.540
1984	228.752	37.535
1985	380.158	49.865
1986	596.685	87.684
1987	687.336	108.830
1988	756.306	117.755
1989	784.700	122.413

a. Draw a scatterplot of revenues (Y axis) versus R&D expenses (X axis).

b. Does a linear relationship exist?
c. Regress revenues on R&D costs.
d. How do revenues change with a 1-unit change in R&D costs?
e. Interpret the p value for the slope coefficient.
f. How much of the variation in revenues is explained by variations in R&D costs and the regression relationship?
g. Complete this paragraph: Based on the data collected from Cray Research, we see (a positive/a negative/no) relationship between R&D expenditure and revenues. A regression of revenues on R&D costs indicates that, for each additional $1 million spent on R&D, revenues can be expected to (increase/decrease) by $_____ million. The p value for the slope coefficient indicates that there (exists/does not exist) a significant relationship at the .05 level. It would appear that R&D costs explain _____% of the variance in revenues. Money spent on R&D (does/does not necessarily) appear to be well spent with regard to revenues.

10.66 Often, company sales are related to industry-wide sales. When there is a strong relationship, the latter can be used to forecast the former. Data for Merck & Company and the health care industry for 1971 to 1985 appear below. (Sales are in millions of dollars, health care expenditures in billions of dollars.)

Year	Sales	Health Care Expenditures
1971	832.4	63.7
1972	942.6	78.3
1973	1104.0	86.7
1974	1260.4	96.4
1975	1402.0	116.8
1976	1561.1	131.8
1977	1724.4	148.7
1978	1981.4	166.7
1979	2384.6	189.7
1980	2734.0	219.7
1981	2929.5	254.7
1982	3063.0	286.5
1983	3246.1	314.7
1984	3559.7	348.1
1985	3547.5	371.4

a. Prepare a scatterplot of the data. Let "Sales" be the response variable.
b. Does a linear relationship exist?
c. Regress sales on health care expenditures.
d. Find the standard error of the estimate.
e. Calculate r and r^2.
f. If health care expenditures increase by $1 billion, what range of increased sales would you predict at a 90% confidence level?
g. Test H_a: $\beta_1 > 0$, using the p value as a measure of significance.
h. If health care expenditures were $400 billion, predict average sales for Merck & Co. with 95% confidence.
i. Complete this paragraph: Merck & Co. sales were found to be (directly/inversely) related to health care expenditures. There (exists/does

not exist) a linear relationship. The regression of Merck & Co.'s sales on health care expenditure appears to fit the data (well/poorly). We can say with 90% confidence that an increase of $1 billion in health care expenditures translates into a potential (increase/decrease) of between $____ and $____ in sales. A test of the slope coefficient suggests that at the .01 level the slope (differs/does not differ) from zero. We can be 95% confident that expenditures of $400 billion on health care would translate to between $____ and $____ in sales.

10.67 Following are data on sales for the Ford Motor Company for 1970 through 1981 and disposable income for the United States for the same period. (Sales are in millions of dollars, U.S. disposable income in billions of dollars.)

Year	Sales	Disposable Income
1970	10.480	715.6
1971	12.360	776.8
1972	15.096	839.6
1973	17.090	949.8
1974	17.098	1038.4
1975	16.129	1142.8
1976	19.858	1252.6
1977	27.762	1379.3
1978	31.019	1551.2
1979	28.191	1729.3
1980	21.882	1918.0
1981	23.482	2127.6

a. Plot sales as a function of disposable income.
b. Does a linear relationship exist?
c. Calculate the least squares regression line.
d. Find r^2 for this model.
e. Interpret the p value for the slope coefficient.
f. Consider the following income figures for 1982–1984:

Year	Disposable Income
1982	2261.4
1983	2425.4
1984	2670.2

Give an interval prediction of sales for these three years with a 95% confidence level.

g. Suppose actual sales figures were as follows:

Year	Sales
1982	24.271
1983	33.107
1984	41.002

How do the actual figures compare with the predictions? What does this suggest about the model?

h. Complete this paragraph: Ford Motor Company sales are a (positive/negative) function of income. A linear relationship (exists/does not exist). A regression of sales on disposable income indicates that a $1 billion increase in income would be associated with a $____ (increase/decrease) in sales. Disposable income explains _____% of the variance in sales. The slope coefficient (differs/does not differ) significantly from zero at the .05 level. The model performs (well/poorly) when predictions are compared with actuals for 1982–1984 period. I (would/would not) use this model for forecasting sales for 1985.

10.68 Advertising plays an important role in the beverage industry. The data below relate to the Coca-Cola Company for 1979–1987. (Sales and advertising costs are in millions of dollars.)

Years	Sales	Advertising Expenditures
1979	4961	73.7
1980	5913	71.1
1981	5889	73.5
1982	6250	125.0
1983	6829	119.7
1984	7364	237.4
1985	7904	393.6
1986	8669	370.4
1987	7658	364.7

a. Plot sales as a function of advertising costs.
b. Does a linear relationship exist?
c. Regress sales on advertising costs.
d. Given a $1 million change in advertising expenditures, find the 95% confidence interval for the change in sales.
e. Interpret the p value for the slope coefficient.
f. How much variation in sales is explained by advertising?
g. Given that Coca-Cola spent $385.4 million on advertising in 1988, find an interval that predicts average annual sales with 95% confidence.
h. Actual sales in 1988 were $8338 million. How does this amount compare with your prediction?
i. Complete this paragraph: Sales are (positively/negatively) related to advertising. There (exists/does not exist) a linear relationship between Coca-Cola Company's advertising expenditures and sales. It is estimated that a $1 million increase in advertising would be associated with a $_____ (increase/decrease) in sales. An analysis of the significance of the slope coefficient indicates that the slope (differs/does not differ) significantly from zero at the .05 level. Advertising appears to explain _____% of the variance observed in sales. A forecast for 1988 using this model fell within _____% of actual sales.

j. In 1982, advertising expenditures increased tremendously. What historical events accounted for that jump? There was another jump in 1985. Can you explain what happened? Prepare a brief report.

10.69 Examine the following hypothesis: The more routes an airline has, the higher its revenue will be. The data for United Airlines for 1974 through 1985 are given below. (Revenues are in millions of dollars.)

Year	Revenues	No. of Routes
1974	2698.3	77
1975	2244.5	81
1976	2517.0	86
1977	2891.9	89
1978	3423.9	92
1979	2847.7	94
1980	3880.7	99
1981	3937.9	103
1982	4695.3	121
1983	5372.6	128
1984	6218.7	135
1985	5291.6	161

a. Plot revenues versus routes. Let "No. of Routes" be the X variable.
b. Does a linear relationship exist?
c. Perform a regression analysis.
d. Is the slope coefficient greater than zero? Test at the .05 level. State the results of your test.
e. The following data represent UAL routes for 1986–1988:

Year	No. of Routes
1986	174
1987	190
1988	203

Predict revenues for 1986–1988.

f. Evaluate your equation's performance. Actual revenues were as follows:

Year	Revenues
1986	7105.1
1987	6855.8
1988	7123.1

g. Complete this paragraph: A test of the hypothesis that airline revenues increase with an increase in the number of routes revealed (a direct/an inverse/no) relationship. The relationship (appears/does not appear) to be linear. A 1-unit increase in the number of routes appears to be linked with a $_____ (increase/decrease) in revenues. Comparing our forecasts with actuals indicates that the model performs (reasonably well/rather poorly).

FORMULAS

The Normal Equations

$$\sum y_i = nb_0 + b_1 \sum x_i$$

$$\sum x_i y_i = b_0 \sum x_i + b_1 \sum x_i^2$$

Shortcut Formulas for b_1 and b_0

$$b_1 = \frac{SS_{xy}}{SS_{xx}}$$

$$b_0 = \bar{y} - b_1 \bar{x}$$

where

$$SS_{xy} = \sum (x - \bar{x})(y - \bar{y}) = \sum xy - \frac{\left(\sum x\right)\left(\sum y\right)}{n}$$

$$SS_{xx} = \sum (x - \bar{x})^2 = \sum x^2 - \frac{\left(\sum x\right)^2}{n}$$

Standard Error of the Estimate

$$s_{y|x} = \sqrt{\frac{\sum (y - y_c)^2}{n - 2}}$$

Shortcut Formulas for $s_{y|x}$

$$SSE = SS_{yy} - b_1 SS_{xy}$$

$$SS_{yy} = \sum y^2 - \frac{\left(\sum y\right)^2}{n}$$

$$s_{y|x} = \sqrt{\frac{SSE}{n - 2}}$$

Coefficient of Correlation

$$r = \frac{SS_{xy}}{\sqrt{SS_{xx} SS_{yy}}}$$

The fraction of variance explained is given by r^2, the coefficient of determination.

Prediction Interval for Y

$$y_c \pm t_{\alpha/2,n-2}s_{y|x}\sqrt{1 + \frac{1}{n} + \frac{(x_0 - \bar{x})^2}{SS_{xx}}}$$

Confidence Interval for the Mean of Y

$$y_c \pm t_{\alpha/2,n-2}s_{y|x}\sqrt{\frac{1}{n} + \frac{(x_0 - \bar{x})^2}{SS_{xx}}}$$

Standard Error of b_1

$$s_{b_1} = \frac{s_{y|x}}{\sqrt{SS_{xx}}}$$

Test Statistic for $\boldsymbol{\beta}_1$

$$t_{\text{test}} = \frac{b_1}{s_{b_1}}$$

Confidence Interval for $\boldsymbol{\beta}_1$

$$P(b_1 \pm t_{\alpha/2,n-2}s_{b_1}) = 1 - \alpha$$

NEW STATISTICAL TERMS

coefficient of correlation
coefficient of determination
confidence interval
dependent variable
direct relationship
errors in measurement
errors in specification
independent variable
intercept
inverse relationship
least squares regression line
mathematical model
negative relationship
positive relationship
prediction interval
predictor variable
residuals
response variable
sampling error
scatterplot
simple linear regression line
standard error of the estimate
stochastic relationship
trend line

EXERCISES 10.70–10.74

Supplementary Problems

10.70 An economist working for General Motors Corporation wished to describe the relationship between sales of GM vehicles (in millions) and the unemployment rate (a percent). Data for the 1970–1985 period appear below:

Year	Sales	Unemployment Rate
1970	18,752.4	5.4
1971	28,263.9	6.4
1972	30,435.3	6.0
1973	35,798.3	4.9
1974	31,550.0	5.6
1975	35,725.0	8.5
1976	47,181.0	7.7
1977	54,961.3	7.0
1978	63,221.1	6.0
1979	66,311.3	5.8
1980	57,728.5	7.1
1981	62,698.5	7.6
1982	60,025.6	9.7
1983	74,581.6	9.6
1984	83,889.9	7.5
1985	96,373.7	7.2

a. Draw a scatterplot of the data. Let "Sales" be the response variable.

b. Is there a linear relationship?

c. Compute the least squares regression of sales on the unemployment rate.

d. Find the standard error of the estimate.

e. What is the value of r^2?

f. Test H_a: $\beta_1 \neq 0$ against H_0: $\beta_1 = 0$ at the .05 level, using the p value to explain your conclusion.

g. For 10% unemployment, estimate the average value of sales with 90% confidence.

h. Complete this paragraph: There appears to be (no/a linear/a nonlinear) relationship between unemployment and sales. A (positive/negative/insignificant) relationship exists between the variables. A regression of sales on unemployment reveals that when unemployment increases by 1%, sales (increase/decrease) by $_____. It would appear that unemployment explains about _____% of the variation in sales. At a 10% level of unemployment, it is expected with 90% confidence that, on average, sales will lie between $_____ and $_____.

10.71 An article in *U.S. News & World Report* (April 5, 1993) identified the 25 cities where homesellers were most successful. In each market, the price for a starter home (in thousands of dollars) and average days on the market were given. Is it true that the lower the price, the faster the home will move?

City	Price	Average Number of Days on Market
Philadelphia	75	75
Honolulu	200	70
Fort Myers	60	90
Chicago	130	63
Portland	100	60
Sacramento	140	30
Youngstown	50	125
Pittsburgh	80	40
Kalamazoo	65	50
Shreveport	50	153
El Paso	75	122
Lansing	75	92
Madison	80	40
Louisville	60	35
Cleveland	95	65
Oklahoma City	35	120
San Antonio	65	75
Albany	110	160
Omaha	75	82
Akron	60	63
South Bend	60	60
Columbia	85	148
Raleigh	75	60
Corpus Christi	45	115
New Orleans	50	109

a. Draw a scatterplot of the data. Make "Price" the predictor variable.
b. Is there a linear relationship?
c. Compute the least squares regression line of average number of days on market on price.
d. Find the standard error of the estimate for the equation.
e. Find r and r^2.
f. Test the slope coefficient for significance. Use the p value to draw your conclusion.
g. Complete this paragraph: It appears that there (is/is not) a relationship between the time a house is on the market and the length of time it takes to sell it. A plot of the data indicated (a weak/a strong/no particular) relationship. The linear relationship between the variables explains _____% of the variance seen in the average number of days on the market.
h. How could this model be improved? Would more data provide better results, or should you look at other predictors to determine how fast houses will sell in a given area? If the latter, what predictors would you suggest?

Exercises 10.72 through 10.74 refer to the following data set, concerning downtimes for New York City Department of Sanitation trucks serviced by one of the borough shops.

Year: Quarter		No. of Trucks Down	No. of Trucks	Average Truck Age	No. of Mechanics
1984:	I	38	254	3.85	95
	II	44	261	3.98	94
	III	38	268	4.12	95
	IV	45	273	4.13	91
1985:	I	39	277	4.18	90
	II	34	297	3.39	87
	III	45	298	3.63	84
	IV	54	299	3.87	86
1986:	I	42	303	3.43	89
	II	38	292	3.04	91
	III	38	282	3.48	93
	IV	43	284	3.36	89
1987:	I	33	257	3.74	90
	II	32	254	3.64	90
	III	23	261	3.98	90
	IV	56	270	4.15	88
1988:	I	41	270	4.42	83
	II	42	268	4.38	83
	III	40	267	4.35	83
	IV	44	265	4.30	82
1989:	I	40	267	4.54	94

10.72 Use the data on "No. of Trucks Down" as the dependent variable and "No. of Trucks" as the independent variable.

a. Draw a scatterplot of the data.

b. Is there a linear relationship?

c. Compute the least squares regression line.

d. Calculate the value of r^2.

e. Test the slope for significance.

f. How does the addition of 10 more trucks affect the number of trucks down? Give a 95% confidence interval in your prediction.

10.73 Use "No. of Trucks Down" as the dependent variable and "Average Truck Age" as the independent variable.

a. Draw a scatterplot of the data.

b. Is there a linear relationship?

c. Compute the least squares regression line.

d. Calculate the value of r^2.

e. Test the slope for significance.

f. Would increasing the average age by one year seriously affect downtime? Estimate the effect, giving a 95% confidence interval.

10.74 Use "No. of Trucks Down" as the dependent variable and "No. of Mechanics" as the independent variable.

a. Draw a scatterplot of the data.

b. Is there a linear relationship?

c. Compute the least squares regression line.

d. Find the value of r^2.

e. Test the slope for significance.

f. Will reducing the number of mechanics have any effect on downtime? Estimate the effect of reducing the number of mechanics by five, giving a 95% confidence interval.

g. Referring to your results in Exercises 10.72 and 10.73, as well as parts a through f of this exercise, tell which equation you would choose, if any, to explain downtime. Defend your choice.

11

Multiple Regression and Time Series Analysis

OUTLINE

LEARNING OBJECTIVES

1. Understand the mathematical structure of multiple regression.
2. Estimate coefficients in a multiple regression equation.
3. Evaluate the usefulness of a multiple regression model.
4. Use indicator (dummy) variables.
5. Examine the validity of a regression model through residual analysis.
6. Deal with multicollinearity.
7. Forecast using a multiple regression model.
8. Forecast using smoothing techniques.
9. Calculate measures of forecast accuracy.
10. Seasonally adjust data.
11. Forecast using trend and seasonal factors.

BUILD YOUR KNOWLEDGE *OF STATISTICS*

During 1997, the Board of Governors of the Federal Reserve System began to use a new econometric model of the U.S. economy, referred to as the FRB/US model. This econometric model—a system of regression equations describing interactions among economic variables such as inflation, interest rates, and gross domestic product—is one of the tools employed by the board in economic forecasting and analysis of macroeconomic policy ("The Role of Expectations in the FRB/US Macroeconomic Model," *Federal Reserve Bulletin,* April 1997).

If every response variable could be forecast with one predictor variable—for example, if we could accurately predict company sales just from disposable income—then simple linear regression would suffice for model building. Unfortunately, model building is usually not that easy. To predict sales, we probably need to look at factors in addition to disposable income. For automobile sales, for example, the interest rate on auto loans would be important. And even after considering both disposable income and interest rate, we would still find some unexplained variation in company sales.

In this chapter, we will focus on adding explanatory variables to construct a multiple regression model, which often can explain a large percentage of the observed variation. We will look at ways to ensure that a model can be used to draw valid conclusions about the relationship between variables.

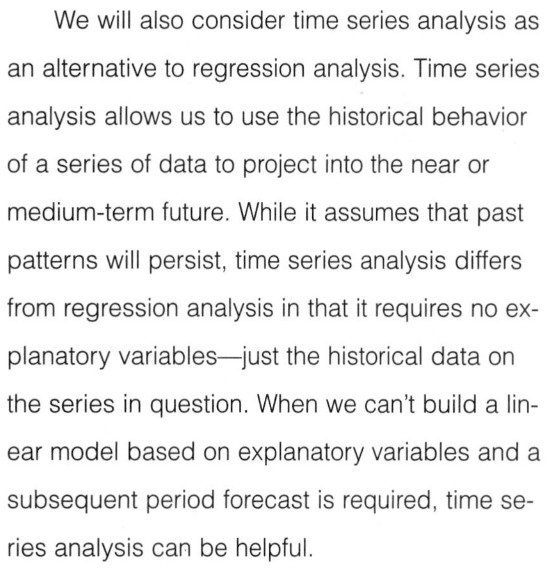

We will also consider time series analysis as an alternative to regression analysis. Time series analysis allows us to use the historical behavior of a series of data to project into the near or medium-term future. While it assumes that past patterns will persist, time series analysis differs from regression analysis in that it requires no explanatory variables—just the historical data on the series in question. When we can't build a linear model based on explanatory variables and a subsequent period forecast is required, time series analysis can be helpful.

SECTION 1 ■ THE MATHEMATICAL STRUCTURE OF MULTIPLE REGRESSION

When more than one factor significantly influences the response variable, simple regression analysis is not sufficient. In such cases, simple linear regression is just the first step in building a mathematical model. Recall from Chapter 10 that a mathematical model is a simplified representation of the relationships between variables, expressed through the use of equations.

A **multiple regression model** allows us to evaluate the influence of more than one predictor variable on a response variable. With two predictor variables, X_1 and X_2, we produce a model for the response variable, Y, of the form

$$Y = b_0 + b_1X_1 + b_2X_2 + \text{Error}$$

The **error terms** for given values x_1 and x_2 of the predictor variables are independent normally distributed random variables, all with the same standard deviations and all with a mean of zero. As Figure 11.1 shows, the equation $Y_{\text{predicted}} = b_0 + b_1X_1 + b_2X_2$ defines a plane. Its height above point (x_1, x_2) in the horizontal plane is the predicted value for Y, corresponding to $X_1 = x_1$ and $X_2 = x_2$. That is, Predicted $Y = b_0 + b_1X_1 + b_2X_2$. The observed value is given by the data point $(x_1, x_2, y) = (x_1, x_2, \text{Observed } Y)$. The error term for this data point and the regression given is Error = Observed $Y -$ Predicted $Y = y - (b_0 + b_1x_1 + b_2x_2)$.

In working out the details of a multiple regression, we make estimates for the b's that will minimize the sum of the squares of these error terms. In the model, the b's represent the population parameters. In practice, b's calculated from the data are statistical estimates of population parameters in some underlying process not directly accessible to us.

Multiple regression is a direct extension of simple linear regression. The b's are derived from systems of normal equations determined by the data. These equations

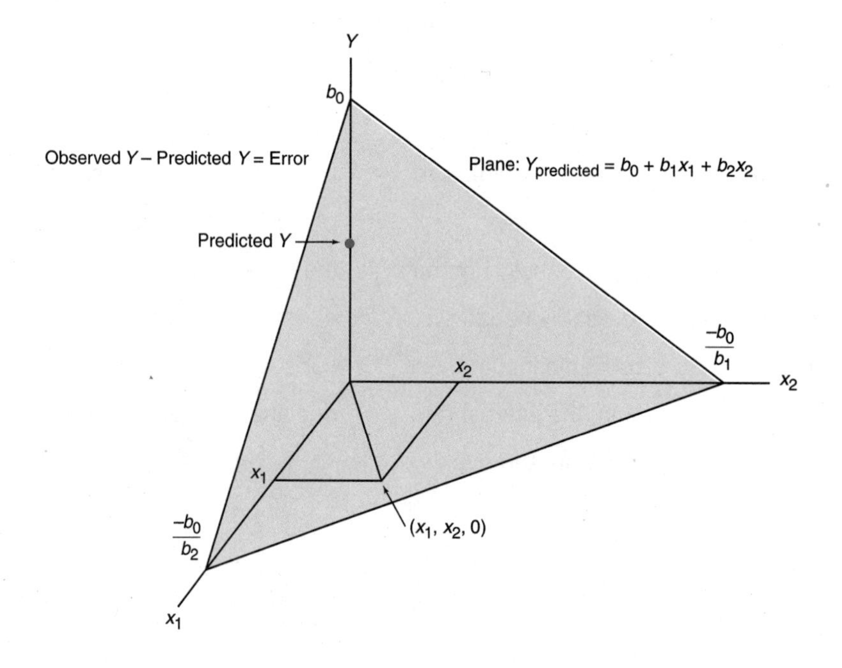

Figure 11.1
Multiple regression: predicted values and errors

minimize the sum of the squared errors: (Observed value – Predicted value)^2. The normal equations for multiple regression are similar to the normal equations for simple regression except that there is one equation for each of $b_0, b_1, \ldots, b_n$. Hence, there are $n + 1$ equations if there are n variables in the regression.

Given the amount of calculation involved, computer software is almost a necessity for working out the coefficients. As the necessary equations are built into the software, we will not go into them here. We will look at how to use the tools of multiple regression and how to interpret the results of multiple regression analysis.

SECTION 2 ▪ CALCULATING AND INTERPRETING MULTIPLE REGRESSION EQUATIONS

To answer the question "What factors influence ridership on the New York City subway system?" data over a 23-year period were examined. These data show a decline in ridership from over 1.2 billion riders in 1970 to about 1 billion in 1992. Figure 11.2 illustrates the decline in ridership; Figures 11.3 through 11.5 show the variations in related variables over this same period.

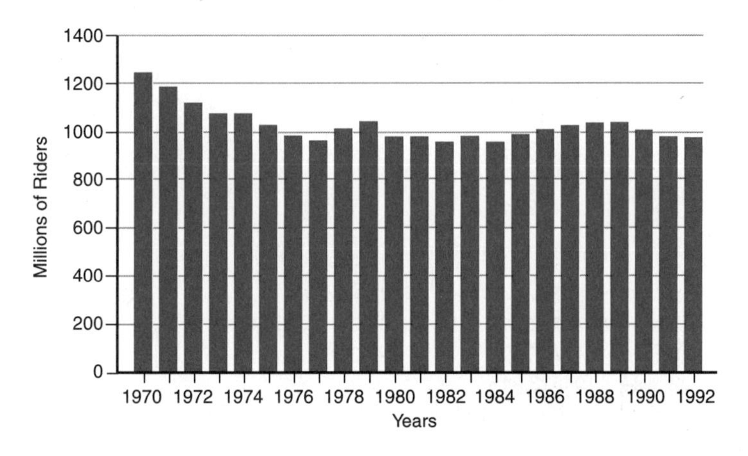

Figure 11.2
New York City subway ridership, 1970–1992

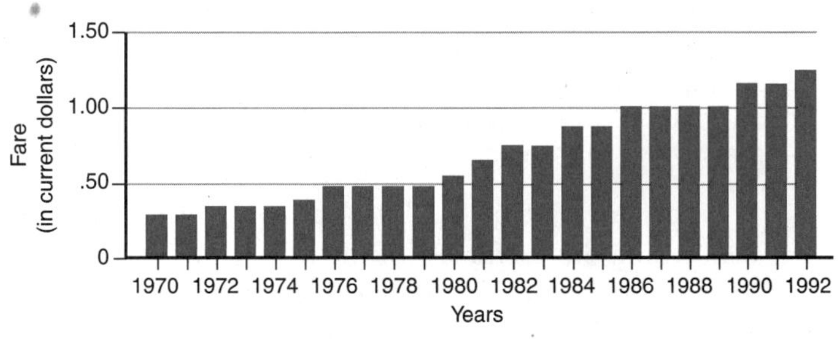

Figure 11.3
New York City subway fares, 1970–1992

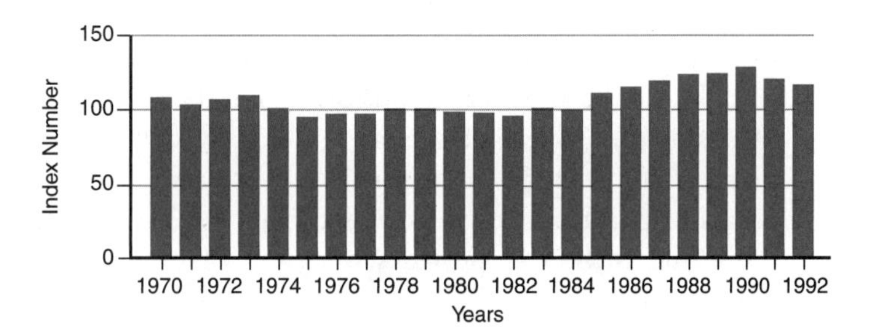

Figure 11.4
New York City business activity, 1970–1992

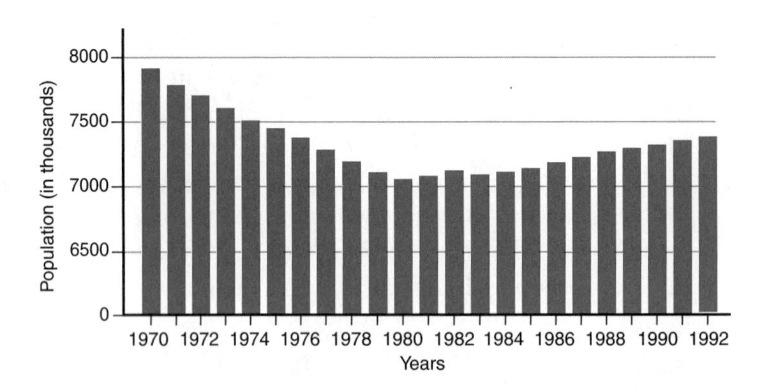

Figure 11.5
New York City population, 1970–1992

Using Excel

Using Excel, a regression was calculated in which ridership was looked at as a function of three variables: fare, business activity, and population. It was expected that ridership would have a negative relationship to fare (economic demand theory), while its relationships to business activity and population would be positive. The Excel printout follows.

Excel multiple regression output

SUMMARY OUTPUT

Regression Statistics	
Multiple R	0.90759644
R Square	0.82373129
Adjusted R Square	0.79589939
Standard Error	31.1846588
Observations	23

ANOVA

	df	SS	MS	F	Significance F
Regression	3	86346.73713	28782.25	29.59666	2.28456E-07
Residual	19	18477.17592	972.4829		
Total	22	104823.913			

(Annotations: Regression df 3 — number of coefficients − 1 $(k - 1)$; Total df 22 — number of observations − 1 $(n - 1)$; Significance F — *p* value of 2.28456E-07 is very small, so regression has "global utility.")

	Coefficients	Standard Error	t Stat	P-value	Lower 95%	Upper 95%
Intercept	−241.83512	279.8483446	−0.86416	0.398276	−827.564619	343.8943787
BUS_ACT	4.74279848	1.528656997	3.102592	0.005861	1.543281624	7.942315344
POP	0.12362071	0.047106264	2.624294	0.016696	0.025026133	0.222215283
FARE	−1.8580907	0.539044313	−3.44701	0.002701	−2.986323755	−0.729857628

The first part of the Excel printout provides a number of summary statistics concerning the usefulness of the regression. "Multiple R" tells us how strongly related the independent variables, as a whole, are to the dependent one. In this instance, Multiple R has a value of nearly .91 on a scale from 0 to 1, indicating a strong relationship. "R Square" indicates the degree to which the independent variables, taken together, explain the dependent variable. Here, the independent variables can be said to account for 82% of the variation in the dependent variable. The value of "Adjusted R Square," which we will discuss in more detail later in this chapter, is most useful in building a regression model; it should be examined as new variables are introduced into an equation.

While R Square provides one way of evaluating a regression equation, an equally valid way of doing so is through the use of **analysis of variance** (or

ANOVA). The ANOVA portion of the printout is designed to evaluate the global utility of the multiple regression model.

> The **global utility** of a model is its overall statistical significance. If a model has global utility, at least one of its coefficients differs significantly from zero.

How probable would our observations be under the null hypothesis that all the coefficients in the model equal zero? The ANOVA table examines variation in the response variable explained by the regression and compares it with variation unexplained by the response variable.

An F test can be used to measure the global utility of a multiple regression model. From the ANOVA table, we see that the three predictor variables in our model account for 86,347 units of variation in the dependent variable. There are 18,477 units of variation still unexplained; this is the error term. Dividing the sum of the squares (SS) by the appropriate degrees of freedom gives the result in the mean square (MS) column—the mean square regression (MSR) for the regression and the mean square residual, or mean square error (MSE), for the residual. The square root of the MSE is s, the standard error of the estimate for the equation. The F statistic is the MSR divided by the MSE, with 3 degrees of freedom for the regression and 19 degrees of freedom for the error. The number of degrees of freedom for the regression is the number of coefficients used (in this case, 4) less 1. The number of degrees of freedom for the error is the number of observations less 1 less the number of degrees of freedom for the regression (in this case, $23 - 1 - 3 = 19$). The p value for this regression (significance of F) is 2.28456E-07, or 0.000000228456. If we were testing at the .05 level, we would conclude that the overall regression model fits the data much too well to be accounted for by chance alone because the p value is less than the α level.

The last portion of the printout contains the regression equation. In this case, the terms of the equation are labeled Intercept, BUS_ACT, POP, and FARE. The value of -241 is the y intercept. It is the predicted response when all the predictor variables are zero. Here, the predicted response is far beyond the range of the data—a ridership of -241 million is meaningless. However, in some cases, having all the predictor variables equal to zero makes sense and is within the range of validity for the relationship; then the intercept can be interpreted in the model. We see that b_1, business activity, is 4.74. This means that, for each unit increase in the business activity index, we can expect that ridership will increase by 4.74 million (when variability due to fare and population are held constant). Here, b_2, population, equals .124. For each increase of 1000 in the New York City population, the number of riders can be expected to increase by 124,000 (when variability due to fare and business activity are held constant). (This suggests that each new resident takes an average of 2.4 subways rides per week.) Here, b_3, fare, is -1.86. That is, for every 1-unit increase in fare, there is an average loss of 1.86 million riders (when variability due to business activity and population are held constant). This doesn't mean that revenues necessarily decrease, however, because the smaller numbers of riders will be paying more. The coefficients relate the average change in the dependent variable to a 1-unit change in the independent variable.

Each coefficient is presented along with its standard error, *t* statistic, *p* value, and lower and upper 95% confidence interval. Examining the intercept, we see that the standard error is 279.85, with a *t* statistic of −.86. The *p* value of this *t* statistic is .398. The 95% confidence interval for b_0 falls between −827.6 and 343.9, inclusive. While the intercept does not differ significantly from zero at the .05 level, the other coefficients, b_1, b_2, and b_3, all differ significantly from zero. The *p* values are all well below .05, and the coefficients have the expected signs. The coefficient for FARE is −1.858. Its standard error is a comforting .53930, yielding a hardy *t* statistic of −3.447. The *p* value for this *t* statistic is .002701. This means that b_3, the observed decrease in ridership with an increase in fare, is unlikely to be a fluke due to random variation; instead, it represents a real and substantial effect, reflecting a nonzero parameter b_3 in the underlying population. The analysis is similar for the other variables.

To obtain the multiple regression printout for the subway ridership data using Excel, first install the Data Analysis Add-in and then you are ready to fill in the details of the regression. Go to "Regression" under "Data Analysis" and fill in the details in the dialog box. Note that the range, as used in Excel, refers not to the range of the data in the statistical study but to the cells in which the data are to be found. For "Output Range," simply indicate that you wish to open up a new workbook and everything will be put in place. Checking off residuals, standardized residuals, residual plots, line fit plots, and normal probability plot will provide you with a variety of useful charts for further analysis.

Using Minitab

The Minitab printout begins with presentation of the regression equation:

RIDERS = −241 − 1.86 FARE + 4.75 BUS.ACT + 0.124 NYC-POP

The second portion of the Minitab printout, with the headings "Predictor," "Coef," "Stdev," "t-ratio," and "p," respectively, contains the details on the *b*'s in the equation. Beneath the section containing details about the coefficients are three statistics: "s," "R-sq," and "R-sq(adj)." The value of *s* is the standard error for the regression equation as a whole. Using the empirical rule, we can conclude that 95% of the data points fall within approximately 2 standard errors of the values predicted by the regression, assuming the residuals are normally distributed with a constant variance. The smaller the standard error, the better the equation fits the data.

The quantity R-sq, or r^2, the coefficient of determination for the equation as a whole, appears next to *s*. Minitab presents the ANOVA table next. Note its similarity to the Excel ANOVA table; the differences are due to rounding.

Minitab multiple regression output

MTB > Retrieve 'C:\MTBWIN\DATA\SUBWAY.MTW'.
Retrieving worksheet from file: C:\MTBWIN\DATA\SUBWAY.MTW
Worksheet was saved on 1/11/1995
MTB > Regress 'RIDERS' 3 'FARE' – 'NYC-POP'.
The regression equation is
RIDERS = −241 − 1.86 FARE + 4.75 BUS.ACT + 0.124 NYC-POP

Predictor	Coef	Stdev	t-ratio	p
Constant	−241.5	280.1	−0.86	0.399
FARE	−1.8589	0.5393	−3.45	0.003
BUS.ACT	4.745	1.529	3.10	0.006
NYC-POP	0.12354	0.04714	2.62	0.017

s = 31.20   R-sq = 82.4%   R-sq(adj) = 79

Analysis of Variance

SOURCE	DF	SS	MS	F	p
Regression	3	86333	28778	29.57	0.000
Error	19	18491	973		
Total	22	104824			

SOURCE	DF	SEQ SS
FARE	1	33821
BUS.ACT	1	45828
NYC-POP	1	6684

Unusual Observations

Obs.	FARE	RIDERS	Fit	Stdev.Fit	Residual	St.Resid
1	30	1258.00	1185.95	17.14	72.05	2.76R
4	35	1102.00	1159.17	13.98	−57.17	−2.05R

R denotes an obs. with a large st. residual.

The importance of each explanatory variable is noted below the ANOVA table. We see that business activity has the greatest influence on ridership, while fare is the next most influential factor. Fare accounted for 33,821 units of variation in this sequential sum of squares analysis; business activity accounted for 45,828 units of variation.

The last part of the printout notes unusual observations—in this case, observations 1 and 4, which correspond to years 1970 and 1973. It is important to follow up on the "alert" messages provided by the multiple regression software. Was there an error in transcribing the data? Were there some unusual events that account for the unusual results? Here, the answer to the second question is yes. The Arab oil embargo hit the United States in 1973–1974.

EXERCISES 11.1–11.7

Skill Builders

11.1 Using a suitable statistical package, such as Excel, Minitab, SAS, or SPSS, calculate a multiple regression, using the data below. Treat Y as the dependent variable.

Y	2	4	4	5	9	6	5	2	9	3	9
X_1	5	6	2	4	9	0	8	9	1	7	7
X_2	3	4	9	4	8	9	4	7	6	6	2

a. Solve for the regression equation.

b. If b_1 were significant, what would it imply concerning a change in X_1 and its effect on Y?

c. Does b_2 differ significantly from zero? Test at the .05 level. Justify your answer.

d. Examining the ANOVA table, evaluate the regression as a whole. Test at the .05 level. (*Note:* Unless an alpha level is specified, assume .05 as a default value.)

11.2 Using the data below and a suitable statistical package, calculate a multiple regression. Treat Y as the response variable.

Y	−.04	.29	.03	.28	.79	.87	.16	−.91	.80
X_1	−.56	.32	.72	.20	−.27	.01	.33	−.88	.42
X_2	12	34	56	78	90	78	56	34	12
X_3	100	30	212	40	302	20	511	17	0

a. Solve for the regression equation.
b. Assess the global utility of the equation.
c. Do any of the coefficients differ significantly from zero? If so, which ones?
d. Explain the numbers in the standard deviation column.
e. Identify the sum of squares due to the regression.

11.3 Examine the portion of a computer printout displayed below:

```
MTB > Regress 'Y' 3 'X1' – 'X3'.
The regression equation is
Y = 79.2 – 0.742 X1 – 0.100 X2 + 0.017 X3
```

Predictor	Coef	Stdev	t-ratio	p
Constant	79.162	9.683	8.18	0.000
X1	−0.7424	0.1309	−5.67	0.001
X2	−0.0996	0.1610	−0.62	0.559
X3	0.0166	0.1687	0.10	0.925

s = 15.02 R-sq = 86.9% R-sq(adj) = 80.4%

Analysis of Variance

SOURCE	DF	SS	MS	F	p
Regression	3	9021.8	3007.3	13.32	0.005
Error	6	1354.3	225.7		
Total	9	10376.1			

SOURCE	DF	SEQ SS
X1	1	8932.7
X2	1	86.8
X3	1	2.2

a. Identify the t statistic for the first variable. What does it refer to? Write the formula for this t statistic.
b. What does the p value represent for the last variable?
c. What percentage of the variation in Y does this equation explain?

11.4 Multiple regression on a four-variable data set gave the results shown below.

```
MTB > Regress 'Y' 3 'W' – 'X';
SUBC > Predict 200 190 250.
The regression equation is
Y = −151 + 0.996 W – 0.005 Z + 0.913 X
```

Predictor	Coef	Stdev	t-ratio	p
Constant	−150.65	80.69	−1.87	0.104
W	0.9962	0.2670	3.73	0.007
Z	−0.0049	0.2221	−0.02	0.983
X	0.9132	0.3724	2.45	0.044

S = 21.92 R-sq = 85.0% R-sq(adj) = 78.6%

Analysis of Variance

SOURCE	DF	SS	MS	F	p
Regression	3	19062.6	6354.2	13.23	0.003
Error	7	3362.1	480.3		
Total	10	22424.7			

SOURCE	DF	SEQ SS
W	1	15501.2
Z	1	672.5
X	1	2888.9

a. Are any of the coefficients not significantly different from zero?
b. Which variable seems to be most strongly related to the dependent variable?
c. What percentage of variation is explained by the equation?

11.5 Consider the following portion of a computer printout:

Predictor	Coef	Stdev	t-ratio	p
Constant	54.94	22.21	2.47	0.024
W	0.00651	0.02681	0.24	?
V	−0.402	1.623	?	0.807

s = 38.68 R-sq = 0.9% R-sq(adj) = 0.0%

Analysis of Variance

SOURCE	DF	SS	MS	F	p
Regression	2	222	111	0.07	0.929
Error	17	25435	?		
Total	19	25657			

SOURCE	DF	SEQ SS
W	1	131
V	1	92

a. Write the regression equation.
b. From the data in the printout, calculate the value of p for variable W.
c. Calculate the t statistic for variable V.
d. Find the mean square error, MSE. Use at least two different approaches to calculate this result.
e. How many observations were used for this regression analysis?
f. Why is this regression worthless?

Applying the Concepts

11.6 Let's look at the idea that automobile sales can be predicted using the average price of a new car and the annual unemployment rate. The data for the U.S. automobile industry for 1970–1979 appear below. (Sales are in millions of units.)

Year	Sales	Price	Unemployment
1970	8.4	3542	4.8
1971	10.2	3742	4.1
1972	10.9	3879	3.9
1973	11.4	4052	4.1
1974	8.9	4440	7.1
1975	8.6	4950	8.3
1976	10.1	5418	4.2
1977	11.2	5814	4.5
1978	11.3	6379	6.0
1979	10.7	6847	5.8

a. Using a suitable statistical package, solve for the regression equation.
b. Test the price coefficient for significance.
c. What is the effect of a $100 increase in price?
d. Does the coefficient for unemployment differ significantly from zero?
e. Using the ANOVA table, evaluate the regression as a whole.

11.7 A study of the steel industry (1970–1979), relating steel sales to construction industry activity, automotive industry activity, and rail transport activity, yielded the following computer printout:

MTB > Regress 'SALES' 3 'CONST.' – 'RAIL'.
The regression equation is
SALES = 74789 + 6.98 CONST. + 0.291 AUTO. – 8.59 RAIL

Predictor	Coef	Stdev	t-ratio	p
Constant	74789	26208	2.85	0.029
CONST.	6.955	1.885	3.69	0.010
AUTO	0.2914	0.8457	0.34	0.742
RAIL	−8.590	7.021	−1.22	0.267

s = 7089 R-sq = 69.8% R-sq(adj) = 54.4%

Analysis of Variance

SOURCE	DF	SS	MS	F	p
Regression	3	689628160	229876048	4.57	0.054
Error	6	301511776	50251964		
Total	9	991139968			

SOURCE	DF	SEQ SS
CONST	1	614012544
AUTO	1	388572
RAIL	1	75227064

a. Identify the t statistic for construction activity. What does it refer to? Does construction industry activity appear to have a bearing on steel sales?
b. Interpret the p value for rail transport activity.
c. Using R-sq, evaluate the equation as a whole.
d. Complete this paragraph, considering an R-sq above .66 to be evidence of a "strong" relationship and one below .33 to be evidence of a "weak" relationship: Sales in the steel industry during the 1970s were related (strongly/moderately/weakly) to activity in the construction industry. A

1-unit change in construction activity (1 unit = \$1 million) had a \$_____ effect on steel sales. A (strong/moderate/weak) relationship could be said to exist between auto and steel industry activity. The relationship to rail transport activity appears to differ (significantly/insignificantly) from zero at the .05 level. I would conclude that this model (is/is not) useful for forecasting.

SECTION 3 ■ IMPROVING THE REGRESSION MODEL

In Sections 1 and 2, we examined the mathematical structure of a multiple regression model, techniques for calculating regression coefficients and related statistics, and interpretation of computer printouts. Once we have produced a multiple regression model, we seek to improve it. In this section, we will examine two tools that can help improve regression models.

A Variable Selection Criterion: Adjusted r^2

In constructing a multiple regression model, we wish to obtain the highest possible r^2. The higher the r^2 is in value, the more variation in the response variable is explained by the predictor variables. If r^2 is closer to 1 than to 0, the points in the sample space fall near the regression surface. In Figure 11.6, part (a) illustrates a high r^2 value; parts (b) and (c) illustrate low r^2 values.

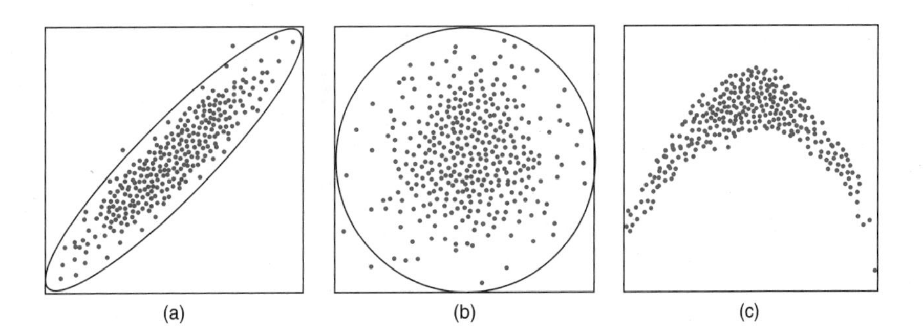

Figure 11.6
Comparison of high and low r^2

It is possible, however, to obtain a perfect r^2 of 1.00 without statistical significance. A simple least squares regression model fitting two points to a straight line will always yield a perfect fit, though the fact that r^2 equals 1 proves nothing. If a regression model has as many variables as observations, the r^2 will always equal 1 for algebraic reasons. To avoid this problem, adjusted r^2 takes into account the number of observations and the number of parameters being estimated (that is, the degrees of freedom). Thus, adjusted r^2 will decrease when a variable that does not carry its weight is added or when a useful variable with a nonlinear relationship is added. The formula for the adjusted r^2 is

Calculating r^2_{adj}

$$r^2_{adj} = 1 - (1 - r^2)\frac{n - 1}{n - p - 1}$$

where r^2 is the coefficient of multiple determination, n is the number of observations, and p is the number of predictor variables.

If r^2_{adj} increases when a new variable is added to the regression, we can be sure that the new variable has improved the model. Example 11.1 shows how adjusted r^2 can be used in model building.

EXAMPLE 11.1 ■ BUILDING A MODEL TO PREDICT GENERAL MOTORS' SALES

Suppose we want to construct a model to predict sales of General Motors' cars. We might wish to consider a number of variables: disposable income, unemployment, the prime rate, and population. To see the incremental effects of these variables, we add them one by one to our model and note how they affect the r^2_{adj}.

Let's try disposable income first. The regression equation is

$$\text{Sales} = 7389 + 27.7 \times \text{Disposable income}$$
$$r^2 = 89.9\% \qquad r^2_{adj} = 89.1\%$$

The second variable we will add is unemployment:

$$\text{Sales} = 19{,}800 + 34.6 \times \text{Disposable income} - 3325 \times \text{Unemployment}$$
$$r^2 = 94.9\% \qquad r^2_{adj} = 94.0\%$$

The third variable is the prime rate:

$$\text{Sales} = 21{,}775 + 35.5 \times \text{Disposable income} - 3330 \times \text{Unemployment} - 342 \times \text{Prime rate}$$
$$r^2 = 95.2\% \qquad r^2_{adj} = 93.9\%$$

Note that r^2 increased as additional variables were added. However, when the prime rate was considered, r^2_{adj} fell, showing that the model was stronger without the prime rate. The p value for the prime rate provides further evidence that the prime rate should be omitted from this model. The p value, as provided by the software package, is .385.

Can this model be improved upon? Let's consider the addition of a population variable. When we add total U.S. population as a regression variable, the results are surprising: (1) Disposable income becomes insignificant, and (2) r^2_{adj} increases to 95.5%.

Predictor	Coef	Stdev	t ratio	p
Constant	−441191	208189	−2.12	0.058
DI	2.93	14.49	0.20	0.843
unemp	−3993.5	900.1	−4.44	0.001
pop	2325	1050	2.21	0.049

s = 4361 R-sq = 96.4% R-sq(adj) = 95.5%

Dropping disposable income further improves the model, as r^2_{adj} increases to 95.8%. The final model relates General Motors' sales to unemployment and population. Disposable income and population are both measures of a general level of economic activity and propensity to spend. But, for this regression, population turns out to be a better measure of this factor than disposable income—and that is why disposable income became superfluous when we introduced population as a variable. The regression equation is

$$\text{Sales} = -482{,}696 - 4033 \times \text{Unemployment} + 2534 \times \text{Population}$$

Predictor	Coef	Stdev	t-ratio	p
Constant	−482696	33289	−14.50	0.000
unemp	−4032.8	843.0	−4.78	0.000
pop	2534.2	168.1	15.07	0.000

s = 4183   R-sq = 96.4%   R-sq(adj) = 95.8%

If forecasting were our only objective, we might be satisfied with using disposable income, but if we want to understand what is going on, we must keep looking and thinking.

Indicator Variables

Some of the conditions influencing response variables are qualitative. In regression, the effects of qualitative variables can be taken into account through the use of **indicator,** or **dummy, variables.** For example, to investigate the effects of a change in the tax code on industrial investment, we would introduce X_2, an indicator variable whose value was defined to be 0 at all times before the change in the tax code and 1 at all times after the change. We would then consider whether X_2 is significant as a predictor of industrial development.

Let's consider the effect of the introduction of a sales tax on taxable sales. If Y represents taxable sales, X_1 represents disposable income, and X_2 represents the status of the sales tax, we have the following model:

Using dummy variables

$$Y = f(X_1, X_2)$$

where

$$X_2 = \begin{cases} 0 & \text{before the sales tax} \\ 1 & \text{after the sales tax} \end{cases}$$

The end result of such a regression is as follows:

for the entire period,

$$Y_{\text{predicted}} = b_0 + b_1X_1 + b_2X_2$$

for the period prior to the change in sales tax, when $X_2 = 0$,

$$Y_{\text{predicted}} = b_0 + b_1X_1 + b_2(0) = b_0 + b_1X_1$$

for the period after the change in sales tax, when $X_1 = 1$,

$$Y_{\text{predicted}} = b_0 + b_1X_1 + b_2(1) = b_0 + b_2 + b_1X_1$$

In essence, the operating level of the response variable is shifted by the amount b_2 after the change in the sales tax. Example 11.2 illustrates the use of dummy variables in analyzing the effects of airline deregulation.

EXAMPLE 11.2 ▪ ASSESSING THE EFFECT OF DEREGULATION ON AN AIRLINE

Suppose an economist working for the Federal Aviation Administration wished to assess the effects of deregulation on the airline industry. The economist could run a regression of revenue passenger miles (RPM) on regulation (represented as a binary, or 0-1, variable) and real GNP. The results of such a regression might be as follows:

$$\text{RPM} = -22{,}699 + 5719 \times \text{Regulation/Deregulation} + 18.4 \times \text{Real GNP}$$

If all the coefficients in the equation prove to differ significantly from zero—and in this case they do—the equation above includes the necessary information on the periods before and after deregulation. The equation reflects the effects of both growth and deregulation. For each 1-unit increase in real GNP, we may expect an 18.4-unit increase in RPM. For the period prior to deregulation, the equation reads

$$\text{RPM} = -22{,}699 + 18.4 \times \text{Real GNP}$$

For the period following deregulation, the equation reads

$$\begin{aligned}\text{RPM} &= -22{,}699 + 5719 + 18.4 \times \text{Real GNP}\\ &= -16{,}980 + 18.4 \times \text{Real GNP}\end{aligned}$$

Interpreting the results, we see that deregulation raised revenue passenger miles (RPM) by 5719 units on average. Note how inclusion of the indicator (dummy) variable allows the analyst to use *all* the data available.

SECTION 4 ▪ RESIDUAL ANALYSIS: SPOTTING POTENTIAL PROBLEMS

Having constructed a regression model, we may use it to draw conclusions about the relationship between response and predictor variables. For our conclusions to be valid, it is not enough simply to get low p values when we perform a t test or an F test. We also need to check that the assumptions underlying these tests are met. In this section, we examine these assumptions and the techniques used to validate them.

Our key assumption in regression analysis is that the error terms are normally distributed, with a mean of zero, and constant—that is, they are independent of the predictor variables and independent of each other. Think of the residuals as being on a spreadsheet, where each row is devoted to a single data point (see Table 11.1). From the information on the predictor variables $X_1, X_2, \ldots, X_n$ and the b's we get the predicted value of the response variable Y for the row: $y_{ci} = b_0 + b_1x_{1i} + b_2x_{2i} + \cdots + b_nx_{ni}$. The observed value Y in the ith row is y_i. The residual value, or ith error term, is $y_i - y_{ci}$, the ith value for Observed Y minus Predicted Y (or Expected y).

Table 11.1

SPREADSHEET FOR RESIDUALS

Row	Observed Y	Predicted Y	Residual (error)
i	y_i	y_{ci}	$y_i - y_{ci} = e_i$

If the e_i do not look like samples from independent, identically distributed normal random variables with a mean of zero, then the significance tests used, coefficients estimated, and forecasts made may be flawed. Such a violation of our regression assumptions can occur even when all the statistics accompanying the regression look good—the t tests appear satisfactory, the F statistic differs significantly from zero, and r^2 is respectably high. To examine the distribution of e_i, we need **residual analysis.**

In order to make residual analysis easier, **standardized residuals** are often used (see Chapter 10). All computer software packages provide standardized residuals as an option in regression analysis.

$$sr_i = \frac{e_i}{s_{y|x}\sqrt{1 + \frac{1}{n} + \frac{(x - \bar{x})^2}{SS_{xx}}}}$$

where sr_i is the standardized residual—a z score with a mean of 0 and a standard deviation of 1, $s_{y|x}$ is the standard error of the estimate for the regression equation, and SS_{xx} is the sum of squares value used in estimating b values.

We will now look at techniques for assessing whether the conditions for a proper regression have been met or violated.

Looking for Linearity

The residuals must exhibit randomness about a linear regression line. A marked departure from patternless scatter indicates that linear regression analysis is not the appropriate tool for describing the relationship between the variables.

EXAMPLE 11.3 ■ TESTING FOR LINEARITY IN THE GENERAL MOTORS' SALES REGRESSION MODEL

Data for the late 1970s and 1980s on the relationship of General Motors' sales to disposable income yielded the following equation:

$$\text{Sales} = 7389 + 27.7 \times \text{Disposable income}$$
$$r^2 = 89.9\%$$

To test for linearity, we calculate the standardized residuals. The Minitab commands are as follows:

Residual analysis: linearity

```
MTB > Regress 'sales' on 1 'DI';
SUBC > Sresiduals 'SRES1'.
```

Figure 11.7 contains a plot of the standardized residuals versus disposable income. The points fall evenly about the center line on the chart. (The center line has to be drawn in by hand.) There should be no discernible relationship between the residuals and the predictor variable. With eight points below the line and seven points above the line, the data appear to be linear.

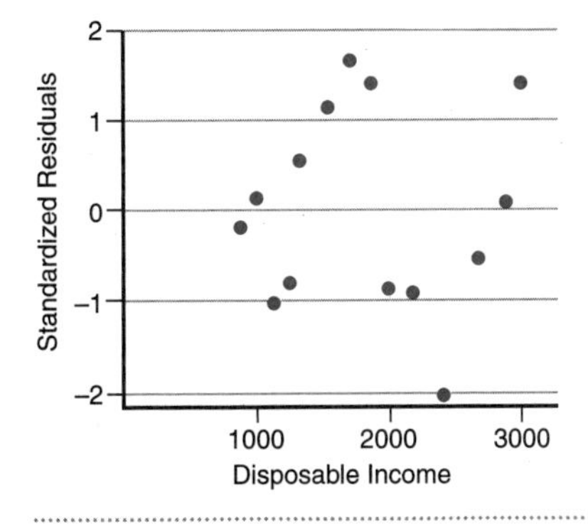

Figure 11.7
Standardized residuals versus disposable income for the General Motors' sales model

With a single predictor variable, X, we can plot the residuals against the predictor variable to see whether they appear normally distributed, with equal variance and a mean of zero. When dealing with multiple regressions, we usually begin by plotting the residuals against the predicted values of y_c. This plot is easy to inspect and can reveal departures from assumptions.

Looking for Homoscedasticity

Constant variance, or **homoscedasticity,** is important if estimates are to be efficient (see Chapter 8). If there is nonconstant variance and the residuals are related to a predictor variable, the estimates of the standard errors of the coefficients will be biased. Any tests done on those coefficients will be inaccurate. Figure 11.7 shows constant variability, as the points on the right side of the plot are no farther from the center line than the points on the left side. We may conclude that the regression exhibits homoscedasticity, as opposed to **heteroscedasticity.**

We see an example of heteroscedasticity when we regress reported taxable income on true personal disposable income (see Figure 11.8). Note that the variance increases with true personal disposable income, as shown by the increasing scatter of the data points about the zero line as we move toward the right.

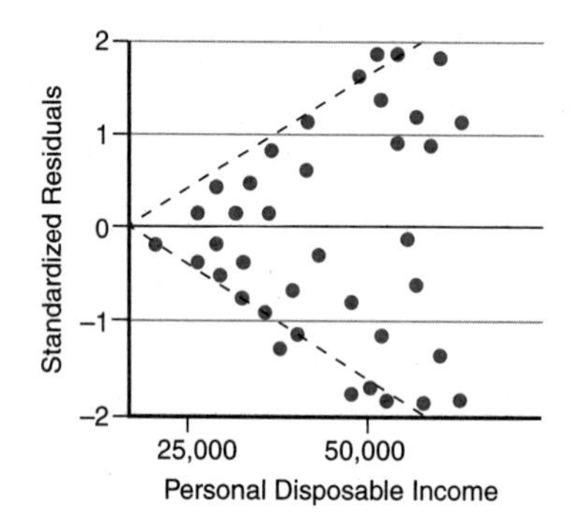

Figure 11.8
Regression of reported taxable income on true personal disposable income

Residual analysis: homoscedasticity

Looking for Normality

The residuals must exhibit normality. Normality is important if inferences about a population are to be drawn from the regression sample.

EXAMPLE 11.4 ▪ TESTING FOR NORMALITY IN THE GENERAL MOTORS' SALES REGRESSION MODEL

If the residuals show an approximately normal distribution, then it is reasonable to assume that the regression satisfies the conditions of normality. In the case of the General Motors model, which has only 15 observations, approximately normal is good enough. If the histogram for the data roughly matches a normal curve, as the one in Figure 11.9 does, we consider the normality criterion satisfied.

Figure 11.9
Histogram for the General Motors' sales model

Residual analysis: normality

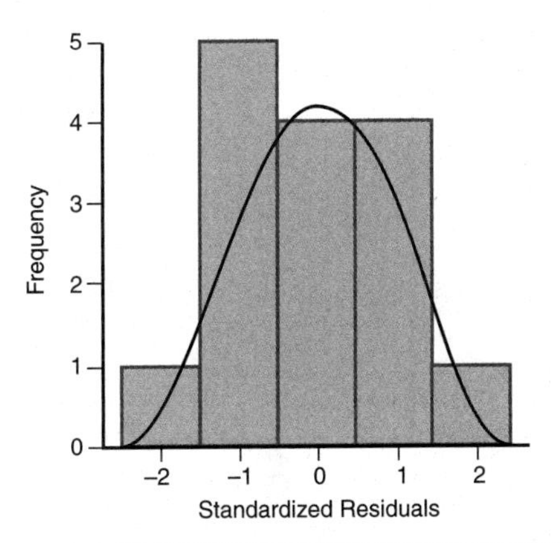

Looking for Serial Correlation

All forecasting must consider the possibility that the response variable is influenced not just by external predictor variables but also by its own past behavior. A prime example of positive serial correlation is provided by the stock market. In a bull market, buyers rush to buy stocks because they see that prices have been rising. High and rising prices bring on more of the same, until a correction takes place. In contrast, other processes show negative serial correlation. Like a pendulum or a spring, they are drawn to one side and then snap back and shoot over to the other side. Figure 11.10 illustrates positive and negative serial correlation, also called **autocorrelation.**

The standard red flag, warning forecasters to beware of serial correlation, is the Durbin-Watson statistic. It is important to look at this statistic even when plots of the residuals against y_c are satisfactory, as serial correlation may exist despite other signs that all is well.

The **Durbin-Watson (DW) statistic** is calculated with this formula:

$$d_{\text{test}} = \frac{(e_i - e_{i-1})^2}{e_i^2}$$

All major computer software packages have a subcommand for calculating the Durbin-Watson statistic. In Minitab, the subcommand for the General Motors' sales model would appear as follows:

```
MTB > Regress 'sales' on 1 'DI';
SUBC > DW.
```

Figure 11.10
Serial correlations: positive and negative

Residual analysis: serial correlation

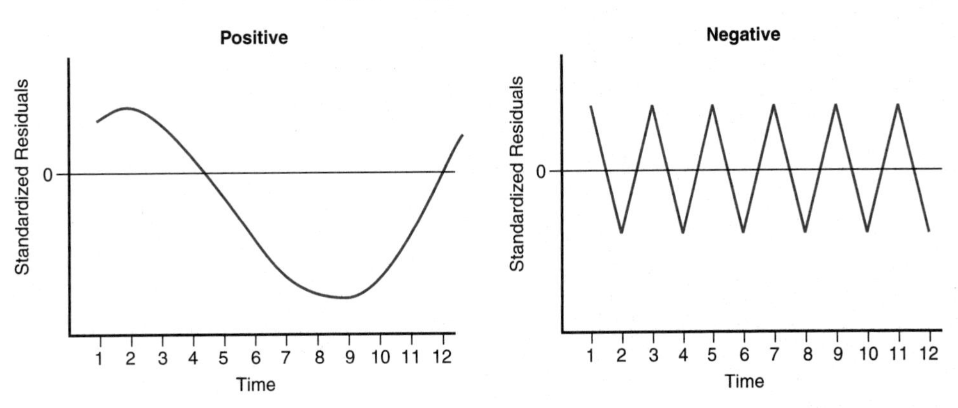

The DW statistic has upper and lower values. To interpret it, we need to set an alpha level and to know the number of observations (minimum of 15) and number of predictor variables. Table 11.2 contains guidelines for the interpretation of the DW statistic.

Table 11.2

GUIDELINES FOR INTERPRETING THE DURBIN-WATSON STATISTIC

HYPOTHESIS		ACTION		
H_0	H_a	Accept H_0	Accept H_a	Inconclusive
No positive serial correlation	Positive serial correlation	$d_{test} > d_U$	$d_{test} < d_L$	$d_L < d_{test} < d_U$
No negative serial correlation	Negative serial correlation	$d_{test} < 4 - d_U$	$d_{test} > 4 - d_L$	$4 - d_U < d_{test} < 4 - d_L$

EXAMPLE 11.5 ▪ TESTING FOR SERIAL CORRELATION IN THE GENERAL MOTORS' SALES MODEL

Let's apply the rules in Table 11.2 to the General Motors model. To test for positive serial correlation, we must compare d_{test} with d_L and d_U. In this instance, $d_L = 1.05$ and $d_U = 1.36$. Minitab tells us that DW = .92. As $d_{test} < d_L$, we accept H_a. We see evidence of a positive serial correlation. This means that we should be cautious about using the regression in forecasting and that we may want to turn the problem over to a specialist for further investigation. A DW of approximately 2.00 would signal an absence of serial correlation and would be reassuring.

Quick Fixes for Regression Problems

Many of the means for remedying violations of an ordinary least squares regression model fall beyond the scope of this textbook. However, there are some quick fixes that address linearity and serial correlation.

When linearity appears to be violated, a plot of the dependent variable may reveal a curvilinear pattern. A transformation of the predictor variable from X to X^2 may solve the problem. In studies of demand and price, the use of logarithms may work. There are many different possibilities for transforming variables.

One possible solution to the serial correlation problem is to use first differences, a special case of a more complex procedure for effectively predicting changes in Y based on the changes in X. With first differences, the model would look like this:

$$Y_i - Y_{i-1} = f(X_i - X_{i-1})$$

Although these techniques do not always work, they may be helpful. If these methods do not correct the specific violation or if they cause another, it is always possible to select new variables and start again.

SECTION 5 ▪ MULTICOLLINEARITY

Checking for multicollinearity

A special problem arising in multiple regression is that two or more predictor variables may move so close together with respect to the dependent variable that it is impossible to disentangle the effects of one from those of the other. The result is a condition called **multicollinearity.**

When two independent variables have a higher correlation with each other than with the dependent variable, the picture that regression presents can be disturbing. Variables take on signs that make no sense. Often, removal of one of the offending variables will correct the problem. Example 11.6 illustrates multicollinearity in the regression of General Motors' sales on four predictor variables.

EXAMPLE 11.6 ■ REMOVING THE EFFECTS OF MULTICOLLINEARITY FROM THE GENERAL MOTORS' SALES REGRESSION MODEL

A total of five variables were considered in the General Motors' sales regression model. Sales were predicted based on disposable income, unemployment, the prime rate, and population. We expected sales to be positively correlated with disposable income and population and negatively correlated with unemployment and the prime rate. The regression, however, yielded the following result:

$$\text{Sales} = -703{,}004 - 12.9 \times \text{Disposable income} - 4395 \times \text{Unemployment} - 881 \times \text{Prime rate} + 3671 \times \text{Population}$$

(Note that population is highly correlated with the other predictor variables.)

Clearly, disposable income has the wrong sign. The simple correlation between disposable income and population is .993, which is higher than any other correlation between variables. The correlation is arrived at through the command

```
MTB > CORR C1-C5
```

The results are as follows:

	Sales	DI	Unemp	Prime
DI	0.948			
Unemp	0.538	0.728		
Prime	0.408	0.484	0.349	
Pop	0.947	0.993	0.750	0.528

All of these simple correlations are positive because they do not hold the effects of other variables constant. In the equation, the beta coefficients show the change in the dependent variable with respect to a particular predictor variable when the other variables in the model are held constant.

If one of the offending variables—either population or disposable income—is dropped, the equation becomes

$$\text{Sales} = 21{,}775 + 35.5 \times \text{Disposable income} - 3330 \times \text{Unemployment} - 342 \times \text{Prime rate}$$

None of the independent variables have a stronger correlation than that between the independent variables and the response variable (sales).

SECTION 6 ■ FORECASTING WITH A MULTIPLE REGRESSION MODEL

When a regression is a good one, as in the case of the subway ridership model in Section 2, the researcher may wish to utilize it for the purposes of forecasting. Example 11.7 continues the subway ridership analysis with projected values for fare, business activity, and New York City population.

EXAMPLE 11.7 ■ FORECASTING FUTURE RIDERSHIP WITH A FARE INCREASE

To use the regression in Section 2 to forecast the effect of a fare increase, we use the Minitab command from Example 11.6, adding a subcommand:

SUBC > Predict 150 122.16 7074.6.

Forecasting with multiple regression

This subcommand, which contains a projected value for each of the independent variables, asks for an answer to the question "What would be the effect of a fare increase from $1.25 to $1.50?" The relevant portions of the Minitab printout appear below.

```
MTB > Retrieve 'C:\MTBWIN\DATA\SUBWAY.MTW'.
Retrieving worksheet from file: C:\MTBWIN\DATA\SUBWAY.MTW
Worksheet was saved on 1/11/1995
MTB > Regress 'RIDERS' 3 'FARE' – 'NYC-POP';
SUBC > Predict 150 122.16 7074.6.

Fit       Stdev.Fit      95% C.I.               95% P.I.
933.39    20.60          (890.27, 976.52)       (855.13, 1011.66)
```

To make the forecast, we need the new fare of $1.50 and new values for business activity and New York City population. With the values given, we can expect a ridership of 933.39 million. On average, we expect ridership to fall between 890.27 million and 976.52 million (the 95% confidence interval). The prediction interval provides a potential range for an individual observation. Will the Metropolitan Transit Authority earn more revenues with the higher fare? The answer is yes. Revenues will increase from $1246.25 million to $1400.09 million. Other variables may influence the status of revenues after the fare increase, but, other things being equal, revenues will increase as stated. This example illustrates the type of work that economic forecasters do every day.

EXERCISES 11.8–11.12

Applying the Concepts

The following problems require the use of a statistical software package. Include a copy of your printout, along with any necessary notes, with your solutions.

11.8 Consider sales of the Coca-Cola Company for the period 1970–1980. Following are data on sales and four potential explanatory variables. (Sales are given in millions of dollars, population in millions, GNP per household in thousands of dollars, CPI with 1982–84 = 100, and price in cents.)

Year	Sales	Population	GNP per Household	CPI (Food and and Beverage)	Price
1970	1606.4	205	32.3	114.7	9.69
1971	1728.8	208	35.4	118.3	9.96
1972	1876.2	210	38.6	123.2	10.03
1973	2145.0	212	42.1	139.4	10.31
1974	2522.2	214	45.8	158.7	12.88
1975	2872.8	216	50.6	172.1	15.00
1976	3032.8	218	55.6	177.4	14.64
1977	3559.9	220	60.5	188.0	14.79

(cont.)

1978	4337.7	223	67.8	206.3	15.93
1979	4961.4	225	75.6	228.5	17.02
1980	5912.6	228	85.3	248.0	19.30

a. Using r^2_{adj}, add each variable into the model. (*Hint:* Do the regression of sales with population first, then add GNP, CPI, etc.) Of course, any variable that does not cause r^2_{adj} to increase should not be added.
b. What percentage of variation in sales is explained by your model?

11.9 To examine the influence of the current president's political party on defense industry sales, data for Martin Marietta Corporation were compiled, along with the variables GNP and political party in the White House, from 1976 to 1986. (Sales are in millions of dollars and GNP in billions of dollars; in the column "Presidential Party," 1 means Republican and 0 means Democrat.)

Year	Sales	GNP	Presidential Party
1976	1213	1786	1
1977	1440	1995	0
1978	1758	2255	0
1979	2061	2521	0
1980	2619	2742	0
1981	3295	3064	1
1982	3526	3180	1
1983	3899	3434	1
1984	3921	3802	1
1985	4411	4054	1
1986	4759	4278	1

a. Using the indicator variable technique, perform the regression of sales on GNP and presidential party.
b. According to the regression, does the party in the White House have an influence on Martin Marietta's sales?
c. Assuming that political party has a bearing on sales, what is the difference in sales between Republican and Democratic administrations?
d. What historical events beyond presidential party might be considered here?

11.10 Look, once again, at the data (in millions of dollars) on revenues and profits for Woolworth Corporation:

Subdivision	Revenues	Profits
Kinney U.S.	2357	261
Kinney Canada & Australia	685	67
Richman Brothers	293	10
Other Specialty Operations	442	12
Woolworth U.S.	2169	123
Woolworth Canada	1867	106
Woolworth Germany	1123	78

a. Compute the linear regression.
b. Using residual analysis, evaluate the regression with respect to linearity, variance, and normality.
c. In view of your results in part b, complete this paragraph: The relationship between revenues, X, and profits, Y, appears to be (linear/nonlinear). The data exhibit (heteroscedasticity/homoscedasticity). The data (appear/do not appear) to follow a normal distribution.

11.11 The third quarter of the year is the heaviest cruise traffic season. Analyze cruise passenger volume with respect to the consumer confidence index (given for four quarters earlier) and a seasonal dummy variable (the third quarter is represented by 1, all other quarters by 0).

Year: Quarter		No. of Cruise Passengers	Consumer Confidence Index	Seasonal Dummy Variable
1988:	I	761,597	90.97	0
	II	820,879	100.86	0
	III	907,621	110.68	1
	IV	711,177	107.94	0
1989:	I	766,929	112.45	0
	II	859,461	117.23	0
	III	903,083	114.68	1
	IV	751,003	116.40	0
1990:	I	799,853	112.88	0
	II	968,820	116.83	0
	III	1,046,024	117.38	1
	IV	842,830	115.03	0

a. Using the seasonal dummy variable and the consumer confidence variable, run the regression of number of cruise passengers.

b. According to your model, does season have a bearing on passenger volume?

c. Assuming season does have a bearing, write the combined equation and the two separate equations determining passenger volume.

11.12 In a study to predict the downtime of New York City Sanitation vehicles, data were collected between the first quarter of 1984 and the first quarter of 1989.

Year: Quarter		No. of Trucks Down	No. of Trucks	Average Truck Age	No. of Mechanics
1984:	I	38	254	3.85	95
	II	44	261	3.98	94
	III	38	268	4.12	95
	IV	45	273	4.13	91
1985:	I	39	277	4.18	90
	II	34	297	3.39	87
	III	45	298	3.63	84
	IV	54	299	3.87	86
1986:	I	42	303	3.43	89
	II	38	292	3.04	91
	III	38	282	3.48	93
	IV	43	284	3.36	89
1987:	I	33	257	3.74	90
	II	32	254	3.64	90
	III	23	261	3.98	90
	IV	56	270	4.15	88
1988:	I	41	270	4.42	83
	II	42	268	4.38	83
	III	40	267	4.35	83
	IV	44	265	4.30	82
1989:	I	40	267	4.54	94

a. Using r^2_{adj}, start with the number of trucks and build a regression model to predict the number of trucks down.
b. What percentage of trucks down does your model explain?
c. Evaluate your model using residual analysis—that is, check your model for linearity, normality, homoscedasticity, and serial correlation.
d. Are there signs of multicollinearity in your model?
e. Is there a seasonal component in your model? If not, how might your model be expanded to take seasons into account?
f. Prepare a brief report on your findings, stating your conclusions clearly.

SECTION 7 ■ TIME SERIES ANALYSIS

History has a tendency to repeat itself, though sometimes with variations. Time series analysis is a tool that relies on history to forecast the future. Reliance on history takes many forms, from the analysis of cycles and trends in business to the use of more modern smoothing procedures. Each technique has its place in the forecaster's tool kit. In this section, we will explore two procedures used to analyze time series data. They are representative of more specialized procedures used in particular situations.

> **Time series data** track the evolution of individuals or populations over time. Examples of time series data include annual gross domestic product data, monthly consumer price index data, weekly unemployment statistics, and the daily closing price of a stock. Time series data are also known as **longitudinal data.**

Time series data play a crucial role in short-term and medium-term business forecasting. It is important to estimate the expected demand for a product so that production can be planned accordingly. No business or government service wants to be caught short-handed when a need for its product or service arises.

In many situations, expected demand can be estimated using a regression model. Even without such a model, we can make a forecast based solely on the behavior of a single data series—a single variable as observed over time.

Components of the Classical Time Series Model

The classical model for time series data is

$$Y_t = T \times C \times S \times I$$

The classical time series model

where Y_t is the current observation, T is the trend component, C is the cyclical component, S is the seasonal component, and I is the irregular or random component. If the data are annual, the model looks like this:

$$Y_t = T \times C \times I$$

where all the terms are the same as in the first model, but the seasonal component is missing. Let's examine the data illustrated in Figure 11.11 in terms of these components.

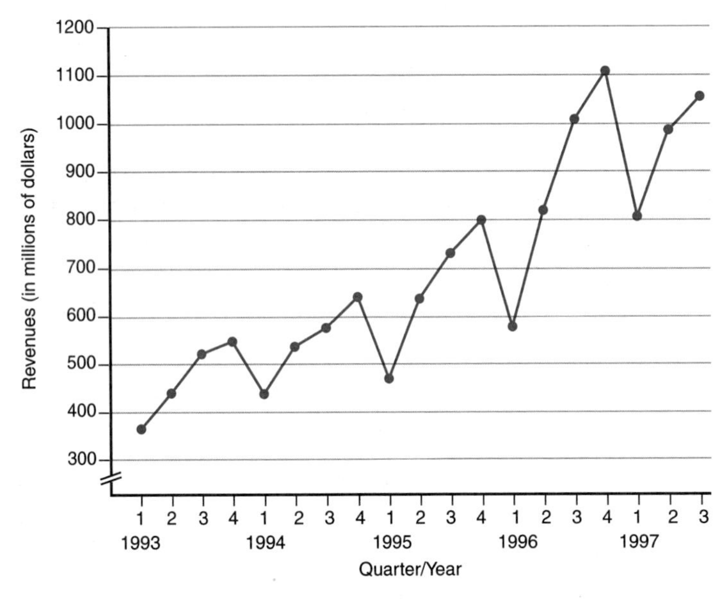

Source: Standard & Poor's Stock Reports.

Figure 11.11
Time series data on Computer Associates International

The data exhibit growth between the first quarter of 1993 and the third quarter of 1997. This growth can be characterized as a trend.

Time series components

> The **trend component** of time series data is the long-term tendency of the series to move in one direction. Examples of time series data that exhibit a trend component include the gross domestic product and your annual income over the course of your career.

Time series data frequently reflect the state of the economy, as it moves into periods of prosperity or recession. In periods of expansion, revenues and sales grow faster than at other times. In periods of contraction, sales growth may slow or cease altogether. These are systematic movements about the trend, as a result of cyclical factors. Business cycles may last from a few months to several years and thus may be difficult to isolate from other movements of a time series.

> The **cyclical component** of time series data is the tendency of the series to follow the phases of the business cycle. This component is tied to indicators of general business activity.

The existence of a cyclical component is not immediately evident in Figure 11.11 because the economy is experiencing expansion during the period shown.

A component that can be seen in Figure 11.11 is the seasonal component. Revenues start out low in the first quarter and climb to a peak in the fourth quarter.

> The **seasonal component** of time series data is the systematic movement that is regular from one year to the next.

Many products and services exhibit a seasonal pattern—lots of greeting cards are sold before major holidays, phone calls increase on Mother's Day, and flowers are in great demand on Valentine's Day. But many less obvious examples of seasonality exist also. To understand a current sales or revenue figure, you must look at it in context. When you compare one quarter's sales with the previous quarter's sales, you should make sure that the figures have been seasonally adjusted.

If a time series could be reduced to just trend, cyclical, and seasonal components, then the line in Figure 11.11 would move uniformly upward. There would be movements about the trend line, reflecting seasonal and cyclical factors, but there wouldn't be any great jumps or long dives in the series. These anomalies exist because economic series may be influenced by other factors, such as strikes, earthquakes, and other nonsystematic events.

The **irregular component** of time series data consists of the nonsystematic, erratic, residual movements remaining after all other causes of variation have been accounted for.

Forecasting with Smoothing Techniques

Smoothing procedures allow the analyst to get an overall impression of the pattern of movement over time. Smoothing techniques tend to minimize the random variation in time series, removing much of the statistical "noise." Two popular smoothing procedures are the moving average technique and simple exponential smoothing.

Moving Average Technique. The **moving average technique** levels out the more pronounced peaks and valleys, which we assume are due to the random component of the process, allowing the analyst to get a more balanced picture of the series. Example 11.8 illustrates how the moving average technique may be used.

EXAMPLE 11.8 ▪ SMOOTHING DATA SETS FOR FORECASTING

Consider the New York State unemployment series shown in Figure 11.12. This series exhibits erratic behavior. It varies from a low of 258, observed at the end of the 1980s, to a high of 789 in 1976.

Figure 11.12 *New York State unemployment series, 1970–1997*

Smoothing techniques

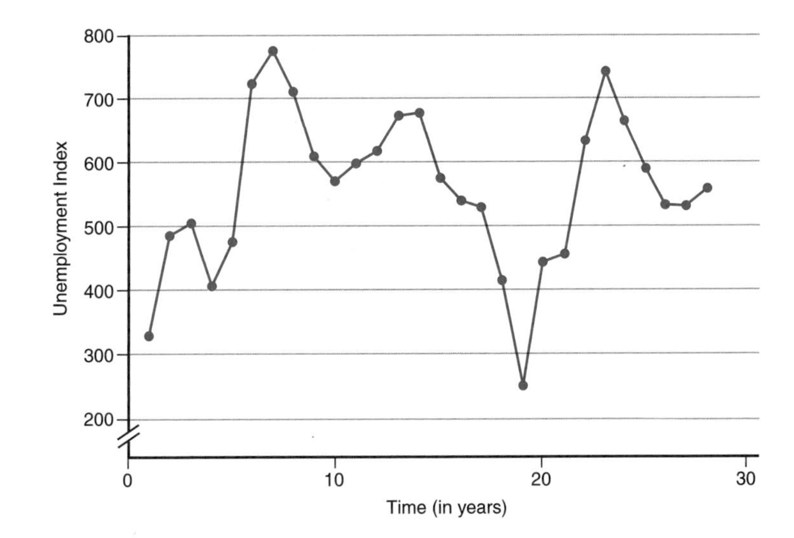

To apply a three-year moving average to the data, we start with this formula:

$$MA_{71} = \frac{Y_{70} + Y_{71} + Y_{72}}{3}$$

where Y_{70} is the observation for 1970, Y_{71} is the observation for 1971, and Y_{72} is the observation for 1972. The next figure in the series is obtained by dropping Y_{70} and adding Y_{73}, yielding

$$MA_{72} = \frac{Y_{71} + Y_{72} + Y_{73}}{3}$$

Table 11.3 contains the original data and the "smoothed" data, corresponding to a three-year moving average. The original series and the smoothed series are both shown in Figure 11.13.

Moving averages

Table 11.3

THE NEW YORK STATE UNEMPLOYMENT SERIES

Year	Original Data	3-Year Moving Average	Year	Original Data	3-Year Moving Average
1970	331		1984	583	605.000
1971	491	441.000	1985	544	551.000
1972	501	465.333	1986	526	494.000
1973	404	462.333	1987	412	398.667
1974	482	537.667	1988	258	370.667
1975	727	666.000	1989	442	389.000
1976	789	740.333	1990	467	515.000
1977	705	699.333	1991	636	616.667
1978	604	626.667	1992	747	686.667
1979	571	590.667	1993	677	673.000
1980	597	593.333	1994	595	603.667
1981	612	631.000	1995	539	557.000
1982	684	661.333	1996	537	547.000
1983	688	651.667	1997	565	

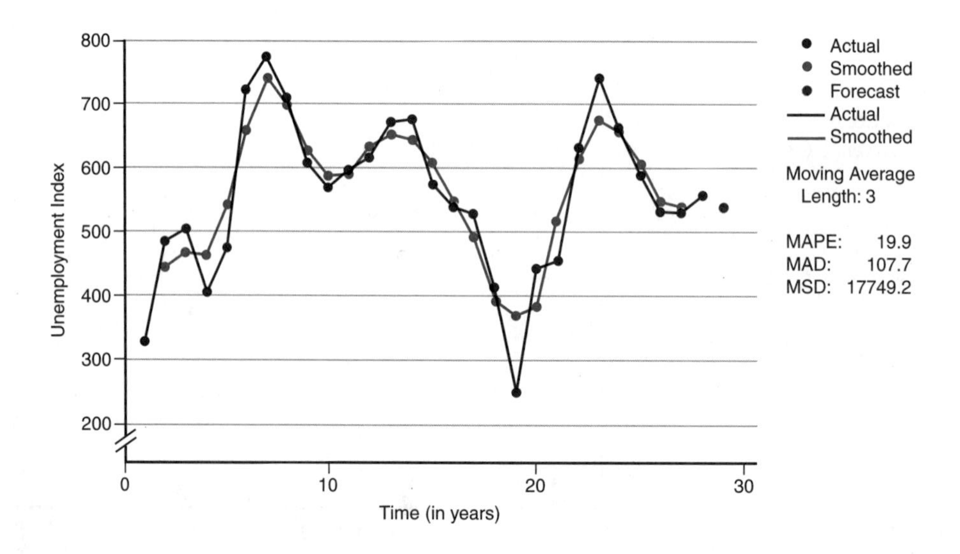

Figure 11.13
Original and smoothed New York State unemployment series, 1970–1997

Note that the first observation and the last observation are lost in this process, so the moving average procedure yields its first result in year 2 and its last result in year $n - 1$. This centered moving average is plotted above the central years of the interval of years.

Forecasting with a smoothing procedure is like predicting tomorrow's high temperature based on the average highs for today, yesterday, and the day before yesterday. To use moving averages in forecasting, we use the average of the past n observations as a predictor for y_{i+1}, the next value.

$$\text{Prediction for the } (i+1)\text{th value, } y_{i+1} = \frac{x_{i-(n-1)} + x_{i-(n-2)} + \cdots + x_i}{n}$$

When $n = 3$, as in Example 11.8,

$$y_{i+1} = \frac{x_{i-2} + x_{i-1} + x_i}{3}$$

The prediction created by the three-year moving average consists of the last observed centered moving average generated by the data. It is the forecast point beyond the lines representing the actual and predicted series. In Example 11.8, the last centered moving average was 547 (see Table 11.3); it becomes the forecast value for 1998. The more time periods you include in the average, the smoother the series will be, but the more data you will lose. Moving average techniques are good for short-term forecasts—say, one or two periods into the future—but no further.

Both Minitab and Excel have moving average routines. It should be noted that Excel plots the three-period moving average opposite the third period, thus producing a graph of forecasted values rather than smoothed values.

Simple Exponential Smoothing. Another short-term forecasting technique is **simple exponential smoothing.** The forecast value for the *i*th term in the series is E_i, and the observed value is Y_i. The values of E_i are determined successively. We start by setting $E_1 = Y_1$. Then $E_2 = wY_1 + (1 - w)E_1$, where w is a weighting factor, between 0 and 1, that is chosen by the forecaster and stays fixed throughout the process. Because $E_1 = Y_1$, we know that

Simple exponential smoothing

$$E_2 = wY_1 + (1 - w)E_1$$

As the forecast is developed, the succeeding values of E_i are determined by

$$E_{i+1} = wY_i + (1 - w)E_i$$

where E_{i+1} is the prediction for the $(i + 1)$th value of the time series, w is the weighting factor, Y_i is the observed *i*th value, and E_i is the predicted value for the *i*th term. In other words, this formula predicts the $(i + 1)$th value by placing weight w on the observed *i*th value, Y_i, and weight $1 - w$ on our *i*th predicted value, E_i. The closer w is to 1, the more weight the predictor places on the most recent observation, Y_i, and the less weight it gives to earlier observations.

The choice of a value for w depends on the objective of the analyst. To simply observe the long-term view, choose a value near zero. For forecasting, a value closer to 1 should be selected. Figure 11.14 on page 422 shows the results when a value of .75 was chosen for the unemployment series. The forecast value for the year 1998 is 559.350. The fitted value at time t is the smoothed value of the previous period.

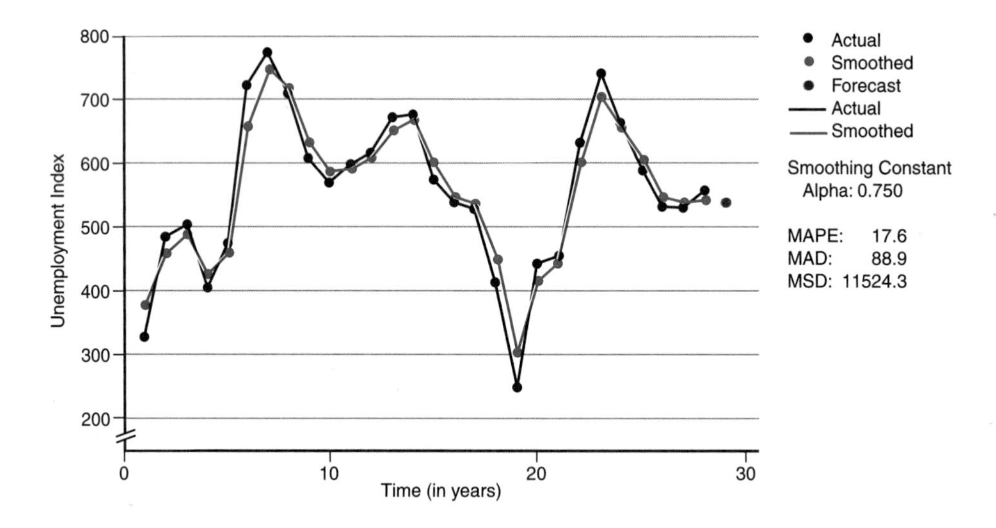

Figure 11.14
Original and exponentially smoothed New York State unemployment series, 1970–1997

Comparing Forecasting Methods

Evaluating forecasts

To determine which method will provide us with a better forecast, we need to check the measures of accuracy produced by each method. Minitab provides three such measures: **MAPE (the mean absolute percentage error), MAD (the mean absolute deviation),** and **MSD (the mean squared deviation).** Each measure assesses the accuracy of the forecast in a different way. The most popular is the mean absolute percentage error (MAPE) because it expresses accuracy as a percentage, allowing the analysts to compare methods without worrying about units of measurement.

$$\text{MAPE} = \frac{\sum |(y_t - \hat{y}_t)/y_t|}{n} \times 100, \quad y_t \neq 0$$

where y_t equals the actual value, $\hat{y}_t$ equals the fitted value, and n equals the number of forecasts.

The mean absolute deviation (MAD) expresses accuracy in the same units as the data, making it easy to visualize the absolute amount of error:

$$\text{MAD} = \frac{\sum |y_t - \hat{y}_t|}{n}$$

where y_t equals the actual value, $\hat{y}_t$ equals the fitted value, and n equals the number of fitted values.

The mean squared deviation (MSD) is similar to the variance. It is the sum of the squared differences between the observations and the fitted values, divided by the number of fitted values. Because the MSD is always computed with the same number of observations for each model, it provides a basis for comparison across models.

$$\text{MSD} = \frac{\sum(y_t - \hat{y}_t)^2}{n}$$

where again y_t equals the actual value, $\hat{y}_t$ equals the fitted value, and n equals the number of fitted values.

A comparison of the three-year moving average and simple exponential smoothing for the New York State unemployment series appears in Table 11.4. Examination of all the measures suggests that, in this instance, simple exponential smoothing provides a more accurate forecast. The mean average percentage error for the three-year moving average is 19.9%, while that for simple exponential smoothing is 17.6%. The mean absolute deviation tells us that, on average, the error is 88.0 thousand unemployed individuals with simple exponential smoothing as opposed to 107.7 thousand unemployed individuals with the three-year moving average. Lastly, the mean squared deviation tells us that simple exponential smoothing is more efficient per fitted value calculated.

Table 11.4

MEASURES OF ACCURACY FOR FORECASTS OF NEW YORK STATE ANNUAL UNEMPLOYMENT SERIES

Measure	Three-Year Moving Average	Simple Exponential Smoothing
MAPE	19.9	17.6
MAD	107.7	88.0
MSD	17,749.2	11,524.2

Adjusting for Seasonal Data

When the data are weekly, monthly, or quarterly and we extrapolate a trend, we can improve our forecasting ability by seasonally adjusting the results. Seasonally adjusting the data allows us to draw more informed conclusions about the behavior of the raw time series data in question.

> **Seasonal adjustment** is a statistical procedure designed to remove the regular periodic effects of seasonal change from a data set. It allows comparison of time series data between weeks, months, or quarters by removing the normally expected changes so that movements away from the norm may be more easily seen.

Consider, once again, the time series data for Computer Associates, illustrated in Figure 11.11. The data show an upward trend, along with seasonal variations. The fourth quarter is always the peak for the year, and the first quarter is always the trough. As a result, a comparison of fourth-quarter revenues and first-quarter revenues will always show a downturn, even though we can recognize an overall upward trend. In comparing the fourth quarter with the succeeding first quarter, what we really want to know is whether the figures are unusually high or low. To answer that question, we use a seasonal index computed from the data. Example 11.9 indicates how a seasonal index may be calculated.

EXAMPLE 11.9 ■ COMPUTING SEASONAL INDICES FOR THE COMPUTER ASSOCIATES DATA

To calculate seasonal indices for the Computer Associates data, we first set up a table using all the quarterly data available. Then we calculate the total and mean for each column, as shown in Table 11.5.

Table 11.5

QUARTERLY REVENUE DATA FOR COMPUTER ASSOCIATES INTERNATIONAL, 1993–1997

Year	Quarter 1	Quarter 2	Quarter 3	Quarter 4
1993	367.0	432.0	501.5	540.0
1994	423.4	517.0	574.4	633.7
1995	426.6	623.3	721.0	802.0
1996	577.5	812.3	1004.0	1110.0
1997	792.1	990.1	1053.0	
Total	2586.6	3374.7	3853.9	3085.7
Mean	517.3	674.9	770.8	771.4

Next we compute the grand mean and seasonal relatives, as defined below. The **grand mean** is the mean of the quarterly means:

$$\frac{517.3 + 674.9 + 770.8 + 771.4}{4} = 683.6$$

Seasonal relatives are given by the quarter's mean divided by the grand mean. The first-quarter relative is

$$\frac{\text{First-quarter mean}}{\text{Grand mean}} = \frac{517.3}{683.6} = .7567$$

The second-quarter relative is

$$\frac{\text{Second-quarter mean}}{\text{Grand mean}} = \frac{674.9}{683.6} = .9873$$

The third-quarter relative is

$$\frac{\text{Third-quarter mean}}{\text{Grand mean}} = \frac{770.8}{683.6} = 1.1276$$

Finally, the fourth-quarter relative is

$$\frac{\text{Fourth-quarter mean}}{\text{Grand mean}} = \frac{771.4}{683.6} = 1.1284$$

These seasonal relatives indicate that first-quarter sales are typically 75.67% of the average quarter's sales and second-quarter sales are at almost the grand average level. Clearly, this business does the majority of its sales in the third and fourth quarters. In the third quarter, we may expect sales to be 12.76% above the average for a quarter; in the fourth quarter, we may expect them to be 12.84% above the average for a quarter.

These seasonal relatives also allow us to estimate annual revenues from quarterly figures. As soon as first-quarter revenues are known, we can divide that figure by .7567 to estimate annual revenues.

By dividing any quarterly figure by its seasonal relative, we get the seasonally adjusted figure for that quarter, which we may compare with any other seasonally adjusted quarterly figure. In Example 11.10, data for Computer Associates are seasonally adjusted.

EXAMPLE 11.10 ■ SEASONALLY ADJUSTING DATA FOR COMPUTER ASSOCIATES

Column 2 of Table 11.6 contains the unadjusted quarterly data for Computer Associates International. Column 3 contains the seasonal adjustment factors developed in Example 11.9. Dividing the number in Column 2 by the number in Column 3, we come up with the seasonally adjusted data in Column 4. (Data are in millions of dollars.)

Table 11.6

SEASONALLY ADJUSTED DATA FOR COMPUTER ASSOCIATES INTERNATIONAL

Year: Quarter		Original Data	Seasonal Relatives	Adjusted Data
1993:	1	367.0	.7567	485.0
	2	432.0	.9873	437.6
	3	501.5	1.1276	444.7
	4	540.0	1.1284	478.6
1994:	1	423.4	.7567	559.5
	2	517.0	.9873	523.7
	3	574.4	1.1276	509.4
	4	633.7	1.1284	561.6
1995:	1	426.6	.7567	563.8
	2	623.3	.9873	631.3
	3	721.0	1.1276	639.4
	4	802.0	1.1284	710.7
1996:	1	577.5	.7567	763.2
	2	812.3	.9873	822.7
	3	1004.0	1.1276	890.4
	4	1110.0	1.1284	983.7
1997:	1	792.0	.7567	1046.6
	2	990.1	.9873	1002.8
	3	1053.0	1.1276	933.8

These data yield a somewhat different picture from that shown in Figure 11.11 on page 418. Once seasonal factors have been accounted for, growth in revenues appears much more regular. Apart from a peak in the first quarter of 1994 and another at the beginning of 1997, growth appears to have followed a steady upward path until the last two quarters plotted in Figure 11.15 on page 426. This downturn suggests that factors other than purely seasonal effects are at work and some policy changes might be in order.

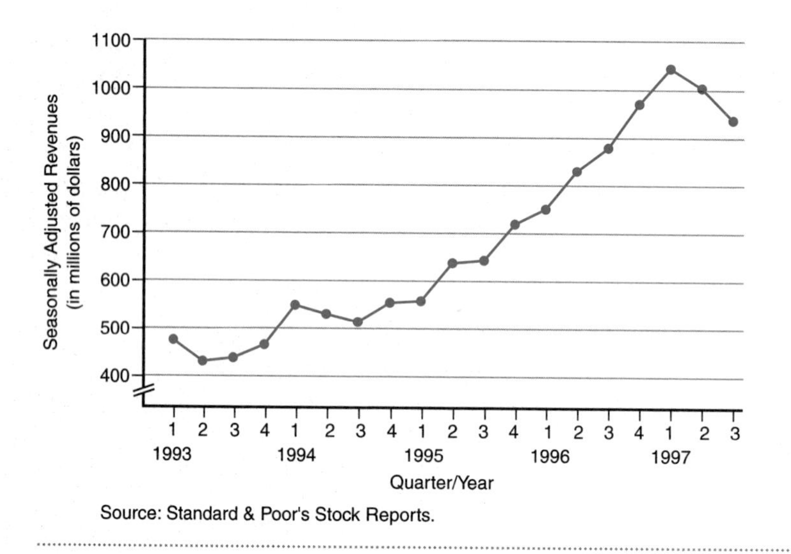

Figure 11.15 *Seasonally adjusted revenues for Computer Associates International*

Source: Standard & Poor's Stock Reports.

Forecasting Using Trend and Seasonal Factors

Making quarterly forecasts that extend a year or more into the future calls for methods beyond exponential smoothing and moving averages. A combination of regression analysis and seasonal adjustment techniques is often used. Consider the data for Home Depot, given in Example 11.11.

EXAMPLE 11.11 ■ USING TRENDS AND SEASONAL ADJUSTMENTS TO FORECAST REVENUE FOR HOME DEPOT

Forecasting with trend analysis and seasonal adjustment techniques

Home Depot operates a chain of over 500 retail warehouse-type stores throughout the United States. Suppose we wish to prepare a set of quarterly revenue forecasts for 1997. Table 11.7 gives quarterly data (in millions of dollars) for the first quarter of 1993 through the fourth quarter of 1996.

Table 11.7

REVENUES FOR HOME DEPOT, 1993–1996

	1993	1994	1995	1996
Quarter 1	1640	2180	2872	4362
Quarter 2	1856	2454	3287	4152
Quarter 3	1834	2317	3240	3998
Quarter 4	1818	2287	3077	3752

To begin, we establish a trend line by regressing revenues on time (see Chapter 10 for details of the regression procedure). The regression analysis yields the following equation:

$$Y_t = 1300.075 + 178.859X$$

where $t = 1$ for the first quarter of 1993.

Next, we calculate the seasonal adjustment factors, which are .9842, 1.0461, .9962, and .9735 for quarters 1, 2, 3, and 4, respectively. To obtain a forecast for each quarter of 1997, we calculate the trend line values Y_{17}, Y_{18}, Y_{19}, and Y_{20}. These values are then modified by the seasonal adjustment factors. Remember that, in seasonally adjusting a quarter's results, we divided the actual by the seasonal relative to get the seasonally adjusted value, a value close to the trend line value. Here, we multiply the trend line value by the seasonal relative to get the forecast for the quarter. Table 11.8 shows the results.

Table 11.8

FORECASTS FOR HOME DEPOT, FIRST QUARTER–FOURTH QUARTER, 1997

	Trend Line Value	Seasonal Relative	Forecast Value	Actual Value
Quarter 1	4341	.9842	4411	4362
Quarter 2	4520	1.0461	4321	5293
Quarter 3	4698	.9962	4716	4922
Quarter 4	4877	.9735	5010	4959

The last column of Table 11.8 contains Home Depot's actual revenues for the four quarters of 1997. While the forecast deviates by nearly 20% for the second quarter of 1997, it falls between 1% and 5% of the actual value in the other three quarters.

EXERCISES 11.13–11.22

Applying the Concepts

You may wish to use a computer to complete these exercises.

11.13 Compaq Computer is a major manufacturer of desktop computers, portable computers, and PC servers. The following data set shows revenues (in millions of dollars) for the period from 1987 to 1996.

Year	Revenues
1987	309
1988	709
1989	1030
1990	1296
1991	1348
1992	1509
1993	1841
1994	2148
1995	2623
1996	3505

Source of Data: Standard & Poor's *100 Best Growth Stocks,* 1998.

a. Use a three-year moving average to prepare a forecast for 1997.
b. Use simple exponential smoothing to prepare a forecast for 1997.
c. Calculate the mean average deviation (MAD) for both methods.
d. Which procedure would you choose in this instance?

11.14 Sales data for Staples, Inc., a leading operator of office product superstores, appear below.

	1993	1994	1995	1996	1997
Quarter 1	187.4	251.0	398.0	668.8	916.8
Quarter 2	192.9	241.3	385.8	605.0	808.1
Quarter 3	240.2	301.1	552.1	818.8	1079
Quarter 4	262.5	328.4	664.7	975.5	1164

Source of Data: Standard & Poor's *100 Best Growth Stocks*, 1998.

a. Compute the seasonal relatives.
b. Which season accounts for the highest sales?
c. It has been reported that revenues for the first quarter of 1998 were $1155 million. Annual revenues for 1997 were $3968 million. Will annual revenues increase in 1998?

11.15 Schering-Plough is a leading producer of pharmaceuticals. Following are the company's annual revenues (in millions of dollars) for the 1987–1996 period.

1987	1988	1989	1990	1991	1992	1993	1994	1995	1996
2699	2969	3158	3323	3616	4056	4341	4657	5104	5656

Source of Data: Standard & Poor's *100 Best Growth Stocks*, 1998.

a. Prepare a forecast for 1997 using a three-year moving average.
b. Prepare a forecast using simple exponential smoothing.
c. Which forecast would you use? Why?

11.16 Gateway 2000, Inc. is a mail-order personal computer marketer. Quarterly revenues (in millions of dollars) for the company appear below.

	1992	1993	1994	1995	1996
Quarter 1	236.9	421.0	615.0	776.0	1142.0
Quarter 2	248.6	364.7	616.5	766.4	1137.0
Quarter 3	267.8	400.0	644.4	888.7	1203.0
Quarter 4	353.7	545.9	824.3	1245.0	1553.0

a. Calculate the trend line for these data.
b. Find the seasonal adjustment factors.
c. Prepare quarterly forecasts for 1997.

11.17 Samsung is South Korea's largest industrial conglomorate. It is into electronics, machinery, petrochemicals, ships, and even a baseball team. Sales (in billions of won) for the company between 1984 and 1993 appear below.

Year	Sales
1984	2,568
1985	3,802
1986	4,275
1987	5,670
1988	6,811
1989	7,613
1990	7,952
1991	10,199
1992	12,055
1993	13,321

Source of Data: *Hoover's Handbook of World Business*, 1995–1996.

a. Use a three-year moving average to prepare a forecast for 1994.

b. Prepare a forecast for 1994, using simple exponential smoothing.
c. Compute the mean average percentage error, and compare your results in parts a and b.
d. Find current data on Samsung and evaluate your forecast. Construct a forecast for next year.

11.18 Coca-Cola Enterprises is the world's largest bottler of Coca-Cola beverage products. Revenues (in millions of dollars) are presented below for the first quarter of 1993 to the fourth quarter of 1996.

	1993	1994	1995	1996
Quarter 1	1208	1320	1462	1600
Quarter 2	1448	1610	1827	2016
Quarter 3	1487	1595	1841	2187
Quarter 4	1322	1486	1643	2118

Source of Data: Standard & Poor's *100 Best Growth Stocks,* 1998.

a. Prepare seasonal indices for Coca-Cola.
b. When are highest sales expected?
c. Compute the trend for these data.
d. Prepare a forecast for the first quarter of 1997.

11.19 As a labor economist, you are interested in forecasting New York State unemployment for 1998. The numbers of unemployed (in thousands) for 1970–1997 appear below.

Year	No. of Unemployed	Year	No. of Unemployed
1970	331	1984	583
1971	491	1985	544
1972	501	1986	526
1973	404	1987	412
1974	482	1988	258
1975	727	1989	442
1976	789	1990	467
1977	705	1991	636
1978	604	1992	747
1979	571	1993	677
1980	597	1994	595
1981	612	1995	539
1982	684	1996	537
1983	688	1997	565

Source of Data: New York State Department of Labor, *Employment Review,* August 1998.

a. Why would applying a trend line be inappropriate in this situation?
b. What procedure would be most useful here? Why?
c. Apply the procedure you recommend.
d. Complete this paragraph: Unemployment in 1998 is expected to (increase/decrease) from 1997 levels. There will be an estimated _____ thousand individuals in New York State seeking employment during the 1998 calendar year. Assuming a civilian labor force of 8,243,000, the average unemployment rate for 1998 will be _____%. (*Hint:* The unemployment rate is the number unemployed divided by the total civilian labor force.)

11.20 Of great concern to economic planners is the domestic level of energy consumption and production. The following data show U.S. energy consumption and production (in quadrillion BTU) for 1970–1991.

Year	Production	Consumption	Year	Production	Consumption
1970	62.1	66.4	1981	64.4	74.0
1971	61.3	67.9	1982	64.0	70.8
1972	62.4	71.3	1983	61.3	70.5
1973	62.1	74.3	1984	65.9	74.1
1974	60.8	72.5	1985	64.8	74.0
1975	59.9	70.5	1986	64.3	74.2
1976	59.9	74.4	1987	64.9	76.8
1977	60.2	76.3	1988	66.1	80.2
1978	61.1	78.1	1989	66.1	81.3
1979	63.8	78.9	1990	67.4	81.3
1980	64.8	76.0	1991	67.5	81.5

Source of Data: *Statistical Abstract of the United States,* 1993–1994.

a. Using a three-year moving average, forecast 1992 energy production.
b. Using a three-year moving average, forecast 1992 energy consumption.
c. According to your forecasts for 1992, how will energy purchases affect the country's net trade (the difference between domestic consumption and production) in 1992?
d. Repeat parts a through c using simple exponential smoothing.
e. Which procedure yields a better estimate?

f. Using the latest edition of the *Statistical Abstract of the United States,* evaluate your forecast. Then repeat the process to forecast next year's energy supply and its disposition. Write a brief paragraph stating your findings. What are your recommendations concerning energy policy?

11.21 Oxford Health operates managed health care plans in the New York metropolitan area. Since 1992, revenues have grown from $151 million to over $3 billion per annum. Quarterly revenues have risen from $57.10 million in the first quarter of 1993 to $987.3 million in the first quarter of 1997.

	1993	1994	1995	1996	1997
Quarter 1	57.10	125.5	331.1	658.1	987.3
Quarter 2	68.50	156.4	409.1	725.3	
Quarter 3	85.80	199.6	480.2	811.3	
Quarter 4	94.30	232.1	537.4	880.3	

Source of Data: Standard & Poor's *100 Best Growth Stocks,* 1998.

a. Compute the seasonal relatives.
b. Estimate a trend line for these data.
c. Forecast revenues for the second quarter of 1997 to the first quarter of 1998.
d. Write a brief paragraph about your findings in parts a through c.
e. Compare your forecasts with the actual results for Oxford.

11.22 Lowe's Companies is a retailer of building supplies, with over 400 stores in 24 states. Revenues (in millions of dollars) exhibit much seasonality.

	1993	1994	1995	1996	1997	1998
Quarter 1	883	992	1397	1635	1907	2401
Quarter 2	1062	1242	1647	1978	2459	
Quarter 3	991	1158	1579	1766	2193	
Quarter 4	910	1146	1487	1697	2042	

Source of Data: Standard & Poor's *100 Best Growth Stocks*, 1998.

a. Compute the seasonal relatives.
b. Estimate a trend line for these data.
c. Forecast revenues for the second quarter of 1998 to the first quarter of 1999.
d. Write a brief paragraph about your findings in parts a through c.

FORMULAS

Calculation of Adjusted r^2

$$r_{\text{adj}}^2 = 1 - (1 - r^2)\frac{n - 1}{n - p - 1}$$

Definition of the ith Error, or ith Residual

$$e_i = y_i - y_{ci}$$

Standardized Residual

$$\text{sr}_i = \frac{e_i}{s_{y|x}\sqrt{1 + \frac{1}{n} + \frac{(x - \bar{x})^2}{\text{SS}_{xx}}}}$$

The Durbin-Watson Statistic

$$d_{\text{test}} = \frac{(e_i - e_{i-1})^2}{e_i^2}$$

First Differences

$$Y_i - Y_{i-1} = f(X_i - X_{i-1})$$

Three-Period Moving Average (Centered)

$$\text{MA}_t = \frac{Y_{t-1} + Y_t + Y_{t+1}}{3}$$

Simple Exponential Smoothing Formulas

$$E_{i+1} = wY_i + (1 - w)E_i$$

Note that in the Excel version the weighting factor w is denoted by α and $\alpha = 1 - w$.

Mean Average Percentage Error

$$\text{MAPE} = \frac{\sum |(y_t - \hat{y}_t)/y_t|}{n} \times 100, \quad y_t \neq 0$$

where y_t is the actual value, $\hat{y}_t$ is the forecast value, and n is the number of observations.

Mean Absolute Deviation

$$\text{MAD} = \frac{\sum |y_t - \hat{y}_t|}{n}$$

Mean Squared Deviation

$$\text{MSD} = \frac{\sum (y_t - \hat{y}_t)^2}{n}$$

Seasonal Adjustment

$$\text{Adjusted data} = \frac{\text{Original data}}{\text{Seasonal relative}}$$

$$\text{Quarterly estimate} = \text{Trend line value} \times \text{Seasonal relative}$$

NEW STATISTICAL TERMS

analysis of variance (ANOVA)
autocorrelation
cyclical component
dummy variable
Durbin-Watson (DW) statistic
error terms
global utility
grand mean
heteroscedasticity
homoscedasticity
indicator variable
irregular component
linearity
longitudinal data
mean absolute deviation (MAD)
mean average percentage error (MAPE)
mean squared deviation (MSD)
moving average technique
multicollinearity
multiple regression model
normality
residual analysis
seasonal adjustment
seasonal component
seasonal relative
serial correlation
simple exponential smoothing
standardized residuals
time series data
trend component

EXERCISES 11.23–11.30

Supplementary Problems

11.23 A study of the New York State lottery (A. C. Ovedovitz, "Factors Influencing Public Lottery Revenues," in William R. Eadington, ed., *Gambling Research: Proceedings of the Seventh International Conference on Gambling and Risk Taking,* vol. 2, 1988, pp. 68–82) analyzed the effect of jackpot size (in thousands of dollars), players' weekly earnings, and the unemployment rate in New York State on lottery sales (in thousands of dollars). A sample of 25 observations was taken from the data set.

Observation	Sales	Jackpot	Earnings per Week	Unemployment Rate
1	$ 6,204	$ 3,000	$361.30	7.3%
2	4,015	3,000	361.30	7.3
3	5,916	3,046	366.73	7.0
4	5,409	7,362	366.73	7.0
5	5,460	4,100	366.73	7.0
6	4,158	3,083	366.73	8.1
7	7,764	8,488	361.82	8.1
8	6,215	7,823	361.82	8.1
9	6,743	3,204	365.04	8.1
10	4,316	3,197	365.09	8.1
11	6,137	3,218	365.09	8.1
12	4,681	5,111	362.34	7.7
13	10,095	10,253	362.34	7.7
14	4,517	3,000	362.34	7.7
15	6,829	3,407	364.57	6.5
16	4,606	3,000	364.57	6.5
17	6,779	6,000	364.57	6.5
18	24,431	22,854	363.37	7.2
19	5,862	3,008	363.37	7.2
20	7,998	4,010	363.37	7.2
21	5,003	3,000	364.45	7.2
22	10,520	11,409	364.19	8.2
23	4,701	3,000	364.19	8.2
24	6,807	3,500	364.19	8.2
25	18,651	20,000	363.13	7.5

a. Construct a multiple regression model to predict sales. Start with sales as a function of jackpot. Use r^2_{adj} in building your model.

b. Use the model to predict sales with a $5 million jackpot, a $10 million jackpot, and a $20 million jackpot. Assume that earnings are $400 per week and the unemployment rate is 6.0%. (*Hint:* Use only the data you need for your final model.)

c. What percentage of variation is explained by the model?

d. Assess the global utility of the model.

e. Answer the following questions only if the model allows you to do so (if it doesn't, state why you cannot answer the question and suggest how the situation might be remedied): What is the effect of an increase of $25 in weekly earnings? What is the effect of an increase of $1 million in jackpot size? What is the effect of an increase of .5% in the unemployment rate?

f. Use residual analysis to test the assumptions of normality, homoscedasticity, serial correlation, and linearity.

g. Is there any problem of multicollinearity in this model? Give your reasoning.

h. Does jackpot size bear on sales? Summarize your findings, using terms suitable for a presentation to the nonstatistician who is lottery manager.

11.24 To the data in Exercise 11.23, add an indicator variable for day of the week. The values for the indicator variable are given below; Wednesday = 1, Saturday = 0.

Observation	Value	Observation	Value
1	0	14	1
2	1	15	0
3	0	16	1
4	1	17	0
5	0	18	0
6	1	19	1
7	0	20	0
8	1	21	1
9	0	22	0
10	1	23	1
11	0	24	0
12	1	25	1
13	0		

a. Construct a model with sales as a function of jackpot and day of the week.

b. Assess the global utility of the model.

c. Does day of the week have any effect on sales?

d. Assuming the day of the week is significant, write the main equation and the two equations it gives rise to describing sales of lottery tickets.

11.25 The Long Island Railroad is the largest commuter railroad in the United States. The data below, on ridership (REVPAS), population (BICTYPOP, short for bicounty population), and fares (HICKFARE, as represented by the fare from Hicksville station), were collected over a 27-year period.

Year	REVPAS	BICTYPOP	HICKFARE
1954	78.0	1.4	25.66
1955	76.4	1.5	25.66
1956	75.3	1.6	25.66
1957	73.9	1.7	29.66
1958	73.6	1.8	29.66
1959	73.9	1.9	33.16
1960	64.4	2.0	35.45
1961	69.1	2.0	35.45
1962	71.3	2.1	35.45
1963	71.5	2.2	35.45
1964	77.8	2.2	37.75
1965	74.0	2.3	37.75
1966	73.8	2.3	37.75
1967	74.2	2.4	37.75
1968	73.7	2.4	42.35
1969	72.4	2.5	42.35

(cont.)

1970	70.1	2.6	46.95
1971	69.6	2.6	46.95
1972	59.8	2.6	54.85
1973	57.0	2.6	54.85
1974	66.2	2.6	63.84
1975	67.2	2.7	63.84
1976	67.4	2.7	63.84
1977	69.5	2.7	63.84
1978	72.5	2.7	63.84
1979	77.4	2.7	63.84
1980	80.8	2.6	76.60

a. Develop a multiple regression model to predict revenue passengers.
b. What percentage of variation is explained by your model?
c. Assess the global utility of the model.
d. Evaluate the model with respect to the assumptions of normality, homoscedasticity, serial correlation, and linearity.
e. The correlations for these variables are shown in the Minitab printout below:

MTB > CORR C1-C4

	YEAR	REVPAS	BICTYPOP
REVPAS	−0.258		
BICTYPOP	0.952	−0.383	
HICKFARE	0.965	−0.262	0.867

What does this information tell you about multicollinearity?
f. Summarize your findings about your model. How might you go about improving it?

Exercises 11.26, 11.27, and 11.28 present data typical of those used by company forecasters to provide managers with target numbers up to three years in the future.

11.26 John Deere and Company are major producers of farm and industrial equipment. To better understand company sales, analysts gathered data on sales, total acreage planted nationally, livestock, and prices received by farmers.

Year	Sales	Acreage	Livestock	Prices
1980	264.4	1039	60.6	134
1981	315.0	1023	52.2	134
1982	332.0	1018	48.7	142
1983	345.4	1012	48.8	128
1984	359.6	1005	45.5	123
1985	375.5	999	47.1	127
1986	398.8	995	57.2	138
1987	419.4	991	61.5	148
1988	440.0	987	65.5	150

a. Solve for the regression equation.
b. Assess the global utility of the equation.
c. Does total acreage have a bearing on sales? (*Hint:* Use the t test to answer this question.)
d. What is the standard error of the estimate for the equation? What does it tell you?
e. What percentage of variation in sales is explained by this model?

f. Complete this paragraph: A model relating sales of John Deere and Company to agricultural sector activity proved to explain ____% of the variation in sales. This model (exhibits/does not exhibit) global utility. The model is characterized by a (large/small) standard error of the estimate. Prices received by farmers (did/did not) play a role in overall sales.

11.27 Two chains—Red Lobster and Olive Garden—make up the General Mills Restaurant Group. To explain sales, a multiple regression model was developed. Data for the model included sales (in millions of dollars) and disposable income (in billions of dollars).

Year	Sales	No. of Restaurants	Disposable Income
1973	54.6	67	914
1974	77.8	95	998
1975	113.6	134	1092
1976	180.7	181	1194
1977	240.9	270	1314
1978	354.9	312	1474
1979	436.3	345	1650
1980	525.7	379	1900
1981	704.0	447	2042
1982	839.4	508	2181

a. Solve for the multiple regression equation.
b. Does the equation exhibit global utility?
c. Test the variable "No. of Restaurants" for significance at the .05 level.
d. Test the variable "Disposable Income" for significance at the .05 level.
e. What is the effect of a $1 billion change in disposable income on sales?
f. What is the effect of an increase of 10 restaurants on the level of sales?
g. Predict sales with 525 restaurants and a disposable income of $2300 billion.
h. Write a paragraph summarizing your results. (*Hint:* Use the paragraph in Exercise 11.26 as a model.)

11.28 The Coca-Cola Company is the world's largest soft drink company. Data on sales (in millions of dollars), disposable personal income (in billions of dollars), and U.S. population (in millions) appear below.

Year	Sales	Disposable Income	Population
1970	1606.4	695.3	205
1971	1728.8	751.8	208
1972	1876.2	810.3	210
1973	2145.0	914.5	212
1974	2522.2	998.3	214
1975	2872.8	1096.1	216
1976	3032.8	1194.4	218
1977	3559.9	1314.0	220
1978	4337.9	1474.0	223
1979	4961.4	1650.2	225
1980	5912.6	1828.9	228
1981	5889.0	2047.6	230
1982	6249.7	2176.5	232
1983	6829.0	2335.6	234

a. Construct a forecasting equation for Coca-Cola's sales.
b. Is the coefficient for disposable personal income significantly different from zero?
c. Does the model have global significance?
d. How much variation in sales does the model explain?
e. If disposable personal income were $2500 billion and population were 240 million, predict the level of sales with 95% confidence.
f. Write a paragraph summarizing your results, for presentation to a long-range planning group at Coca-Cola.

11.29 The Seagram Company Ltd. is the second-largest distiller in the world. Sales for the period between 1985 and 1994 grew at a rate of 8.8% per annum. Annual data for the 1985–1994 period are reproduced below. (Sales are in millions of dollars.)

Year	Sales
1985	2821
1986	2971
1987	3345
1988	3815
1989	5056
1990	5582
1991	6127
1992	6345
1993	6101
1994	6038

Source of Data: *Hoover's Handbook of World Business,* 1995–1996.

a. Prepare a forecast for 1995 using a three-year moving average.
b. Prepare a forecast for 1995 using simple exponential smoothing.
c. Calculate MAPE.
d. Which forecast would you use? Why?

e. Update the series above, and check your forecast. Prepare a forecast for next year based on current data.

11.30 Microsoft Corporation develops and markets systems and applications software. Quarterly revenues (in millions of dollars) from the first quarter of 1993 to the third quarter of 1997 are presented below.

	1993	1994	1995	1996	1997
Quarter 1	$ 818	$ 983	$1247	$2016	$2295
Quarter 2	938	1129	1482	2195	2680
Quarter 3	958	1244	1587	2205	3208
Quarter 4	1039	1293	1621	2255	

a. Calculate a trend line for these data.
b. Find the seasonal adjustment factors.
c. Prepare quarterly forecasts for the fourth quarter of 1997 through the third quarter of 1998.

d. Update these data using current information from Standard and Poor's, Value Line, or Moody's. Evaluate your forecasts, and prepare a new forecast.

Appendix A

Data Sets

Data Set I MERCURY LEASE SURVEY

Data Set II TAMMY'S RESTAURANT SURVEY

Note: These data sets are available in Minitab and Excel format on the CD-ROM bound in the back of this book.

DATA SET I ■ MERCURY LEASE SURVEY: DATA ON DEMOGRAPHICS OF MERCURY LEASE HOLDERS

Explanation of Columns:

C1 Vehicle code

Capri	20200
Cougar	20400
Grand Marquis	20504
Topaz	20700
Sable	21000
Tracer	22000
Villager	61000

C2 Owner's state

State	Code	State	Code	State	Code
Alaska	01	Louisiana	18	Ohio	35
Alabama	02	Massachusetts	19	Oklahoma	36
Arkansas	03	Maryland	20	Oregon	37
Arizona	04	Maine	21	Pennsylvania	38
California	05	Michigan	22	Rhode Island	39
Colorado	06	Minnesota	23	South Carolina	40
Connecticut	07	Missouri	24	South Dakota	41
Delaware	08	Mississippi	25	Tennessee	42
Florida	09	Montana	26	Texas	43
Georgia	10	North Carolina	27	Utah	44
Hawaii	11	North Dakota	28	Virginia	45
Iowa	12	Nebraska	29	Vermont	46
Idaho	13	New Hampshire	30	Washington	47
Illinois	14	New Jersey	31	Wisconsin	48
Indiana	15	New Mexico	32	West Virginia	49
Kansas	16	Nevada	33	Wyoming	50
Kentucky	17	New York	34		

C3 Zip code

C4 Ride quality

Better than expected	3
About what expected	2
Worse than expected	1

C5 Quietness of the engine
(See choices under C4.)

C6 Power and pickup
(See choices under C4.)

C7 Satisfaction with salesperson's concern for your vehicle needs

Completely satisfied	10
	9
Very satisfied	8
	7
Fairly satisfied	6
	5
Somewhat dissatisfied	4
	3
Very dissatisfied	2
	1

C8 Satisfaction with experience of taking delivery of your vehicle
(See choices under C7.)

C9 Did salesperson use a delivery checklist?

Yes	3
No	2
Don't know	1

C10 Did anyone explain what to do if you need service?

Yes	3
No	2
Don't know	1

C11 Gender

Male	1
Female	2

C12 Age

C13 Education

< HS	10
High School Grad	12
Some College	14
College Graduate	16
Post Graduate	18

C14 Race

Black	1
Hispanic	2
Oriental	3
White	4
Other	5
N/A	0

```
Worksheet size:   100000 cells

MTB > Retrieve 'A: \MERCURY.MTW'
Retrieving worksheet from file: A:\MERCURY.MTW
Worksheet was saved on 7/5/1994
MTB > info
```

Column	Name	Count	Missing
C1	VEHICLE	250	0
C2	STATE	250	1
C3	ZIP	250	0
C4	VRQ	250	3
C5	VQE	250	4
C6	VPP	250	3
C7	SSVN	250	0
C8	SEV	250	0
C9	SUDC	250	8
C10	WTDFS	250	1
C11	SEX	250	19
C12	AGE	250	7
C13	EDU	250	19
C14	RACE	250	16

```
MTB > print C1 - C14
```

Row	C1	C2	C3	C4	C5	C6	C7	C8	C9	C10	C11	C12	C13	C14
1	22000	6	81001	2	3	2	10	10	2	3	1	42	12	2
2	22000	22	49001	2	3	3	8	9	3	3	2	57	18	4
3	21000	34	11001	2	2	2	10	7	1	3	2	67	12	4
4	20700	6	81001	3	3	3	10	10	*	3	1	27	14	4
5	20400	34	14001	3	3	2	10	5	3	3	2	37	16	4
6	21000	35	44001	3	3	3	10	10	1	3	1	32	16	4
7	20400	38	19001	2	2	3	8	4	3	3	1	27	14	0
8	21000	38	15001	3	3	3	10	10	3	3	*	67	10	4
9	21000	50	82001	3	2	3	9	9	3	3	1	67	12	4
10	20400	15	47001	3	3	1	9	9	3	3	2	27	14	4
11	20504	35	44001	*	*	*	0	0	3	3	*	67	18	*
12	20400	35	44001	2	2	3	10	10	3	3	2	42	18	4
13	21000	50	82001	3	3	3	10	10	3	3	1	37	14	4
14	22000	5	96001	3	2	3	7	8	3	3	2	67	14	4
15	21000	31	7004	1	2	2	9	9	2	3	1	57	12	4
16	21000	35	43004	3	3	2	10	9	3	3	2	57	14	4
17	20504	14	60004	3	3	3	10	10	3	3	1	67	14	4
18	21000	38	19004	2	2	2	7	7	3	3	1	47	18	4
19	20504	29	68004	3	3	3	10	10	1	3	1	62	12	4
20	22000	5	93004	2	2	2	10	10	3	3	2	67	10	4
21	21000	17	40004	2	3	3	10	10	3	3	1	42	16	4
22	61000	14	60004	2	3	3	10	10	3	3	1	37	18	4
23	61000	14	60004	1	2	2	8	6	3	3	2	42	18	4
24	61000	47	98004	2	2	2	6	3	2	3	2	37	16	3
25	61000	34	14004	1	1	2	9	9	3	3	1	42	16	4
26	61000	22	49004	3	2	3	9	9	1	3	2	32	16	0
27	22000	6	81005	3	3	2	10	10	1	3	2	27	12	4
28	21000	6	81005	3	3	3	10	10	3	3	1	67	12	4
29	20504	31	7005	3	3	3	10	10	3	3	1	47	12	4
30	20504	22	49008	2	2	2	9	9	3	3	1	67	16	4
31	20504	43	77008	3	3	3	10	10	3	3	2	*	*	4
32	20504	5	92008	3	3	2	10	10	1	2	1	62	16	4
33	61000	38	19008	3	3	3	10	10	3	3	2	32	12	4
34	61000	5	90008	3	3	3	8	8	3	3	1	27	14	1
35	61000	22	49008	2	2	2	10	10	1	3	2	42	16	4
36	61000	22	49008	2	3	3	7	8	3	3	1	47	18	4
37	61000	5	95008	2	2	2	8	7	3	3	1	*	16	3
38	61000	5	95008	2	2	3	8	9	1	3	2	42	18	4
39	61000	5	95008	2	3	2	10	10	3	3	2	37	*	4
40	61000	22	49008	3	3	3	10	10	3	3	1	32	14	4
41	21000	22	48009	3	3	3	10	10	1	3	2	62	16	4
42	20504	9	33009	2	1	1	0	9	3	3	*	62	14	*
43	20504	9	33009	3	3	3	10	10	3	3	1	67	16	4
44	21000	22	49009	2	2	2	8	8	1	3	1	62	18	4
45	20504	5	91011	3	3	3	8	8	3	3	*	57	16	4
46	20400	35	45011	*	*	*	10	10	2	3	1	42	10	4
47	22000	6	80011	3	3	3	7	7	3	3	2	42	12	4
48	21000	*	20011	2	3	3	10	10	3	3	2	27	10	1
49	20700	24	63011	2	2	2	10	10	3	3	2	52	14	4
50	20504	36	74011	2	2	2	10	10	3	3	1	42	16	4

Row	C1	C2	C3	C4	C5	C6	C7	C8	C9	C10	C11	C12	C13	C14
51	20202	24	63011	3	2	2	10	10	3	3	2	22	16	4
52	61000	35	44011	3	3	3	8	8	3	3	2	42	16	4
53	61000	5	91011	2	3	1	10	10	3	3	2	37	18	4
54	61000	35	45011	1	2	3	10	10	3	3	2	37	14	4
55	21000	38	15012	3	3	3	10	10	3	3	1	37	14	4
56	21000	22	49012	2	2	2	6	10	3	3	1	67	18	4
57	21000	48	53012	2	3	2	10	9	3	3	1	47	14	*
58	20504	6	80012	2	2	2	10	10	3	3	1	57	16	4
59	21000	48	53012	1	2	2	10	10	3	3	1	67	10	4
60	22000	5	94014	3	3	3	10	10	3	3	2	37	*	1
61	20700	35	45014	3	3	2	10	10	3	3	1	57	16	4
62	20504	33	89014	3	3	2	10	10	3	3	*	62	14	4
63	20504	38	19014	3	3	3	10	10	3	3	2	62	12	4
64	21000	33	89014	2	2	2	10	2	3	3	1	42	18	*
65	20400	33	89014	3	3	2	10	10	*	3	1	57	12	5
66	20400	36	74014	2	3	2	10	10	3	3	1	67	12	4
67	20504	48	53014	3	3	3	10	10	3	3	1	67	10	4
68	61000	20	21014	3	3	3	10	10	1	3	2	27	16	4
69	61000	20	21014	3	3	3	7	10	1	3	2	32	14	4
70	61000	5	95014	3	3	2	9	9	3	3	2	37	18	3
71	61000	9	33014	3	3	3	10	10	3	3	2	37	16	0
72	61000	9	33014	3	3	3	10	10	1	3	1	32	*	4
73	61000	43	77014	3	3	3	10	10	3	3	1	42	16	4
74	61000	24	64014	2	3	3	9	9	2	3	1	67	18	4
75	21000	24	63017	3	3	3	10	10	3	3	1	42	18	4
76	21000	22	48017	2	2	3	9	9	3	3	*	62	12	*
77	21000	22	49017	2	3	3	10	10	3	3	*	67	18	1
78	21000	35	44017	3	3	3	10	10	3	3	2	27	16	4
79	22000	38	18017	3	2	3	10	10	3	3	*	47	14	4
80	21000	34	10017	2	2	3	10	10	3	3	1	62	18	4
81	61000	22	48017	2	3	3	6	3	1	3	2	42	16	4
82	61000	6	80017	2	*	2	10	10	3	3	1	37	16	*
83	61000	36	74017	2	3	3	10	10	3	3	1	42	14	*
84	61000	22	49017	2	2	2	10	10	1	3	1	32	16	4
85	61000	24	63017	2	2	2	10	10	3	3	2	37	16	4
86	61000	22	49017	2	3	3	9	9	3	3	1	27	14	4
87	61000	35	43017	2	2	3	10	10	1	3	1	37	16	4
88	61000	22	48017	3	3	2	10	9	3	3	1	37	16	4
89	21000	38	19018	3	1	2	9	9	3	3	1	62	12	4
90	20504	19	2021	2	2	2	9	9	3	2	1	52	12	4
91	20504	34	10021	3	3	2	10	10	3	3	1	27	18	4
92	20400	31	8021	3	3	3	10	10	3	3	1	52	16	4
93	22000	22	48021	3	3	3	10	10	2	3	1	22	16	4
94	21000	9	33021	3	3	3	10	10	3	3	1	67	16	4
95	20504	34	13021	2	2	2	5	6	2	3	1	67	12	4
96	21000	22	48021	3	3	3	10	10	3	3	1	37	12	4
97	21000	34	13021	2	2	2	9	9	*	3	1	62	18	4
98	61000	12	50021	2	2	2	10	10	3	3	2	47	18	4
99	61000	34	13021	3	3	3	8	9	3	3	2	32	14	4
100	20504	5	94022	2	2	2	10	6	3	3	1	67	14	4

Row	C1	C2	C3	C4	C5	C6	C7	C8	C9	C10	C11	C12	C13	C14
101	21000	5	94022	3	3	2	9	9	3	3	1	47	18	4
102	21000	6	80022	1	2	2	10	10	1	3	*	*	*	*
103	20700	22	49022	3	3	2	10	10	3	3	2	67	12	4
104	20504	47	98022	3	3	3	10	10	*	3	1	67	12	*
105	20400	5	92025	3	3	3	6	5	2	3	*	67	*	4
106	20504	22	48025	3	3	3	10	10	3	3	1	67	16	4
107	20700	29	68025	3	3	2	10	10	*	3	*	67	12	4
108	21000	43	75025	2	3	3	3	6	3	3	1	47	14	4
109	61000	5	94025	3	3	2	10	8	3	3	1	67	14	4
110	61000	29	68025	2	2	2	10	10	1	3	1	47	12	4
111	61000	15	47025	3	3	2	10	10	3	3	1	42	14	4
112	61000	5	92025	3	2	3	10	10	1	3	2	42	16	4
113	61000	22	48025	2	3	3	10	10	1	3	2	42	18	4
114	61000	29	68025	3	3	3	10	10	3	3	1	32	16	4
115	61000	9	33025	2	2	2	10	10	2	3	2	32	16	4
116	61000	38	19025	2	2	2	10	10	1	3	1	32	16	4
117	61000	5	94025	3	3	3	10	10	3	3	1	52	18	4
118	61000	14	60025	2	2	3	5	4	2	2	1	37	18	4
119	61000	38	15025	3	2	2	10	10	3	3	1	32	*	4
120	21000	33	89030	2	2	2	10	10	3	3	2	67	14	4
121	20400	34	11030	3	3	3	10	10	3	3	*	*	*	*
122	21000	34	11030	3	3	3	10	10	3	3	*	*	*	*
123	20504	24	64030	3	3	3	10	10	*	3	1	67	12	4
124	20202	5	95030	1	2	2	9	9	3	3	1	52	16	4
125	20504	5	95030	3	3	3	10	10	3	2	1	67	*	4
126	20400	33	89030	2	2	2	8	10	3	3	2	22	14	1
127	20400	35	45030	2	2	2	10	10	3	3	1	67	10	4
128	21000	5	95030	3	3	3	10	10	3	3	1	57	16	4
129	61000	5	94030	3	3	3	10	10	3	3	2	37	16	4
130	61000	14	60030	3	3	3	10	10	3	3	2	32	18	4
131	61000	5	91030	2	3	2	9	6	3	3	1	62	14	3
132	61000	6	80030	2	3	1	8	8	3	3	1	47	16	1
133	61000	5	94030	3	3	3	6	8	2	2	2	37	14	3
134	61000	5	93030	2	3	3	10	10	3	3	2	47	14	2
135	20504	22	49033	2	3	2	9	5	1	3	1	67	12	4
136	20504	45	22033	2	2	2	8	8	3	3	*	67	16	4
137	21000	7	6033	3	3	3	8	6	1	3	2	47	14	4
138	21000	38	19033	3	3	3	10	10	3	3	2	42	12	4
139	61000	38	15033	3	3	2	10	10	3	3	1	37	16	4
140	61000	15	46033	2	2	2	10	9	3	3	2	32	14	4
141	61000	45	22033	1	3	3	8	6	3	3	2	52	18	4
142	61000	22	48033	2	2	2	10	10	3	3	2	22	14	4
143	61000	15	46033	3	3	3	10	10	3	3	2	37	16	4
144	61000	38	19033	3	3	3	10	10	3	3	2	42	14	4
145	20202	15	46034	2	3	3	10	10	3	3	2	37	16	4
146	20504	31	8034	2	3	3	10	10	3	3	1	57	*	4
147	21000	22	48034	2	2	2	10	10	3	3	1	62	12	4
148	20504	22	48034	2	2	3	8	8	3	3	1	67	14	4
149	20504	31	8034	2	2	2	10	10	3	3	1	42	14	4
150	21000	7	6037	2	2	2	10	10	3	3	2	42	16	4

Row	C1	C2	C3	C4	C5	C6	C7	C8	C9	C10	C11	C12	C13	C14
151	21000	47	98037	2	2	2	10	0	3	3	2	37	14	4
152	20202	20	21037	2	2	3	10	10	1	3	2	42	18	4
153	61000	31	8037	3	3	3	7	8	3	3	2	42	16	4
154	61000	15	47037	2	2	2	10	9	1	3	1	32	14	4
155	61000	36	74037	3	3	2	7	10	3	2	2	37	16	4
156	61000	16	67037	1	3	2	10	10	3	3	1	52	18	4
157	61000	5	92037	2	2	2	10	10	3	3	1	62	14	4
158	61000	5	92037	2	2	2	7	2	2	3	2	52	14	*
159	20504	22	48038	3	3	3	10	10	3	3	1	67	*	4
160	21000	22	48038	3	3	3	10	10	3	3	1	52	18	4
161	21000	35	44038	2	2	2	10	10	2	2	1	42	16	4
162	20400	22	48038	2	1	2	10	10	3	3	1	57	16	4
163	22000	38	17038	2	2	1	10	10	3	3	1	32	16	4
164	20504	47	98038	2	2	2	10	10	3	3	*	57	14	4
165	61000	7	6040	3	3	3	10	10	3	3	1	67	10	4
166	61000	7	6040	3	3	3	10	10	3	3	2	32	16	4
167	61000	24	63040	3	3	2	8	4	2	2	2	32	16	4
168	61000	47	98040	2	2	2	9	10	1	3	2	37	18	4
169	21000	33	89041	3	3	3	10	10	3	3	2	47	14	4
170	21000	12	51041	3	3	3	10	10	3	3	1	57	14	4
171	22000	44	84041	2	2	2	0	9	3	3	*	*	*	*
172	21000	38	19041	3	3	3	10	10	3	3	1	32	18	4
173	20700	35	44041	3	3	3	10	10	3	3	2	57	10	4
174	20400	33	89041	3	3	3	10	10	3	3	1	67	*	4
175	20504	43	77041	2	2	2	7	6	3	3	2	47	14	4
176	22000	43	77041	2	2	2	10	10	3	3	1	47	16	4
177	22000	38	19041	3	3	3	10	10	*	3	2	67	12	4
178	22000	44	84041	2	2	2	10	10	3	3	2	27	14	4
179	20202	5	93041	1	3	2	9	10	3	3	1	62	16	4
180	21000	22	48044	2	2	2	10	10	3	3	1	62	14	4
181	21000	35	44044	1	2	3	10	10	3	3	2	47	14	4
182	22000	20	21044	3	2	3	10	9	3	3	2	42	18	4
183	20700	22	48044	2	2	2	10	10	3	3	1	52	16	*
184	20504	36	74044	2	3	3	10	9	3	3	1	67	12	4
185	21000	22	48044	3	3	3	10	10	3	3	*	37	*	4
186	21000	14	60044	3	3	3	10	10	3	3	1	67	18	4
187	61000	38	15044	3	3	3	10	10	3	3	1	52	18	4
188	61000	31	7044	2	2	2	10	10	3	3	1	42	16	4
189	61000	24	63044	3	3	3	10	10	3	3	1	62	18	4
190	61000	22	48044	2	3	3	10	10	1	3	1	32	16	4
191	22000	5	90045	3	3	2	10	10	3	3	1	62	16	4
192	21000	38	15045	3	3	3	10	10	3	3	2	57	14	4
193	21000	22	48045	3	3	3	10	10	3	3	1	57	16	4
194	22000	5	95050	2	2	3	6	8	2	3	2	57	10	5
195	21000	14	60050	3	3	3	10	10	3	3	2	42	14	4
196	22000	6	81050	3	2	2	9	10	3	3	*	27	*	4
197	22000	35	44050	3	3	3	10	10	3	3	1	32	16	4
198	21000	22	48050	3	3	3	10	10	3	3	2	37	18	4
199	21000	24	64050	3	2	2	10	10	3	3	1	47	14	4
200	20504	14	60050	2	2	3	10	10	3	3	1	67	12	4

Row	C1	C2	C3	C4	C5	C6	C7	C8	C9	C10	C11	C12	C13	C14
201	61000	38	19050	3	3	3	10	10	3	3	2	27	16	4
202	61000	35	44050	2	2	2	10	10	3	3	1	32	*	4
203	61000	22	48050	1	3	3	9	8	3	3	1	42	16	4
204	61000	35	44050	2	2	2	10	10	3	3	1	52	16	4
205	61000	43	75050	3	3	3	6	4	3	3	1	42	18	4
206	20504	34	14051	2	2	2	10	10	3	3	2	57	14	4
207	22000	34	12051	2	2	2	10	10	3	3	1	67	16	4
208	22000	31	8051	2	2	2	9	9	1	3	2	32	14	4
209	20400	43	76054	2	2	2	10	10	3	3	1	52	12	4
210	22000	44	84054	2	2	3	10	10	3	3	2	42	14	4
211	21000	31	7054	1	2	2	6	6	2	3	2	42	12	4
212	22000	30	3054	2	2	2	8	8	3	3	2	52	12	4
213	20700	35	44054	3	3	2	10	10	3	3	2	37	14	4
214	21000	30	3054	3	3	3	10	10	3	3	2	37	12	4
215	20400	5	92054	3	3	3	10	9	1	3	1	67	16	4
216	21000	31	8054	3	3	3	10	10	3	3	1	47	18	4
217	20400	5	92054	2	2	2	10	10	3	3	1	62	16	4
218	20400	35	43054	3	3	3	10	10	3	3	1	47	18	4
219	20504	5	92054	3	2	3	10	10	3	3	1	62	18	4
220	20202	31	8054	3	3	3	10	10	3	3	1	27	14	4
221	21000	22	48054	2	2	3	10	10	1	3	1	32	*	4
222	1000	31	8054	3	3	3	10	10	3	3	2	37	18	4
223	21000	34	12058	2	2	3	10	10	3	3	2	57	16	4
224	22000	44	84058	1	2	3	10	9	3	3	1	67	16	4
225	21000	34	12058	3	3	2	10	10	3	3	1	67	10	4
226	20504	10	30058	2	3	2	10	10	3	3	1	47	14	4
227	22000	47	98058	3	3	2	9	8	3	3	2	42	12	4
228	61000	10	30058	3	3	2	10	10	1	3	1	37	18	4
229	61000	47	98058	2	1	2	9	10	1	3	1	57	16	4
230	20700	17	40059	1	2	2	10	10	3	3	2	47	16	4
231	61000	17	40059	2	2	2	10	10	3	3	1	47	16	4
232	22000	5	95060	2	2	2	6	9	3	2	2	22	14	3
233	20202	35	44060	2	2	3	10	4	3	3	2	52	14	4
234	22000	15	46060	2	2	3	10	10	3	3	1	47	12	4
235	22000	15	46060	1	2	3	10	10	3	3	2	52	12	4
236	20400	23	55060	2	2	2	10	8	3	3	2	37	14	4
237	20400	22	48060	*	*	*	10	10	*	3	2	52	*	4
238	20700	9	33063	2	2	2	3	5	1	2	1	*	12	4
239	21000	5	93063	3	3	2	10	10	3	3	1	52	16	4
240	20504	9	33063	3	3	3	10	10	3	3	*	67	14	*
241	20202	6	81063	3	3	3	10	10	2	*	2	47	14	*
242	21000	22	48063	3	2	3	10	10	3	3	1	42	16	4
243	21000	30	3063	2	2	2	10	10	3	2	1	62	18	4
244	61000	5	92063	1	2	2	7	7	3	2	1	67	12	4
245	20504	9	33064	3	3	3	10	10	1	3	1	57	12	4
246	21000	38	19064	3	3	3	10	10	3	3	1	42	16	4
247	20400	9	33064	2	3	3	9	7	3	2	1	37	12	4
248	21000	29	68064	2	2	3	10	10	3	3	1	47	16	4
249	22000	14	60064	2	2	2	8	7	3	3	2	42	10	1
250	21000	5	92064	2	2	2	10	10	3	3	1	52	12	4

Note: Many thanks go to the people in the Ford Marketing Research Office for being so kind as to provide these data.

DATA SET II ▪ TAMMY'S RESTAURANT SURVEY: SIMULATED DATA ON THE CUSTOMERS OF TAMMY'S RESTAURANT

Explanation of Columns:

C1 Amount

C2 How many times per month do you eat here?

C3 Condition of outside of restaurant

Excellent	5
Good	4
Fair	3
Poor	2
Very poor	1

C4 Condition of inside of restaurant
(See choices under C3.)

C5 Condition of bathrooms
(See choices under C3.)

C6 Employee appearance
(See choices under C3.)

C7 Did you feel welcome?
(See choices under C3.)

C8 Were we working together?
(See choices under C3.)

C9 How do you feel the service was?
(See choices under C3.)

C10 Food quality: hot, fresh, seasoned, etc.
(See choices under C3.)

C11 Order accuracy: Did you get what you paid for?
(See choices under C3.)

Worksheet size: 100000 cells

MTB > Retrieve 'A:\TAMMYS.MTW'
Retriving worksheet from file: A:\TAMMYS.MTW
Worksheet was saved on 6/6/1996
MTB > INFO

Column	Name	Count
C1	AMOUNT	40
C2	TIMES/MO	40
C3	OUTSIDE	40
C4	INSIDE	40
C5	BATHROOM	40
C6	EMPLOYEE	40
C7	HSPTALTY	40
C8	TEAMWORK	40
C9	SERVICE	40
C10	FOODQUAL	40
C11	ACCURACY	40

MTB > PRINT C1 - C11

Row	C1	C2	C3	C4	C5	C6	C7	C8	C9	C10	C11
1	5.10	1	4	3	3	3	3	3	3	3	5
2	5.22	3	2	4	4	3	4	1	4	4	4
3	6.24	3	4	3	5	4	3	3	4	5	5
4	5.42	3	4	2	3	4	3	4	4	3	1
5	10.37	4	3	3	3	4	2	4	4	5	5
6	5.77	4	3	3	1	4	3	3	3	3	3
7	6.99	4	3	4	3	3	4	4	4	4	4
8	6.96	1	4	5	3	4	2	5	4	5	5
9	5.89	4	4	2	3	4	2	2	4	3	1
10	10.85	4	4	4	3	4	3	4	4	2	2
11	4.28	3	2	2	3	4	4	4	4	4	5
12	9.63	4	4	4	1	3	4	5	5	5	5
13	10.09	4	3	2	4	4	3	3	3	4	3
14	5.10	1	4	4	3	2	3	4	4	4	4
15	10.07	4	4	4	4	4	3	4	3	4	4
16	8.61	1	4	4	3	4	3	3	3	4	4
17	6.02	4	4	4	2	3	3	4	3	4	4
18	5.01	4	5	4	2	4	2	3	4	4	4
19	5.52	3	4	4	3	3	5	4	3	3	2
20	6.07	2	4	5	4	3	2	4	4	5	4
21	4.51	4	4	3	3	4	4	4	4	4	5
22	5.67	4	3	4	3	4	3	4	4	4	4
23	5.1	5	3	3	3	4	5	4	4	4	4
24	5.55	4	3	3	2	3	2	3	3	3	3
25	3.76	4	4	4	2	3	3	3	4	4	4
26	6.11	4	4	4	4	3	3	4	4	3	4
27	8.63	4	4	4	2	3	3	3	2	4	4
28	3.79	4	4	4	4	4	4	4	5	4	4
29	5.48	4	4	4	3	4	4	4	4	3	5
30	4.35	3	4	4	2	2	2	3	3	5	4
31	2.11	4	5	5	3	4	2	3	4	3	5
32	0.79	3	4	4	3	4	4	4	4	4	2
33	12.31	1	3	3	3	5	4	3	4	2	5
34	6.39	1	4	4	1	4	3	4	3	5	5
35	7.00	4	4	4	2	4	2	3	3	5	1
36	6.81	1	3	3	3	4	2	4	3	1	4
37	5.45	4	3	3	3	3	3	5	3	3	4
38	6.42	5	3	4	4	3	3	3	4	4	1
39	2.68	3	4	4	4	4	4	3	4	3	2
40	5.66	2	4	4	4	4	1	3	2	2	3

Appendix B

Statistical Tables

Table I

RANDOM NUMBERS

57728	16308	27337	53884	60742	61693	39887	81779	36354
63962	45765	75060	46767	28844	32354	91463	25057	91907
51041	22252	38447	71567	95103	11124	34960	35710	91098
84048	53578	67379	42605	59122	39415	82869	86971	64817
17736	34458	67227	97041	77846	20338	52372	34645	56563
82238	83763	45464	18493	98489	72138	38942	97661	95788
28853	61793	44664	69427	68144	71949	57192	25592	49835
22251	73098	68108	87626	76724	56495	87357	83065	95316
66236	46591	69225	29867	60815	51931	40507	52568	47097
50006	91666	86406	92778	51232	38761	21861	98596	42673
68328	12840	61206	64298	27378	61452	13349	27223	79637
83039	25015	95983	82835	67268	23355	44647	25542	10536
53158	82329	81756	81429	54366	97530	51447	11324	49939
46802	61720	97508	73339	29277	17964	35421	39880	38180
25162	78468	44303	14425	42587	37212	58866	39008	91938
65957	15171	22417	95571	90679	54774	43979	71017	49647
10876	36062	91375	90128	14906	81447	49158	14703	89517
35354	66633	62311	58185	67310	95474	21878	89101	38299
70822	69983	23726	97422	46713	20340	42807	10859	26897
64299	12987	60370	70165	43306	14417	79261	53891	72816
74007	61658	86698	31571	75098	11676	35867	39764	47504
70909	68300	55074	42093	55745	80364	18488	47981	18702
67898	98830	97705	10723	82370	45586	19013	60915	84961
59386	25440	92441	14265	26123	85453	57326	72790	55243
71469	49833	95737	84195	78444	32104	89917	88361	35344
34064	12993	23818	28197	33755	96438	84223	10400	36797
86492	25367	65712	81581	89579	31759	56108	24476	47696
86914	87565	20344	39027	98338	95171	75562	54283	35342
88418	58064	13624	32978	90704	56218	84064	69990	45354
87948	83451	96217	40534	40775	74376	43157	74856	13950
13049	85293	32747	17728	50495	34617	73707	33976	86177
86544	52703	74990	98288	61833	48803	75258	83382	79099
77295	70694	97326	35430	53881	94007	70471	66815	73042
54637	32831	59063	72353	87365	15322	33156	40331	93942
50938	12004	18585	23896	62559	44470	27701	66780	56157
80999	49724	76745	25232	74291	74184	91055	58903	18172
71303	36255	77310	95847	30282	77207	34439	47763	99697
79264	16901	55814	89734	30255	87209	31629	19328	42532
30235	69368	38685	32790	58980	42159	88577	18427	73504
59110	69783	93713	29151	34933	95745	72271	38684	15426
28094	19560	27259	82736	49700	37876	52322	69562	75837
40341	20666	26662	16422	76351	70520	36890	86559	89160
30117	68850	28319	44992	68110	47007	22243	72813	60934
62287	44957	47690	79484	69449	27981	34770	34228	81686
96976	77830	61746	67846	15584	28070	79200	12663	63273
82584	34789	33494	55533	25040	84187	14479	26286	10665
35728	87881	70271	13115	35745	99145	92717	74357	16716
88458	63625	59577	92037	99012	40836	58817	30757	37934
49789	20873	53858	91356	11387	75208	33643	88210	42440
49131	34078	45396	56884	81416	46292	36012	30806	65220

Table II

TABLE OF BINOMIAL PROBABILITIES

		p										
n	*x*	.05	.10	.20	.30	.40	.50	.60	.70	.80	.90	.95
1	0	.9500	.9000	.8000	.7000	.6000	.5000	.4000	.3000	.2000	.1000	.0500
	1	.0500	.1000	.2000	.3000	.4000	.5000	.6000	.7000	.8000	.9000	.9500
2	0	.9025	.8100	.6400	.4900	.3600	.2500	.1600	.0900	.0400	.0100	.0025
	1	.0950	.1800	.3200	.4200	.4800	.5000	.4800	.4200	.3200	.1800	.0950
	2	.0025	.0100	.0400	.0900	.1600	.2500	.3600	.4900	.6400	.8100	.9025
3	0	.8574	.7290	.5120	.3430	.2160	.1250	.0640	.0270	.0080	.0010	.0001
	1	.1354	.2430	.3840	.4410	.4320	.3750	.2880	.1890	.0960	.0270	.0071
	2	.0071	.0270	.0960	.1890	.2880	.3750	.4320	.4410	.3840	.2430	.1354
	3	.0001	.0010	.0080	.0270	.0640	.1250	.2160	.3430	.5120	.7290	.8574
4	0	.8145	.6561	.4096	.2401	.1296	.0625	.0256	.0081	.0016	.0001	.0000
	1	.1715	.2916	.4096	.4116	.3456	.2500	.1536	.0756	.0256	.0036	.0005
	2	.0135	.0486	.1536	.2646	.3456	.3750	.3456	.2646	.1536	.0486	.0135
	3	.0005	.0036	.0256	.0756	.1536	.2500	.3456	.4116	.4096	.2916	.1715
	4	.0000	.0001	.0016	.0081	.0256	.0625	.1296	.2401	.4096	.6561	.8145
5	0	.7738	.5905	.3277	.1681	.0778	.0312	.0102	.0024	.0003	.0000	.0000
	1	.2036	.3280	.4096	.3602	.2592	.1562	.0768	.0284	.0064	.0005	.0000
	2	.0214	.0729	.2048	.3087	.3456	.3125	.2304	.1323	.0512	.0081	.0011
	3	.0011	.0081	.0512	.1323	.2304	.3125	.3456	.3087	.2048	.0729	.0214
	4	.0000	.0004	.0064	.0283	.0768	.1562	.2592	.3601	.4096	.3281	.2036
	5	.0000	.0000	.0003	.0024	.0102	.0312	.0778	.1681	.3277	.5905	.7738
6	0	.7351	.5314	.2621	.1176	.0467	.0156	.0041	.0007	.0001	.0000	.0000
	1	.2321	.3543	.3932	.3025	.1866	.0937	.0369	.0102	.0015	.0001	.0000
	2	.0305	.0984	.2458	.3241	.3110	.2344	.1382	.0595	.0154	.0012	.0001
	3	.0021	.0146	.0819	.1852	.2765	.3125	.2765	.1852	.0819	.0146	.0021
	4	.0001	.0012	.0154	.0595	.1382	.2344	.3110	.3241	.2458	.0984	.0305
	5	.0000	.0001	.0015	.0102	.0369	.0937	.1866	.3025	.3932	.3543	.2321
	6	.0000	.0000	.0001	.0007	.0041	.0156	.0467	.1176	.2621	.5314	.7351
7	0	.6983	.4783	.2097	.0824	.0280	.0078	.0016	.0002	.0000	.0000	.0000
	1	.2573	.3720	.3670	.2471	.1306	.0547	.0172	.0036	.0004	.0000	.0000
	2	.0406	.1240	.2753	.3177	.2613	.1641	.0774	.0250	.0043	.0002	.0000
	3	.0036	.0230	.1147	.2269	.2903	.2734	.1935	.0972	.0287	.0026	.0002
	4	.0002	.0026	.0287	.0972	.1935	.2734	.2903	.2269	.1147	.0230	.0036
	5	.0000	.0002	.0043	.0250	.0774	.1641	.2613	.3177	.2753	.1240	.0406
	6	.0000	.0000	.0004	.0036	.0172	.0547	.1306	.2471	.3670	.3720	.2573
	7	.0000	.0000	.0000	.0002	.0016	.0078	.0280	.0824	.2097	.4783	.6983
8	0	.6634	.4305	.1678	.0576	.0168	.0039	.0007	.0001	.0000	.0000	.0000
	1	.2793	.3826	.3355	.1977	.0896	.0312	.0079	.0012	.0001	.0000	.0000
	2	.0515	.1488	.2936	.2965	.2090	.1094	.0413	.0100	.0011	.0000	.0000
	3	.0054	.0331	.1468	.2541	.2787	.2187	.1239	.0467	.0092	.0004	.0000
	4	.0004	.0046	.0459	.1361	.2322	.2734	.2322	.1361	.0459	.0046	.0004
	5	.0000	.0004	.0092	.0467	.1239	.2187	.2787	.2541	.1468	.0331	.0054

Table II (continued)

		p										
n	*x*	.05	.10	.20	.30	.40	.50	.60	.70	.80	.90	.95
	6	.0000	.0000	.0011	.0100	.0413	.1094	.2090	.2965	.2936	.1488	.0515
	7	.0000	.0000	.0001	.0012	.0079	.0312	.0896	.1977	.3355	.3826	.2793
	8	.0000	.0000	.0000	.0001	.0007	.0039	.0168	.0576	.1678	.4305	.6634
9	0	.6302	.3874	.1342	.0404	.0101	.0020	.0003	.0000	.0000	.0000	.0000
	1	.2985	.3874	.3020	.1556	.0605	.0176	.0035	.0004	.0000	.0000	.0000
	2	.0629	.1722	.3020	.2668	.1612	.0703	.0212	.0039	.0003	.0000	.0000
	3	.0077	.0446	.1762	.2668	.2508	.1641	.0743	.0210	.0028	.0001	.0000
	4	.0006	.0074	.0661	.1715	.2508	.2461	.1672	.0735	.0165	.0008	.0000
	5	.0000	.0008	.0165	.0735	.1672	.2461	.2508	.1715	.0661	.0074	.0006
	6	.0000	.0001	.0028	.0210	.0743	.1641	.2508	.2668	.1762	.0446	.0077
	7	.0000	.0000	.0003	.0039	.0212	.0703	.1612	.2668	.3020	.1722	.0629
	8	.0000	.0000	.0000	.0004	.0035	.0176	.0605	.1556	.3020	.3874	.2985
	9	.0000	.0000	.0000	.0000	.0003	.0020	.0101	.0404	.1342	.3874	.6302
10	0	.5987	.3487	.1074	.0282	.0060	.0010	.0001	.0000	.0000	.0000	.0000
	1	.3151	.3874	.2684	.1211	.0403	.0098	.0016	.0001	.0000	.0000	.0000
	2	.0746	.1937	.3020	.2335	.1209	.0439	.0106	.0014	.0001	.0000	.0000
	3	.0105	.0574	.2013	.2668	.2150	.1172	.0425	.0090	.0008	.0000	.0000
	4	.0010	.0112	.0881	.2001	.2508	.2051	.1115	.0368	.0055	.0001	.0000
	5	.0001	.0015	.0264	.1029	.2007	.2461	.2007	.1029	.0264	.0015	.0001
	6	.0000	.0001	.0055	.0368	.1115	.2051	.2508	.2001	.0881	.0112	.0010
	7	.0000	.0000	.0008	.0090	.0425	.1172	.2150	.2668	.2013	.0574	.0105
	8	.0000	.0000	.0001	.0014	.0106	.0439	.1209	.2335	.3020	.1937	.0746
	9	.0000	.0000	.0000	.0001	.0016	.0098	.0403	.1211	.2684	.3874	.3151
	10	.0000	.0000	.0000	.0000	.0001	.0010	.0060	.0282	.1074	.3487	.5987
11	0	.5688	.3138	.0859	.0198	.0036	.0005	.0000	.0000	.0000	.0000	.0000
	1	.3293	.3835	.2362	.0932	.0266	.0054	.0007	.0000	.0000	.0000	.0000
	2	.0867	.2131	.2953	.1998	.0887	.0269	.0052	.0005	.0000	.0000	.0000
	3	.0137	.0710	.2215	.2568	.1774	.0806	.0234	.0037	.0002	.0000	.0000
	4	.0014	.0158	.1107	.2201	.2365	.1611	.0701	.0173	.0017	.0000	.0000
	5	.0001	.0025	.0388	.1321	.2207	.2256	.1471	.0566	.0097	.0003	.0000
	6	.0000	.0003	.0097	.0566	.1471	.2256	.2207	.1321	.0388	.0025	.0001
	7	.0000	.0000	.0017	.0173	.0701	.1611	.2365	.2201	.1107	.0158	.0014
	8	.0000	.0000	.0002	.0037	.0234	.0806	.1774	.2568	.2215	.0710	.0137
	9	.0000	.0000	.0000	.0005	.0052	.0269	.0887	.1998	.2953	.2131	.0867
	10	.0000	.0000	.0000	.0000	.0007	.0054	.0266	.0932	.2362	.3835	.3293
	11	.0000	.0000	.0000	.0000	.0000	.0005	.0036	.0198	.0859	.3138	.5688
12	0	.5404	.2824	.0687	.0138	.0022	.0002	.0000	.0000	.0000	.0000	.0000
	1	.3413	.3766	.2062	.0712	.0174	.0029	.0003	.0000	.0000	.0000	.0000
	2	.0988	.2301	.2835	.1678	.0639	.0161	.0025	.0002	.0000	.0000	.0000
	3	.0173	.0852	.2362	.2397	.1419	.0537	.0125	.0015	.0001	.0000	.0000
	4	.0021	.0213	.1329	.2311	.2128	.1208	.0420	.0078	.0005	.0000	.0000
	5	.0002	.0038	.0532	.1585	.2270	.1934	.1009	.0291	.0033	.0000	.0000
	6	.0000	.0005	.0155	.0792	.1766	.2256	.1766	.0792	.0155	.0005	.0000
	7	.0000	.0000	.0033	.0291	.1009	.1934	.2270	.1585	.0532	.0038	.0002
	8	.0000	.0000	.0005	.0078	.0420	.1208	.2128	.2311	.1329	.0213	.0021
	9	.0000	.0000	.0001	.0015	.0125	.0537	.1419	.2397	.2362	.0852	.0173

Table II (continued)

		p										
n	x	.05	.10	.20	.30	.40	.50	.60	.70	.80	.90	.95
	10	.0000	.0000	.0000	.0002	.0025	.0161	.0639	.1678	.2835	.2301	.0988
	11	.0000	.0000	.0000	.0000	.0003	.0029	.0174	.0712	.2062	.3766	.3413
	12	.0000	.0000	.0000	.0000	.0000	.0002	.0022	.0138	.0687	.2824	.5404
13	0	.5133	.2542	.0550	.0097	.0013	.0001	.0000	.0000	.0000	.0000	.0000
	1	.3512	.3672	.1787	.0540	.0113	.0016	.0001	.0000	.0000	.0000	.0000
	2	.1109	.2448	.2680	.1388	.0453	.0095	.0012	.0001	.0000	.0000	.0000
	3	.0214	.0997	.2457	.2181	.1107	.0349	.0065	.0006	.0000	.0000	.0000
	4	.0028	.0277	.1535	.2337	.1845	.0873	.0243	.0034	.0001	.0000	.0000
	5	.0003	.0055	.0691	.1803	.2214	.1571	.0656	.0142	.0011	.0000	.0000
	6	.0000	.0008	.0230	.1030	.1968	.2095	.1312	.0442	.0058	.0001	.0000
	7	.0000	.0001	.0058	.0442	.1312	.2095	.1968	.1030	.0230	.0008	.0000
	8	.0000	.0000	.0011	.0142	.0656	.1571	.2214	.1803	.0691	.0055	.0003
	9	.0000	.0000	.0001	.0034	.0243	.0873	.1845	.2337	.1535	.0277	.0028
	10	.0000	.0000	.0000	.0006	.0065	.0349	.1107	.2181	.2457	.0997	.0214
	11	.0000	.0000	.0000	.0001	.0012	.0095	.0453	.1388	.2680	.2448	.1109
	12	.0000	.0000	.0000	.0000	.0001	.0016	.0113	.0540	.1787	.3672	.3512
	13	.0000	.0000	.0000	.0000	.0000	.0001	.0013	.0097	.0550	.2542	.5133
14	0	.4877	.2288	.0440	.0068	.0008	.0001	.0000	.0000	.0000	.0000	.0000
	1	.3593	.3559	.1539	.0407	.0073	.0009	.0001	.0000	.0000	.0000	.0000
	2	.1229	.2570	.2501	.1134	.0317	.0056	.0005	.0000	.0000	.0000	.0000
	3	.0259	.1142	.2501	.1943	.0845	.0222	.0033	.0002	.0000	.0000	.0000
	4	.0037	.0349	.1720	.2290	.1549	.0611	.0136	.0014	.0000	.0000	.0000
	5	.0004	.0078	.0860	.1963	.2066	.1222	.0408	.0066	.0003	.0000	.0000
	6	.0000	.0013	.0322	.1262	.2066	.1833	.0918	.0232	.0020	.0000	.0000
	7	.0000	.0002	.0092	.0618	.1574	.2095	.1574	.0618	.0092	.0002	.0000
	8	.0000	.0000	.0020	.0232	.0918	.1833	.2066	.1262	.0322	.0013	.0000
	9	.0000	.0000	.0003	.0066	.0408	.1222	.2066	.1963	.0860	.0078	.0004
	10	.0000	.0000	.0000	.0014	.0136	.0611	.1549	.2290	.1720	.0349	.0037
	11	.0000	.0000	.0000	.0002	.0033	.0222	.0845	.1943	.2501	.1142	.0259
	12	.0000	.0000	.0000	.0000	.0005	.0056	.0317	.1134	.2501	.2570	.1229
	13	.0000	.0000	.0000	.0000	.0001	.0009	.0073	.0407	.1539	.3559	.3593
	14	.0000	.0000	.0000	.0000	.0000	.0001	.0008	.0068	.0440	.2288	.4877
15	0	.4633	.2059	.0352	.0047	.0005	.0000	.0000	.0000	.0000	.0000	.0000
	1	.3658	.3432	.1319	.0305	.0047	.0005	.0000	.0000	.0000	.0000	.0000
	2	.1348	.2669	.2309	.0916	.0219	.0032	.0003	.0000	.0000	.0000	.0000
	3	.0307	.1285	.2501	.1700	.0634	.0139	.0016	.0001	.0000	.0000	.0000
	4	.0049	.0428	.1876	.2186	.1268	.0417	.0074	.0006	.0000	.0000	.0000
	5	.0006	.0105	.1032	.2061	.1859	.0916	.0245	.0030	.0001	.0000	.0000
	6	.0000	.0019	.0430	.1472	.2066	.1527	.0612	.0116	.0007	.0000	.0000
	7	.0000	.0003	.0138	.0811	.1771	.1964	.1181	.0348	.0035	.0000	.0000
	8	.0000	.0000	.0035	.0348	.1181	.1964	.1771	.0811	.0138	.0003	.0000
	9	.0000	.0000	.0007	.0116	.0612	.1527	.2066	.1472	.0430	.0019	.0000
	10	.0000	.0000	.0001	.0030	.0245	.0916	.1859	.2061	.1032	.0105	.0006
	11	.0000	.0000	.0000	.0006	.0074	.0417	.1268	.2186	.1876	.0428	.0049
	12	.0000	.0000	.0000	.0001	.0016	.0139	.0634	.1700	.2501	.1285	.0307
	13	.0000	.0000	.0000	.0000	.0003	.0032	.0219	.0916	.2309	.2669	.1348
	14	.0000	.0000	.0000	.0000	.0000	.0005	.0047	.0305	.1319	.3432	.3658
	15	.0000	.0000	.0000	.0000	.0000	.0000	.0005	.0047	.0352	.2059	.4633

Table II (continued)

		p										
n	*x*	.05	.10	.20	.30	.40	.50	.60	.70	.80	.90	.95
16	0	.4401	.1853	.0281	.0033	.0003	.0000	.0000	.0000	.0000	.0000	.0000
	1	.3706	.3294	.1126	.0228	.0030	.0002	.0000	.0000	.0000	.0000	.0000
	2	.1463	.2745	.2111	.0732	.0150	.0018	.0001	.0000	.0000	.0000	.0000
	3	.0359	.1423	.2463	.1465	.0468	.0085	.0008	.0000	.0000	.0000	.0000
	4	.0061	.0514	.2001	.2040	.1014	.0278	.0040	.0002	.0000	.0000	.0000
	5	.0008	.0137	.1201	.2099	.1623	.0667	.0142	.0013	.0000	.0000	.0000
	6	.0001	.0028	.0550	.1649	.1983	.1222	.0392	.0056	.0002	.0000	.0000
	7	.0000	.0004	.0197	.1010	.1889	.1746	.0840	.0185	.0012	.0000	.0000
	8	.0000	.0001	.0055	.0487	.1417	.1964	.1417	.0487	.0055	.0001	.0000
	9	.0000	.0000	.0012	.0185	.0840	.1746	.1889	.1010	.0197	.0004	.0000
	10	.0000	.0000	.0002	.0056	.0392	.1222	.1983	.1649	.0550	.0028	.0001
	11	.0000	.0000	.0000	.0013	.0142	.0666	.1623	.2099	.1201	.0137	.0008
	12	.0000	.0000	.0000	.0002	.0040	.0278	.1014	.2040	.2001	.0514	.0061
	13	.0000	.0000	.0000	.0000	.0008	.0085	.0468	.1465	.2463	.1423	.0359
	14	.0000	.0000	.0000	.0000	.0001	.0018	.0150	.0732	.2111	.2745	.1463
	15	.0000	.0000	.0000	.0000	.0000	.0002	.0030	.0228	.1126	.3294	.3706
	16	.0000	.0000	.0000	.0000	.0000	.0000	.0003	.0033	.0281	.1853	.4401
17	0	.4181	.1668	.0225	.0023	.0002	.0000	.0000	.0000	.0000	.0000	.0000
	1	.3741	.3150	.0957	.0169	.0019	.0001	.0000	.0000	.0000	.0000	.0000
	2	.1575	.2800	.1914	.0581	.0102	.0010	.0001	.0000	.0000	.0000	.0000
	3	.0415	.1556	.2393	.1245	.0341	.0052	.0004	.0000	.0000	.0000	.0000
	4	.0076	.0605	.2093	.1868	.0796	.0182	.0021	.0001	.0000	.0000	.0000
	5	.0010	.0175	.1361	.2081	.1379	.0472	.0081	.0006	.0000	.0000	.0000
	6	.0001	.0039	.0680	.1784	.1839	.0944	.0242	.0026	.0001	.0000	.0000
	7	.0000	.0007	.0267	.1201	.1927	.1484	.0571	.0095	.0004	.0000	.0000
	8	.0000	.0001	.0084	.0644	.1606	.1855	.1070	.0276	.0021	.0000	.0000
	9	.0000	.0000	.0021	.0276	.1070	.1855	.1606	.0644	.0084	.0001	.0000
	10	.0000	.0000	.0004	.0095	.0571	.1484	.1927	.1201	.0267	.0007	.0000
	11	.0000	.0000	.0001	.0026	.0242	.0944	.1839	.1784	.0680	.0039	.0001
	12	.0000	.0000	.0000	.0006	.0081	.0472	.1379	.2081	.1361	.0175	.0010
	13	.0000	.0000	.0000	.0001	.0021	.0182	.0796	.1868	.2093	.0605	.0076
	14	.0000	.0000	.0000	.0000	.0004	.0052	.0341	.1245	.2393	.1556	.0415
	15	.0000	.0000	.0000	.0000	.0001	.0010	.0102	.0581	.1914	.2800	.1575
	16	.0000	.0000	.0000	.0000	.0000	.0001	.0019	.0169	.0957	.3150	.3741
	17	.0000	.0000	.0000	.0000	.0000	.0000	.0002	.0023	.0225	.1668	.4181
18	0	.3972	.1501	.0180	.0016	.0001	.0000	.0000	.0000	.0000	.0000	.0000
	1	.3763	.3002	.0811	.0126	.0012	.0001	.0000	.0000	.0000	.0000	.0000
	2	.1683	.2835	.1723	.0458	.0069	.0006	.0000	.0000	.0000	.0000	.0000
	3	.0473	.1680	.2297	.1046	.0246	.0031	.0002	.0000	.0000	.0000	.0000
	4	.0093	.0700	.2153	.1681	.0614	.0117	.0011	.0000	.0000	.0000	.0000
	5	.0014	.0218	.1507	.2017	.1146	.0327	.0045	.0002	.0000	.0000	.0000
	6	.0002	.0052	.0816	.1873	.1655	.0708	.0145	.0012	.0000	.0000	.0000
	7	.0000	.0010	.0350	.1376	.1892	.1214	.0374	.0046	.0001	.0000	.0000
	8	.0000	.0002	.0120	.0811	.1734	.1669	.0771	.0149	.0008	.0000	.0000
	9	.0000	.0000	.0033	.0386	.1284	.1855	.1284	.0386	.0033	.0000	.0000
	10	.0000	.0000	.0008	.0149	.0771	.1669	.1734	.0811	.0120	.0002	.0000
	11	.0000	.0000	.0001	.0046	.0374	.1214	.1892	.1376	.0350	.0010	.0000
	12	.0000	.0000	.0000	.0012	.0145	.0708	.1655	.1873	.0816	.0052	.0002
	13	.0000	.0000	.0000	.0002	.0045	.0327	.1146	.2017	.1507	.0218	.0014

Table II (continued)

		p										
n	*x*	.05	.10	.20	.30	.40	.50	.60	.70	.80	.90	.95
	14	.0000	.0000	.0000	.0000	.0011	.0117	.0614	.1681	.2153	.0700	.0093
	15	.0000	.0000	.0000	.0000	.0002	.0031	.0246	.1046	.2297	.1680	.0473
	16	.0000	.0000	.0000	.0000	.0000	.0006	.0069	.0458	.1723	.2835	.1683
	17	.0000	.0000	.0000	.0000	.0000	.0001	.0012	.0126	.0811	.3002	.3763
	18	.0000	.0000	.0000	.0000	.0000	.0000	.0001	.0016	.0180	.1501	.3972
19	0	.3774	.1351	.0144	.0011	.0001	.0000	.0000	.0000	.0000	.0000	.0000
	1	.3774	.2852	.0685	.0093	.0008	.0000	.0000	.0000	.0000	.0000	.0000
	2	.1787	.2852	.1540	.0358	.0046	.0003	.0000	.0000	.0000	.0000	.0000
	3	.0533	.1796	.2182	.0869	.0175	.0018	.0001	.0000	.0000	.0000	.0000
	4	.0112	.0798	.2182	.1491	.0467	.0074	.0005	.0000	.0000	.0000	.0000
	5	.0018	.0266	.1636	.1916	.0933	.0222	.0024	.0001	.0000	.0000	.0000
	6	.0002	.0069	.0955	.1916	.1451	.0518	.0085	.0005	.0000	.0000	.0000
	7	.0000	.0014	.0443	.1525	.1797	.0961	.0237	.0022	.0000	.0000	.0000
	8	.0000	.0002	.0166	.0981	.1797	.1442	.0532	.0077	.0003	.0000	.0000
	9	.0000	.0000	.0051	.0514	.1464	.1762	.0976	.0220	.0013	.0000	.0000
	10	.0000	.0000	.0013	.0220	.0976	.1762	.1464	.0514	.0051	.0000	.0000
	11	.0000	.0000	.0003	.0077	.0532	.1442	.1797	.0981	.0166	.0002	.0000
	12	.0000	.0000	.0000	.0022	.0237	.0961	.1797	.1525	.0443	.0014	.0000
	13	.0000	.0000	.0000	.0005	.0085	.0518	.1451	.1916	.0955	.0069	.0002
	14	.0000	.0000	.0000	.0001	.0024	.0222	.0933	.1916	.1636	.0266	.0018
	15	.0000	.0000	.0000	.0000	.0005	.0074	.0467	.1491	.2182	.0798	.0112
	16	.0000	.0000	.0000	.0000	.0001	.0018	.0175	.0869	.2182	.1796	.0533
	17	.0000	.0000	.0000	.0000	.0000	.0003	.0046	.0358	.1540	.2852	.1787
	18	.0000	.0000	.0000	.0000	.0000	.0000	.0008	.0093	.0685	.2852	.3774
	19	.0000	.0000	.0000	.0000	.0000	.0000	.0001	.0011	.0144	.1351	.3774
20	0	.3585	.1216	.0115	.0008	.0000	.0000	.0000	.0000	.0000	.0000	.0000
	1	.3774	.2702	.0576	.0068	.0005	.0000	.0000	.0000	.0000	.0000	.0000
	2	.1887	.2852	.1369	.0278	.0031	.0002	.0000	.0000	.0000	.0000	.0000
	3	.0596	.1901	.2054	.0716	.0123	.0011	.0000	.0000	.0000	.0000	.0000
	4	.0133	.0898	.2182	.1304	.0350	.0046	.0003	.0000	.0000	.0000	.0000
	5	.0022	.0319	.1746	.1789	.0746	.0148	.0013	.0000	.0000	.0000	.0000
	6	.0003	.0089	.1091	.1916	.1244	.0370	.0049	.0002	.0000	.0000	.0000
	7	.0000	.0020	.0545	.1643	.1659	.0739	.0146	.0010	.0000	.0000	.0000
	8	.0000	.0004	.0222	.1144	.1797	.1201	.0355	.0039	.0001	.0000	.0000
	9	.0000	.0001	.0074	.0654	.1597	.1602	.0710	.0120	.0005	.0000	.0000
	10	.0000	.0000	.0020	.0308	.1171	.1762	.1171	.0308	.0020	.0000	.0000
	11	.0000	.0000	.0005	.0120	.0710	.1602	.1597	.0654	.0074	.0001	.0000
	12	.0000	.0000	.0001	.0039	.0355	.1201	.1797	.1144	.0222	.0004	.0000
	13	.0000	.0000	.0000	.0010	.0146	.0739	.1659	.1643	.0545	.0020	.0000
	14	.0000	.0000	.0000	.0002	.0049	.0370	.1244	.1916	.1091	.0089	.0003
	15	.0000	.0000	.0000	.0000	.0013	.0148	.0746	.1789	.1746	.0319	.0022
	16	.0000	.0000	.0000	.0000	.0003	.0046	.0350	.1304	.2182	.0898	.0133
	17	.0000	.0000	.0000	.0000	.0000	.0011	.0123	.0716	.2054	.1901	.0596
	18	.0000	.0000	.0000	.0000	.0000	.0002	.0031	.0278	.1369	.2852	.1887
	19	.0000	.0000	.0000	.0000	.0000	.0000	.0005	.0068	.0576	.2702	.3774
	20	.0000	.0000	.0000	.0000	.0000	.0000	.0000	.0008	.0115	.1216	.3585
21	0	.3406	.1094	.0092	.0006	.0000	.0000	.0000	.0000	.0000	.0000	.0000
	1	.3764	.2553	.0484	.0050	.0003	.0000	.0000	.0000	.0000	.0000	.0000

Table II (continued)

		p										
n	*x*	.05	.10	.20	.30	.40	.50	.60	.70	.80	.90	.95
	2	.1981	.2837	.1211	.0215	.0020	.0001	.0000	.0000	.0000	.0000	.0000
	3	.0660	.1996	.1917	.0585	.0086	.0006	.0000	.0000	.0000	.0000	.0000
	4	.0156	.0998	.2156	.1128	.0259	.0029	.0001	.0000	.0000	.0000	.0000
	5	.0028	.0377	.1833	.1643	.0588	.0097	.0007	.0000	.0000	.0000	.0000
	6	.0004	.0112	.1222	.1878	.1045	.0259	.0027	.0001	.0000	.0000	.0000
	7	.0000	.0027	.0655	.1725	.1493	.0554	.0087	.0005	.0000	.0000	.0000
	8	.0000	.0005	.0286	.1294	.1742	.0970	.0229	.0019	.0000	.0000	.0000
	9	.0000	.0001	.0103	.0801	.1677	.1402	.0497	.0063	.0002	.0000	.0000
	10	.0000	.0000	.0031	.0412	.1342	.1682	.0895	.0176	.0008	.0000	.0000
	11	.0000	.0000	.0008	.0176	.0895	.1682	.1342	.0412	.0031	.0000	.0000
	12	.0000	.0000	.0002	.0063	.0497	.1402	.1677	.0801	.0103	.0001	.0000
	13	.0000	.0000	.0000	.0019	.0229	.0970	.1742	.1294	.0286	.0005	.0000
	14	.0000	.0000	.0000	.0005	.0087	.0554	.1493	.1725	.0655	.0027	.0000
	15	.0000	.0000	.0000	.0001	.0027	.0259	.1045	.1878	.1222	.0112	.0004
	16	.0000	.0000	.0000	.0000	.0007	.0097	.0588	.1643	.1833	.0377	.0028
	17	.0000	.0000	.0000	.0000	.0001	.0029	.0259	.1128	.2156	.0998	.0156
	18	.0000	.0000	.0000	.0000	.0000	.0006	.0086	.0585	.1917	.1996	.0660
	19	.0000	.0000	.0000	.0000	.0000	.0001	.0020	.0215	.1211	.2837	.1981
	20	.0000	.0000	.0000	.0000	.0000	.0000	.0003	.0050	.0484	.2553	.3764
	21	.0000	.0000	.0000	.0000	.0000	.0000	.0000	.0006	.0092	.1094	.3406
22	0	.3235	.0985	.0074	.0004	.0000	.0000	.0000	.0000	.0000	.0000	.0000
	1	.3746	.2407	.0406	.0037	.0002	.0000	.0000	.0000	.0000	.0000	.0000
	2	.2070	.2808	.1065	.0166	.0014	.0001	.0000	.0000	.0000	.0000	.0000
	3	.0726	.2080	.1775	.0474	.0060	.0004	.0000	.0000	.0000	.0000	.0000
	4	.0182	.1098	.2108	.0965	.0190	.0017	.0001	.0000	.0000	.0000	.0000
	5	.0034	.0439	.1898	.1489	.0456	.0063	.0004	.0000	.0000	.0000	.0000
	6	.0005	.0138	.1344	.1808	.0862	.0178	.0015	.0000	.0000	.0000	.0000
	7	.0001	.0035	.0768	.1771	.1314	.0407	.0051	.0002	.0000	.0000	.0000
	8	.0000	.0007	.0360	.1423	.1642	.0762	.0144	.0009	.0000	.0000	.0000
	9	.0000	.0001	.0140	.0949	.1703	.1186	.0336	.0032	.0001	.0000	.0000
	10	.0000	.0000	.0046	.0529	.1476	.1542	.0656	.0097	.0003	.0000	.0000
	11	.0000	.0000	.0012	.0247	.1073	.1682	.1073	.0247	.0012	.0000	.0000
	12	.0000	.0000	.0003	.0097	.0656	.1542	.1476	.0529	.0046	.0000	.0000
	13	.0000	.0000	.0001	.0032	.0336	.1186	.1703	.0949	.0140	.0001	.0000
	14	.0000	.0000	.0000	.0009	.0144	.0762	.1642	.1423	.0360	.0007	.0000
	15	.0000	.0000	.0000	.0002	.0051	.0407	.1314	.1771	.0768	.0035	.0001
	16	.0000	.0000	.0000	.0000	.0015	.0178	.0862	.1808	.1344	.0138	.0005
	17	.0000	.0000	.0000	.0000	.0004	.0063	.0456	.1489	.1898	.0439	.0034
	18	.0000	.0000	.0000	.0000	.0001	.0017	.0190	.0965	.2108	.1098	.0182
	19	.0000	.0000	.0000	.0000	.0000	.0004	.0060	.0474	.1775	.2080	.0726
	20	.0000	.0000	.0000	.0000	.0000	.0001	.0014	.0166	.1065	.2808	.2070
	21	.0000	.0000	.0000	.0000	.0000	.0000	.0002	.0037	.0406	.2407	.3746
	22	.0000	.0000	.0000	.0000	.0000	.0000	.0000	.0004	.0074	.0985	.3235
23	0	.3074	.0886	.0059	.0003	.0000	.0000	.0000	.0000	.0000	.0000	.0000
	1	.3721	.2265	.0339	.0027	.0001	.0000	.0000	.0000	.0000	.0000	.0000
	2	.2154	.2768	.0933	.0127	.0009	.0000	.0000	.0000	.0000	.0000	.0000
	3	.0794	.2153	.1633	.0382	.0041	.0002	.0000	.0000	.0000	.0000	.0000
	4	.0209	.1196	.2042	.0818	.0138	.0011	.0000	.0000	.0000	.0000	.0000
	5	.0042	.0505	.1940	.1332	.0350	.0040	.0002	.0000	.0000	.0000	.0000

Table II (continued)

		p										
n	*x*	.05	.10	.20	.30	.40	.50	.60	.70	.80	.90	.95
	6	.0007	.0168	.1455	.1712	.0700	.0120	.0008	.0000	.0000	.0000	.0000
	7	.0001	.0045	.0883	.1782	.1133	.0292	.0029	.0001	.0000	.0000	.0000
	8	.0000	.0010	.0442	.1527	.1511	.0584	.0088	.0004	.0000	.0000	.0000
	9	.0000	.0002	.0184	.1091	.1679	.0974	.0221	.0016	.0000	.0000	.0000
	10	.0000	.0000	.0064	.0655	.1567	.1364	.0464	.0052	.0001	.0000	.0000
	11	.0000	.0000	.0019	.0332	.1234	.1612	.0823	.0142	.0005	.0000	.0000
	12	.0000	.0000	.0005	.0142	.0823	.1612	.1234	.0332	.0019	.0000	.0000
	13	.0000	.0000	.0001	.0052	.0464	.1364	.1567	.0655	.0064	.0000	.0000
	14	.0000	.0000	.0000	.0016	.0221	.0974	.1679	.1091	.0184	.0002	.0000
	15	.0000	.0000	.0000	.0004	.0088	.0584	.1511	.1527	.0442	.0010	.0000
	16	.0000	.0000	.0000	.0001	.0029	.0292	.1133	.1782	.0883	.0045	.0001
	17	.0000	.0000	.0000	.0000	.0008	.0120	.0700	.1712	.1455	.0168	.0007
	18	.0000	.0000	.0000	.0000	.0002	.0040	.0350	.1332	.1940	.0505	.0042
	19	.0000	.0000	.0000	.0000	.0000	.0011	.0138	.0818	.2042	.1196	.0209
	20	.0000	.0000	.0000	.0000	.0000	.0002	.0041	.0382	.1633	.2153	.0794
	21	.0000	.0000	.0000	.0000	.0000	.0000	.0009	.0127	.0933	.2768	.2154
	22	.0000	.0000	.0000	.0000	.0000	.0000	.0001	.0027	.0339	.2265	.3721
	23	.0000	.0000	.0000	.0000	.0000	.0000	.0000	.0003	.0059	.0886	.3074
24	0	.2920	.0798	.0047	.0002	.0000	.0000	.0000	.0000	.0000	.0000	.0000
	1	.3688	.2127	.0283	.0020	.0001	.0000	.0000	.0000	.0000	.0000	.0000
	2	.2232	.2718	.0815	.0097	.0006	.0000	.0000	.0000	.0000	.0000	.0000
	3	.0862	.2215	.1493	.0305	.0028	.0001	.0000	.0000	.0000	.0000	.0000
	4	.0238	.1292	.1960	.0687	.0099	.0006	.0000	.0000	.0000	.0000	.0000
	5	.0050	.0574	.1960	.1177	.0265	.0025	.0001	.0000	.0000	.0000	.0000
	6	.0008	.0202	.1552	.1598	.0560	.0080	.0004	.0000	.0000	.0000	.0000
	7	.0001	.0058	.0998	.1761	.0960	.0206	.0017	.0000	.0000	.0000	.0000
	8	.0000	.0014	.0530	.1604	.1360	.0438	.0053	.0002	.0000	.0000	.0000
	9	.0000	.0003	.0236	.1222	.1612	.0779	.0141	.0008	.0000	.0000	.0000
	10	.0000	.0000	.0088	.0785	.1612	.1169	.0318	.0026	.0000	.0000	.0000
	11	.0000	.0000	.0028	.0428	.1367	.1488	.0608	.0079	.0002	.0000	.0000
	12	.0000	.0000	.0008	.0199	.0988	.1612	.0988	.0199	.0008	.0000	.0000
	13	.0000	.0000	.0002	.0079	.0608	.1488	.1367	.0428	.0028	.0000	.0000
	14	.0000	.0000	.0000	.0026	.0318	.1169	.1612	.0785	.0088	.0000	.0000
	15	.0000	.0000	.0000	.0008	.0141	.0779	.1612	.1222	.0236	.0003	.0000
	16	.0000	.0000	.0000	.0002	.0053	.0438	.1360	.1604	.0530	.0014	.0000
	17	.0000	.0000	.0000	.0000	.0017	.0206	.0960	.1761	.0998	.0058	.0001
	18	.0000	.0000	.0000	.0000	.0004	.0080	.0560	.1598	.1552	.0202	.0008
	19	.0000	.0000	.0000	.0000	.0001	.0025	.0265	.1177	.1960	.0574	.0050
	20	.0000	.0000	.0000	.0000	.0000	.0006	.0099	.0687	.1960	.1292	.0238
	21	.0000	.0000	.0000	.0000	.0000	.0001	.0028	.0305	.1493	.2215	.0862
	22	.0000	.0000	.0000	.0000	.0000	.0000	.0006	.0097	.0815	.2718	.2232
	23	.0000	.0000	.0000	.0000	.0000	.0000	.0001	.0020	.0283	.2127	.3688
	24	.0000	.0000	.0000	.0000	.0000	.0000	.0000	.0002	.0047	.0798	.2920
25	0	.2774	.0718	.0038	.0001	.0000	.0000	.0000	.0000	.0000	.0000	.0000
	1	.3650	.1994	.0236	.0014	.0000	.0000	.0000	.0000	.0000	.0000	.0000
	2	.2305	.2659	.0708	.0074	.0004	.0000	.0000	.0000	.0000	.0000	.0000
	3	.0930	.2265	.1358	.0243	.0019	.0001	.0000	.0000	.0000	.0000	.0000
	4	.0269	.1384	.1867	.0572	.0071	.0004	.0000	.0000	.0000	.0000	.0000
	5	.0060	.0646	.1960	.1030	.0199	.0016	.0000	.0000	.0000	.0000	.0000

Table II (continued)

		p										
n	*x*	**.05**	**.10**	**.20**	**.30**	**.40**	**.50**	**.60**	**.70**	**.80**	**.90**	**.95**
	6	.0010	.0239	.1633	.1472	.0442	.0053	.0002	.0000	.0000	.0000	.0000
	7	.0001	.0072	.1108	.1712	.0800	.0143	.0009	.0000	.0000	.0000	.0000
	8	.0000	.0018	.0623	.1651	.1200	.0322	.0031	.0001	.0000	.0000	.0000
	9	.0000	.0004	.0294	.1336	.1511	.0609	.0088	.0004	.0000	.0000	.0000
	10	.0000	.0001	.0118	.0916	.1612	.0974	.0212	.0013	.0000	.0000	.0000
	11	.0000	.0000	.0040	.0536	.1465	.1328	.0434	.0042	.0001	.0000	.0000
	12	.0000	.0000	.0012	.0268	.1140	.1550	.0760	.0115	.0003	.0000	.0000
	13	.0000	.0000	.0003	.0115	.0760	.1550	.1140	.0268	.0012	.0000	.0000
	14	.0000	.0000	.0001	.0042	.0434	.1328	.1465	.0536	.0040	.0000	.0000
	15	.0000	.0000	.0000	.0013	.0212	.0974	.1612	.0916	.0118	.0001	.0000
	16	.0000	.0000	.0000	.0004	.0088	.0609	.1511	.1336	.0294	.0004	.0000
	17	.0000	.0000	.0000	.0001	.0031	.0322	.1200	.1651	.0623	.0018	.0000
	18	.0000	.0000	.0000	.0000	.0009	.0143	.0800	.1712	.1108	.0072	.0001
	19	.0000	.0000	.0000	.0000	.0002	.0053	.0442	.1472	.1633	.0239	.0010
	20	.0000	.0000	.0000	.0000	.0000	.0016	.0199	.1030	.1960	.0646	.0060
	21	.0000	.0000	.0000	.0000	.0000	.0004	.0071	.0572	.1867	.1384	.0269
	22	.0000	.0000	.0000	.0000	.0000	.0001	.0019	.0243	.1358	.2265	.0930
	23	.0000	.0000	.0000	.0000	.0000	.0000	.0004	.0074	.0708	.2659	.2305
	24	.0000	.0000	.0000	.0000	.0000	.0000	.0000	.0014	.0236	.1994	.3650
	25	.0000	.0000	.0000	.0000	.0000	.0000	.0000	.0001	.0038	.0718	.2774

Table III

TABLE OF POISSON PROBABILITIES

	λ									
x	0.1	0.2	0.3	0.4	0.5	0.6	0.7	0.8	0.9	1.0
0	.9048	.8187	.7408	.6703	.6065	.5488	.4966	.4493	.4066	.3679
1	.0905	.1637	.2222	.2681	.3033	.3293	.3476	.3595	.3659	.3679
2	.0045	.0164	.0333	.0536	.0758	.0988	.1217	.1438	.1647	.1839
3	.0002	.0011	.0033	.0072	.0126	.0198	.0284	.0383	.0494	.0613
4	.0000	.0001	.0003	.0007	.0016	.0030	.0050	.0077	.0111	.0153
5	.0000	.0000	.0000	.0001	.0002	.0004	.0007	.0012	.0020	.0031
6	.0000	.0000	.0000	.0000	.0000	.0000	.0001	.0002	.0003	.0005
7	.0000	.0000	.0000	.0000	.0000	.0000	.0000	.0000	.0000	.0001

	λ									
x	1.1	1.2	1.3	1.4	1.5	1.6	1.7	1.8	1.9	2.0
0	.3329	.3012	.2725	.2466	.2231	.2019	.1827	.1653	.1496	.1353
1	.3662	.3614	.3543	.3452	.3347	.3230	.3106	.2975	.2842	.2707
2	.2014	.2169	.2303	.2417	.2510	.2584	.2640	.2678	.2700	.2707
3	.0738	.0867	.0998	.1128	.1255	.1378	.1496	.1607	.1710	.1804
4	.0203	.0260	.0324	.0395	.0471	.0551	.0636	.0723	.0812	.0902
5	.0045	.0062	.0084	.0111	.0141	.0176	.0216	.0260	.0309	.0361
6	.0008	.0012	.0018	.0026	.0035	.0047	.0061	.0078	.0098	.0120
7	.0001	.0002	.0003	.0005	.0008	.0011	.0015	.0020	.0027	.0034
8	.0000	.0000	.0001	.0001	.0001	.0002	.0003	.0005	.0006	.0009
9	.0000	.0000	.0000	.0000	.0000	.0000	.0001	.0001	.0001	.0002

	λ									
x	2.1	2.2	2.3	2.4	2.5	2.6	2.7	2.8	2.9	3.0
0	.1225	.1108	.1003	.0907	.0821	.0743	.0672	.0608	.0550	.0498
1	.2572	.2438	.2306	.2177	.2052	.1931	.1815	.1703	.1596	.1494
2	.2700	.2681	.2652	.2613	.2565	.2510	.2450	.2384	.2314	.2240
3	.1890	.1966	.2033	.2090	.2138	.2176	.2205	.2225	.2237	.2240
4	.0092	.1082	.1169	.1254	.1336	.1414	.1488	.1557	.1622	.1680
5	.0417	.0476	.0538	.0602	.0668	.0735	.0804	.0872	.0940	.1008
6	.0146	.0174	.0206	.0241	.0278	.0319	.0362	.0407	.0455	.0504
7	.0044	.0055	.0068	.0083	.0099	.0118	.0139	.0163	.0188	.0216
8	.0011	.0015	.0019	.0025	.0031	.0038	.0047	.0057	.0068	.0081
9	.0003	.0004	.0005	.0007	.0009	.0011	.0014	.0018	.0022	.0027
10	.0001	.0001	.0001	.0002	.0002	.0003	.0004	.0005	.0006	.0008
11	.0000	.0000	.0000	.0000	.0000	.0001	.0001	.0001	.0002	.0002
12	.0000	.0000	.0000	.0000	.0000	.0000	.0000	.0000	.0000	.0001

Table III (continued)

	λ									
x	**3.1**	**3.2**	**3.3**	**3.4**	**3.5**	**3.6**	**3.7**	**3.8**	**3.9**	**4.0**
0	.0450	.0408	.0369	.0334	.0302	.0273	.0247	.0224	.0202	.0183
1	.1397	.1304	.1217	.1135	.1057	.0984	.0915	.0850	.0789	.0733
2	.2165	.2087	.2008	.1929	.1850	.1771	.1692	.1615	.1539	.1465
3	.2237	.2226	.2209	.2186	.2158	.2125	.2087	.2046	.2001	.1954
4	.1733	.1781	.1823	.1858	.1888	.1912	.1931	.1944	.1951	.1954
5	.1075	.1140	.1203	.1264	.1322	.1377	.1429	.1477	.1522	.1563
6	.0555	.0608	.0662	.0716	.0771	.0826	.0881	.0936	.0989	.1042
7	.0246	.0278	.0312	.0348	.0385	.0425	.0466	.0508	.0551	.0595
8	.0095	.0111	.0129	.0148	.0169	.0191	.0215	.0241	.0269	.0298
9	.0033	.0040	.0047	.0056	.0066	.0076	.0089	.0102	.0116	.0132
10	.0010	.0013	.0016	.0019	.0023	.0028	.0033	.0039	.0045	.0053
11	.0003	.0004	.0005	.0006	.0007	.0009	.0011	.0013	.0016	.0019
12	.0001	.0001	.0001	.0002	.0002	.0003	.0003	.0004	.0005	.0006
13	.0000	.0000	.0000	.0000	.0001	.0001	.0001	.0001	.0002	.0002
14	.0000	.0000	.0000	.0000	.0000	.0000	.0000	.0000	.0000	.0001

	λ									
x	**4.1**	**4.2**	**4.3**	**4.4**	**4.5**	**4.6**	**4.7**	**4.8**	**4.9**	**5.0**
0	.0166	.0150	.0136	.0123	.0111	.0101	.0091	.0082	.0074	.0067
1	.0679	.0630	.0583	.0540	.0500	.0462	.0427	.0395	.0365	.0337
2	.1393	.1323	.1254	.1188	.1125	.1063	.1005	.0948	.0894	.0842
3	.1904	.1852	.1798	.1743	.1687	.1631	.1574	.1517	.1460	.1404
4	.1951	.1944	.1933	.1917	.1898	.1875	.1849	.1820	.1789	.1755
5	.1600	.1633	.1662	.1687	.1708	.1725	.1738	.1747	.1753	.1755
6	.1093	.1143	.1191	.1237	.1281	.1323	.1362	.1398	.1432	.1462
7	.0640	.0686	.0732	.0778	.0824	.0869	.0914	.0959	.1002	.1044
8	.0328	.0360	.0393	.0428	.0463	.0500	.0537	.0575	.0614	.0653
9	.0150	.0168	.0188	.0209	.0232	.0255	.0281	.0307	.0334	.0363
10	.0061	.0071	.0081	.0092	.0104	.0118	.0132	.0147	.0164	.0181
11	.0023	.0027	.0032	.0037	.0043	.0049	.0056	.0064	.0073	.0082
12	.0008	.0009	.0011	.0014	.0016	.0019	.0022	.0026	.0030	.0034
13	.0002	.0003	.0004	.0005	.0006	.0007	.0008	.0009	.0011	.0013
14	.0001	.0001	.0001	.0001	.0002	.0002	.0003	.0003	.0004	.0005
15	.0000	.0000	.0000	.0000	.0001	.0001	.0001	.0001	.0001	.0002

	λ									
x	**5.1**	**5.2**	**5.3**	**5.4**	**5.5**	**5.6**	**5.7**	**5.8**	**5.9**	**6.0**
0	.0061	.0055	.0050	.0045	.0041	.0037	.0033	.0030	.0027	.0025
1	.0311	.0287	.0265	.0244	.0225	.0207	.0191	.0176	.0162	.0149
2	.0793	.0746	.0701	.0659	.0618	.0580	.0544	.0509	.0477	.0446
3	.1348	.1293	.1239	.1185	.1133	.1082	.1033	.0985	.0938	.0892
4	.1719	.1681	.1641	.1600	.1558	.1515	.1472	.1428	.1383	.1339

Table III (continued)

	λ									
x	5.1	5.2	5.3	5.4	5.5	5.6	5.7	5.8	5.9	6.0
5	.1753	.1748	.1740	.1728	.1714	.1697	.1678	.1656	.1632	.1606
6	.1490	.1515	.1537	.1555	.1571	.1584	.1594	.1601	.1605	.1606
7	.1086	.1125	.1163	.1200	.1234	.1267	.1298	.1326	.1353	.1377
8	.0692	.0731	.0771	.0810	.0849	.0887	.0925	.0962	.0998	.1033
9	.0392	.0423	.0454	.0486	.0519	.0552	.0586	.0620	.0654	.0688
10	.0200	.0220	.0241	.0262	.0285	.0309	.0334	.0359	.0386	.0413
11	.0093	.0104	.0116	.0129	.0143	.0157	.0173	.0190	.0207	.0225
12	.0039	.0045	.0051	.0058	.0065	.0073	.0082	.0092	.0102	.0113
13	.0015	.0018	.0021	.0024	.0028	.0032	.0036	.0041	.0046	.0052
14	.0006	.0007	.0008	.0009	.0011	.0013	.0015	.0017	.0019	.0022
15	.0002	.0002	.0003	.0003	.0004	.0005	.0006	.0007	.0008	.0009
16	.0001	.0001	.0001	.0001	.0001	.0002	.0002	.0002	.0003	.0003
17	.0000	.0000	.0000	.0000	.0000	.0001	.0001	.0001	.0001	.0001

	λ									
x	6.1	6.2	6.3	6.4	6.5	6.6	6.7	6.8	6.9	7.0
0	.0022	.0020	.0018	.0017	.0015	.0014	.0012	.0011	.0010	.0009
1	.0137	.0126	.0116	.0106	.0098	.0090	.0082	.0076	.0070	.0064
2	.0417	.0390	.0364	.0340	.0318	.0296	.0276	.0258	.0240	.0223
3	.0848	.0806	.0765	.0726	.0688	.0652	.0617	.0584	.0552	.0521
4	.1294	.1249	.1205	.1162	.1118	.1076	.1034	.0992	.0952	.0912
5	.1579	.1549	.1519	.1487	.1454	.1420	.1385	.1349	.1314	.1277
6	.1605	.1601	.1595	.1586	.1575	.1562	.1546	.1529	.1511	.1490
7	.1399	.1418	.1435	.1450	.1462	.1472	.1480	.1486	.1489	.1490
8	.1066	.1099	.1130	.1160	.1188	.1215	.1240	.1263	.1284	.1304
9	.0723	.0757	.0791	.0825	.0858	.0891	.0923	.0954	.0985	.1014
10	.0441	.0469	.0498	.0528	.0558	.0588	.0618	.0649	.0679	.0710
11	.0244	.0265	.0285	.0307	.0330	.0353	.0377	.0401	.0426	.0452
12	.0124	.0137	.0150	.0164	.0179	.0194	.0210	.0227	.0245	.0263
13	.0058	.0065	.0073	.0081	.0089	.0099	.0108	.0119	.0130	.0142
14	.0025	.0029	.0033	.0037	.0041	.0046	.0052	.0058	.0064	.0071
15	.0010	.0012	.0014	.0016	.0018	.0020	.0023	.0026	.0029	.0033
16	.0004	.0005	.0005	.0006	.0007	.0008	.0010	.0011	.0013	.0014
17	.0001	.0002	.0002	.0002	.0003	.0003	.0004	.0004	.0005	.0006
18	.0000	.0001	.0001	.0001	.0001	.0001	.0001	.0002	.0002	.0002
19	.0000	.0000	.0000	.0000	.0000	.0000	.0001	.0001	.0001	.0001

	λ									
x	7.1	7.2	7.3	7.4	7.5	7.6	7.7	7.8	7.9	8.0
0	.0008	.0007	.0007	.0006	.0006	.0005	.0005	.0004	.0004	.0003
1	.0059	.0054	.0049	.0045	.0041	.0038	.0035	.0032	.0029	.0027
2	.0208	.0194	.0180	.0167	.0156	.0145	.0134	.0125	.0116	.0107
3	.0492	.0464	.0438	.0413	.0389	.0366	.0345	.0324	.0305	.0286
4	.0874	.0836	.0799	.0764	.0729	.0696	.0663	.0632	.0602	.0573

Table III (continued)

	λ									
x	7.1	7.2	7.3	7.4	7.5	7.6	7.7	7.8	7.9	8.0
5	.1241	.1204	.1167	.1130	.1094	.1057	.1021	.0986	.0951	.0916
6	.1468	.1445	.1420	.1394	.1367	.1339	.1311	.1282	.1252	.1221
7	.1489	.1486	.1481	.1474	.1465	.1454	.1442	.1428	.1413	.1396
8	.1321	.1337	.1351	.1363	.1373	.1381	.1388	.1392	.1395	.1396
9	.1042	.1070	.1096	.1121	.1144	.1167	.1187	.1207	.1224	.1241
10	.0740	.0770	.0800	.0829	.0858	.0887	.0914	.0941	.0967	.0993
11	.0478	.0504	.0531	.0558	.0585	.0613	.0640	.0667	.0695	.0722
12	.0283	.0303	.0323	.0344	.0366	.0388	.0411	.0434	.0457	.0481
13	.0154	.0168	.0181	.0196	.0211	.0227	.0243	.0260	.0278	.0296
14	.0078	.0086	.0095	.0104	.0113	.0123	.0134	.0145	.0157	.0169
15	.0037	.0041	.0046	.0051	.0057	.0062	.0069	.0075	.0083	.0090
16	.0016	.0019	.0021	.0024	.0026	.0030	.0033	.0037	.0041	.0045
17	.0007	.0008	.0009	.0010	.0012	.0013	.0015	.0017	.0019	.0021
18	.0003	.0003	.0004	.0004	.0005	.0006	.0006	.0007	.0008	.0009
19	.0001	.0001	.0001	.0002	.0002	.0002	.0003	.0003	.0003	.0004
20	.0000	.0000	.0001	.0001	.0001	.0001	.0001	.0001	.0001	.0002
21	.0000	.0000	.0000	.0000	.0000	.0000	.0000	.0000	.0001	.0001

	λ									
x	8.1	8.2	8.3	8.4	8.5	8.6	8.7	8.8	8.9	9.0
0	.0003	.0003	.0002	.0002	.0002	.0002	.0002	.0002	.0001	.0001
1	.0025	.0023	.0021	.0019	.0017	.0016	.0014	.0013	.0012	.0011
2	.0100	.0092	.0086	.0079	.0074	.0068	.0063	.0058	.0054	.0050
3	.0269	.0252	.0237	.0222	.0208	.0195	.0183	.0171	.0160	.0150
4	.0544	.0517	.0491	.0466	.0443	.0420	.0398	.0377	.0357	.0337
5	.0882	.0849	.0816	.0784	.0752	.0722	.0692	.0663	.0635	.0607
6	.1191	.1160	.1128	.1097	.1066	.1034	.1003	.0972	.0941	.0911
7	.1378	.1358	.1338	.1317	.1294	.1271	.1247	.1222	.1197	.1171
8	.1395	.1392	.1388	.1382	.1375	.1366	.1356	.1344	.1332	.1318
9	.1255	.1269	.1280	.1290	.1299	.1306	.1311	.1315	.1317	.1318
10	.1017	.1040	.1063	.1084	.1104	.1123	.1140	.1157	.1172	.1186
11	.0749	.0775	.0802	.0828	.0853	.0878	.0902	.0925	.0948	.0970
12	.0505	.0530	.0555	.0579	.0604	.0629	.0654	.0679	.0703	.0728
13	.0315	.0334	.0354	.0374	.0395	.0416	.0438	.0459	.0481	.0504
14	.0182	.0196	.0210	.0225	.0240	.0256	.0272	.0289	.0306	.0324
15	.0098	.0107	.0116	.0126	.0136	.0147	.0158	.0169	.0182	.0194
16	.0050	.0055	.0060	.0066	.0072	.0079	.0086	.0093	.0101	.0109
17	.0024	.0026	.0029	.0033	.0036	.0040	.0044	.0048	.0053	.0058
18	.0011	.0012	.0014	.0015	.0017	.0019	.0021	.0024	.0026	.0029
19	.0005	.0005	.0006	.0007	.0008	.0009	.0010	.0011	.0012	.0014
20	.0002	.0002	.0002	.0003	.0003	.0004	.0004	.0005	.0005	.0006
21	.0001	.0001	.0001	.0001	.0001	.0002	.0002	.0002	.0002	.0003
22	.0000	.0000	.0000	.0000	.0001	.0001	.0001	.0001	.0001	.0001

Table III (continued)

	λ									
x	9.1	9.2	9.3	9.4	9.5	9.6	9.7	9.8	9.9	10
0	.0001	.0001	.0001	.0001	.0001	.0001	.0001	.0001	.0001	.0000
1	.0010	.0009	.0009	.0008	.0007	.0007	.0006	.0005	.0005	.0005
2	.0046	.0043	.0040	.0037	.0034	.0031	.0029	.0027	.0025	.0023
3	.0140	.0131	.0123	.0115	.0107	.0100	.0093	.0087	.0081	.0076
4	.0319	.0302	.0285	.0269	.0254	.0240	.0226	.0213	.0201	.0189
5	.0581	.0555	.0530	.0506	.0483	.0460	.0439	.0418	.0398	.0378
6	.0881	.0851	.0822	.0793	.0764	.0736	.0709	.0682	.0656	.0631
7	.1145	.1118	.1091	.1064	.1037	.1010	.0982	.0955	.0928	.0901
8	.1302	.1286	.1269	.1251	.1232	.1212	.1191	.1170	.1148	.1126
9	.1317	.1315	.1311	.1306	.1300	.1293	.1284	.1274	.1263	.1251
10	.1198	.1209	.1219	.1228	.1235	.1241	.1245	.1249	.1250	.1251
11	.0991	.1012	.1031	.1049	.1067	.1083	.1098	.1112	.1125	.1137
12	.0752	.0776	.0799	.0822	.0844	.0866	.0888	.0908	.0928	.0948
13	.0526	.0549	.0572	.0594	.0617	.0640	.0662	.0685	.0707	.0729
14	.0342	.0361	.0380	.0399	.0419	.0439	.0459	.0479	.0500	.0521
15	.0208	.0221	.0235	.0250	.0265	.0281	.0297	.0313	.0330	.0347
16	.0118	.0127	.0137	.0147	.0157	.0168	.0180	.0192	.0204	.0217
17	.0063	.0069	.0075	.0081	.0088	.0095	.0103	.0111	.0119	.0128
18	.0032	.0035	.0039	.0042	.0046	.0051	.0055	.0060	.0065	.0071
19	.0015	.0017	.0019	.0021	.0023	.0026	.0028	.0031	.0034	.0037
20	.0007	.0008	.0009	.0010	.0011	.0012	.0014	.0015	.0017	.0019
21	.0003	.0003	.0004	.0004	.0005	.0006	.0006	.0007	.0008	.0009
22	.0001	.0001	.0002	.0002	.0002	.0002	.0003	.0003	.0004	.0004
23	.0000	.0001	.0001	.0001	.0001	.0001	.0001	.0001	.0002	.0002
24	.0000	.0000	.0000	.0000	.0000	.0000	.0000	.0001	.0001	.0001

	λ									
x	11	12	13	14	15	16	17	18	19	20
0	.0000	.0000	.0000	.0000	.0000	.0000	.0000	.0000	.0000	.0000
1	.0002	.0001	.0000	.0000	.0000	.0000	.0000	.0000	.0000	.0000
2	.0010	.0004	.0002	.0001	.0000	.0000	.0000	.0000	.0000	.0000
3	.0037	.0018	.0008	.0004	.0002	.0001	.0000	.0000	.0000	.0000
4	.0102	.0053	.0027	.0013	.0006	.0003	.0001	.0001	.0000	.0000
5	.0224	.0127	.0070	.0037	.0019	.0010	.0005	.0002	.0001	.0001
6	.0411	.0255	.0152	.0087	.0048	.0026	.0014	.0007	.0004	.0002
7	.0646	.0437	.0281	.0174	.0104	.0060	.0034	.0019	.0010	.0005
8	.0888	.0655	.0457	.0304	.0194	.0120	.0072	.0042	.0024	.0013
9	.1085	.0874	.0661	.0473	.0324	.0213	.0135	.0083	.0050	.0029
10	.1194	.1048	.0859	.0663	.0486	.0341	.0230	.0150	.0095	.0058
11	.1194	.1144	.1015	.0844	.0663	.0496	.0355	.0245	.0164	.0106
12	.1094	.1144	.1099	.0984	.0829	.0661	.0504	.0368	.0259	.0176
13	.0926	.1056	.1099	.1060	.0956	.0814	.0658	.0509	.0378	.0271
14	.0728	.0905	.1021	.1060	.1024	.0930	.0800	.0655	.0514	.0387

Table III (continued)

	λ									
x	11	12	13	14	15	16	17	18	19	20
15	.0534	.0724	.0885	.0989	.1024	.0992	.0906	.0786	.0650	.0516
16	.0367	.0543	.0719	.0866	.0960	.0992	.0963	.0884	.0772	.0646
17	.0237	.0383	.0550	.0713	.0847	.0934	.0963	.0936	.0863	.0760
18	.0145	.0255	.0397	.0554	.0706	.0830	.0909	.0936	.0911	.0844
19	.0084	.0161	.0272	.0409	.0557	.0699	.0814	.0887	.0911	.0888
20	.0046	.0097	.0177	.0286	.0418	.0559	.0692	.0798	.0866	.0888
21	.0024	.0055	.0109	.0191	.0299	.0426	.0560	.0684	.0783	.0846
22	.0012	.0030	.0065	.0121	.0204	.0310	.0433	.0560	.0676	.0769
23	.0006	.0016	.0037	.0074	.0133	.0216	.0320	.0438	.0559	.0669
24	.0003	.0008	.0020	.0043	.0083	.0144	.0226	.0328	.0442	.0557
25	.0001	.0004	.0010	.0024	.0050	.0092	.0154	.0237	.0336	.0446
26	.0000	.0002	.0005	.0013	.0029	.0057	.0101	.0164	.0246	.0343
27	.0000	.0001	.0002	.0007	.0016	.0034	.0063	.0109	.0173	.0254
28	.0000	.0000	.0001	.0003	.0009	.0019	.0038	.0070	.0117	.0181
29	.0000	.0000	.0001	.0002	.0004	.0011	.0023	.0044	.0077	.0125
30	.0000	.0000	.0000	.0001	.0002	.0006	.0013	.0026	.0049	.0083
31	.0000	.0000	.0000	.0000	.0001	.0003	.0007	.0015	.0030	.0054
32	.0000	.0000	.0000	.0000	.0001	.0001	.0004	.0009	.0018	.0034
33	.0000	.0000	.0000	.0000	.0000	.0001	.0002	.0005	.0010	.0020
34	.0000	.0000	.0000	.0000	.0000	.0000	.0001	.0002	.0006	.0012
35	.0000	.0000	.0000	.0000	.0000	.0000	.0000	.0001	.0003	.0007
36	.0000	.0000	.0000	.0000	.0000	.0000	.0000	.0001	.0002	.0004
37	.0000	.0000	.0000	.0000	.0000	.0000	.0000	.0000	.0001	.0002
38	.0000	.0000	.0000	.0000	.0000	.0000	.0000	.0000	.0000	.0001
39	.0000	.0000	.0000	.0000	.0000	.0000	.0000	.0000	.0000	.0001

Table IV

STANDARD NORMAL DISTRIBUTION TABLE

The entries in the table give the areas under the standard normal curve from 0 to z.

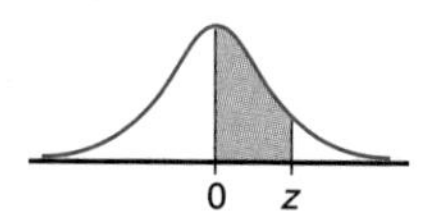

z	.00	.01	.02	.03	.04	.05	.06	.07	.08	.09
0	.0000	.0040	.0080	.0120	.0160	.0199	.0239	.0279	.0319	.0359
0.1	.0398	.0438	.0478	.0517	.0557	.0596	.0636	.0675	.0714	.0753
0.2	.0793	.0832	.0871	.0910	.0948	.0987	.1026	.1064	.1103	.1141
0.3	.1179	.1217	.1255	.1293	.1331	.1368	.1406	.1443	.1480	.1517
0.4	.1554	.1591	.1628	.1664	.1700	.1736	.1772	.1808	.1844	.1879
0.5	.1915	.1950	.1985	.2019	.2054	.2088	.2123	.2157	.2190	.2224
0.6	.2257	.2291	.2324	.2357	.2389	.2422	.2454	.2486	.2517	.2549
0.7	.2580	.2611	.2642	.2673	.2704	.2734	.2764	.2794	.2823	.2852
0.8	.2881	.2910	.2939	.2967	.2995	.3023	.3051	.3078	.3106	.3133
0.9	.3159	.3186	.3212	.3238	.3264	.3289	.3315	.3340	.3365	.3389
1.0	.3413	.3438	.3461	.3485	.3508	.3531	.3554	.3577	.3599	.3621
1.1	.3643	.3665	.3686	.3708	.3729	.3749	.3770	.3790	.3810	.3830
1.2	.3849	.3869	.3888	.3907	.3925	.3944	.3962	.3980	.3997	.4015
1.3	.4032	.4049	.4066	.4082	.4099	.4115	.4131	.4147	.4162	.4177
1.4	.4192	.4207	.4222	.4236	.4251	.4265	.4279	.4292	.4306	.4319
1.5	.4332	.4345	.4357	.4370	.4382	.4394	.4406	.4418	.4429	.4441
1.6	.4452	.4463	.4474	.4484	.4495	.4505	.4515	.4525	.4535	.4545
1.7	.4554	.4564	.4573	.4582	.4591	.4599	.4608	.4616	.4625	.4633
1.8	.4641	.4649	.4656	.4664	.4671	.4678	.4686	.4693	.4699	.4706
1.9	.4713	.4719	.4726	.4732	.4738	.4744	.4750	.4756	.4761	.4767
2.0	.4772	.4778	.4783	.4788	.4793	.4798	.4803	.4808	.4812	.4817
2.1	.4821	.4826	.4830	.4834	.4838	.4842	.4846	.4850	.4854	.4857
2.2	.4861	.4864	.4868	.4871	.4875	.4878	.4881	.4884	.4887	.4890
2.3	.4893	.4896	.4898	.4901	.4904	.4906	.4909	.4911	.4913	.4916
2.4	.4918	.4920	.4922	.4925	.4927	.4929	.4931	.4932	.4934	.4936
2.5	.4938	.4940	.4941	.4943	.4945	.4946	.4948	.4949	.4951	.4952
2.6	.4953	.4955	.4956	.4957	.4959	.4960	.4961	.4962	.4963	.4964
2.7	.4965	.4966	.4967	.4968	.4969	.4970	.4971	.4972	.4973	.4974
2.8	.4974	.4975	.4976	.4977	.4977	.4978	.4979	.4979	.4980	.4981
2.9	.4981	.4982	.4982	.4983	.4984	.4984	.4985	.4985	.4986	.4986
3.0	.4987	.4987	.4987	.4988	.4988	.4989	.4989	.4989	.4990	.4990

Table V

THE *t* DISTRIBUTION TABLE

The entries in the table give the critical values of t for the specified number of degrees of freedom and areas in the right tail.

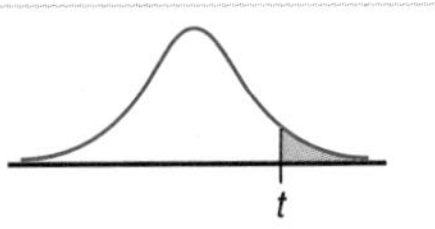

df	Area in the Right Tail under the *t* Distribution Curve					
	.10	.05	.025	.01	.005	.001
1	3.078	6.314	12.706	31.821	63.657	318.309
2	1.886	2.920	4.303	6.965	9.925	22.327
3	1.638	2.353	3.182	4.541	5.841	10.215
4	1.533	2.132	2.776	3.747	4.604	7.173
5	1.476	2.015	2.571	3.365	4.032	5.893
6	1.440	1.943	2.447	3.143	3.707	5.208
7	1.415	1.895	2.365	2.998	3.499	4.785
8	1.397	1.860	2.306	2.896	3.355	4.501
9	1.383	1.833	2.262	2.821	3.250	4.297
10	1.372	1.812	2.228	2.764	3.169	4.144
11	1.363	1.796	2.201	2.718	3.106	4.025
12	1.356	1.782	2.179	2.681	3.055	3.930
13	1.350	1.771	2.160	2.650	3.012	3.852
14	1.345	1.761	2.145	2.624	2.977	3.787
15	1.341	1.753	2.131	2.602	2.947	3.733
16	1.337	1.746	2.120	2.583	2.921	3.686
17	1.333	1.740	2.110	2.567	2.898	3.646
18	1.330	1.734	2.101	2.552	2.878	3.610
19	1.328	1.729	2.093	2.539	2.861	3.579
20	1.325	1.725	2.086	2.528	2.845	3.552
21	1.323	1.721	2.080	2.518	2.831	3.527
22	1.321	1.717	2.074	2.508	2.819	3.505
23	1.319	1.714	2.069	2.500	2.807	3.485
24	1.318	1.711	2.064	2.492	2.797	3.467
25	1.316	1.708	2.060	2.485	2.787	3.450
26	1.315	1.706	2.056	2.479	2.779	3.435
27	1.314	1.703	2.052	2.473	2.771	3.421
28	1.313	1.701	2.048	2.467	2.763	3.408
29	1.311	1.699	2.045	2.462	2.756	3.396
30	1.310	1.697	2.042	2.457	2.750	3.385
31	1.309	1.696	2.040	2.453	2.744	3.375
32	1.309	1.694	2.037	2.449	2.738	3.365
33	1.308	1.692	2.035	2.445	2.733	3.356
34	1.307	1.691	2.032	2.441	2.728	3.348
35	1.306	1.690	2.030	2.438	2.724	3.340
36	1.306	1.688	2.028	2.434	2.719	3.333
37	1.305	1.687	2.026	2.431	2.715	3.326
38	1.304	1.686	2.024	2.429	2.712	3.319

Table V (continued)

	Area in the Right Tail under the *t* Distribution Curve					
df	**.10**	**.05**	**.025**	**.01**	**.005**	**.001**
39	1.304	1.685	2.023	2.426	2.708	3.313
40	1.303	1.684	2.021	2.423	2.704	3.307
41	1.303	1.683	2.020	2.421	2.701	3.301
42	1.302	1.682	2.018	2.418	2.698	3.296
43	1.302	1.681	2.017	2.416	2.695	3.291
44	1.301	1.680	2.015	2.414	2.692	3.286
45	1.301	1.679	2.014	2.412	2.690	3.281
46	1.300	1.679	2.013	2.410	2.687	3.277
47	1.300	1.678	2.012	2.408	2.685	3.273
48	1.299	1.677	2.011	2.407	2.682	3.269
49	1.299	1.677	2.010	2.405	2.680	3.265
50	1.299	1.676	2.009	2.403	2.678	3.261
51	1.298	1.675	2.008	2.402	2.676	3.258
52	1.298	1.675	2.007	2.400	2.674	3.255
53	1.298	1.674	2.006	2.399	2.672	3.251
54	1.297	1.674	2.005	2.397	2.670	3.248
55	1.297	1.673	2.004	2.396	2.668	3.245
56	1.297	1.673	2.003	2.395	2.667	3.242
57	1.297	1.672	2.002	2.394	2.665	3.239
58	1.296	1.672	2.002	2.392	2.663	3.237
59	1.296	1.671	2.001	2.391	2.662	3.234
60	1.296	1.671	2.000	2.390	2.660	3.232
61	1.296	1.670	2.000	2.389	2.659	3.229
62	1.295	1.670	1.999	2.388	2.657	3.227
63	1.295	1.669	1.998	2.387	2.656	3.225
64	1.295	1.669	1.998	2.386	2.655	3.223
65	1.295	1.669	1.997	2.385	2.654	3.220
66	1.295	1.668	1.997	2.384	2.652	3.218
67	1.294	1.668	1.996	2.383	2.651	3.216
68	1.294	1.668	1.995	2.382	2.650	3.214
69	1.294	1.667	1.995	2.382	2.649	3.213
70	1.294	1.667	1.994	2.381	2.648	3.211
71	1.294	1.667	1.994	2.380	2.647	3.209
72	1.293	1.666	1.993	2.379	2.646	3.207
73	1.293	1.666	1.993	2.379	2.645	3.206
74	1.293	1.666	1.993	2.378	2.644	3.204
75	1.293	1.665	1.992	2.377	2.643	3.202
∞	1.282	1.645	1.960	2.326	2.576	3.090

Table VI

CHI-SQUARE DISTRIBUTION TABLE

The entries in the table give the critical values of χ^2 for the specified number of degrees of freedom and areas on the right tail.

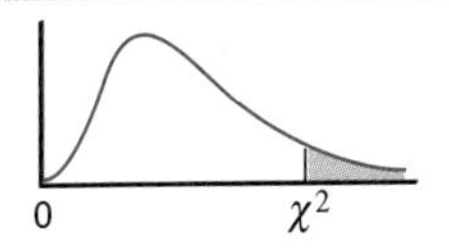

	Area in the Right Tail under the Chi-square Distribution Curve									
df	**.995**	**.990**	**.975**	**.950**	**.900**	**.100**	**.050**	**.025**	**.010**	**.005**
1	0.000	0.000	0.001	0.004	0.016	2.706	3.841	5.024	6.635	7.879
2	0.010	0.020	0.051	0.103	0.211	4.605	5.991	7.378	9.210	10.597
3	0.072	0.115	0.216	0.352	0.584	6.251	7.815	9.348	11.345	12.838
4	0.207	0.297	0.484	0.711	1.064	7.779	9.488	11.143	13.277	14.860
5	0.412	0.554	0.831	1.145	1.610	9.236	11.070	12.833	15.086	16.750
6	0.676	0.872	1.237	1.635	2.204	10.645	12.592	14.449	16.812	18.548
7	0.989	1.239	1.690	2.167	2.833	12.017	14.067	16.013	18.475	20.278
8	1.344	1.646	2.180	2.733	3.490	13.362	15.507	17.535	20.090	21.955
9	1.735	2.088	2.700	3.325	4.168	14.684	16.919	19.023	12.666	23.589
10	2.156	2.558	3.247	3.940	4.865	15.987	18.307	20.483	23.209	25.188
11	2.603	3.053	3.816	4.575	5.578	17.275	19.675	21.920	24.725	26.757
12	3.074	3.571	4.404	5.226	6.304	18.549	21.026	23.337	26.217	28.300
13	3.565	4.107	5.009	5.892	7.042	19.812	22.362	24.736	27.688	29.819
14	4.075	4.660	5.629	6.571	7.790	21.064	23.685	26.119	29.141	31.319
15	4.601	5.229	6.262	7.261	8.547	22.307	24.996	27.488	30.578	32.801
16	5.142	5.812	6.908	7.962	9.312	23.542	26.296	28.845	32.000	34.267
17	5.697	6.408	7.564	8.672	10.085	24.769	27.587	30.191	33.409	35.718
18	6.265	7.015	8.231	9.390	10.865	25.989	28.869	31.526	34.805	37.156
19	6.844	7.633	8.907	10.117	11.651	27.204	30.144	32.852	36.191	38.582
20	7.434	8.260	9.591	10.851	12.443	28.412	31.410	34.170	37.566	39.997
21	8.034	8.897	10.283	11.591	13.240	29.615	32.671	35.479	38.932	41.401
22	8.643	9.542	10.982	12.338	14.041	30.813	33.924	36.781	40.289	42.796
23	9.260	10.196	11.689	13.091	14.848	32.007	35.172	38.076	41.638	44.181
24	9.886	10.856	12.401	13.848	15.659	33.196	36.415	39.364	42.980	45.559
25	10.520	11.524	13.120	14.611	16.473	34.382	37.652	40.646	44.314	46.928
26	11.160	12.198	13.844	15.379	17.292	35.563	38.885	41.923	45.642	48.290
27	11.808	12.879	14.573	16.151	18.114	36.741	40.113	43.195	46.963	49.645
28	12.461	13.565	15.308	16.928	18.939	37.916	41.337	44.461	48.278	50.993
29	13.121	14.256	16.047	17.708	19.768	39.087	42.557	45.722	49.588	52.336
30	13.787	14.953	16.791	18.493	20.599	40.256	43.773	46.979	50.892	53.672
40	20.707	22.164	24.433	26.509	29.051	51.805	55.758	59.342	63.691	66.766
50	27.991	29.707	32.357	34.764	37.689	63.167	67.505	71.420	76.154	79.490
60	35.534	37.485	40.482	43.188	46.459	74.397	79.082	83.298	88.379	91.952
70	43.275	45.442	48.758	51.739	55.329	85.527	90.531	95.023	100.425	104.215
80	51.172	53.540	57.153	60.391	64.278	96.578	101.879	106.629	112.329	116.321
90	59.196	61.754	65.647	69.126	73.291	107.565	113.145	118.136	124.116	128.299
100	67.328	70.065	74.222	77.929	82.358	118.498	124.342	129.561	135.807	140.169

Table VII

THE *F* DISTRIBUTION TABLE

Area in the Right Tail under the Curve = .01

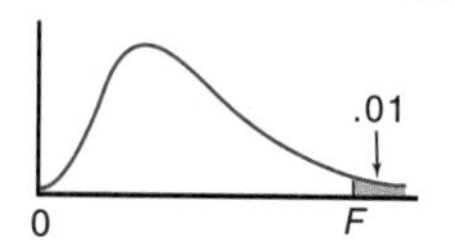

Degrees of Freedom for Numerator

Degrees of Freedom for Denominator	1	2	3	4	5	6	7	8	9
1	4052	5000	5403	5625	5764	5859	5928	5981	6022
2	98.50	99.00	99.17	99.25	99.30	99.33	99.36	99.37	99.39
3	34.12	30.82	29.46	28.71	28.24	27.91	27.67	27.49	27.35
4	21.20	18.00	16.69	15.98	15.52	15.21	14.98	14.80	14.66
5	16.26	13.27	12.06	11.39	10.97	10.67	10.46	10.29	10.16
6	13.75	10.92	9.78	9.15	8.75	8.47	8.26	8.10	7.98
7	12.25	9.55	8.45	7.85	7.46	7.19	6.99	6.84	6.72
8	11.26	8.65	7.59	7.01	6.63	6.37	6.18	6.03	5.91
9	10.56	8.02	6.99	6.42	6.06	5.80	5.61	5.47	5.35
10	10.04	7.56	6.55	5.99	5.64	5.39	5.20	5.06	4.94
11	9.65	7.21	6.22	5.67	5.32	5.07	4.89	4.74	4.63
12	9.33	6.93	5.95	5.41	5.06	4.82	4.64	4.50	4.39
13	9.07	6.70	5.74	5.21	4.86	4.62	4.44	4.30	4.19
14	8.86	6.51	5.56	5.04	4.69	4.46	4.28	4.14	4.03
15	8.68	6.36	5.42	4.89	4.56	4.32	4.14	4.00	3.89
16	8.53	6.23	5.29	4.77	4.44	4.20	4.03	3.89	3.78
17	8.40	6.11	5.18	4.67	4.34	4.10	3.93	3.79	3.68
18	8.29	6.01	5.09	4.58	4.25	4.01	3.84	3.71	3.60
19	8.18	5.93	5.01	4.50	4.17	3.94	3.77	3.63	3.52
20	8.10	5.85	4.94	4.43	4.10	3.87	3.70	3.56	3.46
21	8.02	5.78	4.87	4.37	4.04	3.81	3.64	3.51	3.40
22	7.95	5.72	4.82	4.31	3.99	3.76	3.59	3.45	3.35
23	7.88	5.66	4.76	4.26	3.94	3.71	3.54	3.41	3.30
24	7.82	5.61	4.72	4.22	3.90	3.67	3.50	3.36	3.26
25	7.77	5.57	4.68	4.18	3.85	3.63	3.46	3.32	3.22
30	7.56	5.39	4.51	4.02	3.70	3.47	3.30	3.17	3.07
40	7.31	5.18	4.31	3.83	3.51	3.29	3.12	2.99	2.89
50	7.17	5.06	4.20	3.72	3.41	3.19	3.02	2.89	2.78
100	6.90	4.82	3.98	3.51	3.21	2.99	2.82	2.69	2.59

Table VII (continued)

Degrees of Freedom for Numerator

10	11	12	15	20	25	30	40	50	100
6056	6083	6106	6157	6209	6240	6261	6287	6303	6334
99.40	99.41	99.42	99.43	99.45	99.46	99.47	99.47	99.48	99.49
27.23	27.13	27.05	26.87	26.69	26.58	26.50	26.41	26.35	26.24
14.55	14.45	14.37	14.20	14.02	13.91	13.84	13.75	13.69	13.58
10.05	9.96	9.89	9.72	9.55	9.45	9.38	9.29	9.24	9.13
7.87	7.79	7.72	7.56	7.40	7.30	7.23	7.14	7.09	6.99
6.62	6.54	6.47	6.31	6.16	6.06	5.99	5.91	5.86	5.75
5.81	5.73	5.67	5.52	5.36	5.26	5.20	5.12	5.07	4.96
5.26	5.18	5.11	4.96	4.81	4.71	4.65	4.57	4.52	4.41
4.85	4.77	4.71	4.56	4.41	4.31	4.25	4.17	4.12	4.01
4.54	4.46	4.40	4.25	4.10	4.01	3.94	3.86	3.81	3.71
4.30	4.22	4.16	4.01	3.86	3.76	3.70	3.62	3.57	3.47
4.10	4.02	3.96	3.82	3.66	3.57	3.51	3.43	3.38	3.27
3.94	3.86	3.80	3.66	3.51	3.41	3.35	3.27	3.22	3.11
3.80	3.73	3.67	3.52	3.37	3.28	3.21	3.13	3.08	2.98
3.69	3.62	3.55	3.41	3.26	3.16	3.10	3.02	2.97	2.86
3.59	3.52	3.46	3.31	3.16	3.07	3.00	2.92	2.87	2.76
3.51	3.43	3.37	3.23	3.08	2.98	2.92	2.84	2.78	2.68
3.43	3.36	3.30	3.15	3.00	2.91	2.84	2.76	2.71	2.60
3.37	3.29	3.23	3.09	2.94	2.84	2.78	2.69	2.64	2.54
3.31	3.24	3.17	3.03	2.88	2.79	2.72	2.64	2.58	2.48
3.26	3.18	3.12	2.98	2.83	2.73	2.67	2.58	2.53	2.42
3.21	3.14	3.07	2.93	2.78	2.69	2.62	2.54	2.48	2.37
3.17	3.09	3.03	2.89	2.74	2.64	2.58	2.49	2.44	2.33
3.13	3.06	2.99	2.85	2.70	2.60	2.54	2.45	2.40	2.29
2.98	2.91	2.84	2.70	2.55	2.45	2.39	2.30	2.25	2.13
2.80	2.73	2.66	2.52	2.37	2.27	2.20	2.11	2.06	1.94
2.70	2.63	2.56	2.42	2.27	2.17	2.10	2.01	1.95	1.82
2.50	2.43	2.37	2.22	2.07	1.97	1.89	1.80	1.74	1.60

Table VII (continued)

Area in the Right Tail under the Curve = .05

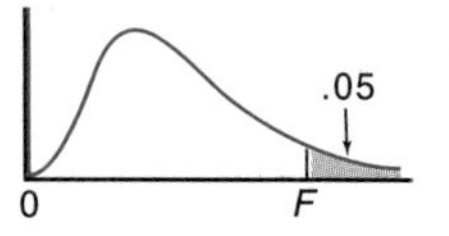

Degrees of Freedom for Numerator

Degrees of Freedom for Denominator	1	2	3	4	5	6	7	8	9
1	161.5	199.5	215.7	224.6	230.2	234.0	236.8	238.9	240.5
2	18.51	19.00	19.16	19.25	19.30	19.33	19.35	19.37	19.38
3	10.13	9.55	9.28	9.12	9.01	8.94	8.89	8.85	8.81
4	7.71	6.94	6.59	6.39	6.26	6.16	6.09	6.04	6.00
5	6.61	5.79	5.41	5.19	5.05	4.95	4.88	4.82	4.77
6	5.99	5.14	4.76	4.53	4.39	4.28	4.21	4.15	4.10
7	5.59	4.74	4.35	4.12	3.97	3.87	3.79	3.73	3.68
8	5.32	4.46	4.07	3.84	3.69	3.58	3.50	3.44	3.39
9	5.12	4.26	3.86	3.63	3.48	3.37	3.29	3.23	3.18
10	4.96	4.10	3.71	3.48	3.33	3.22	3.14	3.07	3.02
11	4.84	3.98	3.59	3.36	3.20	3.09	3.01	2.95	2.90
12	4.75	3.89	3.49	3.26	3.11	3.00	2.91	2.85	2.80
13	4.67	3.81	3.41	3.18	3.03	2.92	2.83	2.77	2.71
14	4.60	3.74	3.34	3.11	2.96	2.85	2.76	2.70	2.65
15	4.54	3.68	3.29	3.06	2.90	2.79	2.71	2.64	2.59
16	4.49	3.63	3.24	3.01	2.85	2.74	2.66	2.59	2.54
17	4.45	3.59	3.20	2.96	2.81	2.70	2.61	2.55	2.49
18	4.41	3.55	3.16	2.93	2.77	2.66	2.58	2.51	2.46
19	4.38	3.52	3.13	2.90	2.74	2.63	2.54	2.48	2.42
20	4.35	3.49	3.10	2.87	2.71	2.60	2.51	2.45	2.39
21	4.32	3.47	3.07	2.84	2.68	2.57	2.49	2.42	2.37
22	4.30	3.44	3.05	2.82	2.66	2.55	2.46	2.40	2.34
23	4.28	3.42	3.03	2.80	2.64	2.53	2.44	2.37	2.32
24	4.26	3.40	3.01	2.78	2.62	2.51	2.42	2.36	2.30
25	4.24	3.39	2.99	2.76	2.60	2.49	2.40	2.34	2.28
30	4.17	3.32	2.92	2.69	2.53	2.42	2.33	2.27	2.21
40	4.08	3.23	2.84	2.61	2.45	2.34	2.25	2.18	2.12
50	4.03	3.18	2.79	2.56	2.40	2.29	2.20	2.13	2.07
100	3.94	3.09	2.70	2.46	2.31	2.19	2.10	2.03	1.97

able VII (continued)

Degrees of Freedom for Numerator

10	11	12	15	20	25	30	40	50	100
241.9	243.0	243.9	246.0	248.0	249.3	250.1	251.1	251.8	253.0
19.40	19.40	19.41	19.43	19.45	19.46	19.46	19.47	19.48	19.49
8.79	8.76	8.74	8.70	8.66	8.63	8.62	8.59	8.58	8.55
5.96	5.94	5.91	5.86	5.80	5.77	5.75	5.72	5.70	5.66
4.74	4.70	4.68	4.62	4.56	4.52	4.50	4.46	4.44	4.41
4.06	4.03	4.00	3.94	3.87	3.83	3.81	3.77	3.75	3.71
6.64	3.60	3.57	3.51	3.44	3.40	3.38	3.34	3.32	3.27
3.35	3.31	3.28	3.22	3.15	3.11	3.08	3.04	3.02	2.97
3.14	3.10	3.07	3.01	2.94	2.89	2.86	2.83	2.80	2.76
2.98	2.94	2.91	2.85	2.77	2.73	2.70	2.66	2.64	2.59
2.85	2.82	2.79	2.72	2.65	2.60	2.57	2.53	2.51	2.46
2.75	2.72	2.69	2.62	2.54	2.50	2.47	2.43	2.40	2.35
2.67	2.63	2.60	2.53	2.46	2.41	2.38	2.34	2.31	2.26
2.60	2.57	2.53	2.46	2.39	2.34	2.31	2.27	2.24	2.19
2.54	2.51	2.48	2.40	2.33	2.28	2.25	2.20	2.18	2.12
2.49	2.46	2.42	2.35	2.28	2.23	2.19	2.15	2.12	2.07
2.45	2.41	2.38	2.31	2.23	2.18	2.15	2.10	1.08	2.02
2.41	2.37	2.34	2.27	2.19	2.14	2.11	2.06	2.04	1.98
2.38	2.34	2.31	2.23	2.16	2.11	2.07	2.03	2.00	1.94
2.35	2.31	2.28	2.20	2.12	2.07	2.04	1.99	1.97	1.91
2.32	2.28	2.25	2.18	2.10	2.05	2.01	1.96	1.94	1.88
2.30	2.26	2.23	2.15	2.07	2.02	1.97	1.94	1.91	1.85
2.27	2.24	2.20	2.13	2.05	2.00	1.96	1.91	1.88	1.82
2.25	2.22	2.18	2.16	2.03	1.97	1.94	1.89	1.86	1.80
2.24	2.20	2.16	2.09	2.01	1.96	1.92	1.87	1.84	1.78
2.16	2.13	2.09	2.01	1.93	1.88	1.84	1.79	1.76	1.70
2.08	2.04	2.00	1.92	1.84	1.78	1.74	1.69	1.66	1.59
2.03	1.99	1.95	1.87	1.78	1.73	1.69	1.63	1.60	1.52
1.93	1.89	1.85	1.77	1.68	1.62	1.57	1.52	1.48	1.39

Table VIII

CRITICAL VALUES FOR THE DURBIN-WATSON STATISTIC

Sample Size = n	Pr = Probability in Lower Tail (Significance Level = α)	k = Number of Regressors (excluding the constant) 1		2		3		4		5	
		D_L	D_U	D_L	D_U	D_L	D_U	D_L	D_U	D_L	D_U
15	.01	.81	1.07	.70	1.25	.59	1.46	.49	1.70	.39	1.96
	.025	.95	1.23	.83	1.40	.71	1.61	.59	1.84	.48	2.09
	.05	1.08	1.36	.95	1.54	.82	1.75	.69	1.97	.56	2.21
20	.01	.95	1.15	.86	1.27	.77	1.41	.68	1.57	.60	1.74
	.025	1.08	1.28	.99	1.41	.89	1.55	.79	1.70	.70	1.87
	.05	1.20	1.41	1.10	1.54	1.00	1.68	.90	1.83	.79	1.99
25	.01	1.05	1.21	.98	1.30	.90	1.41	.83	1.52	.75	1.65
	.025	1.18	1.34	1.10	1.43	1.02	1.54	.94	1.65	.86	1.77
	.05	1.29	1.45	1.21	1.55	1.12	1.66	1.04	1.77	.95	1.89
30	.01	1.13	1.26	1.07	1.34	1.01	1.42	.94	1.51	.88	1.61
	.025	1.25	1.38	1.18	1.46	1.12	1.54	1.05	1.63	.98	1.73
	.05	1.35	1.49	1.28	1.57	1.21	1.65	1.14	1.74	1.07	1.83
40	.01	1.25	1.34	1.20	1.40	1.15	1.46	1.10	1.52	1.05	1.58
	.025	1.35	1.45	1.30	1.51	1.25	1.57	1.20	1.63	1.15	1.69
	.05	1.44	1.54	1.39	1.60	1.34	1.66	1.29	1.72	1.23	1.79
50	.01	1.32	1.40	1.28	1.45	1.24	1.49	1.20	1.54	1.16	1.59
	.025	1.42	1.50	1.38	1.54	1.34	1.59	1.30	1.64	1.26	1.69
	.05	1.50	1.59	1.46	1.63	1.42	1.67	1.38	1.72	1.34	1.77
60	.01	1.38	1.45	1.35	1.48	1.32	1.52	1.28	1.56	1.25	1.60
	.025	1.47	1.54	1.44	1.57	1.40	1.61	1.37	1.65	1.33	1.69
	.05	1.55	1.62	1.51	1.65	1.48	1.69	1.44	1.73	1.41	1.77
80	.01	1.47	1.52	1.44	1.54	1.42	1.57	1.39	1.60	1.36	1.62
	.025	1.54	1.59	1.52	1.62	1.49	1.65	1.47	1.67	1.44	1.70
	.05	1.61	1.66	1.59	1.69	1.56	1.72	1.53	1.74	1.51	1.77
100	.01	1.52	1.56	1.50	1.58	1.48	1.60	1.46	1.63	1.44	1.65
	.025	1.59	1.63	1.57	1.65	1.55	1.67	1.53	1.70	1.51	1.72
	.05	1.65	1.69	1.63	1.72	1.61	1.74	1.59	1.76	1.57	1.78

Answers

to Selected Odd-Numbered Exercises

Note: Because of differences in rounding, answers obtained by students may differ slightly from the ones given here.

CHAPTER 1

1.1. **a.** accounts payable **b.** all accounts payable **c.** random sample
d. an identification for each observation, the date the debt was incurred, the amount of debt

1.3. **a.** all consumer prices
b. Problems in measuring this population would include selection of items, the form of the product, and the place of sale.
c. Random sampling would not be appropriate because it would not correspond to people's purchasing patterns.

1.5. **a.** all those covered by SSA's hospital and supplementary medical care insurance programs
b. Randomly select enrollees from a list of those enrolled in the insurance program.
c. information on types of services rendered and charges for those services

1.7. **a.** tourists to India
b. Randomly select from customs a sample of those leaving the country.
c. how much was spent on gifts, meals, and other activities
d. Example of a typical case:

Amount Spent				
Hotel(s)	Restaurants	Tours	Gifts	Length of Stay
$1500	$600	$250	$200	5 days

1.9. **a.** all parachute cords produced
b. Randomly select a batch of cords from production each hour.
c. tensile strength measurements
d. whether specifications were being met or not being met **e.** yes

1.11. Primary data are collected by the user by direct observation through surveys, censuses, and experimentation. Secondary data come from a number of sources, such as newspapers, magazines, books, journals, and databases.

1.13. **a.** primary data **b.** secondary data **c.** primary data
d. secondary data (except what they collect) **e.** primary data

1.15. **a.** qualitative (It deals with attributes.) **b.** cross-sectional **c.** quantitative
d. continuous **e.** question 1: nominal; question 2: ratio

1.17. ease of shopping: cross-sectional, qualitative, ordinal; most important ranking: cross-sectional, quantitative, ordinal; marital status: cross-sectional, qualitative, nominal

1.19. All the variables are quantitative, insofar as they are scores. All are discrete and ordinal.

CHAPTER 2

2.1. c.w. = 20 units; reasonable class limits: 40–59.9, . . . , 140–159.9

2.3.

Group	Frequency
A	7
B	6
AB	7
O	10

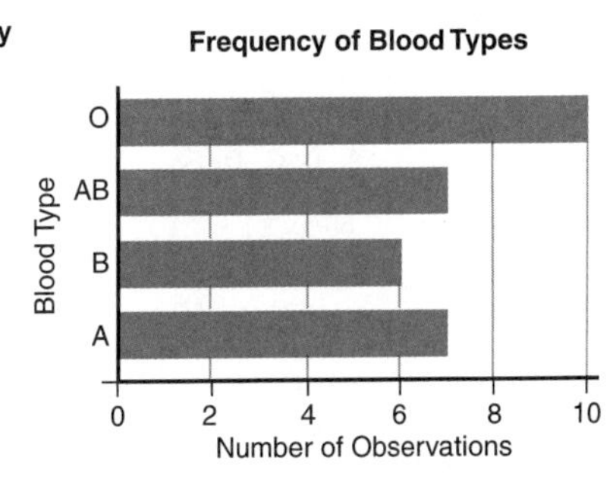

2.5. **a.** overlap, unequal size of amounts being compared, and gaps
b. LCLs should be 800, 850, 900, . . . ; UCLs should be 849, 899, 949, . . . Exact values would be necessary so that data could be properly allocated to classes.

2.7. **a.** Prices of umbrellas:

Price Range	No. of Umbrellas
$ 0–24	27
25–49	10
50–74	2
⋮	⋮
175–199	1

b. A price below $25 would be reasonable; a price below $10 would be a bargain.
c. Prices of umbrellas:

Price Range	No. of Umbrellas
Less than $10	5
$10–19	18
20–29	7
30–39	7
40–49	0
50–59	1
60–69	1

Use the second table to present a picture of the umbrella market because it gives more information about umbrella prices.

2.9. **a.** Winners of the Grand Prix Formula-One Road Race:

Country	No. of Winners
Argentina	5
Australia	4
Austria	4
Brazil	8
Finland	1
France	3
Germany	1
Great Britain	12
Italy	3
New Zealand	1
South Africa	1
United States	2

b. Great Britain has had the most winners. Finland, Germany, New Zealand, and South Africa are tied for fewest winners.

2.11.

Stem	Leaf
4	3 3
5	
6	
7	2 5 8
8	
9	0 1 1 6

b. 40 50 60 70 80 90

c. Possible awarding of grades: 4 A's, 3 C's, and 2 F's

d.

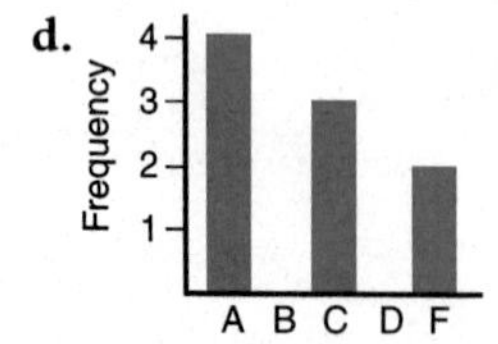

2.17.

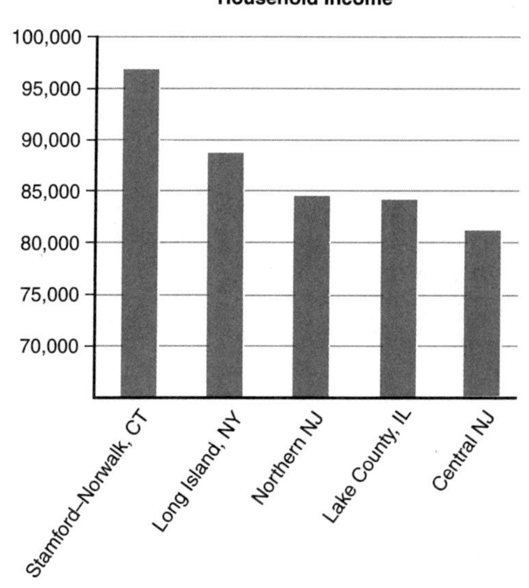

2.19. a.

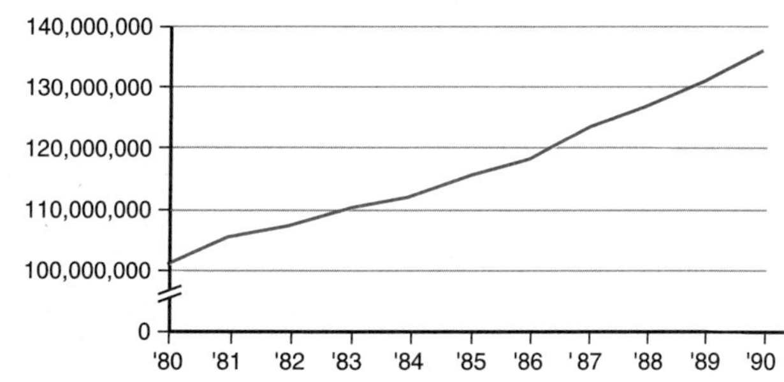

b. There is clearly significant growth between 1980 and 1990. Access lines appear to be growing at a steady rate of about 2 to 3% per annum, suggesting a fairly steady need for switching and line equipment.

2.21. a.

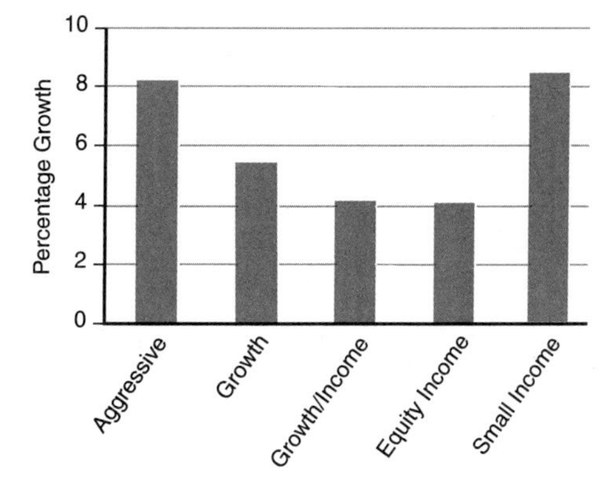

2.21. b.

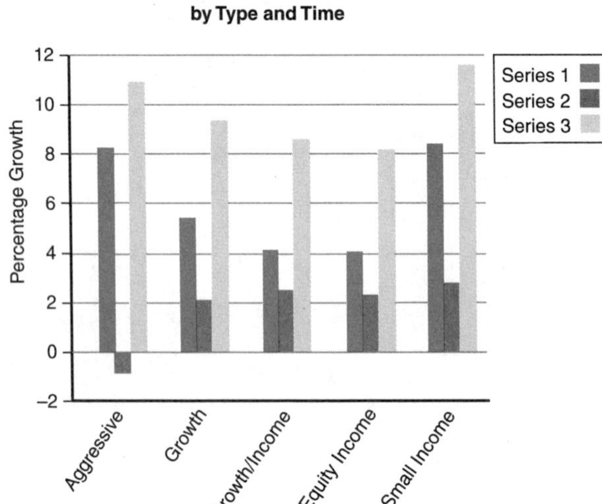

Series 1 is 3-month percentage growth; series 2 is 1-year percentage growth; and series 3 is 5-year percentage growth.

c. You might point out that the type of investment a client wishes to make might well depend on the client's time horizon. You might suggest advertising that the long-term investor should consider aggressive growth or small income funds.

2.23. a. and b.

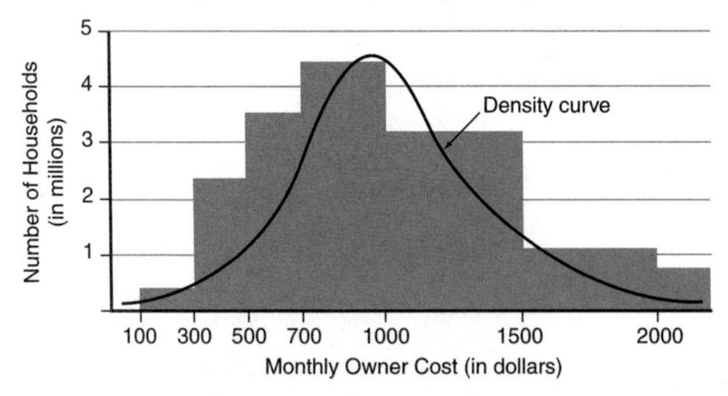

2.25. a.

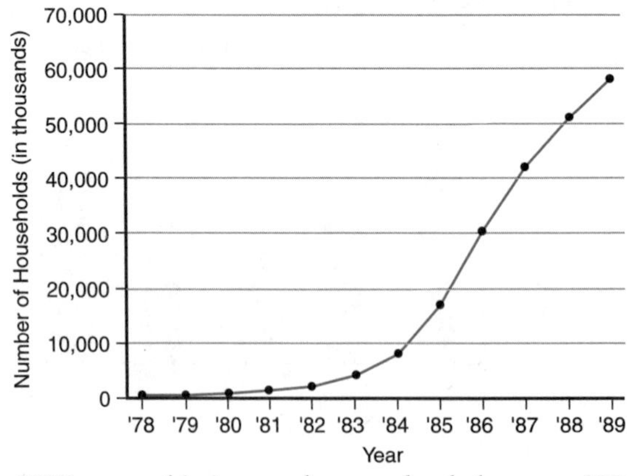

b. VCR ownership increased tremendously between 1978 and 1989. The number of households with VCRs increased almost geometrically in the early years, doubling between 1979 and 1980 and almost doubling again between 1980 and 1981. Growth continued at a dramatic rate, reaching over 58 million households by 1989.

2.27. a.

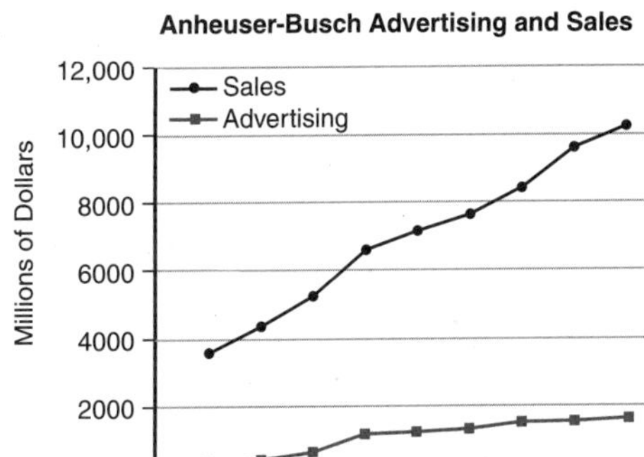

b. Yes; it would appear that increased advertising expenditures have a positive effect on sales.

2.29. a.

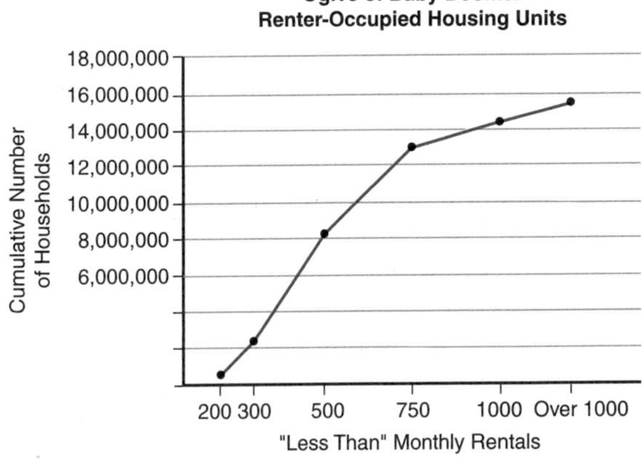

d. The middle 50% of baby boomer renters pay between \$475 and \$750 per month.

2.31. a. Breakdown of energy consumption by end-use sector:

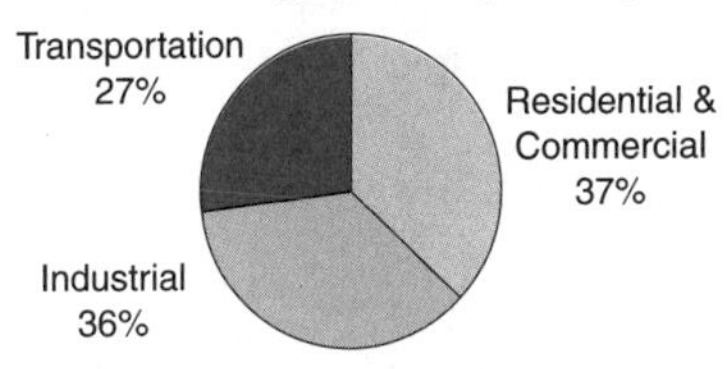

b. The Residential & Commercial and Industrial sectors use about the same share of energy. Both use more than Transportation does. It seems that Residential & Commercial or Industrial would be a better target than Transportation.

2.33. a.

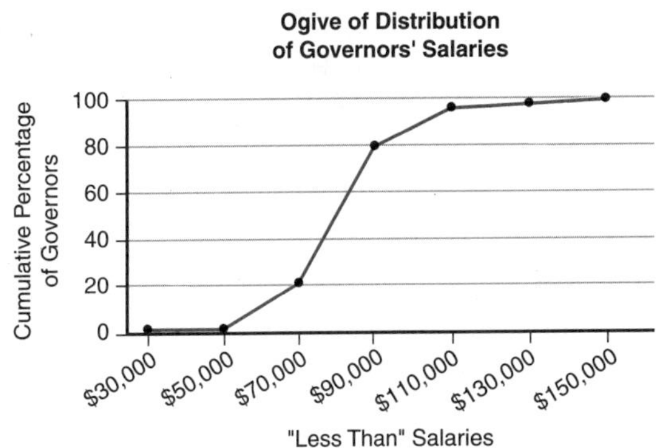

b. The median value is approximately $80,000.
c. Ninety percent of all governors earn less than $105,000.

2.35. a.

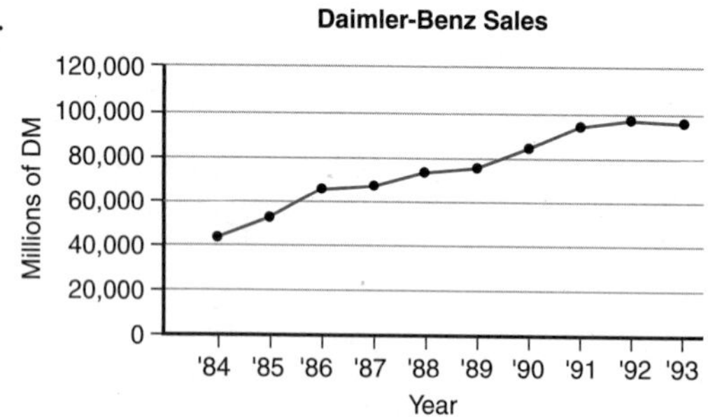

b.

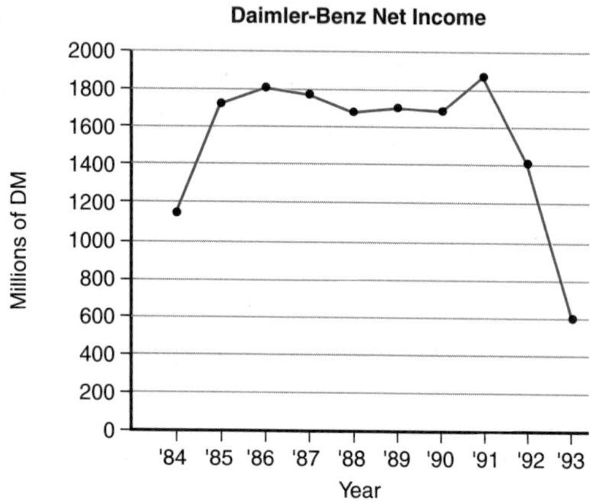

c. Growth in sales has not been accompanied by growth in net income.
d. Increased costs accounted for a drop in net income.

2.37. a. Pie chart of U.S. tableware shipments in 1970:

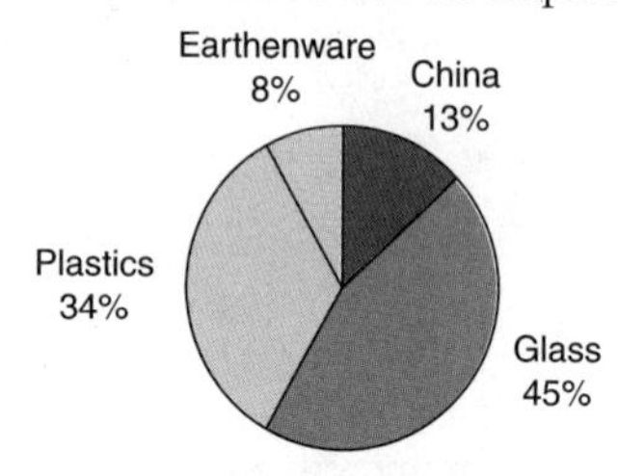

b. Pie chart of U.S. tableware shipments in 1977:

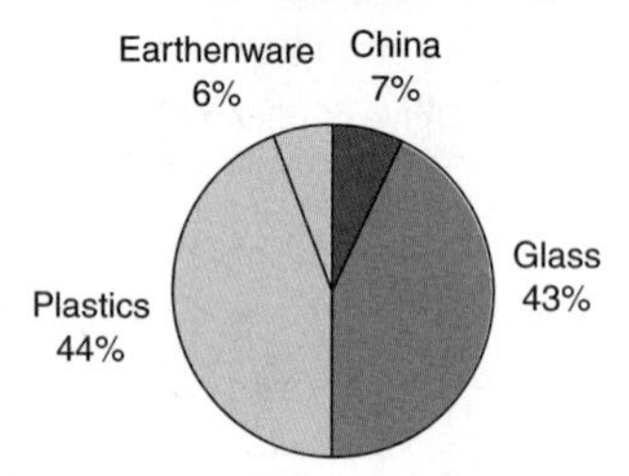

c. Between 1970 and 1977, total tableware shipments grew. All areas grew, but growth was not even. Plastics showed the largest change in market share.

2.39. a.

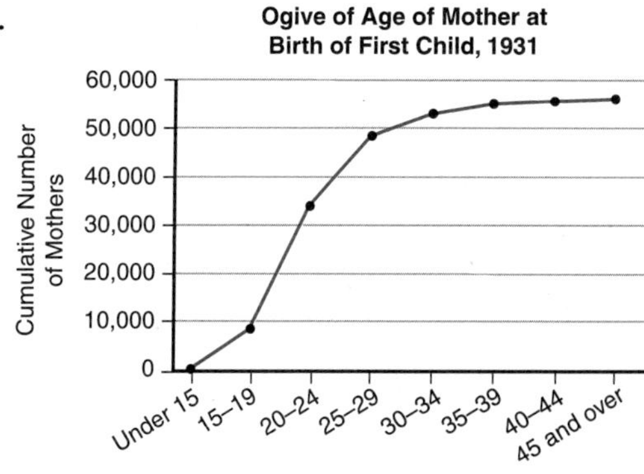

b. The approximate 90th percentile value is 30 years of age.

c.

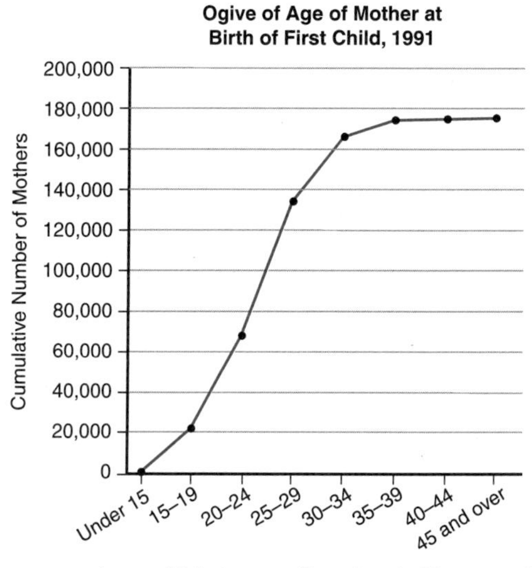

The approximate 90th percentile value is 35 years of age.

CHAPTER 3

3.1. **a.** 42.78 **b.** 32 **c.** 29
d. Mean increases to 113.33; mode and median remain the same.

3.3. **a.** 74.17
b.

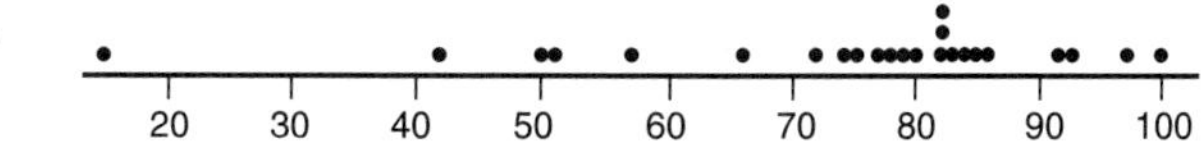

c. 79.5 **d.** 82 **e.** Mean decreases to 73.33; median and mode remain the same.
f. 42 is an outlier that should be removed because it is distant from the rest of the data set.
g. Mean and median increase; mode remains the same.

3.5. **a.** 40,561.22 **b.** 31,099.5 **c.** Mean decreases to 39,922.69; median remains the same.
d. Mean decreases to 37,584.76; median decreases to 30,624.

3.7. **a.** 13 oz **b.** $.19 **c.** $.19 **d.** 7.58 mg **e.** 7.74 mg

3.9. **a.** 5.68 **b.** 5 **c.** 5 **d.** Mean decreases to 5.26; median remains the same.
e. mode

3.11. **a.** mean = 6.72; median = 7.05 **b.** mean = 6.61; median = 6.55 **c.** no **d.** yes

3.13. **a.** $7010 **b.** $6112 **c.** $9755
d. yes, based on the average for all the colleges in the sample

3.15. **a.** $7.58 million **b.** $3.35 million
c. Average decreases to $5.14 million; median decreases to $3.2 million.

3.17. **a.** 66 **b.** 50.0 **c.** 24.31 **d.** 48.62%

3.19. **a.** 48,530.40 **b.** 86,063 **c.** 24,638 **d.** 50.77%

3.21. **a.** 4.0 **b.** 2.31 **c.** .92 **d.** 40.26%

3.23. **a.** $3088
b. c. and d. Northeast averages $8985; std dev. = $3344; c.v. = 37.22% . Others average $5947; std dev. = $2451; c.v. = 41.22%. **e.** Northeast
f. more absolute variation in northeast; more relative variation elsewhere

3.25. **a.** $407 **b.** $75.78 **c.** $320, $24.49 **d.** removable: 18.62%; shaft-mounted: 7.65%
e. removable

3.27. **a.** 87.2% **b.** .99% **c.** 90.63%, .78% **d.** 1.14 % **e.** .86% **f.** quarter slots

3.29. **a.** 26.43 **b.** 36.26 **c.** 7.29 **d.** 20.10%

3.31. **b.** 26.5 **c.** 61 **d.** 34.5 **e.** 67

3.33. **b.** −5.16 **c.** 17.99 **d.** 5.51 **e.** −5.16; −.04; 18.03; 5.51

3.35. **a.** 17 **b.** 35 **c.** 35; 6; 22

3.37. **a.** 10 **b.** 14
c. 4; 14; −27; Sterling; 39; Crompton & Knowles; Great Lake Chemical; 7.5; 14

3.39. **a.** 193.5 **b.** 174 **c.** 216 **d.** 174; 193.5; 216

3.41. **a.** 5 **b.** 4 **c.** yes, skewed to the right

3.43. **a.** median = 20.00, mean = 20.74; because the median and mean are nearly equal, we may conclude that the distribution is not skewed.
b. Median remains at 20.00; mean becomes 20.02; distribution is not skewed.

3.45. **a.** 1.48 **b.** .55 **c.** positively skewed

3.47. **a.** 16.88 **b.** 16.52 **c.** no

3.49. **a.** 146.65 **b.** 146.44 **c.** no

3.51. **b.**

Variable	N	Mean	Median	TrMean	StDev	SE Mean
Income	52	244.0	47.2	177.4	421.8	58.5

Variable	Minimum	Maximum	Q1	Q3
Income	10.4	1538.6	29.6	239.5

c.

d. –599.6; 1087.6; the empirical rule does not fit these data well.
e. 52; $244 million; $10.4 million; Pan American Foundation; Catholic Relief Services; $1538.6 million; $29.6 million; $239.5 million

3.53. **a.** fountain **b.** $29.50 to $83.75 (ballpoint); $45.00 to $124.50 (fountain)

3.55. **a.** mean = 18,857; st. dev. = 48,127
b. five-number summary: 266; 1487; 2265; 8450; 170,000
c.

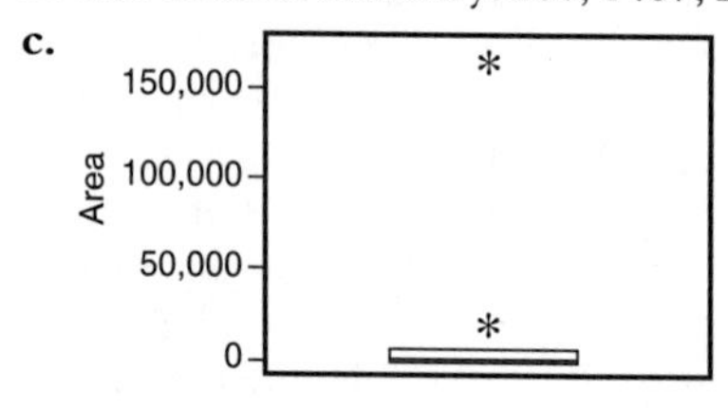

e. mean = 2348; st. dev. = 3400; five-number summary: −1290; −73; 468; 4965; 10,000

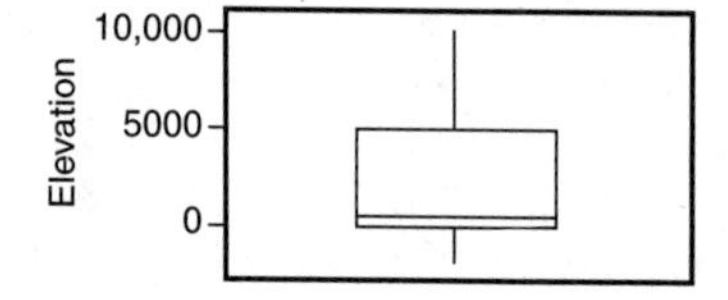

3.57. **a.** mean = 52,127; median = 53,000

b.

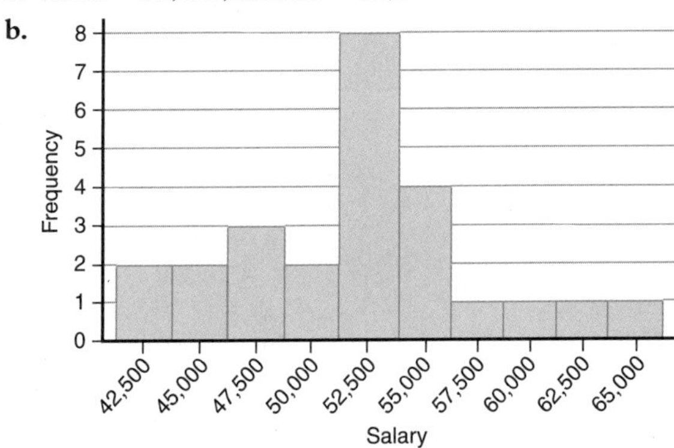

c. IQR = 5750

3.59. **a.** \$58.73 **b.** \$26.70 **c.** \$26 **d.** 45.46% **e.** \$89 **f.** underbudgeting

3.61. 43 **3.63.** 1849 **3.65.** 352; 1720 **3.67.** 83 **3.69.** 38 **3.71.** 33

CHAPTER 4

4.1. $\Omega = \{A, B, \ldots, Z\}$; *P*(Selecting a vowel) = 5/26

4.3. **a.** $\Omega = \{A_1B_1, A_1B_2, A_1B_3, A_1B_4, A_2B_1, A_2B_2, A_2B_3, A_2B_4, A_3B_1, A_3B_2, A_3B_3, A_3B_4\}$
b. {All points given in part a and $A_1A_2, A_1A_3, A_2A_3, B_1B_2, B_1B_3, B_1B_4, B_2B_3, B_2B_4, B_3B_4$} without replacement; {All points given in part a and $A_1A_1, A_2A_2, A_3A_3, B_1B_1, B_2B_2, B_3B_3, B_4B_4$} with replacement

4.5. $\Omega = \{M_1N_1, M_1N_2, M_2N_1, M_2N_2, M_3N_1, M_3N_2\}$

4.7. *P*(Selecting ice tea) = 30/50; classical definition of probability

4.9. **a.** $\Omega = \{R_1, \ldots, R_{25}, G_1, \ldots, G_{25}, Y_1, \ldots, Y_{20}, B_1, \ldots, B_{10}, Bl_1, \ldots, Bl_{10}\}$
b. *P*(Selecting a blue) = 10/90

4.13. **a.** Ω = {Survival, Death} **b.** *P*(Not surviving) = .02

4.15. *P*(Must stop) = 1/3 **4.17.** *P*(Not yes) = .6; no

4.19. **a.** simple events: high income, low income, rewrap, keep
b. joint events: high income and rewrap, high income and keep, low income and rewrap, low income and keep
c. Complement of "rewrap" is "keep."
d. Complement of "high income" is "low income."
e. Disposition and income are not statistically independent.

4.21. **a.** *P*(Not scoring field goal) = .514 **b.** *P*(Successful on two opportunities) = .2362
c. *P*(Failing on two opportunities) = .2642

4.23. **a.** *P*(H and R) = .155 **b.** *P*(L or K) = .845 **c.** P(R | H) = .31

4.25. **a.** simple events: A, B, AB, O, Rh+, Rh−
b. *P*(A) = .42; *P*(B) = .10; *P*(AB) = .04; *P*(O) = .44; *P*(Rh+) = .85; *P*(Rh−) = .15
c. *P*(A and Rh+) = .357 **d.** *P*(O or Rh−) = .524 **e.** *P*(AB | Rh−) = .04 **f.** yes

4.27. **a.** *P*(Does not eat healthy) = .054 **b.** *P*(F | H) = .5412 **c.** *P*(LB | F) = .60 **d.** no

4.29. **a.** *P*(High) = 11/21 **b.** *P*(H and A) = 8/21 **c.** *P*(A | H) = 8/11 **d.** no

CHAPTER 5

5.1. **a.** $E(X) = 34.98$ **b.** $V(X) = 531.74$ **c.** $S(X) = 23.06$

5.3. **a.** $E(X) = 5.5$ **b.** $V(X) = 8.25$ **c.** $S(X) = 2.87$

5.5. **a.** $E(X) = .28$ **b.** *P*(More than 1 breakdown) = .06
c. It does pay to get a service contract.

5.7. **a.** mean time for re-arrest = 1.85 yr **b.** $V(X) = 1.23$ **c.** $S(X) = 1.11$ yr
d. 1.85 yr; 90

5.9. **a.** $E(X) = 8.6$ min **b.** $S(X) = 4.1$ min **c.** *P*(*X* at most 10) = .8

5.11. **a.** $P(3) = .2903$ **b.** $P(X > 3) = .2898$ **c.** $P(X \geq 3) = .5801$
d. $P(X < 3) = .4199$ **e.** $P(X \leq 3) = .7102$

5.13. **a.** $P(X \geq 4) = .3438$ **b.** $P(X \leq 4) = .8906$ **c.** $E(X) = 3$
d. $V(X) = 1.5$; $S(X) = 1.22$

5.15. **a.** $P(X > 20) = .0905$ **b.** $P(X < 20) = .8065$ **c.** $P(X > 15) = .8106$
d. $P(X \leq 14) = .0978$

5.17. **a.** $E(X) = 7.2$ **b.** $P(X \leq 2) = .0143$ **c.** $P(X \geq 2) = .9972$

5.19. **a.** $P(X > 5) = .6331$ **b.** $P(X < 2) = .0017$ **c.** $E(X) = 18$
d. $S(X) = 2.68$ **e.** between 15.32 and 20.68

5.21. **a.** $E(X) = 4.44$ **b.** $P(X = 5) = .1936$ **c.** $P(X \leq 5) = .7268$

5.23. **a.** $P(3) = .0998$ **b.** $P(X \leq 3) = .9569$ **c.** $P(X \geq 3) = .1429$
d. $P(X > 3) = .0431$ **e.** $V(X) = 1.3$

5.25. **a.** $P(1) = .3476$ **b.** $P(X \geq 1) = .5034$ **c.** $P(X < 2) = .8442$
d. $P(X > 2) = .0341$ **e.** $V(X) = .7$

5.27. **a.** $P(0) = .1225$ **b.** $P(X \leq 1) = .3797$ **c.** $P(X > 3) = .1614$

5.29. **a.** $P(X \geq 6) = .3883$ **b.** $P(7 \leq X \leq 9) = .2060$ **c.** $P(X > 9) = .0318$
d. $P(X < 3) = .1246$ **e.** $S(X) = 2.24$

5.31. **a.**

No. of Hits	Actual Hits	*P*(Hit)
0	229	.3976
1	211	.3663
2	93	.1615
3	35	.0608
4	7	.0122
≥5	1	.0017

b. Given $\lambda = .9323$:

No. of Hits	*P*(*k*)
0	.3936
1	.3670
2	.1711
3	.0532
4	.0124
≥5	.0027

c. Data fit the Poisson model.

5.33. **a.** area between *z* score and mean: 1.18, .3810; −2.03, .4788; .47, .1808; −.99, .3389; 3.09, .4990
b. area to the right of *z* score: 1.18, .1190; −2.03, .9788; .47, .3192; −.99, .8389; 3.09, .0010
c. area to the left of *z* score: 1.18, .8810; −2.03, .0212; .47, .6808; −.99, .1611; 3.09, .9990

5.35. **a.** 1.00 **b.** .1587 lies to the right; .8413 falls to the left. **c.** −.52 **d.** .1985
e. .5398

5.37. **a.** $P(X > 25) = .1587$ **b.** $P(19 < X < 27) = .7938$ **c.** $P(23 < X < 29) = .3608$
d. $P_{90} = 25.84$ **e.** 16.12, 27.88

5.39. **a.** $P(X < \$150) = .1170$ **b.** $P(X > \$350) = .0150$ **c.** \$180.72, \$260.62
d. $P_{95} = \$318.76$ **e.** 54.50%

5.41. **a.** $P_{90} = 3.69\%$ **b.** $P_{10} = 2.85\%$ **c.** $P(X \geq 3\%) = .7939$ **d.** $P(X > 4\%) = .0136$

5.43. **a.** $E(X) = 160$ **b.** $S(X) = 10.43$ **c.** normal, because $np \geq 5$ and $nq \geq 5$
d. $P(X > 172) = .1151$ **e.** $P(150 < X < 160) = .3438$

5.45. When $np \geq 5$ and $nq \geq 5$, the normal distribution can be used to approximate the binomial. The Poisson distribution can be used when n is very large and/or p is very small.

5.47. **a.** $E(X) = 440$ **b.** $S(X) = 15.70$ **c.** $P(X > 450) \approx .2514$ **d.** $P(X < 430) \approx .2514$

5.49. **a.** $P(X > 5) = .0046$ **b.** $P(X \leq 3) = .9342$ **c.** $S(X) = 1.2247$
d. .46; .9342; 1.5; 1.2

5.51. **a.** $P(X \geq 15) = .4080$ **b.** $P(X \leq 10) = .0479$ **c.** $E(X) = 13.96$ **d.** 14; 5

5.53. **a.** $E(X) = \$251.85$ **b.** $V(X) = 20{,}843.83$ **c.** $S(X) = \$144.37$
d. $P(X > 195) = .39$ **e.** 251.85; 144.37; 39

5.55. **a.** $P(X = 0) = .51$ **b.** $P(X = 1) = .34$ **c.** $P(X > 1) = .14$

5.57. **a.** $S(X) = 2.14$ **b.** $P(X \leq 2) = .1634$ **c.** $P(X > 5) = .3110$
d. $P(X = 0) = .0101$

5.59. **a.** $E(X) = 4.54$ **b.** $S(X) = 2.20$ **c.** $P(X < 5) = .38$ **d.** $P(X > 6) = .04$
e. 4.54; 38; 4

5.61. **a.** $P(2300 < X < 2500) = .2128$ **b.** $P(X < 1000) = .0228$

CHAPTER 6

6.1. **a.** 50 **b.** 33.3 **c.** 25 **d.** 20
6.3. **a.** 56.36 **b.** 54.90 **c.** 54.18 **d.** 53.74
6.5. **a.** .9525 **b.** .9564 **c.** .3674
6.7. **a.** .0009 **b.** 1.000 **c.** .2676 **d.** .0000 **e.** \$1568.64 to \$1631.36
6.9. **a.** .0000 **b.** .7486 **c.** 100% **d.** 425 min
6.11. **a.** .1562 **b.** .3845 **c.** .8869 **d.** .63
6.13. **a.** 160 **b.** 10.43 **c.** .1251 **d.** .3315
6.15. **a.** fpcp = .8949, $\sigma(p)$ = .0298 **b.** .2514 **c.** .0901 **d.** .0000 **e.** .65
6.17. **a.** .81 **b.** .0000 **c.** 1.0000 **d.** .0000
6.19. **a.** .0681 **b.** .008 **c.** 12; 20
6.29. **a.** .0000 **b.** .2676 **c.** .9306
6.31. **a.** .2709 **b.** .0793 **c.** .3845 **d.** no

CHAPTER 7

7.1. **a.** $\pi = 3/5$ **b.** ABCD, ABCE, ABDE, ACDE, BCDE **c.** $E(p_s) = .60$
7.3. **a.** Samples of size $n = 3$ have a standard error of the mean of .72, while samples of size $n = 4$ have a standard error of .47. Likewise, the ranges for the two samples are 2.33 and 1.10. Both measures demonstrate consistency. **b.** $E(\bar{x}_3) = E(\bar{x}_4) = \mu = 42.52$
7.5. **a.**, **b.**

Sample	Mean	Median
ABCD	73.150	73.15
ABCE	71.925	73.15
BCDE	67.750	65.05
CDEA	67.875	65.05
DEAB	71.800	72.90

c. $\mu = 70.5$; population median = 65.3
d.

Sample	Mean	Median
ABC	75.933	81.0
BCD	70.367	65.3
CDE	63.333	64.8
DEA	68.733	64.8
EAB	74.133	81.0
ACD	70.533	65.3
BDE	68.567	64.8
CEA	68.900	65.3
DAB	75.767	81.0
EBC	68.733	65.3

e. The sample mean is an unbiased, consistent estimator of the population parameter. The sample median is not unbiased; however, it is consistent.
7.7. 90% confidence interval for the proportion: $.022 \leq \pi \leq .060$
7.9. **a.** $\bar{x} = .441$ **b.** $s_x = .291$ **c.** $s_{\bar{x}} = .092$
d. 90% confidence interval for the mean: $.272 \leq \mu \leq .610$
e. We can say with 90% confidence that the true mean for this population falls between .272 and .610.
7.11. **a.** $s_{\bar{x}} = 110.51$ **b.** 95% confidence interval for the mean: \$421.8 thousand $\leq \mu \leq$ \$1035.4 thousand
7.13. **a.** $s_{p(\text{males})} = .014$ **b.** 90% confidence interval for the proportion: $.33 \leq \pi \leq .37$
c. $s_{p(\text{females})} = .010$; 90% confidence interval for the proportion: $.12 \leq \pi \leq .16$
d. It would appear that the true percentage of males who drink at least once a week is between 2 and 3 times the true percentage for females.
7.15. 99% confidence interval for the mean: $15.2 \leq \mu \leq 17.2$
7.17. **a.** 90% confidence interval for the mean: $.40 \leq \pi \leq .46$ **b.** ME = ±3%
c. 90% confidence interval for the proportion : $.24 \leq \pi \leq .28$ **d.** ME = ±2%
e. The percentage of females believing a gun in the home makes it a more dangerous place exceeds the percentage for males.
7.19. 99% confidence interval for the proportion: $.37 \leq \pi \leq .47$
7.21. $n = 44$ **7.23.** ME = ±2.4 units **7.25.** $n = 1{,}036$ **7.27.** $n = 2$
7.29. $n = 19$ **7.31.** $n = 10$ **7.33.** $n = 5$

7.35. The company cannot make the claim, as we can say with 95% confidence that the true average domestic content falls between 74.82% and 92.43%. Dropping the Festiva and Capri lines would improve the percentage.

7.37. The 95% confidence interval for CEOs' average age is 53.5 $\pm$ 4.8.

7.39. Sample size should be 385. Self-selection would lead to biased results insofar as only those who were particularly satisfied or particularly upset would be likely to respond.

7.41. **a.** standard error of the mean = .57
b. 95% confidence interval for the mean: $8.73 \leq \mu \leq 10.97$
c. 9.85%; 36.3%; 8.73; 10.97; better than

7.43. **a.** average number of blackout days: 17.4 **b.** standard error of the mean = 6.75
c. 99% confidence interval for the mean: $0 \leq \mu \leq 48.5$ **d.** 17.4; 0; 48.5

7.45. $n = 226$

7.47. $n = 17$; this information could be used to target ads to the desired age brackets.

CHAPTER 8

8.1. **a.** t test; approximately normal population, small sample **b.** $t = 3.78$, reject H_0

8.3. **a.** z test; standard normal deviation, large sample **b.** $z = -8.86$; reject H_0

8.5. There is sufficient evidence to conclude that the claim is not acceptable; reject H_0. The z test was chosen because the sample is large.

8.7. More difficult times have forced more people into the labor force.

8.9. Do not reject H_0. It does not appear that career satisfaction for women is significantly higher than that for men.

8.11. Reject H_0. The bottler should readjust the bottling process.

8.13. Do not reject H_0. The claim is not justified.

8.15. Reject H_0. The firm's claim appears to be unjustified.

8.17. Reject H_0. There is convincing evidence that the mean of set 2 is significantly greater than the mean of set 1.

8.19. Using the t statistic, do not reject H_0. There does not appear to be a significant difference.

8.21. Using the z statistic, reject H_0. The data provide convincing evidence that males earn significantly more than females.

8.23. Using the t statistic, reject H_0. There is convincing evidence that significant differences in housing costs exist between the two markets.

8.25. Using the t statistic, reject H_0. There is a significant difference in fuel consumption.

8.27. Using the t statistic, do not reject H_0. **8.29.** Using the t statistic, do not reject H_0.

8.31. **a.** UCL = 1.5015; LCL = 1.4985 **b.** UCL = 1.5011; LCL = 1.4989
c. Limits are tighter with $n = 8$.

8.33. The data suggest that operations should go into overtime; they suggest a highly variable process.

8.35. **a.** UCL = 40.75; LCL = 39.25; CCL = 40 **b.** The process is in control.

8.37. Do not reject H_0. There is no statistically significant difference.

8.39. Do not reject H_0. There does not appear to be a significant difference.

8.41. Do not reject H_0. There is not convincing evidence that homes in East Hampton are less expensive than those in Great Neck.

CHAPTER 9

9.1. The z test was used. The critical value is the z value of 1.96.

9.3. The maximum value p may have is .67.

9.5. A z test should be used because this a proportion. Using a one-tailed test with z_{critical} = 1.645, do not reject H_0.

9.7. The shipment should not be accepted.

9.9. Do not reject the hypothesis that at least 10% of all NBA players failed to complete college.

9.11. UCL = .13; CCL = .10; LCL = .07; the process is not in a state of statistical control. On-time performance does not exhibit stability.

9.13. Using a two-tailed z test with a critical value of 1.96, do not reject H_0. There is no evidence that $\pi_a \neq \pi_b$.

9.15. Using a one-tailed z test with a critical value of 1.28, do not reject H_0. There is no evidence that $\pi_a > \pi_b$.

9.17. Using $z_{critical} = 2.33$, do not reject H_0. There is no evidence that $\pi_a < \pi_b$.
9.19. Using the chi-square test, do not reject the null hypothesis. There is no sign of differences between processes.
9.21. Reject H_0; there exists a significant difference between males and females with respect to unhealthy popcorn. Reject H_0; there exists a significant difference between males and females with respect to smuggled snacks.
9.23. Do not reject H_0. There has not been a statistically significant change in the distribution of values. A chi-square goodness-of-fit test is used; the critical value of χ^2 is 7.815.
9.25. Reject H_0. The claimed distribution cannot be verified.
9.27. Do not reject H_0. There has been no change in the distribution of tax returns.
9.29. Reject H_0. There is a relationship between party affiliation and ADA rating.
9.31. Reject H_0. There is an association between gift action and income level.
9.33. Do not reject H_0. There is no relationship between type of vehicle and quality of ride. p value = .05.
9.35. Do not reject H_0. There is no evidence of a relationship between placement and gender. The p value exceeds the alpha level.
9.37. Accept the null hypothesis. There is no evidence to support the claim.
9.39. Do not reject H_0. No more than two-thirds of the smoking public believe they will die from a smoking-related disease.
9.41. Do not reject H_0. There is not a significant difference between markets.
9.43. Do not reject H_0. There is no association between the type of charity and the program ratio.
9.45. Do not reject H_0. It cannot be concluded that over two-thirds of the patrons felt service was either good or excellent.

CHAPTER 10

10.1. **b.** yes **c.** positive
10.3. **b.** There does not appear to be a linear relationship. **c.** not applicable
10.5. **b.** There does not appear to be a linear relationship. **c.** not applicable
10.7. **a.** previous wins **c.** If there is a linear relationship, it is very weak.
d. direct **e.** Previous wins have little, if any, bearing on attendance.
10.9. **a.** sales **c.** yes **d.** direct
e. The accountant can conclude that greater sales are associated with higher payroll.
10.11. **a.** misery index **c.** There appears to be a linear relationship. **d.** negative
e. The lengths of presidential terms differ markedly. Considering values on a per annum basis might change the results.
10.13. **b.** The plot may show a weak linear relationship; the mean at each of the three levels decreases. **c.** no **d.** Rates and points vary independently.
10.15. **b.** yes **c.** direct
10.17. **a.** $Y = 8.76 + .78X$ **c.** yes
10.19. **b.** $Y = .3697 - .0938X$ **d.** The fit is poor. **e.** no
10.21. **a.** $Y = 2093.6 + 4.01X$ **c.** yes **d.** Advertising is positively associated with sales.
10.23. **a.** $Y = 1.3780 + .0096X$ **c.** The line fits poorly.
10.25. **a.** $Y = 12.93 - .28X$
c. There appears to be a weak to moderate relationship between the change in GDP and the misery index.
10.27. **b.** yes **c.** $Y = -980.6 + .07X$ **e.** The line appears to fit well.
10.29. **b.** There does not appear to be a linear relationship. **c.** $Y = 225.963 + 3.904X$
d. no **e.** is not; no
10.31. **a.** $s_{y|x} = 12.08$ **b.** $r^2 = .83$ **c.** 83% of the variation is accounted for by the relationship.
10.33. **a.** $s_{y|x} = .4854$ **c.** $r = .12$; $r^2 = .0144$
d. r gives the strength of the relationship; r^2 gives the percentage of variation explained by the line.
10.35. **a.** $s_{y|x} = 96.30$ **b.** $r = .99$; $r^2 = .98$ **c.** yes **d.** ±96.30; is; .99; .98
e. Increasing the advertising budget should improve sales.

10.37. **b.** There does not appear to be a relationship within this range.
c. $Y = 20.36 - .0009X$ **d.** $s_{y|x} = .589$ **e.** $r = -.41; r^2 = .17$
f. There does not appear to be a relationship within this range; $Y = 11.47 - .19X$; $s_{y|x} = .607$; $r = -.25$; $r^2 = .06$ **g.** The first regression has the smaller standard error.
h. The first regression has the higher r^2.
i. weak; remained about level; 17; weaker; inverse; .19% decrease

10.39. **b.** There appears to be a linear relationship. **c.** $Y = -2.36 + 6.53X$
d. $s_{y|x} = 38.614$ **e.** $r = .99; r^2 = .98$
f. There appears to be a linear relationship; $Y = -426.238 + .0512X$; $s_{y|x} = 156.679$; $r = .83$; $r^2 = .69$ **g.** The first regression appears to provide a better forecast.
h. a linear; positive; increase; 6.53; positive; 69; 512; engineering and development expenditures

10.41. **a.** $s_{y|x} = 4.54$ **c.** Yes, it did fall outside the bands during the Carter administration.
d. $r = -.48$; $r^2 = .23$ **e.** yes

10.43. **b.** $Y = 188.2 + 12.5X$ **c.** $s_{y|x} = 28.88$ **d.** $r = .88$; $r^2 = .78$
e. $Y = 230.33 + 2.24X$; $s_{y|x} = 22.87$; $r = .93$; $r^2 = .86$ **f.** positive; positively; cholesterol content

10.45. **a.** $Y = 5.15$ **b.** 3.25, 7.90 **c.** –.88, 11.18
d. The answer in part a is a point estimate. The answer in part b provides a 95% prediction interval for an observation of Y when $X = 6$. The answer in part c provides a 95% confidence interval for the mean value of Y when $X = 6$.

10.47. **a.** $y = 90.6$. **b.** 29.3, 151.9 **c.** 66.4, 114.8

10.49. **a.** $Y = 295.391$ **b.** –248.6, 839.3 **c.** 85.99, 504.79

10.51. **a.** 277.7 thousand **b.** 198.3 thousand, 357.1 thousand
c. 50.4 thousand, 504.9 thousand
d. 277.7 thousand; 504.9 thousand; 50.4 thousand; 198.3 thousand; 357.1 thousand

10.53. **b.** $Y = 143.0637 + .3459X$ **c.** $s_{y|x} = 8.6378$ **d.** $r = .9975$; $r^2 = .9949$
e. 800.27 ± 51.98 **f.** positive; .3459 million; 8.6378 million; strong; 99.49

10.55. **a.** Do not reject H_0. **b.** $-.30$, .86
c. Because the confidence interval contains zero, it confirms the conclusion in part a that β_1 could, indeed, equal zero.

10.57. **a.** $b_1 = 0.48$ **b.** Do not reject H_0. **c.** $-.32$, 1.28

10.59. **a.** Do not reject H_0. **b.** $-.367$, .525 **c.** The confidence interval contains zero.

10.61. **a.** $H_0: \beta_1 = 0$; $H_a: \beta_1 \neq 0$; reject H_0. **b.** .32, .38 **c.** $1; dividends; .32; .38

10.63. **a.**

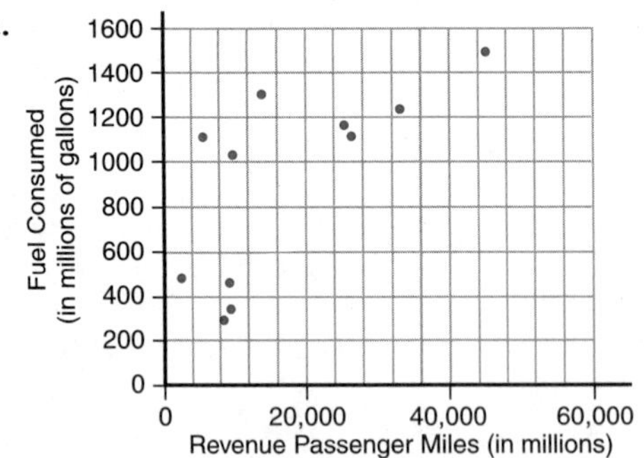

b. $Y = 545.3 + .022238X$ **c.** linear **e.** $s_{y|x} = 310.9$ **f.** $r = .714$; $r^2 = .510$
g. With 95% confidence, we can forecast average consumption of between 23.4 million gallons and 1511.9 million gallons. **h.** Reject H_0. **i.** .0055, .0390
j. appears; 51; 23.4; 1511.9; .0055; .0390

10.65. **a.**

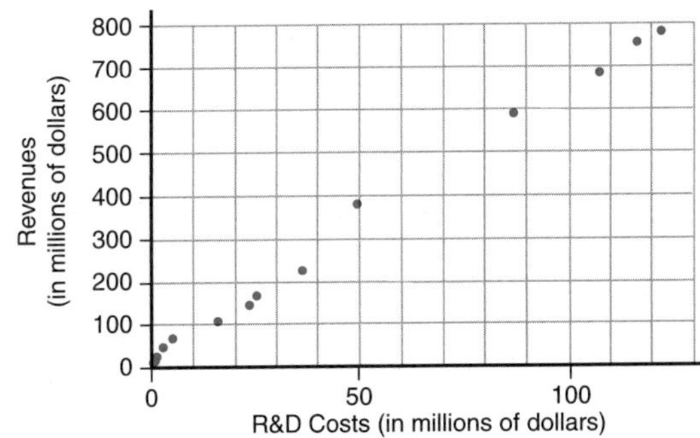

b. yes **c.** Regression analysis using Minitab:

The regression equation is

REVENUE = 11.2 + 6.37 R&D

Predictor	Coef	Stdev	T-ratio	p
Constant	11.204	7.749	1.45	0.174
R&D	6.3745	0.1250	50.98	0.000

s = 20.96 R-sq = 99.5% R-sq(adj) = 99.5%

Analysis of Variance

SOURCE	DF	SS	MS	F	p
Regression	1	1141913	1141913	2599.46	0.000
Residual Error	12	5271	439		
Total	13	1147184			

d. With a 1-unit change in R&D expenditures, there's a 6.37-unit change in revenues.
e. The probability of observing a slope this extreme when the true slope is zero is less than .0005.
f. $r^2 = 99.5\%$; this means that R&D expenditures can explain 99.5% of the change in revenues. **g.** a positive; increase; 6.37; exists; 99.5; does

10.67. **a.**

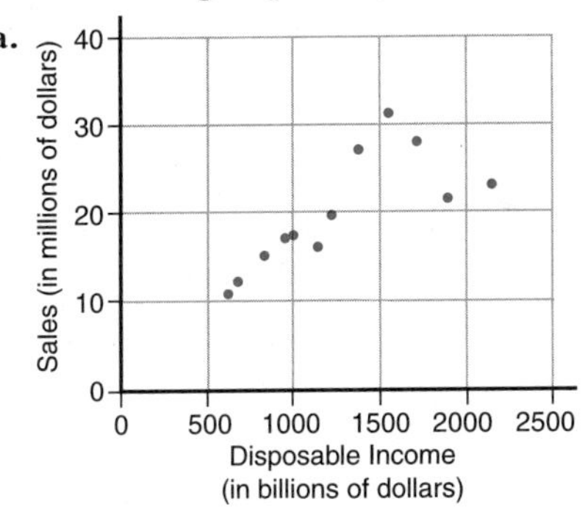

b. yes

c.

SUMMARY OUTPUT

Regression Statistics	
Multiple R	0.755571704
R Square	0.5708886
Adjusted R Square	0.52797746
Standard Error	4.487812837
Observations	12

ANOVA

	df	SS	MS	F	Significance F
Regression	1	267.9481676	267.9482	13.30397	0.004481092
Residual	10	201.4046406	20.14046		
Total	11	469.3528083			

	Coefficients	Standard Error	t Stat	p Value	Lower 95%	Upper 95%
Intercept	6.353247322	3.96903781	1.600702	0.140526	−2.490321558	15.1968162
Income	0.010648339	0.002919384	3.647461	0.004481	0.004143545	0.017153132

d. $r^2 = .57$

e. The p value for this model is .004481, which means that the probability of observing this result or a more extreme one when the null hypothesis is true is less than .005.

f. The prediction and confidence intervals for the sales figures may be obtained using Minitab. The relevant portion of the Minitab printout appears below:

Predicted Values

Fit	Stdev.Fit	95.0% CI	95.0% PI
30.43	3.13	(23.46, 37.41)	(18.24, 42.63)
32.18	3.57	(24.22, 40.14)	(19.40, 44.96) X
34.79	4.25	(25.33, 44.25)	(21.02, 48.55) XX

X denotes a row with X values away from the center

XX denotes a row with very extreme X values

g. For 1982, the actual sales figure was $24.271, while the point estimate for the prediction was $30.43. In 1983, there was only a small difference. However, in 1984, the prediction underestimates the actual by about $6 million.

h. positive; exists; .01 million increase; 57; differs; poorly; would not

10.69. a.

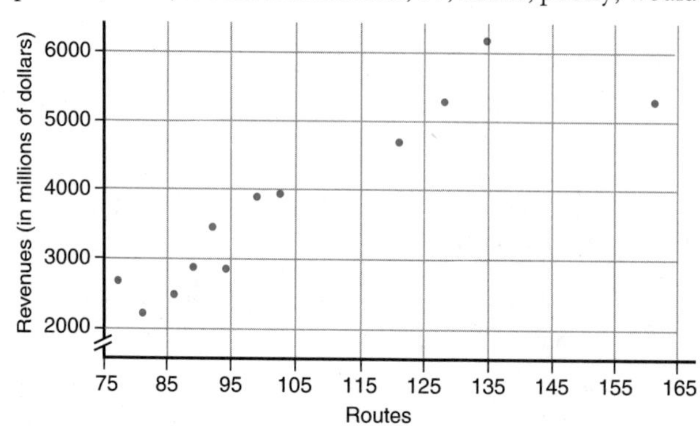

b. A linear relationship appears to exist.

c.

The regression equation is

REVENUE = −1023 + 46.0 ROUTES

Predictor	Coef	Stdev	T-ratio	p
Constant	−1022.7	748.8	−1.37	0.202
ROUTES	46.044	6.915	6.66	0.000

s = 583.7 R-sq = 81.6% R-sq(adj) = 79.8%

Analysis of Variance

SOURCE	DF	SS	MS	F	p
Regression	1	15105581	15105581	44.33	0.000
Residual Error	10	3407238	340724		
Total	11	18512819			

Unusual Observations

Obs	ROUTES	REVENUE	Fit	Stdev.Fit	Residual	St.Resid
12	161	5292	6390	419	−1099	−2.70RX

R denotes an observation with a large standardized residual

X denotes an observation whose X value gives it large influence

d. The slope is significantly greater than zero.

e.

Predicted Values 1986

Fit	Stdev.Fit	95.0% CI	95.0% PI
6989	503	(5869, 8109)	(5273, 8706) X

Predicted Values 1987

Fit	Stdev.Fit	95.0% CI	95.0% PI
7726	608	(6371, 9081)	(5848, 9604) XX

Predicted Values 1988

Fit	Stdev.Fit	95.0% CI	95.0% PI
8324	695	(6776, 9873)	(6302, 10347) XX

X denotes a row with X values away from the center
XX denotes a row with very extreme X values

f. Evaluating the result for 1986 shows that actual revenues fell within the 95% prediction interval. The actual value was $7105.1 million, while the forecast was $6989 million. For 1987 and 1988, actual revenues also fell in the prediction intervals.

g. a direct; appears; 46-unit increase; reasonably well

10.71. a.

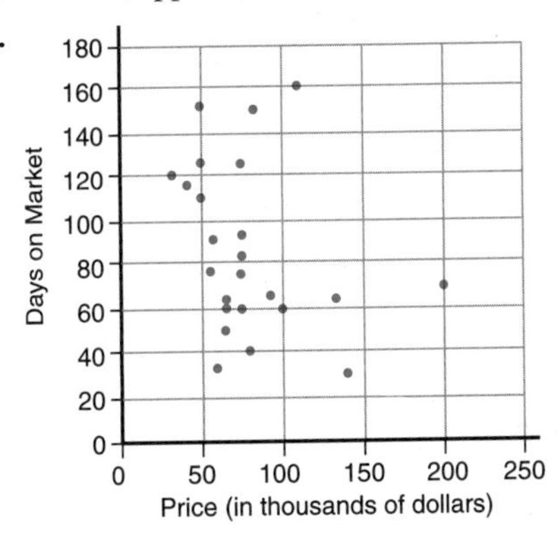

b. There does not appear to be a linear relationship.

c.

SUMMARY OUTPUT

Regression Statistics	
Multiple R	0.264989875
R Square	0.070219634
Adjusted R Square	0.0297944
Standard Error	37.30079372
Observations	25

ANOVA

	df	SS	MS	F	Significance F
Regression	1	2416.808118	2416.808	1.737025	0.200492994
Residual	23	32001.03188	1391.349		
Total	24	34417.84			

	Coefficients	Standard Error	t Stat	p Value	Lower 95%	Upper 95%
Intercept	106.6643468	18.68929628	5.707243	8.22E-06	68.00264445	145.3260491
Price	−0.283011865	0.214734404	−1.31796	0.200493	−0.727223218	0.161199489

d. $s_{y|x} = 37.3$. **e.** $r = .26$; $r^2 = .07$

f. The p value for the slope coefficient is .20. When the p value exceeds α, do not reject H_0. There is no significant linear relationship.

g. is not; no particular; 7 **h.** Adding other variables might improve this model.

10.73. a.

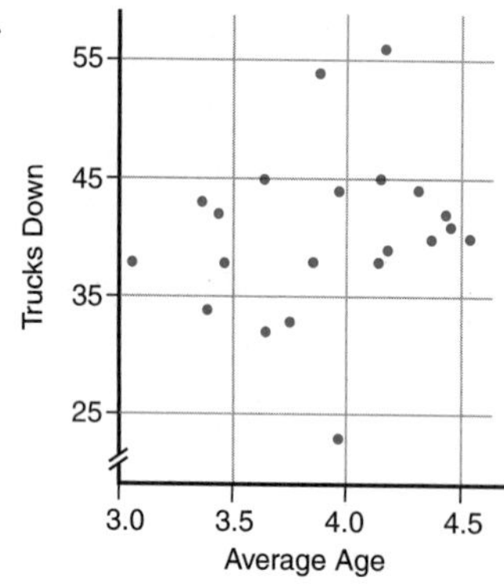

b. There does not appear to be a linear relationship.

c.

The regression equation is
TRUCKS D = 27.7 + 3.26 AGE

Predictor	Coef	Stdev	T-ratio	p
Constant	27.71	15.21	1.82	0.084
AGE	3.259	3.877	0.84	0.411

s = 7.135 R-sq = 3.6% R-sq(adj) = 0.0%

Analysis of Variance

SOURCE	DF	SS	MS	F	p
Regression	1	35.96	35.96	0.71	0.411
Residual Error	19	967.18	50.90		
Total	20	1003.14			

Unusual Observations

Obs	AGE	TRUCKS D	Fit	Stdev.Fit	Residual	St.Resid
15	3.98	23.00	40.68	1.59	−17.68	−2.54R
16	4.15	56.00	41.23	1.83	14.77	2.14R

d. $r^2 = 3.6\%$

e. The p value for the slope is .411, which is greater than α. This means that there does not exist a statistically significant linear relationship.

f. Increasing the average age by another year would have no effect on downtime. The average age of the current fleet is 3.90 years. Plugging 4.90 years into the equation yields

Predicted Values

Fit	Stdev.Fit	95.0% CI	95.0% PI
43.68	4.17	(34.95, 52.40)	(26.38, 60.97) X

X denotes a row with X values away from the center

The average number of trucks down presently is 40.43, which is not significantly different from the predicted result.

CHAPTER 11

11.1. a.

The regression equation is
Y = 6.14 − 0.127 X1 − 0.035 X2

Predictor	Coef	Stdev	T-ratio	p
Constant	6.140	3.377	1.82	0.107
X1	−0.1267	0.3141	−0.40	0.697
X2	−0.0353	0.4109	−0.09	0.934

s = 2.973 R-sq = 2.0% R-sq(adj) = 0.0%

Analysis of Variance

SOURCE	DF	SS	MS	F	p
Regression	2	1.459	0.730	0.08	0.922
Residual Error	8	70.722	8.840		
Total	10	72.182			

SOURCE	DF	Seq SS
X1	1	1.394
X2	1	0.065

b. If b_1 were significant, it would imply that a 1-unit change in X_1 was associated with a $-.127$-unit change in Y.
c. b_2 does not significantly differ from zero.
d. Examining the ANOVA, we find that the p value equals .922, indicating that the regression as a whole exhibits no global significance.

11.3. **a.** $t = -5.67$
b. The p value of .925 indicates that X_3 does not contribute to the explanation of Y, given that X_1 and X_2 are in the model. **c.** 86.9%

11.5. **a.** $Y = 54.94 + .00651W - .402V$ **b.** $p > 0.25$ **c.** $t = -0.25$
d. MSE = 1496 **e.** $n = 20$ **f.** It explains nothing.

11.7. **a.** $t = 3.69$ **b.** $p = .267$
c. R-sq = 69.8%, indicating that this combination of factors can explain 69.8% of the variation in steel sales. **d.** strongly; 6.955 million; weak; insignificantly; is

11.9. **a.**

The regression equation is
SALES = − 1317 + 1.38 GNP + 233 PARTY

Predictor	Coef	Stdev	T-ratio	p
Constant	−1317.5	225.1	−5.85	0.000
GNP	1.38210	0.08654	15.97	0.000
PARTY	233.1	143.0	1.63	0.142

s = 182.3 R-sq = 98.3% R-sq(adj) = 97.8%

Analysis of Variance

SOURCE	DF	SS	MS	F	p
Regression	2	15034878	7517439	226.25	0.000
Residual Error	8	265808	33226		
Total	10	15300687			

SOURCE	DF	Seq SS
GNP	1	14946555
PARTY	1	88324

b. According to this regression, the party in the White House has no influence on sales.
c. If party had a bearing on sales, sales would be $233.1 million more in Republican administrations than in Democratic administrations.

11.11. **a.**

The regression equation is
PASSVOL = 378325 + 3914 CONFID + 126704 SEASON

Predictor	Coef	Stdev	T-ratio	p
Constant	378325	320991	1.18	0.269
CONFID	3914	2908	1.35	0.211
SEASON	126704	50972	2.49	0.035

s = 74251 R-sq = 53.0% R-sq(adj) = 42.6%

Analysis of Variance

SOURCE	DF	SS	MS	F	p
Regression	2	56047652584	28023826292	5.08	0.033
Residual Error	9	49618445821	5513160647		
Total	11	1.05666E+11			

SOURCE	DF	Seq SS
CONFID	1	21981356089
SEASON	1	34066296495

b. Season appears to have a bearing on volume: $p = .035$, and at $\alpha = .05$ it is significant.
c. overall model: $Y = 378{,}325 + 3914 \times \text{Confidence} + 126{,}704 \times \text{Season}$; 3rd quarter: $Y = 505{,}029 + 3914 \times \text{Confidence}$; remainder of year: $Y = 378{,}325 + 3914 \times \text{Confidence}$

11.13. **a.** Three-year moving average for Compaq with 1997 forecast:

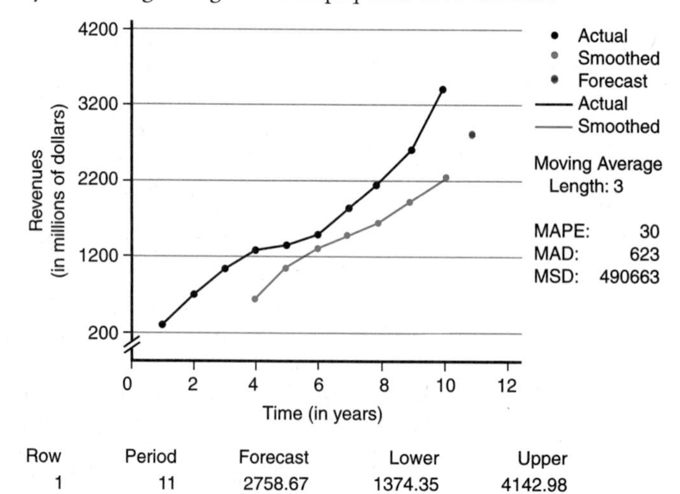

Row	Period	Forecast	Lower	Upper
1	11	2758.67	1374.35	4142.98

b. Single exponential smoothing:

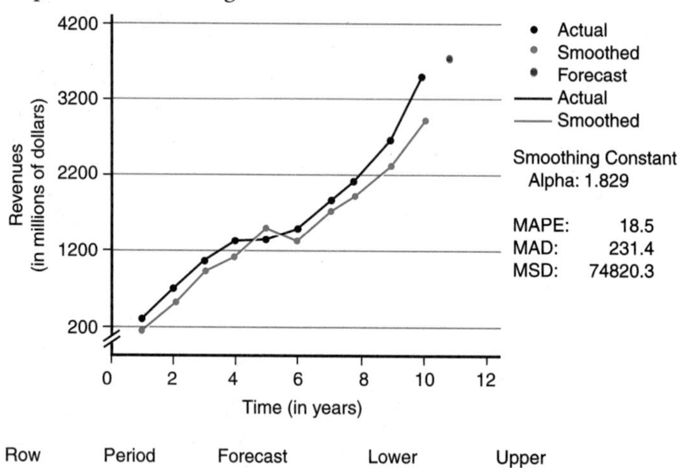

Row	Period	Forecast	Lower	Upper
1	11	3863.72	3180.94	4546.51

c. For the 3-year moving average, MAD is 623; for single exponential smoothing, it is 279.

d. single exponential smoothing

11.15. **a.** Three-year moving average for Schering-Plough with 1997 forecast:

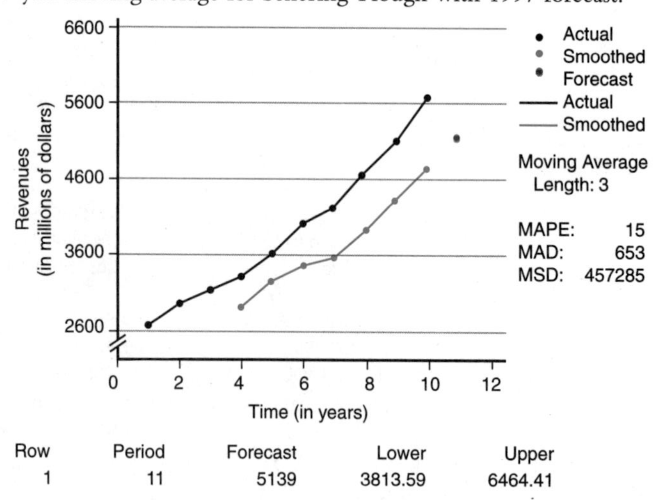

Row	Period	Forecast	Lower	Upper
1	11	5139	3813.59	6464.41

b. Single exponential smoothing:

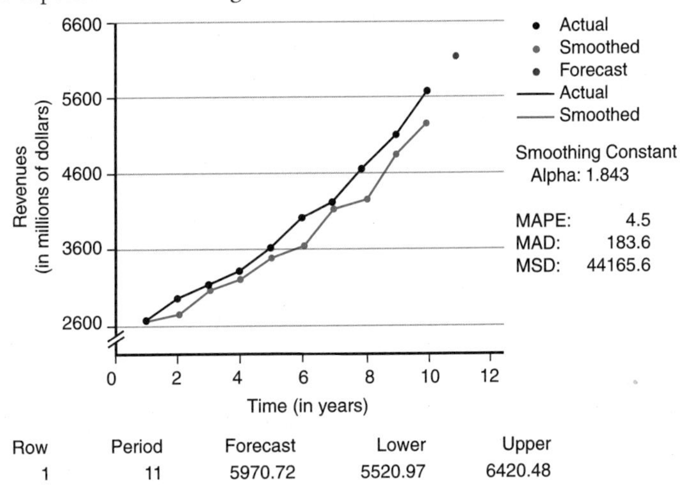

Row	Period	Forecast	Lower	Upper
1	11	5970.72	5520.97	6420.48

c. single exponential smoothing

11.17. a. Three-year moving average for Samsung:

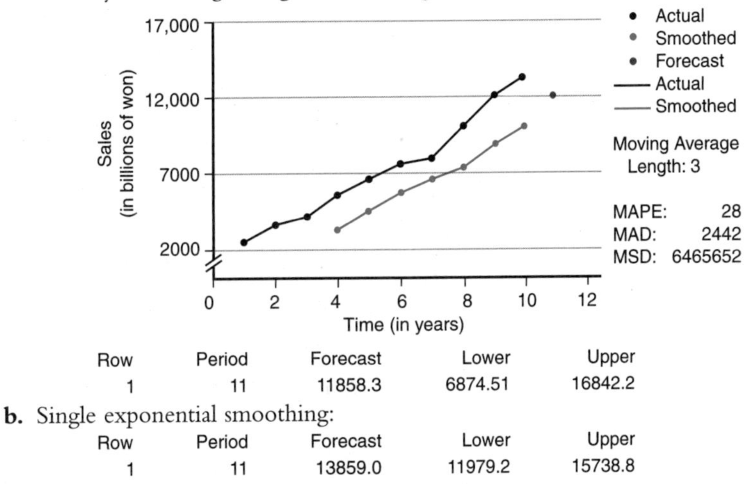

Row	Period	Forecast	Lower	Upper
1	11	11858.3	6874.51	16842.2

b. Single exponential smoothing:

Row	Period	Forecast	Lower	Upper
1	11	13859.0	11979.2	15738.8

11.19. a. The unemployment time series looks like this:

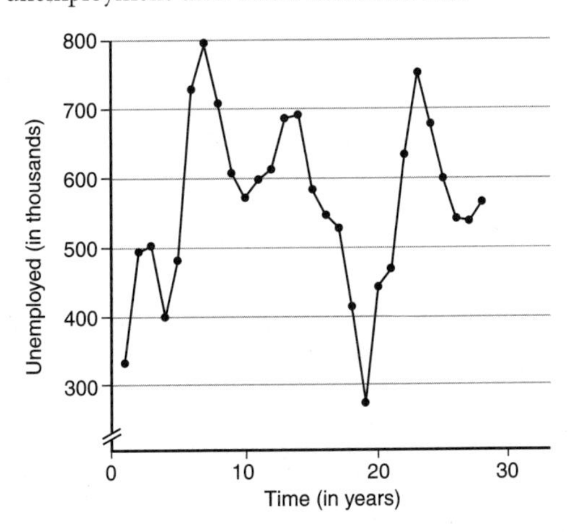

A trend would be inappropriate because of the cyclical nature of the series.

b. Single exponential smoothing would be most useful here, because it traces the series most closely.

c. Applying single exponential smoothing yields the following:

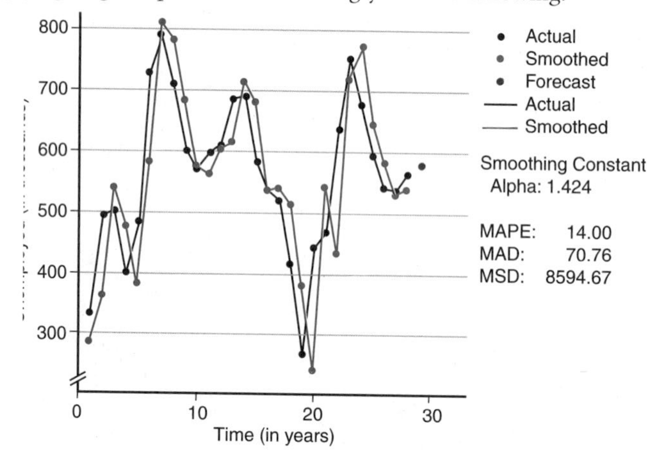

Row	Period	Forecast	Lower	Upper
1	29	574.500	401.132	747.868

d. increase; 589.748; 7.1

11.21. **a.** quarter 1: 1.078273; quarter 2: 0.848557; quarter 3: 0.984397; quarter 4: 1.088773
b. $y = -141.296 + 60.4015t$ **c.** 802.68; 990.63; 1161.43; 1045.32

11.23. **a.**

The regression equation is
SALES = 2296 + 0.837 JACKPOT

Predictor	Coef	Stdev	T-ratio	p
Constant	2295.9	479.2	4.79	0.000
JACKPOT	0.83680	0.06035	13.87	0.000

s = 1555 R-sq = 89.3% R-sq(adj) = 88.9%

Analysis of Variance

SOURCE	DF	SS	MS	F	p
Regression	1	464720392	464720392	192.27	0.000
Residual Error	23	55592428	2417062		
Total	24	520312819			

The other variables do not add anything to the explanatory power of the equation.

b. Just use the jackpot data:

Predicted Values 5 million jackpot

Fit	Stdev.Fit	95.0% CI	95.0% PI
6480	317	(5824, 7136)	(3198, 9762)

Predicted Values 10 million jackpot

Fit	Stdev.Fit	95.0% CI	95.0% PI
10664	392	(9853, 11475)	(7347, 13981)

Predicted Values 20 million jackpot

Fit	Stdev.Fit	95.0% CI	95.0% PI
19032	898	(17175, 20889)	(15318, 22746) X

X denotes a row with X values away from the center

c. 89.3% of the variation in sales

d. The ANOVA table indicates the model has global significance. The p value is less than .05.

e. A $1 million increase in jackpot size will increase sales by .837 million. The coefficients for earnings and unemployment do not differ significantly from zero. It may be said that changes in weekly earnings and unemployment appear to have no effect on lottery sales.

f. The residuals follow a roughly normal distribution. The data indicate a degree of linearity. The assumption of homoscedasticity is not violated. The Durbin-Watson statistic is 2.07, indicating no serial correlation.

g. Because there is only one independent variable, multicollinearity is not a problem.

11.25. a.

The regression equation is
REVPAS = − 5763 + 3.03 YEAR − 35.8 BICTYPOP − 0.810 HICKFARE

Predictor	Coef	Stdev	T-ratio	p
Constant	−5763	1612	−3.57	0.002
YEAR	3.0257	0.8340	3.63	0.001
BICTYPOP	−35.751	8.571	−4.17	0.000
HICKFARE	−0.8097	0.2752	−2.94	0.007

s = 4.183 R-sq = 47.0% R-sq(adj) = 40.1%

Analysis of Variance

SOURCE	DF	SS	MS	F	p
Regression	3	356.94	118.98	6.80	0.002
Residual Error	23	402.47	17.50		
Total	26	759.41			

SOURCE	DF	Seq SS
YEAR	1	50.39
BICTYPOP	1	155.06
HICKFARE	1	151.49

Unusual Observations

Obs	YEAR	REVPAS	Fit	Stdev.Fit	Residual	St.Resid
20	1973	57.000	69.894	1.038	−12.894	−3.18R
27	1980	80.800	73.464	2.648	7.336	2.27R

R denotes an observation with a large standardized residual

b. 47.0% of the variation in revenue passengers

c. The ANOVA table and the *F* statistic indicate the model has global significance. The *p* value for the model is .002, which is less than $\alpha = .05$.

d. The residuals follow a normal distribution. There are no problems with respect to linearity and homoscedasticity. The DW (1.17) is inconclusive with respect to serial correlation.

e. There are problems with respect to multicollinearity. To improve the model, eliminate "Year" as an independent variable, recalculate the model, and re-evaluate the result.

11.27. a.

The regression equation is
SALES = −497 + 0.138 RESTS + 0.550 DI

Predictor	Coef	Stdev	T-ratio	p
Constant	−497.1	123.4	−4.03	0.005
RESTS	0.1376	0.5309	0.26	0.803
DI	0.5503	0.1781	3.09	0.018

s = 43.23 R-sq = 98.0% R-sq(adj) = 97.5%

Analysis of Variance

SOURCE	DF	SS	MS	F	p
Regression	2	647757	323879	173.33	0.000
Residual Error	7	13080	1869		
Total	9	660837			

SOURCE	DF	Seq SS
RESTS	1	629918
DI	1	17839

Unusual Observations

Obs	RESTS	SALES	Fit	Stdev.Fit	Residual	St.Resid
8	379	525.7	600.7	26.7	−75.0	−2.20R

R denotes an observation with a large standardized residual

b. yes **c.** not significant at the .05 level **d.** significant at the .05 level
e. The effect of a $1 billion change in disposable income is a change in sales of $.55 million.
f. An increase of 10 restaurants, if significant, would account for a $1,376,000 increase in sales.
g. With 525 restaurants and disposable income of $2300 billion, sales are predicted to be $840.9 million.

11.29. a. Three-year moving average for Seagram:

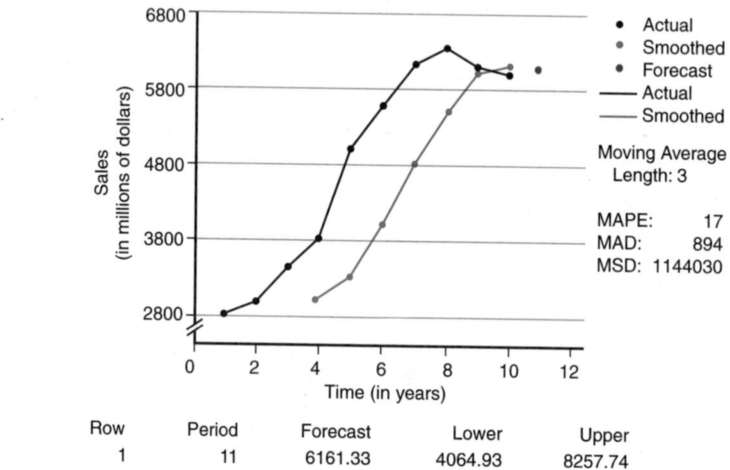

Row	Period	Forecast	Lower	Upper
1	11	6161.33	4064.93	8257.74

b. Single exponential smoothing:

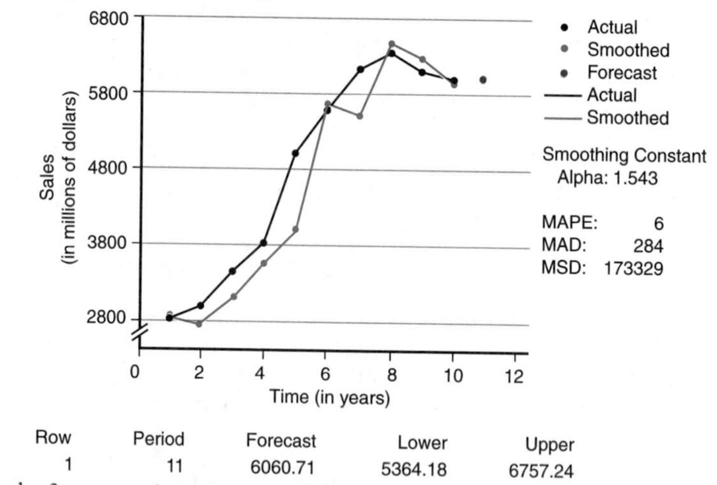

Row	Period	Forecast	Lower	Upper
1	11	6060.71	5364.18	6757.24

c. For the 3-year moving average, MAPE is 17%; for single exponential smoothing, it is 6%.
d. single exponential smoothing, because it provides a potentially more accurate forecast

Index

A

B

C

D

E

H

I

J

K

L

M

N

O

P

Q

R

S

T

U

V

W

X

Y

Z